Excel

MATHEMATICS

STUDY GUIDE

YEAR 7

Get the Results You Want!

John Compton, Allyn Jones & Peter Nicolas

First edition 1996
Reprinted 1998, 2000, 2001
Revised edition 2006
Reprinted 2007, 2009, 2011

Updated in 2013 for the Australian Curriculum

Reprinted 2017, 2018, 2019, 2021 (twice), 2022, 2024

ISBN 978 1 74125 006 0

Pascal Press
PO Box 250
Glebe NSW 2037
(02) 9198 1748
www.pascalpress.com.au

Publisher: Vivienne Joannou
Project Editors: Grant Bailey, Lisa Burrell and Leanne Poll
Edited by Bruce Howarth, Catherine Page and Karen Enkelaar
Indexed by Puddingburn Publishing Services
Answers checked by Peter Little
Typeset by Nikki M Group Pty Ltd
Cover by DiZign Pty Ltd
Printed by Vivar Printing/Green Giant Press

Students
All care has been taken in compiling this study guide, but please check with your teacher about the exact requirements of the course as these can change from year to year.

Acknowledgements
Allyn, Peter and I thank our wives Fiona, Maria and Robyn for their patience over the time that we have spent writing our series of study guides. Without their encouragement and understanding these books would still be but a good idea. Thanks also to our children for their help and understanding.

CONTENTS

HOW TO USE THIS BOOK

- This book covers **the topics in the Year 7 Australian Curriculum (Mathematics)**. It is best to work through it bit by bit over the whole year. However, you can also use it for revision before tests or examinations. You will also find it useful for finding out about particular topics.
- Try to work through the book in **chapter order** (Chapter 1 first, then Chapter 2, Chapter 3, and so on). Your teacher may cover topics in a different order. If you are not sure which chapter of the book relates to your classroom work, ask your teacher for assistance.
- If you want to practise a particular topic, then look up that topic in the Table of contents (the pages before this one) or the Index (at the back of the book).

Chapter by Chapter ...

DECIMALS 3

- Preliminary Ideas
- Converting Decimals to Fractions
- Converting Fractions to Decimals
- Comparing Decimals
- Rounding Off, or Correcting Decimals
- Four Operations Involving Decimals
- Applications of Decimals

- If you want to work through a whole chapter start at the first page. Here you will find an overview of the topics covered in that chapter.

KEYWORDS

Ascending
Correcting
Decimal
Descending
Expanded
Recurring
Repeating
Rounding off
Terminating

- You should also study the **keywords** at the start of the chapter. Look out for these important terms as you work through the chapter.

Converting Decimals to Fractions

The number of digits after the decimal point (number of decimal places) equals the number of zeros in the denominator. ← Explanation

For Example

Convert to simplified fractions: ← Example

1 0.7 **2** 0.41

3 0.04 **4** 0.409

5 0.0078 **6** 1.7

1 $0.7 = \frac{7}{10}$ ← Worked solution

2 $0.41 = \frac{41}{100}$

- Each concept is explained using both an **explanation** and an **example question**. Read the explanation and then work through the **example question**. The worked solution to the example question appears straight after the question in purple type. Make sure you understand how the answer was found. If you are not sure, read through the topic again or ask your teacher for help.
- Use an exercise book to make notes as you go along. This will help you remember what you have learned.

PRACTISE, PRACTISE

1 Write the value of the 3 in: p. 56
a 0.317 **b** 28.031 **c** 104.763

2 Rewrite in expanded notation (form): p. 56
a 3.21 **b** 76.047

3 Write the number of decimal places in: p. 56
a 3.476 **b** 21.5048

4 Write in expanded form: p. 56
a 4.25 **b** 8.017 **c** 0.1022

5 Rewrite as decimals: p. 56
a $4 + \frac{1}{10} + \frac{7}{100}$ **b** $6 + \frac{3}{100} + \frac{7}{1000}$
c $\frac{7}{10} + \frac{3}{100}$ ***d** $74 + \frac{11}{100} + \frac{3}{1000}$

6 Express as fractions in their simplest form: pp. 56–57
a 0.3 **b** 0.5 **c** 0.8
d 0.45 **e** 0.42 **f** 0.125
g 0.56 **h** 6.4 **i** 1.41
j 4.25 **k** 10.35 **l** 1.002

7 Change these fractions to decimals: p. 57
a $\frac{21}{100}$ **b** $\frac{9}{10}$ **c** $\frac{21}{10}$
d $\frac{3}{100}$ **e** $\frac{3}{1000}$ **f** $\frac{3}{5}$
g $\frac{7}{20}$ **h** $\frac{7}{25}$ **i** $\frac{7}{50}$
j $\frac{17}{200}$

Page references to explanations

- When you have worked through all topics in the chapter, try the **Practise, Practise** section. Quick answers and worked solutions to this section are found at the back of the book. The page reference on the right of each question tells you where each concept is explained in the chapter.

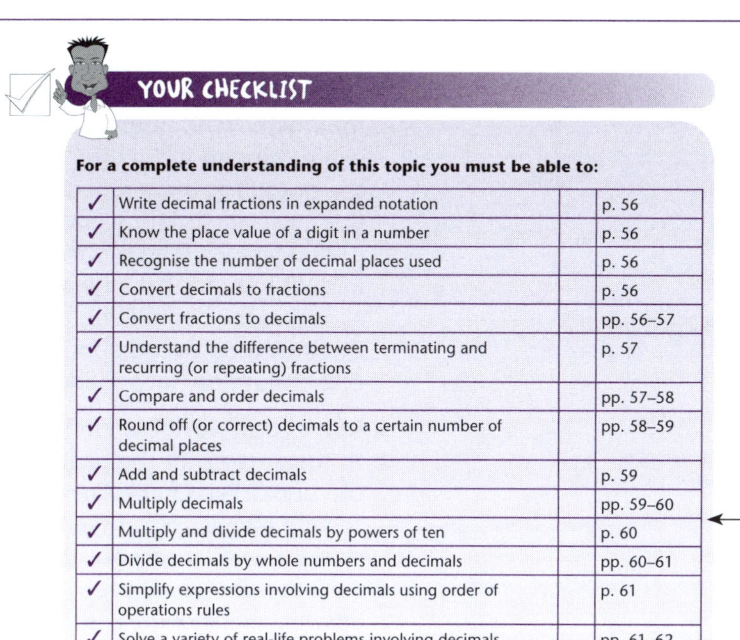

YOUR CHECKLIST

For a complete understanding of this topic you must be able to:

✓	Write decimal fractions in expanded notation	p. 56
✓	Know the place value of a digit in a number	p. 56
✓	Recognise the number of decimal places used	p. 56
✓	Convert decimals to fractions	p. 56
✓	Convert fractions to decimals	pp. 56–57
✓	Understand the difference between terminating and recurring (or repeating) fractions	p. 57
✓	Compare and order decimals	pp. 57–58
✓	Round off (or correct) decimals to a certain number of decimal places	pp. 58–59
✓	Add and subtract decimals	p. 59
✓	Multiply decimals	pp. 59–60
✓	Multiply and divide decimals by powers of ten	p. 60
✓	Divide decimals by whole numbers and decimals	pp. 60–61
✓	Simplify expressions involving decimals using order of operations rules	p. 61
✓	Solve a variety of real-life problems involving decimals.	pp. 61–62

Now you are ready to do the tests!

- Go through the **Checklist** near the end of the chapter to ensure that you understand the key concepts of the chapter. Put a tick next to each checklist item that you have covered. You need to be able to tick every item to be able to do the tests. If you have missed any items, go through the chapter to ensure that you cover the relevant topics. The page reference next to each item tells you where the concept is explained.

Page references

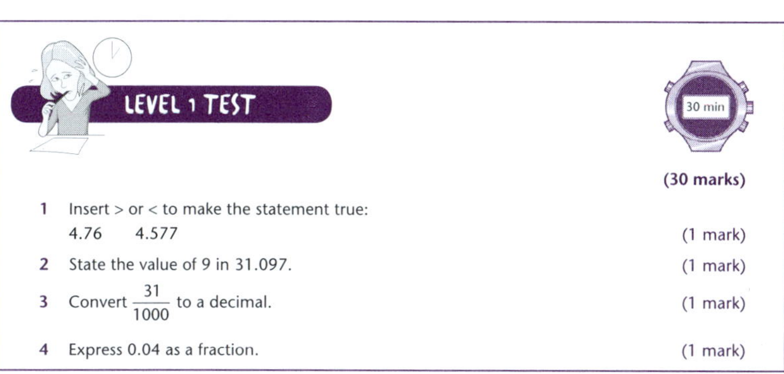

LEVEL 1 TEST

30 min

(30 marks)

1 Insert > or < to make the statement true: 4.76 4.577 (1 mark)

2 State the value of 9 in 31.097. (1 mark)

3 Convert $\frac{31}{1000}$ to a decimal. (1 mark)

4 Express 0.04 as a fraction. (1 mark)

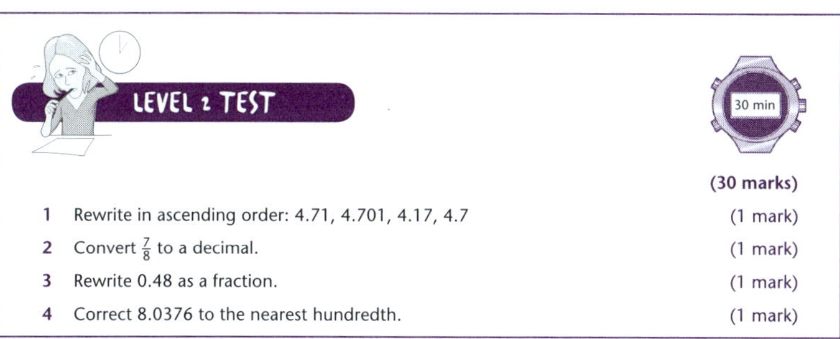

LEVEL 2 TEST

30 min

(30 marks)

1 Rewrite in ascending order: 4.71, 4.701, 4.17, 4.7 (1 mark)

2 Convert $\frac{7}{8}$ to a decimal. (1 mark)

3 Rewrite 0.48 as a fraction. (1 mark)

4 Correct 8.0376 to the nearest hundredth. (1 mark)

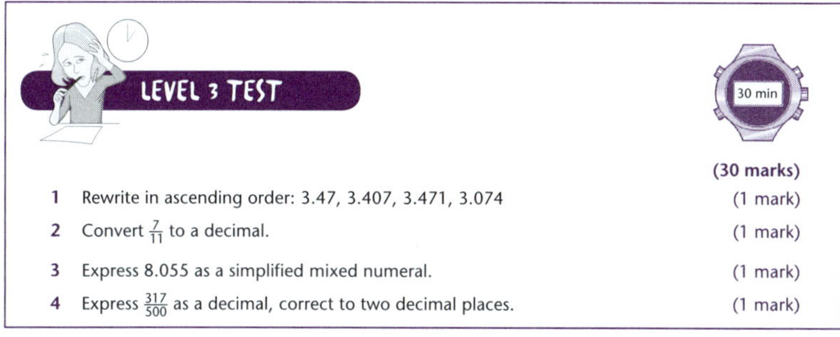

LEVEL 3 TEST

30 min

(30 marks)

1 Rewrite in ascending order: 3.47, 3.407, 3.471, 3.074 (1 mark)

2 Convert $\frac{7}{11}$ to a decimal. (1 mark)

3 Express 8.055 as a simplified mixed numeral. (1 mark)

4 Express $\frac{317}{500}$ as a decimal, correct to two decimal places. (1 mark)

- Next try to do all the **Tests.** The tests are of three levels of difficulty: the Level 1 tests are straightforward, the Level 2 tests are of average difficulty and the Level 3 tests are challenging. Start with the Level 1 test. If you find the first test easy, it will still be good practice before you try the other tests. If you find Test 1 or Test 2 close to your ability, then try the third test and complete all the questions you can—even if you find them difficult. Because there are worked solutions to each question, you will always be able to find out how to obtain each answer.
- The tests are designed to **prepare you for your school tests or examinations** so try to keep to the maximum time allowed for each one. The time appears on the stopwatch at the start of the test. Avoid looking at your textbook or notes while doing the test—this will help you develop your examination technique.

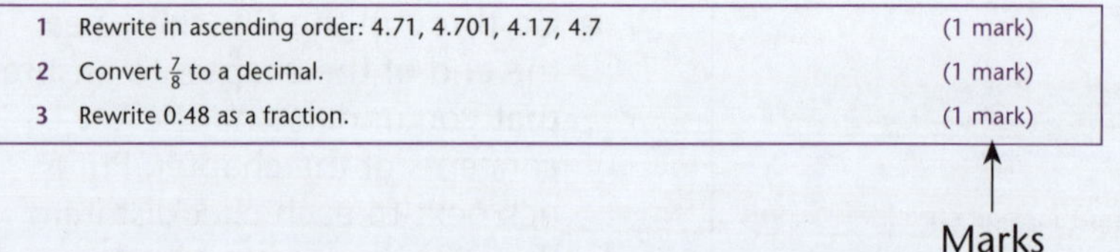

1 Rewrite in ascending order: 4.71, 4.701, 4.17, 4.7 (1 mark)
2 Convert $\frac{7}{8}$ to a decimal. (1 mark)
3 Rewrite 0.48 as a fraction. (1 mark)

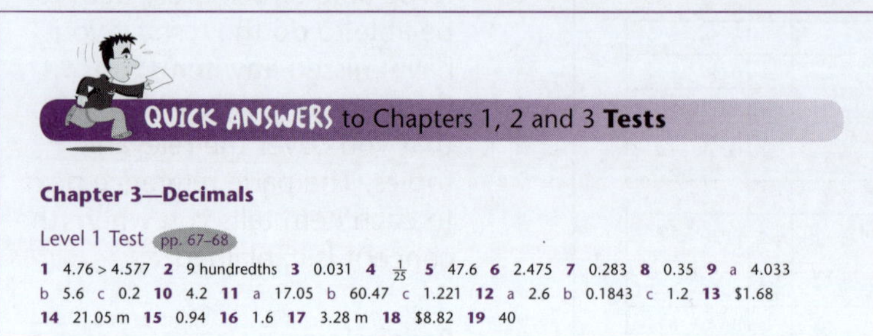

QUICK ANSWERS to Chapters 1, 2 and 3 **Tests**

Chapter 3—Decimals

Level 1 Test pp. 67–68

1 4.76 > 4.577 **2** 9 hundredths **3** 0.031 **4** $\frac{1}{25}$ **5** 47.6 **6** 2.475 **7** 0.283 **8** 0.35 **9** a 4.033 b 5.6 c 0.2 **10** 4.2 **11** a 17.05 b 60.47 c 1.221 **12** a 2.6 b 0.1843 c 1.2 **13** $1.68 **14** 21.05 m **15** 0.94 **16** 1.6 **17** 3.28 m **18** $8.82 **19** 40

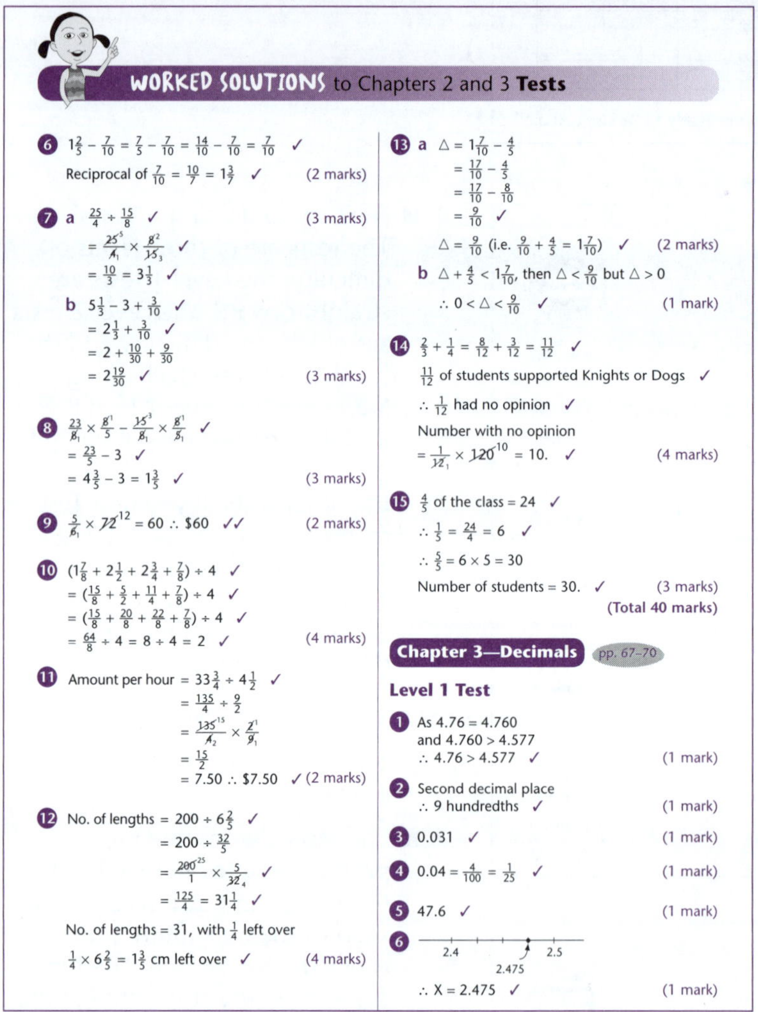

WORKED SOLUTIONS to Chapters 2 and 3 **Tests**

6 $1\frac{2}{5}-\frac{7}{10}=\frac{7}{5}-\frac{7}{10}=\frac{14}{10}-\frac{7}{10}=\frac{7}{10}$ ✓
Reciprocal of $\frac{7}{10}=\frac{10}{7}=1\frac{3}{7}$ ✓ (2 marks)

7 a $\frac{25}{4}\div\frac{15}{8}$ ✓ (3 marks)
$=\frac{\cancel{25}^5}{\cancel{4}_1}\times\frac{\cancel{8}^2}{\cancel{15}_3}$ ✓
$=\frac{10}{3}=3\frac{1}{3}$ ✓

b $5\frac{1}{3}-3+\frac{3}{10}$ ✓
$=2\frac{1}{3}+\frac{3}{10}$ ✓
$=2+\frac{10}{30}+\frac{9}{30}$
$=2\frac{19}{30}$ ✓ (3 marks)

8 $\frac{23}{\cancel{8}_1}\times\frac{\cancel{8}^1}{5}-\frac{\cancel{15}^3}{\cancel{8}_1}\times\frac{\cancel{8}^1}{\cancel{5}_1}$ ✓
$=\frac{23}{5}-3$ ✓
$=4\frac{3}{5}-3=1\frac{3}{5}$ ✓ (3 marks)

9 $\frac{5}{\cancel{6}_1}\times\cancel{72}^{12}=60$ ∴ $60 ✓✓ (2 marks)

10 $(1\frac{7}{8}+2\frac{1}{2}+2\frac{3}{4}+\frac{7}{8})\div 4$ ✓
$=(\frac{15}{8}+\frac{5}{2}+\frac{11}{4}+\frac{7}{8})\div 4$ ✓
$=(\frac{15}{8}+\frac{20}{8}+\frac{22}{8}+\frac{7}{8})\div 4$ ✓
$=\frac{64}{8}\div 4=8\div 4=2$ ✓ (4 marks)

11 Amount per hour $=33\frac{3}{4}\div 4\frac{1}{2}$ ✓
$=\frac{135}{4}\div\frac{9}{2}$
$=\frac{\cancel{135}^{15}}{\cancel{4}_2}\times\frac{\cancel{2}^1}{\cancel{9}_1}$
$=\frac{15}{2}$
$=7.50$ ∴ $7.50 ✓ (2 marks)

12 No. of lengths $=200\div 6\frac{2}{5}$ ✓
$=200\div\frac{32}{5}$
$=\frac{\cancel{200}^{25}}{1}\times\frac{5}{\cancel{32}_4}$ ✓
$=\frac{125}{4}=31\frac{1}{4}$ ✓
No. of lengths = 31, with $\frac{1}{4}$ left over
$\frac{1}{4}\times 6\frac{2}{5}=1\frac{3}{5}$ cm left over ✓ (4 marks)

13 a $\triangle=1\frac{7}{10}-\frac{4}{5}$
$=\frac{17}{10}-\frac{4}{5}$
$=\frac{17}{10}-\frac{8}{10}$
$=\frac{9}{10}$ ✓
$\triangle=\frac{9}{10}$ (i.e. $\frac{9}{10}+\frac{4}{5}=1\frac{7}{10}$) ✓ (2 marks)

b $\triangle+\frac{4}{5}<1\frac{7}{10}$, then $\triangle<\frac{9}{10}$ but $\triangle>0$
$\therefore 0<\triangle<\frac{9}{10}$ ✓ (1 mark)

14 $\frac{2}{3}+\frac{1}{4}=\frac{8}{12}+\frac{3}{12}=\frac{11}{12}$ ✓
$\frac{11}{12}$ of students supported Knights or Dogs ✓
∴ $\frac{1}{12}$ had no opinion ✓
Number with no opinion
$=\frac{1}{\cancel{12}_1}\times\cancel{120}^{10}=10.$ ✓ (4 marks)

15 $\frac{4}{5}$ of the class = 24 ✓
$\therefore\frac{1}{5}=\frac{24}{4}=6$ ✓
$\therefore\frac{5}{5}=6\times 5=30$
Number of students = 30. ✓ (3 marks)
(Total 40 marks)

Chapter 3—Decimals pp. 67–70

Level 1 Test

1 As 4.76 = 4.760
and 4.760 > 4.577
∴ 4.76 > 4.577 ✓ (1 mark)

2 Second decimal place
∴ 9 hundredths ✓ (1 mark)

3 0.031 ✓ (1 mark)

4 $0.04=\frac{4}{100}=\frac{1}{25}$ ✓ (1 mark)

5 47.6 ✓ (1 mark)

6 2.4 2.5 2.475
∴ X = 2.475 ✓ (1 mark)

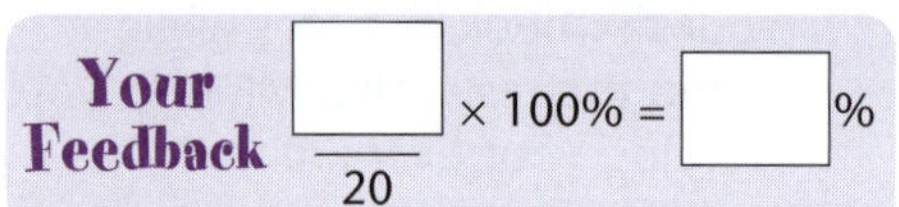

Your Feedback $\frac{\square}{20}\times 100\% = \square\,\%$

- **Marks** are allocated for each question. These are similar to the marks you will be trying for in your school tests and exams. Spend more time on the questions that are worth the most marks. For example, if there are 20 marks in total, and 20 minutes have been allocated for completion of the test, then spend about one minute on each mark.
- **Mark your work** by referring to the answers at the back of the book. The **Quick Answers** section allows you to see straight away if you got the answer right—only the answers are shown. The **Worked Solutions** section sets out the solutions to the question in full, so look at this section if your answer was incorrect. The **ticks** in worked solutions indicate those parts of the working which receive marks. Therefore, even if your answer is wrong, you may be entitled to some marks for the question. Compare the worked solution to your own working to find out if you are entitled to any marks for the question.
- Write your own score in the **Your Feedback** space at the end of the test. From this you will be able to work out your percentage score. If your score for the test was less than 50%, you will need to work through the chapter again. If you are still having trouble, ask your teacher to explain those topics that you can't understand.
- Worked solutions to the tests are found at the back of the book. Check the worked solutions to any test questions that you answered incorrectly.

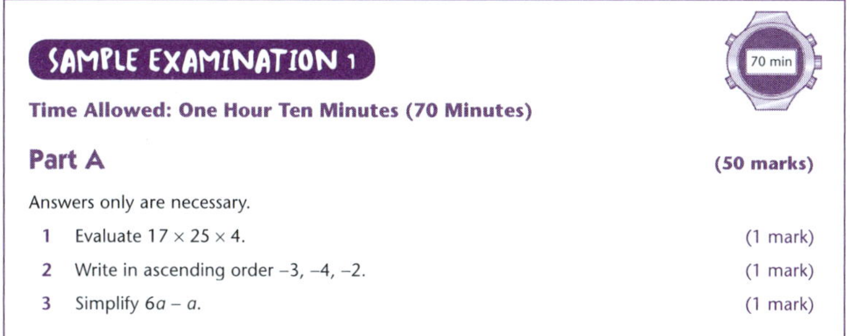

SAMPLE EXAMINATION 1

Time Allowed: One Hour Ten Minutes (70 Minutes)

70 min

Part A **(50 marks)**

Answers only are necessary.

1 Evaluate $17 \times 25 \times 4$. (1 mark)

2 Write in ascending order $-3, -4, -2$. (1 mark)

3 Simplify $6a - a$. (1 mark)

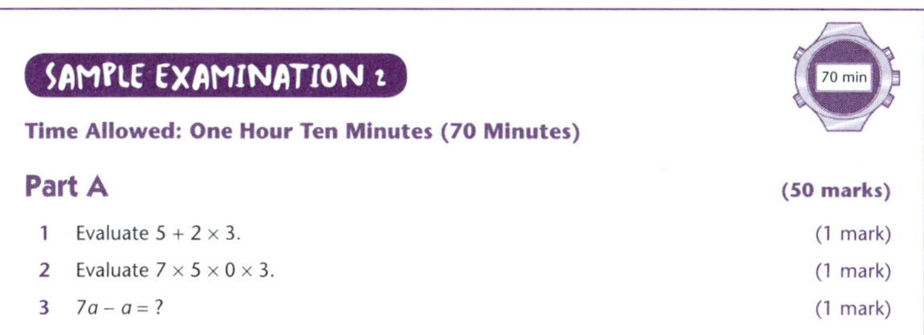

SAMPLE EXAMINATION 2

Time Allowed: One Hour Ten Minutes (70 Minutes)

70 min

Part A **(50 marks)**

1 Evaluate $5 + 2 \times 3$. (1 mark)

2 Evaluate $7 \times 5 \times 0 \times 3$. (1 mark)

3 $7a - a = ?$ (1 mark)

- When you have completed all of the chapters you can try the **Sample Examination Papers**. The Sample Examination Papers cover the entire year's work, so you should attempt the papers if you have covered all the topics in this book. (Ask your teacher if you are not sure.) The Sample Examination Papers will provide you with good practice for your final Year 7 examination. Like the tests, they are of three levels of difficulty. Set aside the time allowed for each paper and complete them under exam conditions.

Some general advice on succeeding in Mathematics

- Mathematics is a subject that **builds on knowledge**. Often you need to master one topic in order to be able to understand a later topic. This is why you should work through the book from beginning to end.
- Many students fail to **read questions carefully**. This accounts for a large number of the mistakes in exams. Although your time is limited, you must still take enough time to read each question carefully. If necessary, read the question a second (or even a third) time. Don't start your answer until you are sure you understand the question. For the longer questions, take a moment or two to **plan** your answer.
- Take care with your working. Write all answers down the page and avoid skipping any steps, as this often leads to errors. Your teachers like to see all your working, so it is wise to get into good habits. For questions that require a lot of working (such as equations), **write down each step of the solution on a separate line, lining up any equals signs** (=). This helps your teachers follow your working and will also help you check your work.
- Set out all of your working, because in Mathematics you may get some marks for your working, even if your answer is wrong.
- Always work through the solutions to any questions you answer incorrectly. This helps you learn. Remember: if you have attempted the questions, you are entitled to look at the answers—this is not cheating. If you don't get feedback, you won't be able to improve!
- You cannot expect to score 100% all the time. Remember that this book has been designed to help you identify your strengths as well as your weaknesses, so it is OK to make mistakes. The key to success is to learn from those mistakes.

NUMBER 1

- Expanded Notation
- Ordering Numbers
- Grouping Symbols
- Order of Operations
- Mathematical Language
- The Unitary Method
- Operations with Whole Numbers
- Problems Involving Different Operations
- Mental Computation
- Odd and Even Numbers
- Figurate Numbers
- Palindromic Numbers
- Fibonacci Numbers
- Multiples
- Factors
- Primes and Composites
- Index Notation
- Prime Factors
- Using a Factor Tree to Find the HCF or LCM
- Squares, Square Roots, Cubes and Cube Roots
- Divisibility
- Problem-Solving Involving LCM

KEYWORDS

Addition	**HCF**
Ascending	**Index**
Composite	**Jump method**
Cube	**Less than**
Cube root	**LCM**
Decimal system	**Multiplication**
Descending	**Odd**
Divisibility	**Order of operations**
Division	**Palindromic**
Even	**Powers of 10**
Expanded notation	**Prime**
Factor tree	**Product**
Fibonacci	**Square**
Figurate	**Square root**
Greater than	**Subtraction**
Grouping symbols	**Triangular**
	Unitary method

Expanded Notation

Our number system is based on ten and is thus known as a **decimal system**. It is called the Hindu–Arabic system because it had its origins in India with the Hindus and was further developed by the Arabs.

For example, because our system is based on **powers of 10**, the number 7469 represents:

$$7000 + 400 + 60 + 9$$

$$\text{or } 7 \times 1000 + 4 \times 100 + 6 \times 10 + 9 \times 1$$

In general, a number can be broken down this way. Let's look at a table that shows the **place value** of numbers on the top line, and their power of 10 value on the bottom line:

Millions	Hundred Thousands	Ten Thousands
$\times 1\,000\,000$	$\times 100\,000$	$\times 10\,000$
$\times 10^6$	$\times 10^5$	$\times 10^4$

Thousands	Hundreds	Tens	Unit
$\times 1000$	$\times 100$	$\times 10$	$\times 1$
$\times 10^3$	$\times 10^2$	$\times 10^1$	$\times 1$

e.g. 7 345 928

$$\begin{aligned} &= 7 \times 1\,000\,000 \\ &\quad + 3 \times 100\,000 + 4 \times 10\,000 \\ &\quad + 5 \times 1000 + 9 \times 100 \\ &\quad + 2 \times 10 + 8 \times 1 \\ &= 7 \times 10^6 + 3 \times 10^5 + 4 \times 10^4 \\ &\quad + 5 \times 10^3 + 9 \times 10^2 + 2 \times 10 \\ &\quad + 8 \times 1 \end{aligned}$$

Breaking a number down this way is known as writing the number in **expanded notation**.

In the number 7 345 928, the 3 has a place value of 300 000, the 4 a place value of 40 000 and so on.

Note: Any number (except 0) that is raised to the power of 0, equals 1.

Write in expanded notation:

1 6345

2 473 829

1 $6345 = 6 \times 1000 + 3 \times 100 + 4 \times 10 + 5 \times 1$
$= 6 \times 10^3 + 3 \times 10^2 + 4 \times 10 + 5 \times 1$

2 $473\,829 = 4 \times 100\,000 + 7 \times 10\,000 + 3 \times 1000 + 8 \times 100 + 2 \times 10 + 9 \times 1$

$= 4 \times 10^5 + 7 \times 10^4 + 3 \times 10^3 + 8 \times 10^2 + 2 \times 10 + 9 \times 1$

For Example

Simplify these numbers:

1 $2 \times 10^3 + 4 \times 10^2 + 5 \times 10 + 2$

2 $5 \times 10^4 + 3 \times 10 + 7$

1 $2 \times 10^3 + 4 \times 10^2 + 5 \times 10 + 2$

$= 2 \times 1000 + 4 \times 100 + 5 \times 10 + 2$

$= 2000 + 400 + 50 + 2$

$= 2452$

2 $5 \times 10^4 + 3 \times 10 + 7$

$= 5 \times 10\,000 + 3 \times 10 + 7$

$= 50\,000 + 30 + 7$

$= 50\,037$

For Example

Write in Hindu–Arabic numerals:

1 $3 \times 10\,000 + 7 \times 100 + 4 \times 10$

2 $7 \times 10^5 + 5 \times 10^4 + 3 \times 10^2 + 6 \times 10 + 2 \times 1$

1 $3 \times 10\,000 + 7 \times 100 + 4 \times 10$

$= 30\,000 + 700 + 40$

$= 30\,740$

2 $7 \times 10^5 + 5 \times 10^4 + 3 \times 10^2 + 6 \times 10 + 2 \times 1$

$= 700\,000 + 50\,000 + 300 + 60 + 2$

$= 750\,362$

Ordering Numbers

These important mathematics symbols are used to show the order of numbers:

$=$	is equal to	$\neq$	is not equal to
$<$	is less than	$>$	is greater than
$\nless$	is not less than	$\ngtr$	is not greater than
$\leq$	is less than or equal to	$\geq$	is greater than or equal to

For Example

1 Place the correct symbol between the two numbers to show their order:

a 97 79 b 4334 4433

2 Indicate true or false:

a $6 + 11 \nless 11 + 7$

b $4 \times 3 = 3 \times 4$

c $17 - 4 \ngtr 14$

d $25 \neq 5 \times 7 - 10$

e $15 + 5 \geq 10 \times 2$

3 Arrange in ascending order:

a 77, 65, 56, 66

b 134, 341, 314, 143, 413, 431

4 Arrange in descending order:

415, 514, 145, 154, 451, 541

1 a $97 > 79$ (97 is greater than 79)

b $4334 < 4433$ (4334 is less than 4433)

2 a $6 + 11 \nless 11 + 7$

$17 \nless 18$

As $17 < 18$ $\therefore$ False

b $4 \times 3 = 3 \times 4$

As $12 = 12$ $\therefore$ True

c $17 - 4 \ngtr 14$

$13 \ngtr 14$

As 13 is not greater than 14 $\therefore$ True

d $25 \neq 5 \times 7 - 10$

$25 \neq 35 - 10$

$25 \neq 25$

As $25 = 25$ $\therefore$ False

e $15 + 5 \geq 10 \times 2$

As $20 \geq 20$ $\therefore$ True

3 **a** 56, 65, 66, 77

b 134, 143, 314, 341, 413, 431

Ascending—going up from smallest to largest

4 541, 514, 451, 415, 154, 145

Descending—going down from largest to smallest

Grouping Symbols

These grouping symbols are used to show the order of work in mathematical expressions:

( ) Parentheses
[ ] Brackets
{ } Braces

Note: Calculations enclosed in grouping symbols are done first.

Order of Operations

Let's look at the signs we use:

Strong signs (with muscles) $\times$ $\div$

Weak signs (puny) 

Follow these rules when evaluating any mathematical expression:

1. Do any calculations within **grouping symbols** first.
2. If there are no grouping symbols and the question contains only strong signs or only weak signs, work **from left to right**, doing each operation as it occurs.
3. If the question contains a mixture of both strong and weak signs, then work from left to right, doing any **strong signs first** and then weak signs next.

This is also referred to as BODMAS: Bracket Order Division Multiplication Addition Subtraction.

For Example

Evaluate:

1 $(6 + 3) \times 4$

2 $17 - [17 - (17 - 9)]$

1 $(6 + 3) \times 4 = 9 \times 4$
$= 36$

2 $17 - [17 - (17 - 9)] = 17 - [17 - 8]$
$= 17 - 9$
$= 8$

For Example

1 $17 + 3 - 18 + 2$

2 $6 + 4 \times 7$

3 $36 - 42 \div 7$

4 $5 \times 7 + 5 \times 4$

5 $19 - 3 \times 4 + 8$

6 $8 + 5 \times (3 + 4) - 1$

7 Insert grouping symbols into these expressions to give the indicated answer:

a $7 + 5 \times 9 = 108$

b $78 - 3 \times 4 + 6 = 48$

1 $17 + 3 - 18 + 2$
$= 20 - 18 + 2$
$= 2 + 2$
$= 4$

[Work from left: 17 + 3 is done first, followed by 20 − 18, and then 2 + 2 = 4]

2 $6 + 4 \times 7 = 6 + 28$
$= 34$

[Strong sign 4 × 7 done first, followed by 6 + 28 = 34]

3 $36 - 42 \div 7 = 36 - 6$
$= 30$

4 $5 \times 7 + 5 \times 4 = 35 + 20$
$= 55$

5 $19 - 3 \times 4 + 8 = 19 - 12 + 8$
$= 7 + 8$
$= 15$

6 $8 + 5 \times (3 + 4) - 1 = 8 + 5 \times 7 - 1$
$= 8 + 35 - 1$
$= 43 - 1$
$= 42$

7 **a** $(7 + 5) \times 9 = 108$

b $78 - 3 \times (4 + 6) = 48$

Mathematical Language

Here are some different words used to indicate mathematical operations:

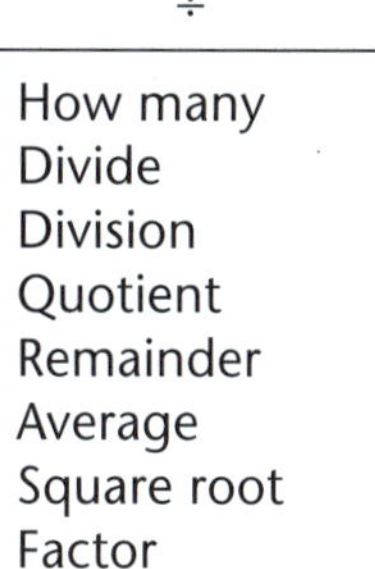

+	−	×	÷
Plus	Minus	Times	How many
Add	Take away	Multiply	Divide
Addition	Subtract	Multiplication	Division
Sum	Exceed	Lots of	Quotient
Total	Difference	Product	Remainder
		Squared	Average
		Power	Square root
			Factor

For Example

1 Find the sum of 17 and 13.

2 By how much does the product of 6 and 8 exceed their sum?

3 Calculate the quotient of 36 and 9.

1 Sum $= 17 + 13$
$= 30$

2 Product $= 6 \times 8$
$= 48$
Sum $= 6 + 8$
$= 14$
Difference $= 48 - 14$
$= 34$
$\therefore$ The product exceeds the sum by 34.

3 Quotient $= 36 \div 9$
$= 4$

The Unitary Method

We use the **unitary method** to find the value of one item and then use this value to answer the question.

For Example

1 If 10 pencils cost $2.00, how much money would I need to buy 15 pencils?

1 10 pencils cost $2.00

$\therefore$ 1 pencil costs $2.00 ÷ 10 = $0.20 or 20c [Find cost of one item]

$\therefore$ 15 pencils cost $0.20 × 15 = $3.00

$\therefore$ I would need $3.00.

For Example

1 45 trucks (all the same size) can carry a total of 450 tonnes. How much can be carried by 20 trucks?

1 45 trucks can carry 450 tonnes

$\therefore$ 1 truck can carry (450 ÷ 45) tonnes
= 10 tonnes

$\therefore$ 20 trucks can carry 20 × 10
= 200 tonnes

Sometimes you can simplify the question by finding the value of 5 or 10 items rather than 1, and then building up to the answer.

For example:

5 trucks (45 ÷ 9) can carry (450 ÷ 9) tonnes
= 50 tonnes

$\therefore$ 20 trucks can carry 4 × 50 tonnes
= 200 tonnes

Operations with Whole Numbers

Addition and Subtraction

When the question involves large numbers it may be easier to rewrite the numbers down rather than across the page.

For Example

Evaluate:

1 72 + 467 + 39 + 7094

2 648 − 399

1
$$\begin{array}{r} 72 + \\ 467 \\ 39 \\ 7094 \\ \hline 7672 \\ \hline \end{array}$$
[Answer: 7672]

2
$$\begin{array}{r} 648 - \\ 399 \\ \hline 249 \\ \hline \end{array}$$
[Answer: 249]

Long Multiplication

When multiplying by large numbers, we multiply by the units on the right first, then the tens, then the hundreds, and so on. This means that when multiplying by 63, we first multiply by 3, then multiply by 60, then add the two results together.

For Example

1 74 × 69

2 804 × 376

1
```
  74 ×
  69
────
 666
4440
────
5106
```

We are multiplying 74 by 9 (666) and then 74 by 60 (4440). This is the reason for putting the zero at the end of the second row. We get our answer by adding the two rows

2
```
    804 ×
    376
───────
   4824
  56280
 241200
───────
 302304
```

Multiply 804 by 6 (4824), then by 70 (56 280), then by 300 (241 200). This time, we have three rows to add up before we get our answer

Division

To divide is to answer the question 'How many … ?' This means that 72 ÷ 8 can be thought of as 'how many 8s in 72?'

For Example

Evaluate:

1 728 ÷ 7 **2** 524 ÷ 5

3 $9\overline{)4635}$ **4** $6\overline{)7215}$

1 $728 \div 7 = 7\overline{)72^28}$ (quotient: 10 4)

How many times does 7 go into 7? 1
How many times does 7 go into 2? 0, carry the 2
How many times does 7 go into 28? 4

Answer: 104

2 $524 \div 5 = 5\overline{)52^24}$ (quotient: $10\,4\frac{4}{5}$)

Answer: $104\frac{4}{5}$

(Note the two alternatives for writing a remainder: R4 or $\frac{4}{5}$)

3 $9\overline{)46^13^45}$ (quotient: 5 1 5)

Answer: 515

4 $6\overline{)7^12 1^15}$ (quotient: $1\,20\,2\frac{3}{6}$)

Answer: $1202\frac{3}{6}$ or $1202\frac{1}{2}$

Long Division

A long division method records the steps on paper that are mentally calculated if using a short division method:

How many
Multiply
Subtract
Bring down

For Example

Evaluate:

1 7259 ÷ 37

2 46830 ÷ 59

1
```
     196
37)7259
   37
   355
   333
    229
    222
      7
```

How many 37s in 72? 1
Multiply 1 × 37 = 37
Subtract 72 − 37 = 35
Bring down the next number—5
How many 37s in 355? 9
Multiply 9 × 37 = 333
Subtract 355 − 333 = 22
Bring down next number—9
How many 37s in 229? 6
Multiply 6 × 37 = 222
Subtract 229 − 222 = 7

Answer: $196\frac{7}{37}$

2
```
      793
59)46830
   413
    553
    531
     220
     177
      43
```

Answer: $793\frac{43}{59}$

Problems Involving Different Operations

Problems can be solved by using addition, subtraction, multiplication and division.

For Example

1 Allyn keeps 6 dogs, 8 cats, 5 birds and 3 snakes as pets. How many pets does Allyn keep?

2 Greg packs 12 eggs in each carton. If Greg has 15 cartons, how many eggs does he pack?

3 Maria has to transport 25 children to the museum by car. Each car can carry 4 passengers. How many cars are needed?

4 A syndicate of 42 won $463 000 in Lotto. The winning amount was divided evenly between the 42 and each person received a 'whole number' of dollars. The remaining amount was donated to charity. How much was donated to charity?

1 Number of pets $= 6 + 8 + 5 + 3$

$= 22$

$\therefore$ Allyn has 22 pets.

2 Number of eggs $= 12 \times 15$

$$\begin{array}{r} 12\times \\ 15 \\ \hline 60 \\ 120 \\ \hline 180 \\ \hline \end{array}$$

$\therefore$ Greg packs 180 eggs.

3 Number of cars $= 25 \div 4$

$$4\overline{)25}\quad = 6\tfrac{1}{4}$$

$\therefore$ Seven cars would be needed (unless we leave one student behind).

4 Each share $= \$463\,000 \div 42$

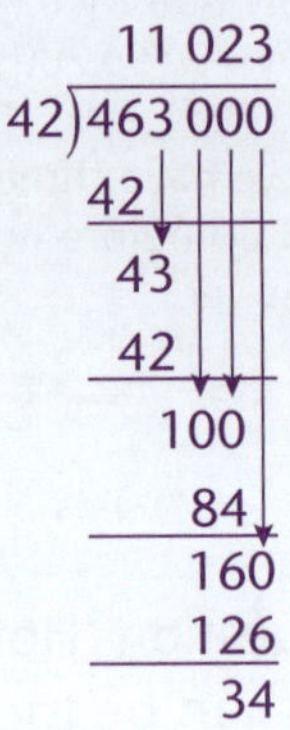

$$\begin{array}{r} 11\,023 \\ 42\overline{)463\,000} \\ 42 \\ \hline 43 \\ 42 \\ \hline 100 \\ 84 \\ \hline 160 \\ 126 \\ \hline 34 \end{array}$$

The remainder is $34.
A charity will receive $34.
(Each winner received $11 023.)

Mental Computation

There are many simple ideas that help people do calculations mentally or just quickly. This is called mental computation. Some of these ideas are included here.

Jumping when Doing Simple Additions or Subtractions

Here the numbers to be added are separated into tens and units.

1 Evaluate $67 + 24$.

1 [number line: 67 —20→ 87 —4→ 91]

$\therefore 67 + 24 = 91$

For Example

1 Evaluate $83 - 47$.

1 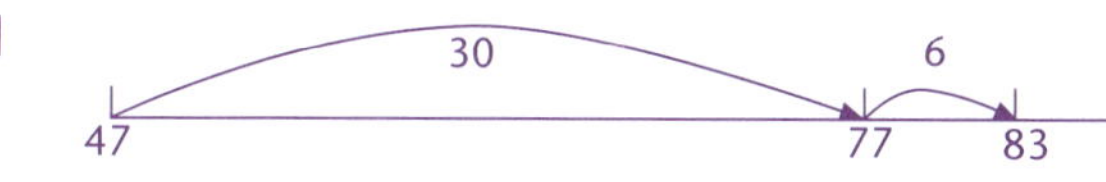

Start from 47 and jump to 83.

The jumps total 36.

$\therefore 83 - 47 = 36$

Another Addition and Subtraction Method

Again, each number can be separated into units, tens, hundreds etc.

For Example

1 Evaluate $67 + 24$.

1 Break up each number:

$$\begin{aligned} 60 + 20 + 7 + 4 &= 80 + 11 \\ &= 91 \end{aligned}$$

For Example

1 Evaluate $83 - 47$.

1 Break up each number:

$$\begin{aligned} 83 - 40 - 7 &= 43 - 7 \\ &= 36 \end{aligned}$$

Multiplication and Division of Whole Numbers by 10, 100 etc.

The tables below give you hints on multiplying and dividing by powers of 10.

To multiply a whole number by:	Do this:
10	Add one zero e.g. $75 \times 10 = 750$
100	Add two zeros e.g. $75 \times 100 = 7500$
1000	Add three zeros e.g. $75 \times 1000 = 75\,000$

To divide a multiple of 10 by:	Do this:
10	Remove the final zero e.g. $6800 \div 10 = 680$
100	Remove the final two zeros e.g. $6800 \div 100 = 68$

A similar process is used for all powers of 10.

Note: Multiplying and dividing decimals by powers of 10 will be treated in Chapter 3.

Multiplying by 10 or 100 after an Order Change

Remember that $7 \times 5 \times 2$ is the same as $7 \times 2 \times 5$ or $2 \times 5 \times 7$. The order of multiplication does not affect the answer.

For Example

1 Evaluate $5 \times 17 \times 2$.

1
$$\begin{aligned} 5 \times 17 \times 2 &= 5 \times 2 \times 17 \\ &= 10 \times 17 \\ &= 170 \end{aligned}$$

1 Evaluate $25 \times 19 \times 4$.

1 $25 \times 19 \times 4 = 25 \times 4 \times 19$
$= 100 \times 19$
$= 1900$

Multiplication by 9 or 11

To multiply a number by 9, multiply by 10 and then subtract the number.

1 Evaluate 47×9.

1 $47 \times 9 = 47 \times 10 - 47$
$= 470 - 47$
$= 423$

To multiply a number by 11, multiply by 10 and then add the number.

1 Evaluate 47×11.

1 $47 \times 11 = 47 \times 10 + 47$
$= 470 + 47$
$= 517$

Similarly, to multiply a number by 8, multiply by 10 and then subtract twice the number. To multiply a number by 12, multiply by 10 and then add twice the number.

How would you multiply by 7 or 13?

Multiplication and Division by Multiples of 10

To multiply a number by 20, double the number and then multiply by 10.

To divide a number by 20, halve the number and then divide by 10.

Evaluate:

1 36×20 2 $360 \div 20$

1 $36 \times 2 = 72$
$72 \times 10 = 720$
$\therefore\ 36 \times 20 = 720$

2 $360 \div 2 = 180$
$180 \div 10 = 18$
$\therefore\ 360 \div 20 = 18$

A similar process can be used to multiply or divide by 30, 40, 50 etc. Multiplying or dividing by 200, 300, 400 etc. can also be done using a similar process.

Shortcuts when Numbers Add to 10, 100

The order in which numbers are added does not affect the answer.

For example, $7 + 11 + 9 = 7 + 9 + 11$
$= 11 + 7 + 9$

1 Evaluate $7 + 91 + 3$.

1 Rearrange the question:
$7 + 91 + 3 = 7 + 3 + 91$
$= 10 + 91$
$= 101$

1 Evaluate 19 + 37 + 81.

1 Rearrange the question:

$$19 + 37 + 81 = 19 + 81 + 37$$
$$= 100 + 37$$
$$= 137$$

(This shortcut relies upon your seeing pairs of numbers that add up to 10 or 100. This process can be used for sums other than 10 or 100 such as 20, 30, 200 etc.)

Note: The complete working has been shown for all of these shortcuts. However, once these methods have been practised and understood you should be able to do most of these calculations mentally.

Odd and Even Numbers

Odd numbers: 1, 3, 5, 7, 9, 11, 13, 15 …
All odd numbers end in 1, 3, 5, 7 or 9.

Even numbers: 2, 4, 6, 8, 10, 12, 14, 16 …
All even numbers end in 0, 2, 4, 6 or 8.

Note: The number zero, 0, is neither even nor odd.

Figurate Numbers

1 Square numbers

1 4 9 16 25 36 …

Square numbers:
1, 4, 9, 16, 25, 36, 49 …

or $1^2, 2^2, 3^2, 4^2, 5^2, 6^2$ …

2 Triangular numbers

1 3 6 10 15 21 28

Triangular numbers: 1, 3, 6, 10, 15, 21, 28, 36, 45 …

Note:
$3 = 1 + 2$
$6 = 1 + 2 + 3$
$10 = 1 + 2 + 3 + 4$
$15 = 1 + 2 + 3 + 4 + 5$
$21 = 1 + 2 + 3 + 4 + 5 + 6$

Palindromic Numbers

Palindromic numbers have the same value when read either from the beginning or the end.

Examples are 626, 575, 1661, 7007 or 51 515. Palindromic numbers can be generated from non-Palindromic numbers by an addition process.

1 Consider 47.

2 Consider 947.

1

47	
74	Reverse the digits
121	Add

121 is Palindromic.

2

947	
749	Reverse the digits
1696	Add
6961	Reverse the digits
8657	Add
7568	Reverse the digits
16225	Add
52261	Reverse the digits
68486	

68486 is Palindromic.

Fibonacci Numbers

Here the next number in the sequence is obtained by adding the previous two numbers of the sequence.

1, 1, 2, 3, 5, 8, 13, 21, 34, 55 ...

i.e. $1 + 1 = 2$
$1 + 2 = 3$
$2 + 3 = 5$
$3 + 5 = 8$
$5 + 8 = 13$

Multiples

The first 10 multiples of 6 are 6, 12, 18, 24, 30, 36, 42, 48, 54, 60:

i.e. $6 \times 1, 6 \times 2, 6 \times 3, 6 \times 4, 6 \times 5, 6 \times 6, 6 \times 7, 6 \times 8, 6 \times 9, 6 \times 10$

The multiples of 5 are 5, 10, 15, 20, 25, 30, 35, 40, 45, 50 ...

i.e. $5 \times 1, 5 \times 2, 5 \times 3, 5 \times 4, 5 \times 5, 5 \times 6, 5 \times 7$ ...

1 Write down the first 6 multiples of 4.

2 Find all the multiples of 7 between 10 and 30.

3 Write down all the multiples of 3 less than 20.

1 $4 \times 1, 4 \times 2, 4 \times 3, 4 \times 4, 4 \times 5, 4 \times 6$
$= 4, 8, 12, 16, 20, 24$

2 Multiples of 7 are 7, 14, 21, 28, 35, 42 ...
Multiples between 10 and 30 are 14, 21, 28.

3 Multiples of 3 are 3, 6, 9, 12, 15, 18.

Common Multiples

1 Find the common multiples of 3 and 4.

2 Find any common multiples of 3 and 5 that are less than 50.

1 Multiples of 3: 3, 6, 9, (12), 15, 18, 21, (24), 27, 30, 33, (36)...

Multiples of 4: 4, 8, (12), 16, 20, (24), 28, 32, (36), 40 ...

Common multiples of 3 and 4 are 12, 24, 36 ...

Note: The common multiples are multiples of both 3 and 4; that is, multiples of 12.

2 Multiples of 3: 3, 6, 9, 12, (15), 18, 21, 24, 27, (30), 33, 36, 39, 42, (45), 48 ...

Multiples of 5: 5, 10, (15), 20, 25, (30), 35, 40, (45), 50 ...

Common multiples of 3 and 5 less than 50 are 15, 30, 45.

Lowest Common Multiple (LCM)

This is the smallest number that is a common multiple of two or more numbers.

The Lowest Common Multiple of 3 and 5 (from example 2 above) is 15.

For the LCM of 4 and 6, consider the multiples of 4 and 6:

Multiples of 4: 4, 8, (12), 16, 20, 24 …
Multiples of 6: 6, (12), 18, 24 …
LCM of 4 and 6 = 12

Generally, the quickest method of calculating the LCM is to write down (or mentally count through) the multiples of the larger number until the first multiple of the other number is reached.

For example, to find the LCM of 6 and 8, write down the multiples of 8:

8, 16, (24) … until the first multiple of 6 occurs. This is 24.

∴ LCM of 6 and 8 = 24.

For another example, find the lowest common multiple of 2, 5 and 6:

Multiples of 6: 6, 12, 18, 24, (30) …

The first multiple of 6 that is also a multiple of both 2 and 5 is the answer. This is 30 (a multiple of 6, 5 and 2).

∴ LCM of 2, 5 and 6 = 30.

Factors

Any whole number that divides exactly into another number is called a **factor** of that number.

For example, as	$2 \times 3 = 6$
then	2 is a factor of 6
	3 is a factor of 6
But,	$6 \times 1 = 6$ also
∴	6 is a factor of 6
	1 is a factor of 6

∴ The factors of 6 are 1, 2, 3, 6.

One and the **number itself** are always factors of any number.

1 Write down all the factors of 24.

1 Now $1 \times 24 = 24$
$2 \times 12 = 24$
$3 \times 8 = 24$
$4 \times 6 = 24$

∴ Factors of 24 are 1, 2, 3, 4, 6, 8, 12, 24.

Common Factors

When two (or more) numbers have the same factor, that factor is called a **common factor**:

Factors of 24 are (1), (2), (3), 4, (6), 8, 12, 24

Factors of 18 are (1), (2), (3), (6), 9, 18

∴ The common factors of 24 and 18 are 1, 2, 3 and 6.

1 Find all the common factors of 16 and 12.

2 Find all the factors that are common to 9, 15 and 18.

1 Factors of 16: (1), (2), (4), 8, 16

Factors of 12: (1), (2), 3, (4), 6, 12

∴ Common factors of 16 and 12 are 1, 2, 4.

2 Factors of 9: (1), (3), 9

Factors of 15: (1), (3), 5, 15

Factors of 18: (1), 2, (3), 6, 9, 18

∴ Common factors of 9, 15 and 18 are 1 and 3.

Note: One is always a common factor of any two (or more) numbers.

Highest Common Factor (HCF)

The largest number that divides exactly into two (or more) numbers is known as the **Highest Common Factor** (HCF).

It is the largest number that is a common factor.

For example, consider 12 and 16:

Factors of 12: (1), (2), 3, (4), 6, 12

Factors of 16: (1), (2), (4), 8, 16

Common factors are 1, 2, 4

∴ Highest Common Factor is 4.

1 Find the HCF of 21 and 24.

2 Find the Highest Common Factor of 8, 12 and 16.

1 Factors of 21: (1), (3), 7, 21

Factors of 24: (1), 2, (3), 4, 6, 8, 12, 24

Common factors are 1 and (3)

∴ HCF of 21 and 24 is 3.

2 Factors of 8: (1), (2), (4), 8

Factors of 12: (1), (2), 3, (4), 6, 12

Factors of 16: (1), (2), (4), 8, 16

Common factors are 1, 2, (4)

∴ HCF of 8, 12 and 16 is 4.

In reality (and after practise) you write down (or think of) the factors of the smallest number. The largest of these factors that is also a factor of the other numbers is the Highest Common Factor.

Primes and Composites

A **prime** number has only two factors, the number itself and one (except for 0 and 1, which are special cases).

e.g. $5 = 1 \times 5$
5 is a prime number

$7 = 1 \times 7$
7 is a prime number

A **composite** number is any whole number that is not prime (except for 0 and 1).

e.g. 6 has factors 1, 2, 3, 6
∴ 6 is not prime
i.e. 6 is a composite number

The prime numbers less than 20 are 2, 3, 5, 7, 11, 13, 17, 19.

Hence, the composite numbers less than 20 are 4, 6, 8, 9, 10, 12, 14, 15, 16, 18.

The Sieve of Eratosthenes

A neat method for generating primes is to use the Sieve of Eratosthenes. This method was devised in the 3rd century BC by Eratosthenes.

The counting numbers are written down in groups of 6.

Cross out 1. Circle 2 as it is prime and cross out all multiples of 2. Circle 3—it is prime. Cross out all multiples of 3. Circle 5—next prime. Cross out all multiples of 5. Circle 7, cross out multiples of 7. Continue circling the next prime and continue this process until only primes remain:

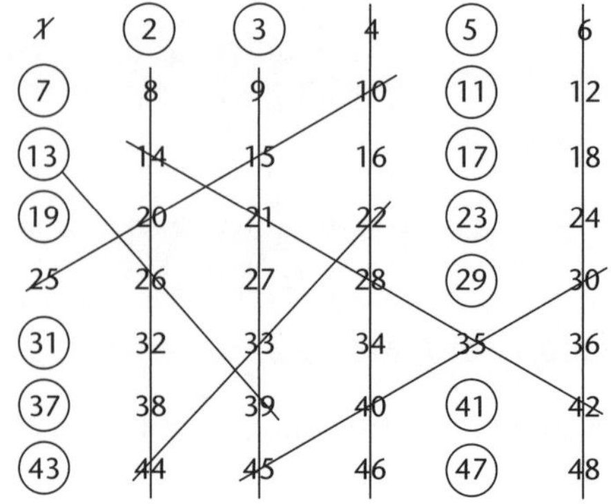

Primes sieved out then are 2, 3, 5, 7, 11, 13, 17, 19, 23, 29, 31, 37, 41, 43, 47 …

This process can be operated up to any limit. Notice how all the multiples occur in lines when groups of six are used.

For Example

1 Write down the prime numbers between 10 and 20.

2 Write 18 as the sum of two primes.

3 Twin primes are primes separated by a single even number. Write down a pair of twin primes.

4 Find the set of composite numbers between 10 and 20.

1 From the Eratosthenes diagram, 11, 13, 17, 19 are primes.

2 $18 = 13 + 5$ or $18 = 11 + 7$

3 (3), 4, (5) … (11), 12, (13) … (17), 18, (19)

Some sets of twin primes are 3 and 5, 11 and 13, 17 and 19.

4 Composite numbers are 12, 14, 15, 16, 18.

Index Notation

Index notation is a method of recording **products** containing the **same number** in a concise manner. (Remember, a product is the result of multiplying two or more numbers together.)

e.g.		
	5×5 can be written as	5^2
	$5 \times 5 \times 5$	5^3
	$5 \times 5 \times 5 \times 5$	5^4

For the term 5^3, 5 is the **base** number and 3 is the **index**. The index tells you how many times to write the base number in the product:

e.g. 5^4 means write the product of four 5s
$5^4 = 5 \times 5 \times 5 \times 5 = 625$

Remember, $5^0 = 1$.

For Example

1 Write 5^7 in expanded form.

1 $5^7 = 5 \times 5 \times 5 \times 5 \times 5 \times 5 \times 5$

[The number 5 is written as a product 7 times.]

For Example

1 Write the product of $3 \times 3 \times 3 \times 3 \times 7 \times 7 \times 7 \times 7 \times 7$ in index form.

1 $\underbrace{3 \times 3 \times 3 \times 3}_{4} \times \underbrace{7 \times 7 \times 7 \times 7 \times 7}_{5} = 3^4 \times 7^5$

Prime Factors

The prime factors of a composite number are the factors of that number that are primes.

For example, factors of 18 are 1, (2), (3), 6, 9, 18.

Of these factors, the primes are 2 and 3.
∴ Prime factors of 18 are 2 and 3.

Any composite number can be written as the product of its prime factors.

The simplest method for doing this is a **factor tree**.

For Example

1 Write 54 as the product of its prime factors.

1

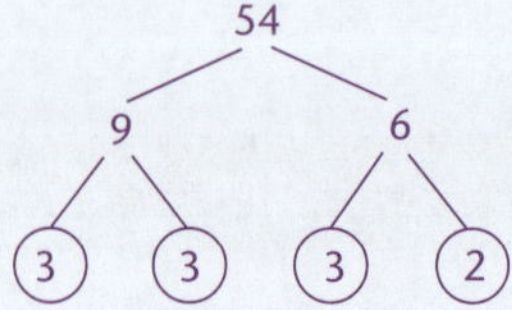

Use any two factors of the number, such as 9×6 (it does not matter which choice of factors you start with). Then write down the factors of both 9 and 6. As you obtain a prime, circle it—this is the end of that branch.

$\therefore 54 = 3 \times 3 \times 3 \times 2$

or $54 = 3^3 \times 2$ (using index notation)

A branch ends only when a prime is obtained.

For Example

1 Find the prime factors of 96.

1

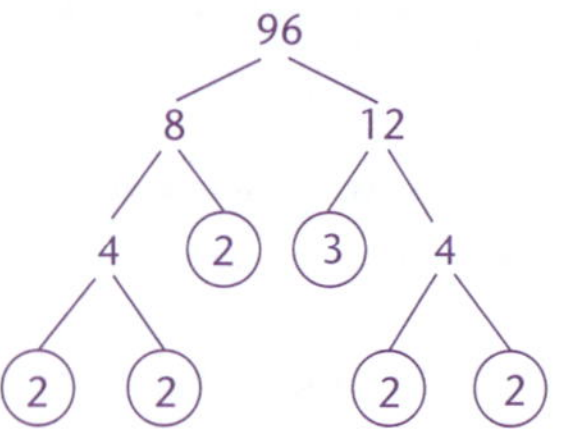

$96 = 2 \times 2 \times 2 \times 2 \times 2 \times 3$

or $96 = 2^5 \times 3$ (index notation)

Suppose we had begun with a different choice of initial factors:

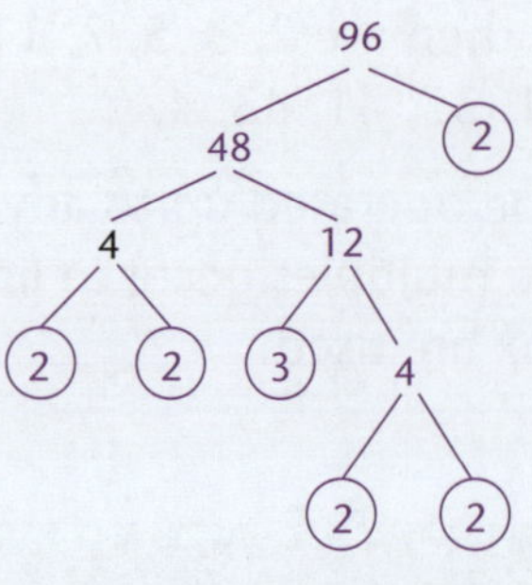

$96 = 2 \times 2 \times 2 \times 2 \times 2 \times 3$

or $96 = 2^5 \times 3$

The results obtained in each case are identical!

For Example

(Note the difference between these questions.)

1 Write down the prime factors of 36.

2 Express 36 as the product of its prime factors.

1 Factors of 36: 1, (2), (3), 4, 6, 9, 12, 18, 36

Primes are 2 and 3

$\therefore$ Prime factors of 36 are 2 and 3.

2

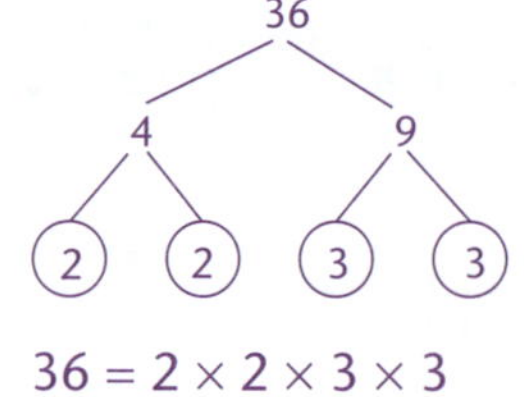

$36 = 2 \times 2 \times 3 \times 3$

$= 2^2 \times 3^2$

Using a Factor Tree to Find the HCF or LCM

When we are attempting to find the HCF or LCM of large numbers, a factor tree is useful.

For Example

1 Find the HCF of 144 and 220.

2 Find the LCM of 144 and 220.

3 Find the HCF and LCM of 150 and 240.

1 The HCF of 144 and 220 is the *largest* number that is a factor of both numbers:

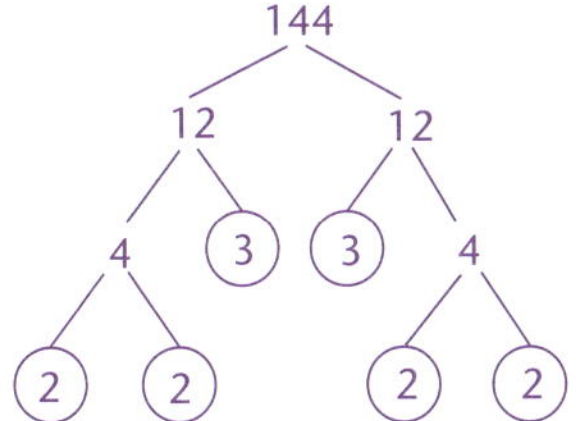

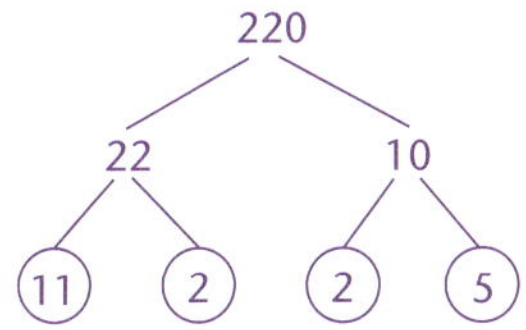

$144 = \underline{2 \times 2} \times 3 \times 3 \times 2 \times 2$

$220 = 11 \times \underline{2 \times 2} \times 5$

$\text{HCF} = 2 \times 2 = 4$

$\therefore$ HCF of 144 and 220 is 4.

2 The LCM is the *smallest* number that both 144 and 220 divide into exactly. First, write down all the factors of the largest number, and then multiply by the factors of the other number that are not already included:

$\text{LCM} = 11 \times 2 \times 2 \times 5 \times (3 \times 3 \times 2 \times 2)$
$= 7920$

Check: $7920 \div 144 = 55$
$7920 \div 220 = 36$

No remainders, so divisions are exact.

55 $(= 5 \times 11)$ and 36 $(= 2 \times 2 \times 3 \times 3)$ have no common factors, so 7920 is the LCM.

We have to include the initial 2×2 only once to obtain the LCM.

3 The HCF and LCM of 150 and 240:

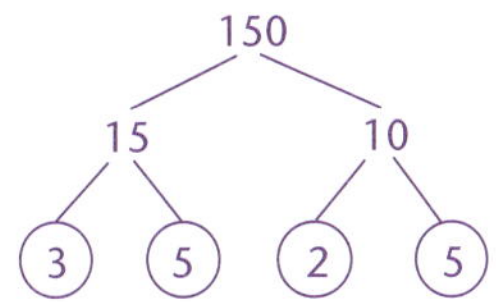

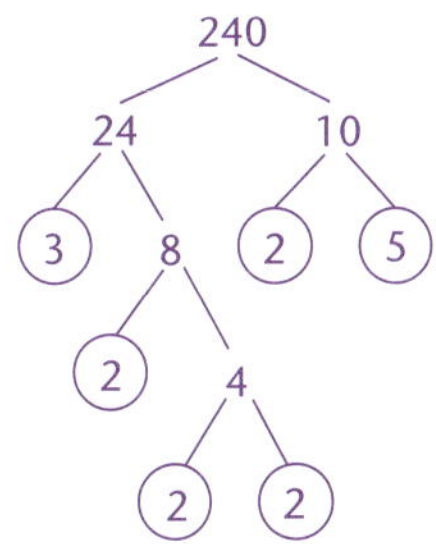

$150 = \underline{2 \times 3 \times 5} \times 5$

$240 = 2 \times 2 \times 2 \times \underline{2 \times 3 \times 5}$

OR	$150 = \underline{2 \times 3 \times 5} \times 5$
	$240 = \underline{2 \times 3 \times 5} \times 2 \times 2 \times 2$

$\text{HCF} = 2 \times 3 \times 5$
$= 30$

$\text{LCM} = \underbrace{2 \times 3 \times 5}_{\text{Include once}} \times \underbrace{2 \times 2 \times 2}_{\text{Other factors from 240}} \times \underset{\text{Other factor from 150}}{5}$

$\therefore$ HCF of 150 and 240 = 30
LCM of 150 and 240 = 1200.

Squares, Square Roots, Cubes and Cube Roots

When you multiply a number by itself, the result is the **square** of the number. The original number is then called the **square root** of the product.

For example:

[$\Rightarrow$ means 'implies'] [$\sqrt{\ }$ means square root]

$$5^2 = 5 \times 5$$
$$= 25$$
$$\Rightarrow \sqrt{25} = 5$$

$$
\begin{aligned}
7^2 &= 7 \times 7 \\
&= 49 \\
\Rightarrow \quad \sqrt{49} &= 7
\end{aligned}
$$

$$
\begin{aligned}
15^2 &= 15 \times 15 \\
&= 225 \\
\Rightarrow \quad \sqrt{225} &= 15
\end{aligned}
$$

$$
\begin{aligned}
20^2 &= 20 \times 20 \\
&= 400 \\
\Rightarrow \quad \sqrt{400} &= 20
\end{aligned}
$$

When you multiply a number by itself twice, the result is the **cube** of the number.

The original number is then called the **cube root** of the product.

[$\sqrt[3]{}$ means cube root]

$$
\begin{aligned}
2^3 \text{ (2 cubed)} &= 2 \times 2 \times 2 \\
&= 8 \\
\Rightarrow \quad \sqrt[3]{8} &= 2
\end{aligned}
$$

$$
\begin{aligned}
3^3 &= 3 \times 3 \times 3 \\
&= 27 \\
\Rightarrow \quad \sqrt[3]{27} &= 3
\end{aligned}
$$

$$
\begin{aligned}
5^3 &= 5 \times 5 \times 5 \\
&= 125 \\
\Rightarrow \quad \sqrt[3]{125} &= 5
\end{aligned}
$$

$$
\begin{aligned}
10^3 &= 10 \times 10 \times 10 \\
&= 1000 \\
\Rightarrow \quad \sqrt[3]{1000} &= 10
\end{aligned}
$$

More examples:

$\sqrt{81} = 9$ (as 9×9 or $9^2 = 81$)

$\sqrt{144} = 12$ (as 12×12 or $12^2 = 144$)

$\sqrt[3]{64} = 4$ (as $4 \times 4 \times 4$ or $4^3 = 64$)

Factor trees can be used to find the square roots or cube roots of large numbers.

For Example

Use factor trees to find:

1 $\sqrt{324}$

2 $\sqrt[3]{3375}$

1

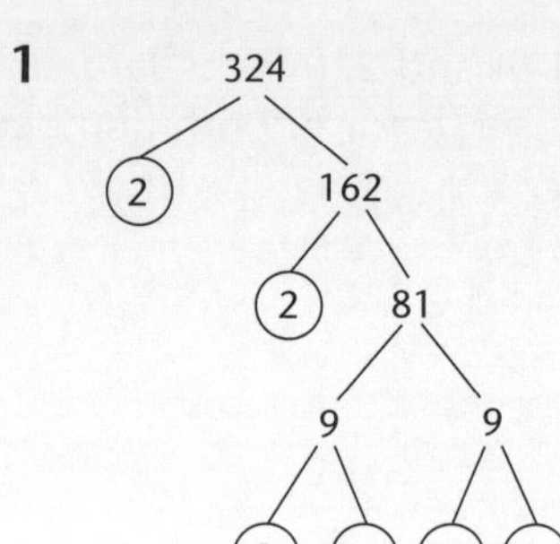

$$
\begin{aligned}
324 &= 2 \times 2 \times 3 \times 3 \times 3 \times 3 \\
&= \underline{2 \times 3 \times 3} \times \underline{2 \times 3 \times 3}
\end{aligned}
$$

$[(2 \times 3 \times 3)^2]$

$$
\begin{aligned}
\therefore \sqrt{324} &= 2 \times 3 \times 3 \\
&= 18
\end{aligned}
$$

2

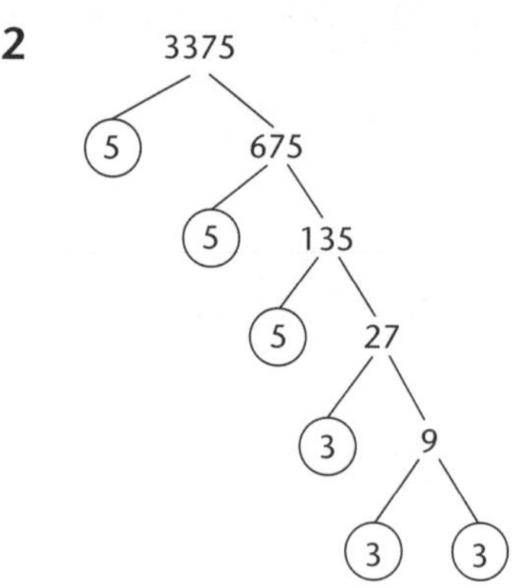

$$
\begin{aligned}
3375 &= 5 \times 5 \times 5 \times 3 \times 3 \times 3 \\
&= \underline{5 \times 3} \times \underline{5 \times 3} \times \underline{5 \times 3}
\end{aligned}
$$

$[(5 \times 3)^3]$

$$
\begin{aligned}
\therefore \sqrt[3]{3375} &= 5 \times 3 \\
&= 15
\end{aligned}
$$

Divisibility

Remember, a factor is a whole number that divides exactly into another whole number. It is helpful when looking for factors to know which numbers divide exactly into others.

Here are some tests for divisibility.

A number is divisible by:

2 if the last digit is zero or even;

3 if the sum of the digits of the number is divisible by 3;

4 if the last two digits are divisible by 4;

5 if the last digit is zero or 5;

6 if it is divisible by 2 and 3;

8 if the last three digits are divisible by 8;

9 if the sum of the digits is divisible by 9;

10 if the last digit is zero;

11 if the sum of the digits in the odd places differs from the sum of the digits in the even places by zero or a multiple of 11.

There is no simple test for divisibility by 7.

For Example

1 Find which of the numbers 234, 525, 744 and 417 208 are divisible by 2, 3, 5, 9 or 11.

1 Consider 234—it is even thus divisible by 2
Also $2 + 3 + 4 = 9$—sum of digits divisible by 3 and 9

$\therefore$ 234 is divisible by 2, 3 and 9.

Consider 525—obviously divisible by 5
$5 + 2 + 5 = 12$—sum of digits divisible by 3

$\therefore$ 525 is divisible by 5 and 3.

Consider 744—obviously divisible by 2
$7 + 4 + 4 = 15$—sum of digits divisible by 3

$\therefore$ 744 is divisible by 2 and 3.

(Note $\therefore$ 744 is divisible by 6.)

Consider 417 208—obviously divisible by 2

Also $4 + 7 + 0 = 11$
and $1 + 2 + 8 = 11$
Now, $11 - 11 = 0$

[Difference between even-placed and odd-placed digits is 0—divisible by 11]

$\therefore$ 417 208 is divisible by 2 and 11.

Problem-Solving Involving LCM

This type of question occurs frequently in mathematics competitions.

For Example

1 Greg, Lee and Maria share out their marbles and find that they each receive the same number and there are none left over. Antonie joins the group so they now share out all the marbles again, but this time between the four friends. Once again there are none left over. What is the least number of marbles they must have started with?

1 The question is really asking you to look at numbers that 3 divides into and also 4 divides into exactly. The smallest such number is the answer needed. In other words, find the LCM of 3 and 4.

LCM of 3 and 4 = 12
The friends had at least 12 marbles.

For Example

1 Suppose the problem begins in the same manner as the above example but then along came Vido. So the marbles were again shared out evenly among the five friends. Once again there were none remaining after the division. Find the least number of marbles the friends could have started with.

1 The question is asking you to find the LCM of 3, 4 and 5.

Consider the multiples of 5:

5, 10, 15, 20, 25, 30, 35, 40, 45, 50, 55, (60), 65, 70 …

60 is the first multiple of 5 that both 3 and 4 divide into exactly.

LCM of 3, 4 and 5 = 60

The friends must have begun with 60 marbles.

For Example

1 Three girls shared out a number of marbles so that each received the same number. There were 2 marbles left over. A fourth friend arrived so the marbles were shared out again, this time four ways. Once again there were 2 left over. A fifth friend was called over. The marbles were shared out once again and amazingly there were 2 left over. Find the least number of marbles that could have been available.

1 Once again the question is asking you to find the LCM of 3, 4 and 5, but the required answer will be 2 more than the LCM.

LCM of 3, 4 and 5 = 60
(from previous example)

$\therefore$ Required number is 60 + 2 = 62

Check: $62 \div 3 = 20$ remainder 2
$62 \div 4 = 15$ remainder 2
$62 \div 5 = 12$ remainder 2

The least number of marbles available is 62.

PRACTISE, PRACTISE

Go to p. 254 for quick answers, or to pp. 268–277 for worked solutions.

1 Write in expanded notation: p. 2

a 6147 **b** 74 291 **c** 70 707
d 317 429 **e** 40 030

2 Simplify these numbers: p. 2

a $4 \times 10^3 + 7 \times 10^2 + 3 \times 10 + 2 \times 1$
b $6 \times 10^3 + 9 \times 1$

3 Write in Hindu–Arabic numerals: p. 2

a $5 \times 1000 + 8 \times 100 + 9 \times 10 + 5 \times 1$
b $9 \times 10^4 + 8 \times 10^2 + 7 \times 10$

4 By how much does the place value of the first 3 exceed that of the second 3 in these numbers? p. 2

a 393 **b** 3037 **c** 33 204
d 3213 **e** 137 302

5 Place a symbol between the pair of numbers to show their order: p. 3

a $74 \square 47$ **b** $235 \square 325$ **c** $1021 \square 1201$
d $3102 \square 3012$ **e** $56174 \square 56147$

6 Indicate true or false: p. 3

a $426 < 462$ **b** $8 \times 9 = 72$ **c** $46 \neq 6 + 14 + 16$
d $42 \div 6 < 42 \div 7$ **e** $5 + 17 > 17 + 5$ **f** $13 \not< 10 + 4$
g $24 \not> 96 \div 4$ **h** $5 + 3 \leq 3 + 5$ **i** $20 - 17 < 17 - 14$
j $3 \times 2 \times 8 = 6 \times 2 \times 4$

7 Write in ascending order: p. 3

a 47, 74, 36, 63 **b** 3112, 2113, 1123, 1132
c 7415, 7514, 7145, 7154, 5147, 5174

8 Write in descending order: p. 3

a 93, 39, 68, 86 **b** 4848, 4884, 8448, 8844
c 3192, 2139, 2193, 2319, 2391, 3291, 3129, 3219

9 Evaluate the following: p. 4

a $(4 + 6) \times 5$ **b** $(16 - 8) \div 4$
c $3 \times (2 + 3) \times 4$ **d** $(3 \times 5) + 5 \times (3 + 2)$
e $[25 \div (3 + 2)] + 7$ **f** $100 - \{80 - [60 - (40 - 20)]\}$
g $(5 + 3) \times [21 - (9 + 8)]$

10 Evaluate these expressions: p. 4

a $27 - 8 - 9 - 10$ **b** $8 + 14 - 8 + 14$ **c** $27 - 14 + 26 - 13$
d $6 + 4 \times 5$ **e** $27 - 36 \div 4$ **f** $5 + 3 \times 2 + 7$
g $6 \times 4 + 4 \times 4$ **h** $36 \div 9 + 49 \div 7$ **i** $21 - 42 \div 7 - 15$
j $14 - 16 \div 8 \times 4 + 4$

11 Insert grouping symbols so that each expression gives the desired answer: p. 5

a $4 + 3 \times 7 = 49$ **b** $12 - 5 \times 3 = 21$ **c** $7 \times 3 + 2 \times 2 = 70$

d $3 \times 2 + 3 + 4 = 19$ **e** $5 + 4 + 3 \times 2 = 19$ **f** $25 \div 3 + 2 + 7 = 12$

12 Calculate: p. 5

a The sum of 15 and 14.

b The product of 15 and 4.

c The difference between 15 and 4.

d How much 15 exceeds 4.

e The quotient of 42 and 6.

f The remainder when 73 is divided by 9.

g The average of 19, 21, 36, 40, 14.

h The total of 11, 28, 14, 9, 23, 5.

i How much the sum of 19 and 13 exceeds the difference between 19 and 13.

j The product of the sum of 7 and 9 and the difference between 17 and 9.

13 If 8 T-shirts are priced at $72, how much would Mario pay for 9 T-shirts? pp. 5–6

14 Michelle received $78.75 for 7 hours' work. How much would she have received if she had worked only 4 hours? pp. 5–6

15 If 6 similar-sized cartons contain a total of 150 packages, how many packages would be contained in 8 similar cartons? pp. 5–6

16 Complete the following operations: pp. 6–7

a $72 + 27 + 98 + 89$

b $407 - 198$

c 72×27

d $465 \div 9$

e $75 + 871 + 4 + 637 + 92$

f $7004 - 989$

g 532×286

h $4237 \div 45$

i $50102 \div 69$

17 Bronwyn collects chess pieces. Bronwyn has 17 bishops, 11 queens, 9 kings and 12 pawns. How many chess pieces does Bronwyn have? pp. 8–11

18 Nicolas bought 15 packets of basketball cards. Each packet contained 20 cards. How many cards did Nicolas buy? pp. 8–11

19 Gregory scored 17, 23, 42, 8 and 10 over 5 innings. Calculate his average score. pp. 8–11

20 Gary paid $10.56 for 8 packets of screws. How much did each packet cost? How much would Gary pay for 10 packets? pp. 8–11

21 Lee had 87 apple trees to plant. He planted 14 trees to each row. How many trees were planted in the row that had less than 14 trees? pp. 8–11

22 Mitchell has 765 pheasants to pluck. He can pluck 45 pheasants every hour. How many hours will Mitchell need? pp. 8–11

23 Evaluate: pp. 8–11

a $26 + 35$ **b** $48 + 37$ **c** $27 + 56$

d $33 - 16$ **e** $48 - 29$ **f** $82 - 48$

24 Evaluate: pp. 8–11

a $33 + 68$ **b** $76 + 55$ **c** $83 + 19$

d $46 - 27$ **e** $53 - 36$ **f** $48 - 39$

25 Evaluate: pp. 8–11

a $5 \times 19 \times 2$ **b** $25 \times 27 \times 4$ **c** $5 \times 16 \times 20$
d $10 \times 31 \times 10$ **e** $20 \times 13 \times 5$ **f** $18 \times 25 \times 4$

26 Evaluate: pp. 8–11

a 38×9 **b** 57×9 **c** 46×9
d 46×11 **e** 78×11 **f** 83×11

27 Evaluate: pp. 8–11

a 58×20 **b** 62×20 **c** 48×30
d $630 \div 30$ **e** $480 \div 40$ **f** $360 \div 90$

28 Evaluate: pp. 8–11

a $4 + 17 + 6$ **b** $43 + 18 + 57$ **c** $36 + 57 + 64$

29 **a** Write down the set of even numbers less than 20. p. 11
b Write down the set of odd numbers between 17 and 27.

30 Nominate which of the following numbers are even: p. 11

a 2374 **b** 7234 **c** 3427
d 4372 **e** 4723 **f** 40 690

31 Write down the first ten: p. 11

a Triangular numbers **b** Square numbers.

32 Decide which of the following numbers are Palindromic: pp. 11–12

a 727 **b** 4848 **c** 4884
d 1 001 001 **e** 43 234

33 By using the addition process, generate Palindromic numbers from the following: pp. 11–12

a 58 **b** 74 **c** 117
d 724 **e** 512

34 Finish the first 20 terms of the Fibonacci sequence 1, 1, 2, 3, 5, 8, 13 … p. 12

35 Find the next three terms in the following sequences of numbers: pp. 11–12

a 5, 7, 9, 11 … **b** 6, 8, 10, 12 …
c 2, 4, 8, 16 … **d** 2, 5, 8, 11 …
e 2, 3, 5, 8, 12 … **f** 2, 6, 18 …
g 2, 4, 7, 11 … **h** 2, 4, 12, 48 …
i 2, 1, 4, 3, 6, 5 … **j** 4, 16, 64 …
k 1, 9, 25 … **l** 2, 3, 5, 6, 8, 9 …
m 17, 18, 16, 19, 15, 20 … **n** 1, 2, 3, 6, 11, 20 …
o 2, 11, 56, 281 … **p** 5, 6, 7, 4, 9, 2 …

36 Consider the consecutive terms of the Fibonacci sequence from Question 34. Calculate the values when each term is divided by the previous term. p. 12

e.g. $\frac{1}{1} = 1$, $\frac{2}{1} = 2$, $\frac{3}{2} = 1.5$, $\frac{5}{3} =$, $\frac{8}{5} =$ etc.

Continue this process using all of the terms you have generated. Write answers correct to 6 decimal places. Do the values approach a single value? If so, what is this value?

37 Consider the Fibonacci sequence from Question 34. Square each term of the sequence. Add pairs of consecutive terms; that is, $(\text{Term one})^2 + (\text{Term two})^2$, $(\text{Term two})^2 + (\text{Term three})^2$, … Discuss the result obtained. p. 12

38 Again, consider the sequence from Question 34: p. 12

a Sum the first 10 terms. **b** Sum the first 8 terms.
c Sum the first 6 terms. **d** Sum the first 12 terms.
e Can you see a pattern emerging? Briefly describe the result you have observed.
f Use this observed pattern to sum the first 18 terms and the first 20 terms.

39 **a** Write down the first 10 multiples of 7. p. 12
b Find the multiples of 7 between 10 and 40.
c How many multiples of 7 are less than 20?
d Write down all the multiples of 7 that are greater than 14 but less than 49.
e Find all the multiples of 7 from 42 up to 63.

40 **a** How many multiples of 8 lie between 20 and 50? p. 12
b Find all the multiples of 8 between 20 and 40.
c Write down all the multiples of 8 less than 80.
d Write down any multiples of 8 less than 100 that are also multiples of 5.

41 **a** Find the first three common multiples of 2 and 7. p. 12
b Find any common multiples of 2 and 7 between 40 and 60.
c How many common multiples of 2 and 7 are less than 50?

42 **a** Find the first two common multiples of 5 and 7. p. 12
b How many common multiples of 5 and 7 are less than 150?
c Find the smallest common multiple of 2, 5 and 7.

43 Find the Lowest Common Multiple (LCM) of the following numbers: pp. 12–13

a 5, 6 **b** 3, 5 **c** 4, 7
d 4, 9 **e** 4, 8 **f** 4, 12
g 6, 9 **h** 9, 12 **i** 2, 12
j 3, 12 **k** 6, 15 **l** 5, 12
m 8, 12 **n** 10, 12 **o** 2, 4, 7
p 3, 5, 6 **q** 3, 4, 5 **r** 3, 4, 6
s 2, 5, 6, 10 **t** 3, 4, 5, 6

44 Four monkeys shared evenly among themselves a number of coconuts. There were none left over. Another two monkeys arrived, so they put all the coconuts back on the ground and shared them out equally among the six monkeys. Once again there were none left over. Calculate the minimum number of coconuts that would make this possible. pp. 12–13

45 Four monkeys shared out a number of coconuts equally. There were none left over. Two more monkeys arrived so the coconuts were handed back and divided up again. There were none left over. A further monkey turned up, so once again the coconuts were handed back and divided evenly pp. 12–13 among the seven monkeys. There were none left over. Calculate the minimum number of coconuts that would make this possible.

46 Fifteen children at a party share out the jelly beans between them equally. There was one jelly bean left over. Three children left early and did not take their jelly beans. As none of the jelly beans had been eaten, all were placed on the table and divided equally between the remaining 12 children. Again there was one bean remaining. Calculate the least number of jelly beans that could possibly have been there at the start of the party. pp. 12–13

47 As for Question 46, but before any jelly beans were consumed another two children had to leave and the beans were again placed on the table and divided evenly among the 10 remaining children. Once again there was one left over. Find the least number of jelly beans that would make this possible. pp. 12–13

48 Five shipwrecked pirates starving on a deserted island collected some bananas. They divided them evenly among themselves and there was one banana left over. They gave this banana to a passing orangutang. Another survivor crawled up the beach so the honourable pirates collected the bananas again (the orangutang unfortunately had devoured his). They divided them up again, this time between the six pirates. Once again there was one left over, which they gave to the orangutang. Find the minimum number of bananas possible for this to happen. pp. 12–13

49 In Question 48, a week later another survivor crawled up the beach. The seven surviviors collected another bunch of bananas and divided them up evenly. This time there was no banana remaining after the division. Calculate the smallest number of bananas needed to make this possible. pp. 12–13

50 **a** Write down all the factors of 48. p. 13
b Write down all the factors of 16.
c Write down all the common factors of 16 and 48.

51 **a** Write down all the factors of 54. p. 13
b Write down all the factors of 36.
c Write down all the common factors of 36 and 54.

52 **a** Write down all the factors of 35. p. 13
b Write down all the factors of 45.
c Write down all the common factors of 35 and 45.

53 Write down the prime factors of: p. 13
a 48 **b** 16
c 54 **d** 35
e 45 **f** 36
g 42 **h** 66
i 39 **j** 51

54 Find the Highest Common Factor (HCF) for the following: p. 14

a 10, 15 **b** 4, 8
c 8, 12 **d** 9, 12
e 10, 12 **f** 5, 7
g 8, 16 **h** 4, 9
i 4, 12 **j** 7, 28
k 16, 24 **l** 15, 25
m 9, 15 **n** 9, 21
o 17, 34 **p** 15, 18, 24
q 45, 15, 50 **r** 15, 60, 45
s 16, 24, 48 **t** 16, 48, 64

55 By drawing up a table of the whole numbers from 1 to 100 in groups of 6, complete the Sieve of Eratosthenes for these numbers. Hence, write down all the prime numbers less than 100. pp. 14–15

56 Write down the: pp. 14–15

a Prime numbers between 20 and 30.
b Composite numbers less than 30.
c Smallest prime number.
d Only even prime number.
e Prime number just less than 100.
f Prime number just greater than 100.
g Odd composite numbers between 40 and 60.
h Composite numbers from 25 up to and including 45.
i Prime numbers less than 100 that contain a 3.

57 Write down all the sets of twin primes less than 100. pp. 14–15

58 Siamese twin primes are two prime numbers that are consecutive whole numbers. Write down the only set of Siamese twin primes. pp. 14–15

59 Express 48 as the sum of two primes in as many different ways as possible. pp. 14–15

60 Repeat Question 59 with the following composite numbers: pp. 14–15

a 24 **b** 36
c 54 **d** 40

61 Express the following numbers as the product of their prime factors (using a factor tree or otherwise): pp. 15–16

a 24 **b** 36
c 42 **d** 48
e 72 **f** 56
g 84 **h** 63
i 60 **j** 78
k 150 **l** 200
m 64 **n** 128
o 81 **p** 162
q 225 **r** 105
s 300 **t** 256

62 Use factor trees (or otherwise) to calculate the HCF and LCM of: pp. 16–17

a 240, 504 **b** 396, 360
c 440, 480 **d** 544, 595

63 Simplify: p. 17

a 12^2 **b** 14^2
c 21^2 **d** 30^2
e 5^3 **f** 7^3
g $\sqrt{256}$ **h** $\sqrt{49}$
i $\sqrt{169}$ **j** $\sqrt{81}$
k $\sqrt[3]{216}$ **l** $\sqrt[3]{512}$

64 Use factor trees (or otherwise) to evaluate: pp. 17–18

a $\sqrt{576}$ **b** $\sqrt{441}$
c $\sqrt{729}$ **d** $\sqrt{1296}$
e $\sqrt[3]{1728}$ **f** $\sqrt[3]{5832}$
g $\sqrt[3]{13824}$

65 Find which of the following numbers are divisible by 2, 3, 5, 9 or 11: p. 19

a 324 **b** 645 **c** 4173
d 7491 **e** 5967

66 The three-digit number $74a$ is divisible by 6. Find the value that a must take. p. 19

67 Four kindergarten classes have 16, 17, 18 and 19 students, respectively. When each teacher divides out the only set of centicubes between the children in their class, there is always one cube left over. (Obviously only one class can use them at the one time.) Find the least number of cubes that could be in the set. p. 19

68 Nine friends shared out a set of marbles between them equally. There were none left over. Three more friends arrived so the marbles were again divided up among the 12 friends. There were none left over. What was the total number of marbles? p. 19

69 Evaluate: p. 15

a 2^3 **b** 10^2 **c** 3^2
d 3^3 **e** 4^3 **f** 2^4
g 6^0 **h** $(2 \times 3)^2$ **i** $(4^2 - 3^2)^2$
j $(3^2 - 7)^3$

70 **a** Evaluate $(2 \times 3)^2$ and $2^2 \times 3^2$ and state whether $(2 \times 3)^2 = 2^2 \times 3^2$. p. 15

b Write true (T) or false (F) for the following:

i $(3 \times 4)^2 = 3^2 \times 4^2$ **ii** $(2 + 3)^2 = 2^2 + 3^2$
iii $\sqrt{9 \times 4} = \sqrt{9} \times \sqrt{4}$ **iv** $2^3 = 8$
v $\sqrt{9 + 4} = \sqrt{9} + \sqrt{4}$

Go to p. 254 for quick answers, or to pp. 268–277 for worked solutions.

YOUR CHECKLIST

For a complete understanding of this topic you must be able to:

✓	Understand and use expanded notation		p. 2
✓	Order whole numbers		p. 3
✓	Understand the meaning of arithmetic symbols		p. 3
✓	Arrange numbers in ascending or descending order		p. 3
✓	Evaluate expressions containing grouping symbols		p. 4
✓	Insert grouping symbols into numerical expressions to obtain a particular answer		p. 4
✓	Apply order of operation rules to evaluate expressions		p. 4
✓	Evaluate expressions involving various words that represent operations		p. 5
✓	Use the unitary method		p. 5
✓	Apply the four operations to whole numbers		p. 6
✓	Solve problems involving the four operations		p. 8
✓	Use mental computation methods appropriately		pp. 8–10
✓	Recognise special numbers such as odd, even, square, triangular, Palindromic and Fibonacci		pp. 11–12
✓	Complete patterns using the above special numbers		pp. 11–12
✓	Find multiples of numbers		p. 12
✓	Calculate the LCM of two or more numbers		p. 12
✓	Find factors of numbers		p. 13
✓	Calculate the HCF of two or more numbers		p. 14
✓	Determine whether numbers are prime or composite		p. 14
✓	Use index notation		p. 15
✓	Write a number as a product of its prime factors (with or without factor trees)		pp. 15–16
✓	Use factor trees to find LCM, HCF, square roots and cube roots of large numbers		pp. 16–18
✓	Determine proper divisors of numbers using divisibility tests.		p. 19

Now you are ready to do the tests!

LEVEL 1 TEST

(40 marks)

1 Evaluate $16 + 17 + 4$. (1 mark)

2 Evaluate $4 \times 17 \times 25$. (1 mark)

3 Find the sum of 8 and 9. (1 mark)

4 Calculate the product of 8 and 9. (1 mark)

5 Find the number that is 8 larger than 12. (1 mark)

6 Calculate the difference between 12 and 8. (1 mark)

7 By how many is 12 bigger than 8? (1 mark)

8 Evaluate $16 + 14 - 12 + 8$. (1 mark)

9 Evaluate $16 + 4 \times 7$. (1 mark)

10 Evaluate $(16 + 4) \times 7$. (1 mark)

11 What is the value of the 7 in 87 034? (1 mark)

12 Arrange the numbers 7302, 3702, 3207, 7203 in ascending order. (1 mark)

13 Vary the sets of grouping symbols in each expression to give the indicated answer:

a $6 + 4 \times 7 - 5 = 20$ **b** $6 + 4 \times 7 - 5 = 14$

c $6 + 4 \times 7 - 5 = 65$ (3 marks)

14 Total 17, 11 and 13. (1 mark)

15 Complete the calculations:

a 64×35 **b** $389 \div 27$ (2 marks)

16 Jim sorts nails into packets of 75. He fills 14 packets and has 5 nails left over. How many nails did Jim sort? (2 marks)

17 Write down the set of:

a Even numbers between 12 and 24

b Multiples of 4 less than 40

c Factors of 12

d Square numbers between 10 and 50. (4 marks)

18 Find the next term in the sequence:

a 7, 13, 19, ____ **b** 3, 12, 48, ____ (2 marks)

19 Find the HCF of 8 and 12. (1 mark)

20 Find the LCM of 8 and 12. (1 mark)

21 Find the set of prime factors of 12. (1 mark)

22 a Write in index form $6 \times 6 \times 6 \times 6 \times 6 \times 6 \times 6$.

b True (T) or False (F): $(3 \times 2)^2 = 3^2 \times 2^2$. (2 marks)

23 Evaluate $\sqrt{81}$. (1 mark)

24 Use a factor tree to express these numbers as products of their prime factors:

a 48

b 54

c Hence, or otherwise, find the HCF and the LCM of 48 and 54. (6 marks)

25 Determine if these numbers are divisible by 3:

a 72 **b** 89 (2 marks)

☞ **Quick answers on page 261**

☞ **Worked solutions on page 312**

Your Feedback $\dfrac{\square}{40} \times 100\% = \square\ \%$

LEVEL 2 TEST

(40 marks)

1 Evaluate:

a $6 + 11 + 4 + 9$ **b** $8 \times 19 \times 125$ (2 marks)

2 Find the quotient of 42 and 7. (1 mark)

3 By how many is 11 less than 17? (1 mark)

4 Calculate the average of 9, 15, 8, 6 and 12. (1 mark)

5 By how much is the sum of 19 and 13 greater than 24? (2 marks)

6 Find the value of the 4 in 649 725. (1 mark)

7 Arrange the numbers 42 169, 21 496, 62 169, 49 612, 46 129 in descending order. (1 mark)

8 Evaluate:

a $(17 - 9) \times 6$ **b** $5 + 15 \times 4$ **c** $7 \times 8 + 7 \times 12$ (3 marks)

9 Complete the calculations:

a 76×47 **b** $894 \div 36$ **c** 47×100 **d** $7530 \div 10$

e 43×30 **f** 27×9 **g** 27×11 (7 marks)

10 If 9 pens cost $14.40, how much would 5 pens cost? (2 marks)

11 Write down the set of:

a Odd numbers from 11 to 23 **b** Multiples of 5 between 10 and 60

c Factors of 36 **d** Triangular numbers less than 30. (4 marks)

12 Find the HCF of 12 and 18. (1 mark)

13 Find the LCM of 12 and 18. (1 mark)

14 Use a factor tree to express these numbers as products of their prime factors:

a 72 **b** 84

c Hence, or otherwise, find the HCF and LCM of 72 and 84. (6 marks)

15 Write in expanded form 3^8. (1 mark)

16 Write in index notation $3 \times 3 \times 3 \times 3 \times 3 \times 4 \times 4 \times 4$. (1 mark)

17 Evaluate 3^4. (1 mark)

18 Evaluate:

a $\sqrt{49}$ **b** $\sqrt[3]{64}$ (2 marks)

19 Evaluate $5^2 - 4^2$ and $\sqrt{5^2 - 4^2}$. Is $\sqrt{5^2 - 4^2} = 5^2 - 4^2$? (2 marks)

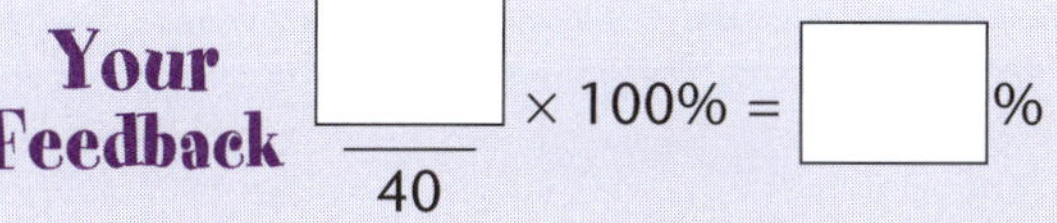

☞ **Quick answers on page 261**
☞ **Worked solutions on page 313**

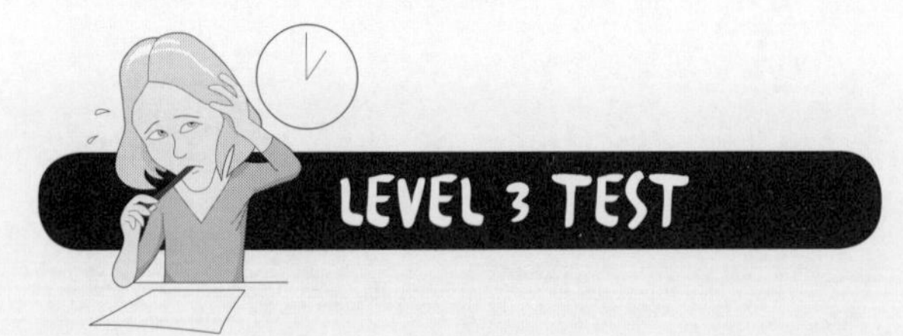

(40 marks)

1 True (T) or false (F): $\sqrt{25 \times 4} = \sqrt{25} \times \sqrt{4}$. (1 mark)

2 By how much does 104 exceed 77? (1 mark)

3 Calculate the total of 17, 177, 117 and 717. (1 mark)

4 Evaluate:

a $123 + 7 - 122 + 8$ b $17 + 3 \times 9$

c $(32 - 29) \times (32 - 19)$ d $17 \times 19 + 17 \times 11$ (4 marks)

5 Place grouping symbols in the expression $4 + 8 \times 3 + 7$ to give an answer of 120. (1 mark)

6 Calculate the difference between the product of 15 and 8 and the sum of 15 and 8. (2 marks)

7 Evaluate the expression:

$\{35 - [49 - (52 - 28) - 15]\} \times 4$ (1 mark)

8 A fence is to be built around three sides of a block of land 120 metres long and 45 metres wide. The fence will be along the two longer sides and one of the shorter sides. A post is to be placed in each corner and another post every 5 metres. How many posts are needed? (3 marks)

9 Sally, batter extraordinaire, has an average of 26 after batting eight times. How many runs must Sally score in the next two innings to raise her average to 32? (3 marks)

10 Write down the prime number closest to 100. (1 mark)

11 Find the HCF of 18, 24 and 27. (1 mark)

12 Find the LCM of 18, 24 and 27. (1 mark)

13 Use a factor tree to express these numbers as products of their prime factors:

a 96 b 108

c Hence, or otherwise, find the HCF and LCM of 96 and 108. (6 marks)

14 Write in expanded form $3^5 \times 5^3$. (1 mark)

15 Write in index form $7 \times 7 \times 7 \times 7 \times 7 \times 7 \times 7 \times 8 \times 8 \times 8 \times 8 \times 8$. (1 mark)

16 Evaluate:

a $\sqrt{121}$ **b** 2^8 **c** $\sqrt{5^2 + 12^2}$ (1 + 1 + 2 marks)

17 Determine whether 47 616 is divisible by:

a 8 **b** 3 **c** 11 (3 marks)

d Hence, or otherwise, determine a number greater than 12 that also divides 47 616 exactly and justify your answer. (2 marks)

18 Use a factor tree to determine $\sqrt[3]{9261}$. (3 marks)

☞ **Quick answers on page 261**
☞ **Worked solutions on page 314**

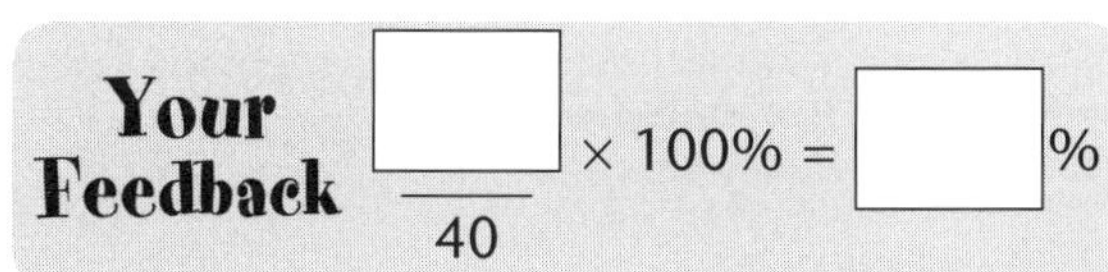

FRACTIONS 2

- Some Introductory Ideas
- Equivalent Fractions
- Comparing Fractions
- Improper Fractions and Mixed Numerals
- Reciprocals
- Four Operations Involving Fractions
- Some Applications
- Using Your Calculator with Fractions
- Mixed Operations Involving Fractions
- Problems Involving Fractions

KEYWORDS

Denominator	Mixed numeral
Equivalent	Numerator
Fraction	Proper
Improper	Reciprocal
Lowest Common Denominator	Simplest

Some Introductory Ideas

A fraction is a part of something. The fraction $\frac{1}{5}$ means 'one out of five equal parts'. The fraction $\frac{2}{3}$ means 'two out of three equal parts'.

A simple diagram can be used to show a fraction:

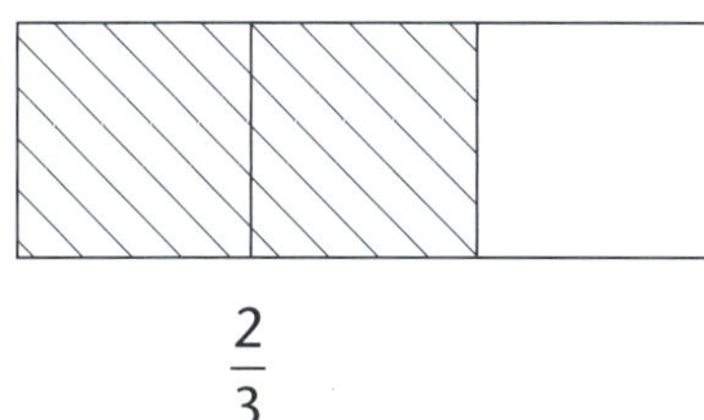

$\frac{2}{3}$

The top number is called the **numerator**. The bottom number is called the **denominator**. In $\frac{2}{3}$, 2 is the numerator while 3 is the denominator.

For Example

Write down the fraction that is shaded in these diagrams:

1 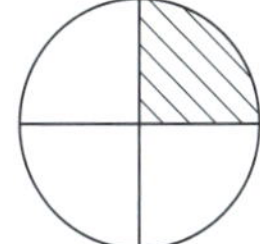

2 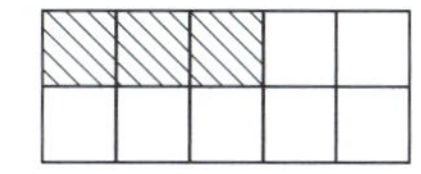

3 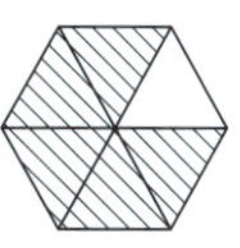

4 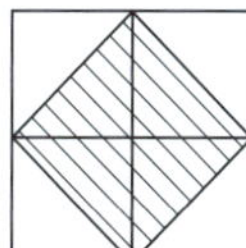

1 $\frac{1}{4}$

2 $\frac{3}{10}$

3 $\frac{5}{6}$

4 $\frac{1}{2}$ $\left[\text{4 out of 8 shaded, i.e. } \frac{4}{8} = \frac{1}{2}\right]$

Equivalent Fractions

A fraction can be expressed in many different but equal ways:

e.g. $\frac{2}{3} = \frac{4}{6} = \frac{6}{9} = \frac{8}{12} = \frac{10}{15} = \ldots$

An equivalent fraction can be formed by multiplying the numerator and denominator by the same number:

e.g. $\frac{2}{3} = \frac{10}{15}$ (numerator ×5, denominator ×5)

For Example

Complete these equivalent fractions:

1 $\frac{3}{5} = \frac{\square}{10}$

2 $\frac{7}{10} = \frac{\square}{40}$

3 $\frac{3}{4} = \frac{15}{\square}$

1 $\frac{3}{5} = \frac{\square}{10}$ (denominator × 2)

$\therefore \frac{3}{5} = \frac{6}{10}$ (numerator × 2)

2 $\dfrac{7}{10} = \dfrac{\square}{40}$ (× 4) $\therefore \dfrac{7}{10} = \dfrac{28}{40}$ (× 4)

3 $\dfrac{3}{4} = \dfrac{15}{\square}$ (× 5) $\therefore \dfrac{3}{4} = \dfrac{15}{20}$ (× 5)

Equivalent fractions can also be formed by **cancelling**; that is, by dividing both parts of the fraction by the same number. This is known as simplifying the fraction, or writing the fraction in **simplest form**. The Highest Common Factor of the numerator and denominator is used for the division:

e.g. (a) $\dfrac{10}{15} = \dfrac{2 \times \cancel{5}_1}{3 \times \cancel{5}_1} = \dfrac{2}{3}$ [Both parts are divided by HCF of 10 and 15, i.e. 5.]

(b) $\dfrac{15}{20} = \dfrac{3 \times 5}{4 \times 5} = \dfrac{3}{4}$ [Both parts are divided by the common factor, 5.]

(c) $\dfrac{8}{20} = \dfrac{2 \times 4}{5 \times 4} = \dfrac{2}{5}$ [Both parts are divided by the common factor, 4.]

For Example

Write in simplest form:

1 $\dfrac{3}{12}$ **2** $\dfrac{8}{12}$ **3** $\dfrac{20}{50}$

1 $\dfrac{3}{12} = \dfrac{\cancel{3}_1}{\cancel{12}_4} = \dfrac{1}{4}$ [Both parts are divided by 3.]

2 $\dfrac{8}{12} = \dfrac{\cancel{8}_2}{\cancel{12}_3} = \dfrac{2}{3}$ [4 divides into 8 and 12.]

3 $\dfrac{20}{50} = \dfrac{\cancel{20}_2}{\cancel{50}_5} = \dfrac{2}{5}$

Comparing Fractions

Fractions can be compared easily if they have the same denominator:

For example, which is the larger: $\dfrac{1}{3}$ or $\dfrac{1}{4}$?

Express both fractions as twelfths:

$\dfrac{1}{3} = \dfrac{4}{12}$ $\dfrac{1}{4} = \dfrac{3}{12}$

Obviously $\dfrac{4}{12} > \dfrac{3}{12}$

$\therefore \dfrac{1}{3} > \dfrac{1}{4}$ [Once the denominators are the same we can compare the numerators.]

For Example

1 Arrange $\dfrac{2}{3}, \dfrac{1}{4}, \dfrac{3}{5}$ in ascending order.

2 Insert either > or < between $\dfrac{3}{4}$ and $\dfrac{2}{3}$ so that the statement is correct.

3 Place $\dfrac{1}{2}$ and $\dfrac{2}{3}$ correctly on a number line.

1 The common denominator is the LCM of 3, 4 and 5, which is 60.

Then $\frac{2}{3} = \frac{40}{60}$ $(2 \times 20 = 40)$
$\times 20$

$\frac{1}{4} = \frac{15}{60}$ $(1 \times 15 = 15)$
$\times 15$

$\frac{3}{5} = \frac{36}{60}$ $(3 \times 12 = 36)$
$\times 12$

$\therefore$ Order is $\frac{15}{60}, \frac{36}{60}, \frac{40}{60}$

(Ascending means going up.)

i.e. $\frac{1}{4}, \frac{3}{5}, \frac{2}{3}$

2 $\frac{3}{4}$ and $\frac{2}{3}$ have a Lowest Common Denominator of 12:

$\frac{3}{4} = \frac{9}{12}$ $\frac{2}{3} = \frac{8}{12}$

Now $\frac{9}{12} > \frac{8}{12}$

$\therefore \frac{3}{4} > \frac{2}{3}$

3 $\frac{1}{2}$ and $\frac{2}{3}$ can be expressed as sixths:

$\frac{1}{2} = \frac{3}{6}$ $\frac{2}{3} = \frac{4}{6}$

Divide the number line between 0 and 1 into sixths:

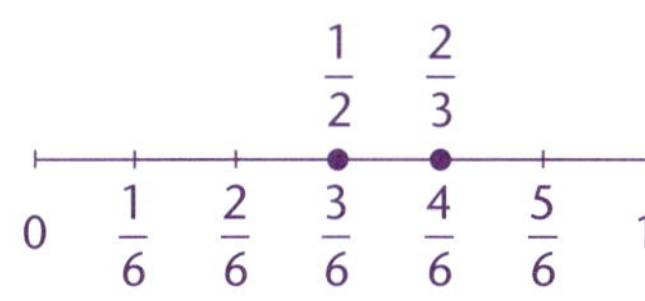

Improper Fractions and Mixed Numerals

Proper fraction: Numerator < Denominator

e.g. $\frac{2}{3}, \frac{3}{4}, \frac{2}{7}, \frac{1}{2}$ etc.

Improper fraction: Numerator > Denominator

e.g. $\frac{7}{2}, \frac{4}{3}, \frac{5}{4}, \frac{9}{7}$ etc.

Mixed numeral: Whole number plus fraction

e.g. $3\frac{1}{2}, 2\frac{1}{4}, 3\frac{3}{5}, 2\frac{7}{3}$ etc.

Converting Mixed Numerals to Improper Fractions

Consider $1\frac{2}{3}$.

There are 3 thirds in the 1 whole, plus 2 thirds in the $\frac{2}{3}$, giving 5 thirds altogether:

i.e. $1\frac{2}{3} = \frac{5}{3}$

or

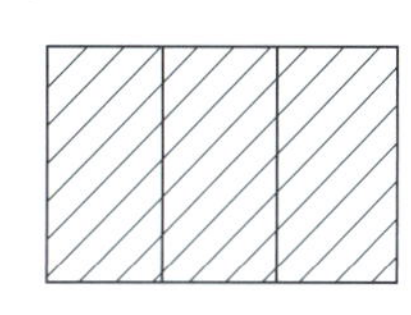 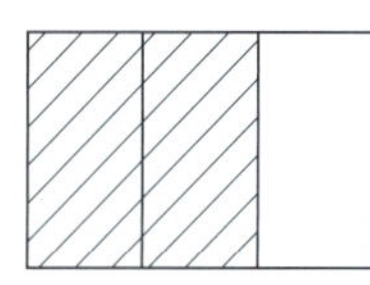

$1\frac{2}{3} = \frac{5}{3}$

or $1 \; \frac{2}{3}$ (+, $\times$) $3 \times 1 + 2 = 5$

$\therefore 1\frac{2}{3} = \frac{5}{3}$

Write as improper fractions:

1 $2\frac{3}{4}$ **2** $4\frac{3}{5}$

1 $2\frac{3}{4}$

$4 \times 2 + 3 = 11$

$\therefore\ 2\frac{3}{4} = \frac{11}{4}$

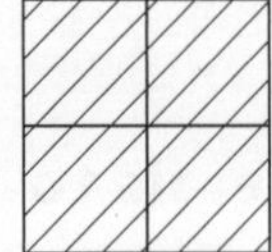 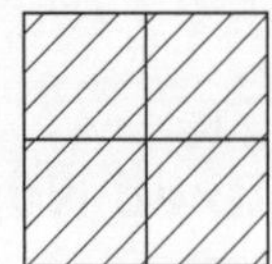 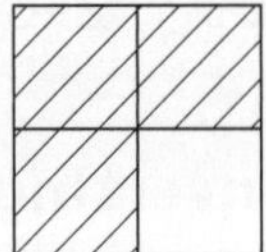

$2\frac{3}{4} = \frac{11}{4}$

2 $4\frac{3}{5} = \frac{5 \times 4 + 3}{5}$ (+, ×)

$= \frac{23}{5}$

Converting Improper Fractions to Mixed Numerals

Divide the numerator by the denominator and put the remainder over denominator:

e.g. $\frac{17}{4} = 4\frac{1}{4}$ $\quad 4\overline{)17}$ gives $4\frac{1}{4}$

Write as mixed numerals:

1 $\frac{17}{5}$ **2** $\frac{15}{2}$

1 $\frac{17}{5} = 3\frac{2}{5}$ $\quad 5\overline{)17}$ gives $3\frac{2}{5}$

2 $\frac{15}{2} = 7\frac{1}{2}$ $\quad 2\overline{)15}$ gives $7\frac{1}{2}$

Reciprocals

To find the reciprocal of a fraction we switch the denominator and the numerator (i.e. we flip the fraction upside down):

e.g. reciprocal of $\frac{2}{3}$ is $\frac{3}{2}$ or $1\frac{1}{2}$

Find the reciprocals of:

1 $\frac{2}{7}$ **2** 5 **3** $3\frac{2}{5}$

1 $\frac{2}{7}$ reciprocal is $\frac{7}{2} = 3\frac{1}{2}$

2 $5 = \frac{5}{1}$ reciprocal is $\frac{1}{5}$

3 $3\frac{2}{5} = \frac{17}{5}$ reciprocal is $\frac{5}{17}$

Four Operations Involving Fractions

In this section we look at adding, subtracting, multiplying and dividing fractions without using a calculator. Calculator usage is discussed later.

Addition of Fractions

We can add fractions only if they have the same denominators (we look for the **Lowest Common Denominator**). We then add the numerators.

For Example

Find the value of:

1 $\frac{2}{5}+\frac{1}{5}$ **2** $\frac{3}{5}+\frac{1}{10}$

3 $\frac{2}{3}+\frac{3}{4}$ **4** $1\frac{1}{10}+3\frac{1}{4}$

5 $2\frac{1}{5}+1\frac{1}{3}+2\frac{3}{4}$

1 $\frac{2}{5}+\frac{1}{5}=\frac{3}{5}$ [by adding numerators]

2 $\frac{3}{5}+\frac{1}{10}=\frac{6}{10}+\frac{1}{10}$ $\left[\text{as } \frac{3}{5}=\frac{6}{10}\right]$

$=\frac{7}{10}$

3 $\frac{2}{3}+\frac{3}{4}=\frac{8}{12}+\frac{9}{12}$ $\left[\frac{2}{3}=\frac{8}{12},\ \frac{3}{4}=\frac{9}{12}\right]$

$=\frac{17}{12}$

$=1\frac{5}{12}$

4 $1\frac{1}{10}+3\frac{1}{4}=(1+3)+\left(\frac{1}{10}+\frac{1}{4}\right)$

$=4+\left(\frac{2}{20}+\frac{5}{20}\right)$ [20 is Lowest Common Denominator]

$=4\frac{7}{20}$

5 $2\frac{1}{5}+1\frac{1}{3}+2\frac{3}{4}=(2+1+2)+\left(\frac{1}{5}+\frac{1}{3}+\frac{3}{4}\right)$

$=5+\left(\frac{12}{60}+\frac{20}{60}+\frac{45}{60}\right)$

$=5+\frac{77}{60}$

$=5+1\frac{17}{60}$

$=6\frac{17}{60}$

Subtracting Fractions

As with adding fractions, we can subtract fractions only if they have the same denominators.

For Example

Find the value of:

1 $1-\frac{2}{5}$ **2** $4-\frac{7}{10}$

3 $6-2\frac{3}{4}$ **4** $\frac{7}{10}-\frac{3}{10}$

5 $\frac{4}{5}-\frac{1}{2}$ **6** $2\frac{1}{2}-1\frac{1}{3}$

7 $5\frac{1}{4}-3\frac{7}{10}$ **8** $3\frac{2}{3}-2\frac{4}{5}$

1 $1-\frac{2}{5}=\frac{5}{5}-\frac{2}{5}$ $\left[1=\frac{5}{5}\right]$

$=\frac{3}{5}$

2 $4-\frac{7}{10}=3+\left(1-\frac{7}{10}\right)$

$=3+\left(\frac{10}{10}-\frac{7}{10}\right)$ $\left[1=\frac{10}{10}\right]$

$=3\frac{3}{10}$

3 $6-2\frac{3}{4}=6-2-\frac{3}{4}$

$=4-\frac{3}{4}$

$=3\frac{1}{4}$

4 $\frac{7}{10}-\frac{3}{10}=\frac{4}{10}$

$=\frac{2}{5}$

5 $\frac{4}{5}-\frac{1}{2}=\frac{8}{10}-\frac{5}{10}$

$=\frac{3}{10}$

6 $2\frac{1}{2}-1\frac{1}{3}=(2-1)+\left(\frac{1}{2}-\frac{1}{3}\right)$

$=1+\left(\frac{3}{6}-\frac{2}{6}\right)$

$=1+\frac{1}{6}$

$=1\frac{1}{6}$

7 $5\frac{1}{4} - 3\frac{7}{10} = 2 + \left(\frac{1}{4} - \frac{7}{10}\right)$

[Our problem is $\frac{1}{4} < \frac{7}{10}$]

$= 2 + \frac{1}{4} - 1 + \frac{3}{10}$

$= 1 + \frac{1}{4} + \frac{3}{10}$

[One is taken away and then $\frac{3}{10}$ added back on because only $\frac{7}{10}$ is to be subtracted.]

$= 1 + \frac{5}{20} + \frac{6}{20}$

$= 1\frac{11}{20}$

Alternatively:

$5\frac{1}{4} - 3\frac{7}{10} = 2 + \left(\frac{1}{4} - \frac{7}{10}\right)$

$= 2 + \left(\frac{5}{20} - \frac{14}{20}\right)$

[In Chapter 5, $5 - 14 = -9$]

$= 2 - \frac{9}{20}$

$= 1\frac{11}{20}$

Alternatively:

$5\frac{1}{4} - 3\frac{7}{10} = 2 + \left(\frac{1}{4} - \frac{7}{10}\right)$

$= 2 + \left(\frac{5}{20} - \frac{14}{20}\right)$

[Take 1 from the 2 and add it to the $\frac{5}{20}$]

$= 1 + \left(1\frac{5}{20} - \frac{14}{20}\right)$

$= 1 + \left(\frac{25}{20} - \frac{14}{20}\right)$

$= 1\frac{11}{20}$

Alternatively:

$5\frac{1}{4} - 3\frac{7}{10} = \frac{21}{4} - \frac{37}{10}$

[Using improper fractions]

$= \frac{105}{20} - \frac{74}{20}$

$= \frac{31}{20}$

$= 1\frac{11}{20}$

Alternatively:

$5\frac{1}{4} - 3\frac{7}{10} = \left(5\frac{1}{4} - 4\right) + \frac{3}{10}$

[Take off the next whole number and then add back the extra bit that was taken away.]

$= 1\frac{1}{4} + \frac{3}{10}$

$= 1 + \left(\frac{5}{20} + \frac{6}{20}\right)$

$= 1\frac{11}{20}$

8 $3\frac{2}{3} - 2\frac{4}{5} = \left(3\frac{2}{3} - 3\right) + \frac{1}{5}$

$= \frac{2}{3} + \frac{1}{5}$

[$\frac{1}{5}$ too much was subtracted]

$= \frac{10}{15} + \frac{3}{15}$

$= \frac{13}{15}$

Multiplying Fractions

Multiply the numerators and multiply the denominators:

e.g. $\frac{7}{9} \times \frac{4}{5} = \frac{7 \times 4}{9 \times 5} = \frac{28}{45}$

$1\frac{1}{2} \times \frac{3}{4} = \frac{3}{2} \times \frac{3}{4} = \frac{9}{8} = 1\frac{1}{8}$

[Convert mixed numerals to improper fractions then multiply.]

Some questions can be simplified by cancelling within the question. Remember, only top numbers can cancel bottom numbers, and vice versa:

e.g. $\frac{7}{9} \times \frac{3}{14} = \frac{\cancel{7}_1}{\cancel{9}_3} \times \frac{\cancel{3}_1}{\cancel{14}_2} = \frac{1 \times 1}{3 \times 2} = \frac{1}{6}$

$\frac{5}{8} \times \frac{16}{25} = \frac{\cancel{5}_1}{\cancel{8}_1} \times \frac{\cancel{16}_2}{\cancel{25}_5} = \frac{1 \times 2}{1 \times 5} = \frac{2}{5}$

For Example

Evaluate:

1 $\frac{3}{7} \times \frac{3}{4}$ **2** $\frac{5}{9} \times 1\frac{1}{3}$

3 $\frac{2}{3} \times \frac{2}{5} \times \frac{1}{3}$ **4** $\frac{3}{4} \times \frac{2}{5}$

5 $\frac{7}{10} \times \frac{5}{7}$ **6** $\frac{3}{4}$ of $\frac{2}{9}$

7 $\frac{5}{6} \times 1\frac{2}{5}$ **8** $\frac{7}{10}$ of $1\frac{3}{7}$

9 $\left(\frac{3}{5}\right)^2$ **10** $6\frac{1}{4} \times 8$

11 $1\frac{1}{2} \times \frac{3}{4} \times \frac{8}{9}$ **12** $\left(2\frac{1}{2}\right)^2$

1 $\frac{3}{7} \times \frac{3}{4} = \frac{9}{28}$

2 $\frac{5}{9} \times 1\frac{1}{3} = \frac{5}{9} \times \frac{4}{3}$
$= \frac{20}{27}$

3 $\frac{2}{3} \times \frac{2}{5} \times \frac{1}{3} = \frac{4}{45}$

4 $\frac{3}{\cancel{4}_2} \times \frac{\cancel{2}_1}{5} = \frac{3 \times 1}{2 \times 5}$
$= \frac{3}{10}$

5 $\frac{\cancel{7}_1}{\cancel{10}_2} \times \frac{\cancel{5}_1}{\cancel{7}_1} = \frac{1 \times 1}{2 \times 1}$
$= \frac{1}{2}$

6 $\frac{3}{4}$ of $\frac{2}{9} = \frac{\cancel{3}_1}{\cancel{4}_2} \times \frac{\cancel{2}_1}{\cancel{9}_3}$ ['of' implies multiply]
$= \frac{1}{6}$

7 $\frac{5}{6} \times 1\frac{2}{5} = \frac{\cancel{5}_1}{6} \times \frac{7}{\cancel{5}_1}$ $\left[1\frac{2}{5} = \frac{7}{5}\right]$
$= \frac{7}{6}$
$= 1\frac{1}{6}$

8 $\frac{7}{10}$ of $1\frac{3}{7} = \frac{\cancel{7}_1}{\cancel{10}_1} \times \frac{\cancel{10}_1}{\cancel{7}_1}$ $\left[1\frac{3}{7} = \frac{10}{7}\right]$
$= 1$

9 $\left(\frac{3}{5}\right)^2 = \frac{3}{5} \times \frac{3}{5}$
$= \frac{9}{25}$

10 $6\frac{1}{4} \times 8 = \frac{25}{\cancel{4}_1} \times \frac{\cancel{8}_2}{1}$
$= \frac{50}{1}$
$= 50$

11 $1\frac{1}{2} \times \frac{3}{4} \times \frac{8}{9} = \frac{\cancel{3}_1}{\cancel{2}_1} \times \frac{\cancel{3}_1}{\cancel{4}_1} \times \frac{\cancel{8}_1}{\cancel{9}_1}$
$= \frac{1}{1}$
$= 1$

12 $\left(2\frac{1}{2}\right)^2 = \left(\frac{5}{2}\right)^2$
$= \frac{5}{2} \times \frac{5}{2}$
$= \frac{25}{4}$
$= 6\frac{1}{4}$

Division of Fractions

We do not actually divide fractions. Instead we multiply by the reciprocal of the fraction that follows the division sign:

e.g. $\frac{7}{10} \div \frac{3}{5} = \frac{7}{\cancel{10}_2} \times \frac{\cancel{5}_1}{3}$

$= \frac{7}{6}$

$= 1\frac{1}{6}$

Find:

1 $\frac{3}{5} \div \frac{1}{10}$ **2** $\frac{3}{4} \div \frac{7}{8}$

3 $1\frac{1}{2} \div \frac{3}{4}$ **4** $\frac{3}{4} \div 1\frac{1}{2}$

5 $\frac{3}{8} \div 4$

1 $\frac{3}{5} \div \frac{1}{10} = \frac{3}{\cancel{5}_1} \times \frac{\cancel{10}_2}{1}$

$= \frac{6}{1}$

$= 6$

2 $\frac{3}{4} \div \frac{7}{8} = \frac{3}{\cancel{4}_1} \times \frac{\cancel{8}_2}{7}$

$= \frac{6}{7}$

3 $1\frac{1}{2} \div \frac{3}{4} = \frac{\cancel{3}_1}{\cancel{2}_1} \times \frac{\cancel{4}_2}{\cancel{3}_1}$

$= \frac{2}{1}$

$= 2$

4 $\frac{3}{4} \div 1\frac{1}{2} = \frac{3}{4} \div \frac{3}{2}$

$= \frac{\cancel{3}_1}{\cancel{4}_2} \times \frac{\cancel{2}_1}{\cancel{3}_1}$

$= \frac{1}{2}$

5 $\frac{3}{8} \div 4 = \frac{3}{8} \div \frac{4}{1}$

$= \frac{3}{8} \times \frac{1}{4}$

$= \frac{3}{32}$

Some Applications

Finding Fractions of Quantities

To find a fraction of a quantity, the word 'of' can be replaced with a multiplication symbol. This means two-thirds of 6 is $\frac{2}{3} \times 6$.

Find:

1 $\frac{3}{4}$ of \$2

2 $\frac{4}{5}$ of 6 litres

3 $\frac{7}{10}$ of 2 km

1 $\frac{3}{4}$ of \$2 $= \frac{3}{\cancel{4}_1} \times \frac{\cancel{200}_{50}}{1}$

$= 150$ [\$2 = 200c]

∴ \$1.50 is the answer.

2 $\frac{4}{5}$ of 6 L $= \frac{4}{\cancel{5}_1} \times \frac{\cancel{6000}_{1200}}{1}$

$= 4800$ [6 L = 6000 mL]

∴ 4800 mL is the answer.

3 $\frac{7}{10}$ of 2 km $= \frac{7}{1\cancel{0}} \times \frac{200\cancel{0}}{1}$

$= 1400$ [2 km = 2000 m]

∴ 1400 m is the answer.

Expressing One Quantity as a Fraction of Another

After ensuring that both quantities are expressed in the same units, place the first over the second in fraction form, then simplify if possible.

For Example

1 What fraction is \$4 of \$6?

2 What fraction is 20c of \$4?

3 Express 20 cm as a fraction of 1 metre.

1 $\frac{\$4}{\$6} = \frac{4}{6} = \frac{2}{3}$

2 $\frac{20c}{400c} = \frac{20}{400} = \frac{1}{20}$ [Numerator and denominator must be same units: \$4 = 400c]

3 $\frac{20}{100} = \frac{1}{5}$ [1 m = 100 cm]

Using Your Calculator with Fractions

You should refer to the instruction booklet for your calculator. All operations, simplification, cancelling and many conversions are possible on most popular brand calculators.

However, it is important for students who aspire to do well in mathematics competitions or to be able to manipulate fractions at higher stages of mathematics to be capable of working with fractions without significant reliance on the calculator.

Mixed Operations Involving Fractions

Remember to deal with brackets first, then the stronger symbols (× and ÷) and then the weaker symbols (+ and −).

For Example

Simplify:

1 $\frac{3}{4} - \frac{1}{2} \times \frac{2}{5}$

2 $\left(\frac{3}{4} - \frac{1}{2}\right) \times \left(\frac{4}{5} + \frac{1}{10}\right)$

3 $3 - \frac{2}{5} \div \frac{3}{10}$

4 $1\frac{1}{2} - \frac{2}{5} \times \frac{5}{6}$

5 $\dfrac{\frac{3}{4} - \frac{1}{2}}{\frac{3}{4} + \frac{1}{2}}$

1 $\frac{3}{4} - \left(\frac{1}{\cancel{2}_1} \times \frac{\cancel{2}_1}{5}\right) = \frac{3}{4} - \frac{1}{5}$

$= \frac{15}{20} - \frac{4}{20}$

$= \frac{11}{20}$

2 $\left(\frac{3}{4} - \frac{1}{2}\right) \times \left(\frac{4}{5} + \frac{1}{10}\right)$

$= \left(\frac{3}{4} - \frac{2}{4}\right) \times \left(\frac{8}{10} + \frac{1}{10}\right)$

$= \frac{1}{4} \times \frac{9}{10}$

$= \frac{9}{40}$

3 $3 - \frac{2}{5} \div \frac{3}{10} = 3 - \left(\frac{2}{\cancel{5}_1} \times \frac{\cancel{10}_2}{3}\right)$

$= 3 - \frac{4}{3}$

$= \frac{9}{3} - \frac{4}{3}$ [$3 = \frac{9}{3}$]

$= \frac{5}{3}$

$= 1\frac{2}{3}$

4 $1\frac{1}{2} - \left(\frac{\cancel{2}_1}{\cancel{5}_1} \times \frac{\cancel{5}_1}{\cancel{6}_3}\right) = 1\frac{1}{2} - \frac{1}{3}$

$= 1 + \left(\frac{3}{6} - \frac{2}{6}\right)$

$= 1\frac{1}{6}$

5 $\dfrac{\frac{3}{4} - \frac{1}{2}}{\frac{3}{4} + \frac{1}{2}} = \left(\frac{3}{4} - \frac{1}{2}\right) \div \left(\frac{3}{4} + \frac{1}{2}\right)$

$= \left(\frac{3}{4} - \frac{2}{4}\right) \div \left(\frac{3}{4} + \frac{2}{4}\right)$

$= \frac{1}{4} \div \frac{5}{4}$

$= \frac{1}{\cancel{4}_1} \times \frac{\cancel{4}_1}{5}$

$= \frac{1}{5}$

Problems Involving Fractions

Problems involving fractions can be solved using addition, subtraction, multiplication or division.

1 A French stick is cut into 12 equal pieces. If I eat a third of the French stick, how many pieces will I eat?

2 Our basket of fruit contains 4 bananas, 3 apples and 3 oranges. What fraction of the fruit in the basket is bananas?

3 Adam's brother has lost 4 of his 24 'baby teeth'. What fraction remains?

4 Daniel takes three-quarters of a minute to swim fifty metres. How long would it take for him to swim two-hundred metres at the same speed?

5 A petrol tank has a capacity of 60 litres. If it shows '$\frac{1}{4}$ full', how much petrol must be added to the tank to fill it?

6 From a reel of ribbon, $10\frac{1}{2}$ metres in length, a piece $1\frac{3}{4}$ metres long is removed. How much will remain on the roll?

7 A 'Pizza Party' comprising twelve pizzas was first prize in a competition. How many people could attend the party, if it was anticipated each person would eat:

a 2 pizzas? **b** $\frac{1}{2}$ pizza?

c $\frac{3}{4}$ pizza?

1 Fraction eaten $= \frac{1}{3} \times 12 = \frac{1}{\cancel{3}_1} \times \frac{\cancel{12}_4}{1} = \frac{4}{1}$

$= 4$

$\therefore$ I'll eat four pieces.

2 Fraction bananas $= \frac{4}{10}$ $[4 + 3 + 3 = 10]$

$= \frac{2}{5}$

$\therefore \frac{2}{5}$ are bananas.

3 Lost = 4

No. remaining = 24 − 4

= 20

$\therefore$ Fraction remaining $= \frac{20}{24}$

$= \frac{5}{6}$

$\therefore \frac{5}{6}$ teeth remain.

4 $200 \div 50 = 4$

$\therefore$ Number of 50 metres = 4

$\therefore$ Time $= 4 \times \frac{3}{4}$ minutes

= 3 minutes

$\therefore$ Daniel takes 3 minutes.

5 Tank is $\frac{1}{4}$ full

$\therefore \frac{3}{4}$ required to fill the tank.

$\therefore$ Petrol required

$= \frac{3}{4}$ of 60 L

$= \frac{3}{\cancel{4}_1} \times \frac{\cancel{60}_{15}}{1}$

$= 45$ L

$\therefore$ 45 litres is needed to fill the tank.

6 Remaining ribbon $= 10\frac{1}{2} - 1\frac{3}{4}$

$= 10\frac{1}{2} - 2 + \frac{1}{4}$

$= 8\frac{1}{2} + \frac{1}{4}$

$= 8 + \frac{2}{4} + \frac{1}{4}$

$= 8\frac{3}{4}$

$\therefore 8\frac{3}{4}$ m remaining.

7 **a** No. of people $= 12 \div 2$

$= 6$

$\therefore$ 6 people could attend.

b No. of people $= 12 \div \frac{1}{2}$

$= 12 \times \frac{2}{1}$

$= 24$

$\therefore$ 24 people could attend.

c No. of people $= 12 \div \frac{3}{4}$

$= \frac{\cancel{12}_4}{1} \times \frac{4}{\cancel{3}_1}$

$= 16$

$\therefore$ 16 people could attend.

PRACTISE, PRACTISE

Go to p. 255 for quick answers, or to pp. 277–280 for worked solutions.

1 Write down the fraction that is shaded: p. 35

a 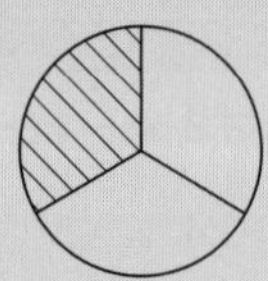

b 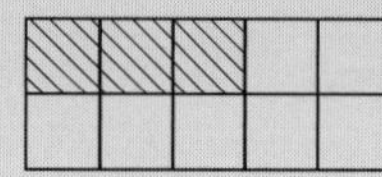

2 Complete these equivalent fractions: p. 35

a $\frac{2}{3} = \frac{\square}{12}$ **b** $\frac{4}{5} = \frac{12}{\square}$

3 Simplify these fractions: pp. 35–36

a $\frac{6}{9}$ **b** $\frac{25}{35}$ **c** $\frac{16}{36}$

4 Arrange in ascending order: pp. 36–37

a $\frac{1}{3}, \frac{1}{2}, \frac{2}{5}$ **b** $\frac{3}{8}, \frac{2}{5}, \frac{1}{4}$

5 On the same number line, plot $\frac{1}{4}, \frac{1}{2}$ and $\frac{2}{3}$. pp. 36–37

6 Arrange in descending order: pp. 36–37

a $\frac{2}{3}, \frac{5}{6}, \frac{1}{2}$ **b** $1\frac{1}{3}, 1\frac{7}{8}, 1\frac{3}{4}$

7 Place the correct symbol (< or >) between these fractions: pp. 36–37

a $\frac{7}{10} \quad \frac{2}{3}$ **b** $\frac{3}{4} \quad \frac{5}{6}$

8 Write as improper fractions: pp. 37–38

a $4\frac{1}{2}$ **b** $3\frac{2}{5}$ **c** $5\frac{7}{10}$

9 Change to mixed numerals: p. 38

a $\frac{21}{5}$ **b** $\frac{13}{3}$ **c** $\frac{47}{10}$

10 Write down the reciprocals of: p. 38

a $\frac{4}{3}$ **b** $1\frac{1}{2}$ **c** $\frac{2}{5}$

11 Evaluate: pp. 38–39

a $\frac{2}{3} + \frac{1}{3}$ **b** $\frac{7}{10} - \frac{3}{10}$ **c** $\frac{2}{5} + \frac{4}{5}$

d $1\frac{1}{4} - \frac{3}{4}$ **e** $2\frac{2}{3} + 1\frac{1}{3}$ **f** $4\frac{3}{5} + 2\frac{4}{5}$

12 Evaluate: pp. 38–40

a $\frac{3}{4}+\frac{3}{4}$ b $\frac{4}{5}-\frac{1}{2}$ c $\frac{7}{10}-\frac{3}{5}$

d $\frac{2}{3}+\frac{1}{2}-\frac{1}{4}$ e $3\frac{1}{2}+2\frac{1}{5}$ f $4\frac{1}{4}-2\frac{1}{5}$

g $3\frac{1}{2}-2\frac{2}{3}$ h $5\frac{1}{3}-3\frac{4}{5}$ i $5-2\frac{4}{5}$

j $7-3\frac{2}{3}-1\frac{9}{10}$

13 Evaluate: pp. 40–41

a $\frac{2}{3}\times\frac{2}{3}$ b $\frac{3}{4}\times\frac{5}{8}$ c $\frac{9}{10}\times\frac{3}{5}$

d $\frac{3}{4}\times\frac{1}{2}\times\frac{3}{5}$ e $2\frac{1}{2}\times1\frac{1}{4}$ f $20\times1\frac{1}{3}$

14 Evaluate: pp. 40–41

a $\frac{3}{4}\times\frac{4}{5}$ b $\frac{2}{3}\times12$ c $\frac{4}{5}$ of 20

d $\left(\frac{4}{5}\right)^2$ e $1\frac{1}{2}\times\frac{2}{3}$ f $\left(3\frac{1}{2}\right)^2$

g $2\frac{1}{2}\times1\frac{3}{5}$ h $6\frac{1}{4}\times1\frac{3}{5}$ i $60\times1\frac{1}{3}$

j $20\times2\frac{1}{2}\times\frac{4}{5}$ k $\frac{1}{2}$ of $\frac{1}{2}$ of $\frac{1}{2}$

15 Evaluate: pp. 41–42

a $\frac{4}{5}\div\frac{3}{4}$ b $\frac{7}{10}\div\frac{5}{7}$ c $\frac{4}{5}\div\frac{1}{5}$

d $1\frac{1}{2}\div\frac{2}{3}$ e $40\div\frac{3}{4}$ f $40\div1\frac{1}{4}$

g $3\frac{1}{2}\div1\frac{3}{4}$ h $2\frac{1}{2}\div1\frac{1}{4}$

16 Evaluate: pp. 43–44

a $\frac{1}{2}+\frac{3}{4}\times\frac{2}{3}$ b $\frac{4}{5}+\frac{2}{3}\div\frac{1}{6}$ c $\dfrac{\frac{3}{4}-\frac{1}{2}}{\frac{1}{4}}$

d $\dfrac{\frac{1}{2}-\frac{1}{4}}{\frac{1}{2}+\frac{1}{4}}$ e $\left(\frac{3}{4}+\frac{3}{5}\right)\div\left(\frac{3}{4}-\frac{3}{5}\right)$ f $\frac{3}{5}+\frac{2}{5}\times\frac{3}{4}$

g $\frac{2}{3}+\frac{3}{5}\times1\frac{1}{3}+\frac{5}{6}$ h $1\frac{1}{3}+1\frac{1}{4}-\frac{3}{4}+\frac{2}{3}$ i $6-2\frac{1}{2}-2\frac{3}{4}$

17 Find: p. 42

a $\frac{4}{5}$ of \$6 b $\frac{3}{4}$ of \$2.40 c $\frac{7}{10}$ of 2 litres

d $\frac{2}{3}$ of 2 minutes e $\frac{3}{5}$ of $1\frac{1}{2}$ metres f $\frac{9}{10}$ of $2\frac{1}{2}$ km

18 What fraction is: pp. 42–43

a \$4 of \$20? b 20 cents of \$2?

c 3 days of 3 weeks? d 21 out of 35?

19 Write the next term in the sequences: pp. 38–42

a $\frac{1}{4}, \frac{1}{2}, \frac{3}{4} \ldots$ b $1, \frac{1}{2}, \frac{1}{4} \ldots$

20 Name the fraction mid-way between $\frac{1}{2}$ and $\frac{1}{3}$. pp. 38–42

21 It took six-and-a-half minutes to wash up and four-and-a-quarter minutes to wipe up. How long did the 'wash and wipe up' take? pp. 44–45

22 In each one hour of television there is twelve minutes of advertising: pp. 44–45

a What fraction of each hour is devoted to advertising?

b If each advertising break takes three minutes, how many advertising breaks occur in each 24-hour period?

23 pp. 44–45

350 MOREE	TAREE 250

When John reached this signpost on his way to Taree after leaving Moree, what fraction of his journey did he yet have to travel?

24 Farmer Jones has an orchard containing 600 trees. One-third of the trees are peach, one-fifth are nectarine and the remainder are plum. How many plum trees are there? pp. 44–45

25 Oranges, cut into quarters, are consumed by a junior soccer team at half time. If the eight players each eat three pieces of orange, how many oranges were required? pp. 44–45

26 From Monday to Friday, Adam watches two-and-a-half hours of television before dinner each afternoon: pp. 44–45

a What fraction of each day does Adam watch television?

b Adam watches television for a total of $4\frac{1}{2}$ hours over the weekend. How many hours of television does Adam watch during the week?

27 In a certain Year 7 class, half of the twenty-four students are girls. If three-quarters of the girls have brown hair, how many brown-haired girls are there in the class? pp. 44–45

Go to p. 255 for quick answers, or to pp. 277–280 for worked solutions.

YOUR CHECKLIST

For a complete understanding of this topic you must be able to:

✓	Use the correct terminology for the topic		p. 35
✓	Understand the basic concept of a fraction		p. 35
✓	Calculate equivalent fractions		pp. 35–36
✓	Compare and order fractions		pp. 36–37
✓	Convert mixed numerals to improper fractions		pp. 37–38
✓	Convert improper fractions to mixed numerals		p. 37
✓	Find the reciprocal of a fraction		p. 38
✓	Apply the four operations to fractions		pp. 38–42
✓	Find fractions of quantities		p. 42
✓	Expressing one quantity as a fraction of another		pp. 42–43
✓	Apply order of operation rules to fractions questions involving mixed operations		pp. 43–44
✓	Use fractions in word problems.		pp. 44–45

Now you are ready to do the tests!

(35 marks)

1 Simplify $\frac{12}{18}$ (1 mark)

2 Write as a mixed number $\frac{5}{3}$ (1 mark)

3 Write as an improper fraction $2\frac{1}{4}$ (1 mark)

4 Place the correct sign (<, >, =) between these fractions:

a $\frac{1}{4} \quad \frac{1}{3}$ b $\frac{2}{5} \quad \frac{4}{10}$ (2 marks)

5 Find the reciprocal of $\frac{4}{5}$ (1 mark)

6 Complete the following additions or subtractions (write answer in simplest form):

a $\frac{9}{10} - \frac{7}{10}$ b $\frac{3}{4} + \frac{2}{5}$ c $4 - 2\frac{1}{4}$ (3 marks)

d $6\frac{1}{4} + 3\frac{2}{3}$ (2 marks)

7 Evaluate:

a $\frac{3}{5} \times \frac{3}{4}$ b $\frac{3}{5} \times \frac{5}{9}$ (2 marks)

c $1\frac{1}{2} \times 1\frac{1}{3}$ d $\frac{7}{10} \div \frac{5}{7}$ e $\frac{5}{8} \div \frac{3}{4}$ (6 marks)

8 Evaluate $\frac{7}{10} + \frac{1}{2} \times \frac{1}{5}$ (2 marks)

9 Find:

a $\frac{1}{4}$ of \$44 b $\frac{3}{4}$ of \$44 (2 marks)

10 What fraction is \$5 of \$25? (1 mark)

11 Find the fraction mid-way between $\frac{1}{4}$ and $\frac{3}{4}$ (2 marks)

12 James watches television for $\frac{1}{3}$ of each day. How many hours does James watch television each week? [7 days] (3 marks)

13 Jenni scored 36 out of 40 in her mathematics test. Convert this to a mark out of 100. (3 marks)

14 Min was given a pay rise of $8 per week. This was an increase of $\frac{1}{10}$ over her previous wage. What was Min's original wage? (3 marks)

☞ Quick answers on page 261
☞ Worked solutions on page 315

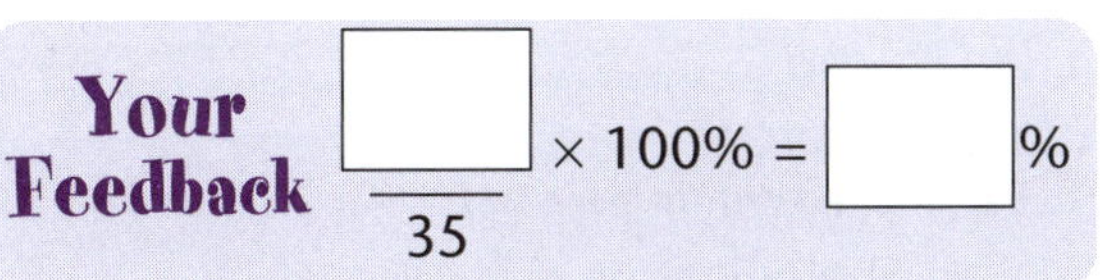

LEVEL 2 TEST

(35 marks)

1 Simplify the fraction $\frac{27}{45}$ (1 mark)

2 Write as a mixed number $\frac{17}{4}$ (1 mark)

3 Write as an improper fraction $5\frac{4}{5}$ (1 mark)

4 Write in ascending order $\frac{3}{4}, \frac{2}{3}, \frac{1}{2}$ (1 mark)

5 Find the reciprocal of $1\frac{2}{3}$ (2 marks)

6 Evaluate:

a $\frac{2}{3} + \frac{1}{2}$ b $5 - 3\frac{1}{3}$ c $2\frac{1}{3} + 1\frac{4}{5}$ d $5\frac{1}{2} - \frac{7}{10}$ (6 marks)

7 Evaluate:

a $\frac{2}{7} \times \frac{3}{4}$ b $\frac{4}{5} \times 1\frac{7}{8}$ c $\frac{7}{9} \div \frac{5}{6}$ d $30 \div 1\frac{1}{5}$ (8 marks)

8 Evaluate $\frac{3}{4} + \frac{7}{10} \times \frac{5}{7}$ (2 marks)

9 Find $\frac{2}{3}$ of \$24 (2 marks)

10 Calculate the average of $\frac{2}{5}$ and $\frac{7}{10}$ (2 marks)

11 In class 7M at Mt Voo school, $\frac{5}{8}$ of the class are girls. If there are 24 students in 7M, how many are girls? (3 marks)

12 Convert a mark of 18 out of 30 to a mark out of 100. (2 marks)

13 Jacqui runs the 800 metres. Jacqui's best time was 2 minutes 30 seconds. If Jacqui reduces her time by 40 seconds, by what fraction has she reduced her time? (3 marks)

14 Melissa earns \$8 an hour. How much does she receive if she works $7\frac{1}{2}$ hours in the week? (1 mark)

Your Feedback $\frac{\square}{35} \times 100\% = \square\%$

☞ **Quick answers on page 261**
☞ **Worked solutions on page 315**

LEVEL 3 TEST

(40 marks)

1 Write $\frac{38}{8}$ in simplest form (1 mark)

2 Write as an improper fraction $8\frac{4}{7}$ (1 mark)

3 Write in descending order $1\frac{7}{12}$, $1\frac{3}{4}$, $\frac{8}{5}$ (1 mark)

4 Find the sum of $\frac{5}{8}$ and $\frac{5}{6}$ (2 marks)

5 Find the product of $\frac{3}{8}$ and $1\frac{7}{9}$ (2 marks)

6 Calculate the reciprocal of $\left(1\frac{2}{5} - \frac{7}{10}\right)$ (2 marks)

7 Evaluate:

a $6\frac{1}{4} \div 1\frac{7}{8}$ b $5\frac{1}{3} - 2\frac{7}{10}$ (6 marks)

8 Evaluate $2\frac{7}{8} \times 1\frac{3}{5} - 1\frac{7}{8} \times 1\frac{3}{5}$ (3 marks)

9 Find $\frac{5}{6}$ of \$72 (2 marks)

10 Find the average of $1\frac{7}{8}$, $2\frac{1}{2}$, $2\frac{3}{4}$ and $\frac{7}{8}$ (4 marks)

11 Stacey worked an extra $4\frac{1}{2}$ hours during the week. She was paid \$33.75 for the extra hours. Calculate the amount Stacey earns each hour. (2 marks)

12 How many $6\frac{2}{5}$ cm lengths can be cut from a length of timber 2 metres long, and what length of timber will be left over? (4 marks)

13 a If $\Delta + \frac{4}{5} = 1\frac{7}{10}$, find the value of Δ. (2 marks)

b However, if $\Delta + \frac{4}{5} < 1\frac{7}{10}$ and it is known that Δ is positive, find the range of values for Δ. (1 mark)

14 Of 120 students interviewed, $\frac{2}{3}$ supported the Knights, $\frac{1}{4}$ supported the Bulldogs and the remainder had no opinion. How many students had no opinion? (4 marks)

15 Four-fifths of my class purchased the new 'Healthy Kids Lunch' at the canteen. Twenty-four students purchased the 'Healthy Kids Lunch'. How many students are in my class? (3 marks)

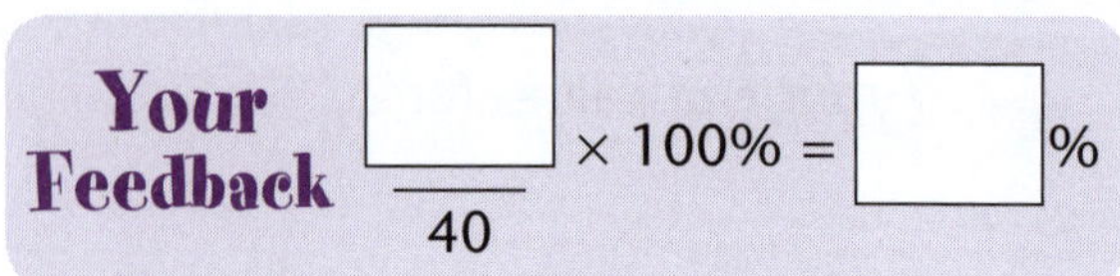

☞ **Quick answers on page 261**
☞ **Worked solutions on page 316**

3 DECIMALS

- Preliminary Ideas
- Converting Decimals to Fractions
- Converting Fractions to Decimals
- Comparing Decimals
- Rounding Off, or Correcting Decimals
- Four Operations Involving Decimals
- Applications of Decimals

KEYWORDS

Ascending	**Recurring**
Correcting	**Repeating**
Decimal	**Rounding off**
Descending	**Terminating**
Expanded	

Preliminary Ideas

A decimal fraction is one with a denominator that uses some power of 10; that is, 10, 100, 1000 etc:

e.g. $0.4 = \frac{4}{10}$ [0.4 has one decimal place]

$0.41 = \frac{41}{100}$ [0.41 has two decimal places]

$0.411 = \frac{411}{1000}$ [0.411 has three decimal places]

A decimal can be written in expanded notation:

$$5.73 = (5 \times 1) + \left(7 \times \frac{1}{10}\right) + \left(3 \times \frac{1}{100}\right)$$

The first number after the point represents tenths, second number hundredths and so on:

$$4.862 = (4 \times 1) + \left(8 \times \frac{1}{10}\right) + \left(6 \times \frac{1}{100}\right) + \left(2 \times \frac{1}{1000}\right)$$

For Example

1 What is the value of 2 in:

a 3.247 b 14.0328

2 Rewrite in expanded notation (form):

a 0.47 b 3.0407

3 How many decimal places in 8.2476?

1 a 2 tenths b 2 thousandths

2 a $0.47 = \left(4 \times \frac{1}{10}\right) + \left(7 \times \frac{1}{100}\right)$

b $3.0407 = (3 \times 1) + \left(4 \times \frac{1}{100}\right) + \left(7 \times \frac{1}{10000}\right)$

3 Four decimal places.

Converting Decimals to Fractions

The number of digits after the decimal point (number of decimal places) equals the number of zeros in the denominator.

For Example

Convert to simplified fractions:

1 0.7 2 0.41

3 0.04 4 0.409

5 0.0078 6 1.7

1 $0.7 = \frac{7}{10}$

2 $0.41 = \frac{41}{100}$

3 $0.04 = \frac{4}{100} = \frac{1}{25}$

4 $0.409 = \frac{409}{1000}$

5 $0.0078 = \frac{78}{10000} = \frac{39}{5000}$

6 $1.7 = 1\frac{7}{10}$

Converting Fractions to Decimals

Some fractions can be converted to decimals by changing the fraction to an equivalent fraction with a denominator of 10 or 100 or 1000 etc:

e.g. $\frac{7}{10} = 0.7$ $\frac{17}{100} = 0.17$

$\frac{4}{5} = \frac{8}{10} = 0.8$ $\frac{13}{20} = \frac{65}{100} = 0.65$

When this method is not possible, we divide the numerator by the denominator, remembering that we can add zeros at the end of the decimal without affecting its value:

e.g. $\frac{5}{8} = 8\overline{)5.{}^{5}0{}^{2}0{}^{4}0}$ with quotient $0.\ 6\ 2\ 5$

The decimal point is after the whole number.

Convert these fractions to decimals:

1 $\frac{9}{10}$ **2** $\frac{18}{100}$ **3** $\frac{41}{50}$

4 $\frac{11}{25}$ **5** $\frac{25}{1000}$ **6** $\frac{2}{3}$

7 $\frac{1}{6}$ **8** $\frac{7}{11}$ **9** $\frac{7}{27}$

1 $\frac{9}{10} = 0.9$

2 $\frac{18}{100} = 0.18$

3 $\frac{41}{50} = \frac{82}{100} = 0.82$

4 $\frac{11}{25} = \frac{44}{100} = 0.44$

5 $\frac{25}{1000} = 0.025$

6 $\frac{2}{3} = 2 \div 3$ $3\overline{)2.00\ldots}$ with quotient $0.66\ldots$

$= 0.\dot{6}$

The 6 continues or repeats ∴ place a dot above the number that repeats.

7 $\frac{1}{6} = 1 \div 6$ $6\overline{)1.000\ldots}$ with quotient $0.166\ldots$

$= 0.1\dot{6}$

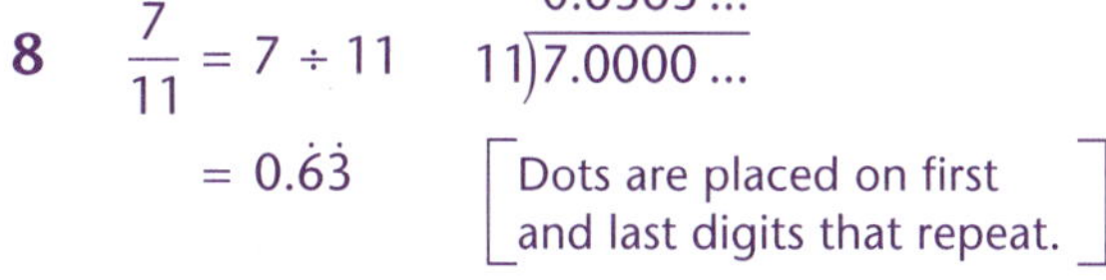

8 $\frac{7}{11} = 7 \div 11$ $11\overline{)7.0000\ldots}$ with quotient $0.6363\ldots$

$= 0.\dot{6}\dot{3}$

Dots are placed on first and last digits that repeat.

9 $\frac{7}{27} = 7 \div 27$ $27\overline{)7.00000\ldots}$ with quotient $0.25925\ldots$

$= 0.\dot{2}5\dot{9}$

We can see that some decimals terminate (e.g. 0.9, 0.82) but other decimals repeat or recur (e.g. $0.\dot{6}$, $0.\dot{6}\dot{3}$).

Comparing Decimals

To compare decimals, give them the same number of decimal places. Remember that:

2.1 = 2.10 = 2.100 etc.

3 = 3.0 = 3.000 etc.

To compare 2.45 with 2.453 and 2.4, write them all with three places of decimals:

i.e. 2.450, 2.453, 2.400

Obviously now

$2.400 < 2.450 < 2.453$

This shows 400 is less than 450 and is less than 453.

Then, in ascending order, the numbers are 2.4, 2.45, 2.453.

Or, to place the correct sign ($<$ or $>$) between 1.161 and 1.16, write the numbers as 1.161 and 1.160.

Then $1.161 > 1.160$ (as $161 > 160$).

1 Insert either $<$ or $>$ between the decimals:

a 3.14 3.2 **b** 0.307 0.31

c 2.3 $2\frac{1}{4}$

2 Rearrange in ascending order (smallest to largest):

a 2, 0.22, 0.021

b 1.04, 0.14, 0.401

c $1\frac{1}{5}$, 1.02, $1\frac{3}{10}$

1 a $3.14 < 3.20$ (as $14 < 20$)

$\therefore 3.14 < 3.2$

b $0.307 < 0.310$ (as $307 < 310$)

$\therefore 0.307 < 0.31$

c $2\frac{1}{4} = 2.25$

$\therefore 2.30 > 2.25$

i.e. $2.3 > 2.25$

2 a Firstly, add zeros: 2.000, 0.220, 0.021

$\therefore$ order is 0.021, 0.22, 2

b Again, firstly add zeros:

1.040, 0.140, 0.401

$\therefore$ order is 0.14, 0.401, 1.04

c $1\frac{1}{5} = 1.2$, $1\frac{3}{10} = 1.3$

$\therefore$ 1.02, 1.20, 1.30

$\therefore$ order is 1.02, $1\frac{1}{5}$, $1\frac{3}{10}$

Rounding Off, or Correcting Decimals

Rounding, or correcting to a number of decimal places, approximates the decimal or makes an answer sensible. For example, to round off to 2 decimal places, we check the digit in the third decimal place. If it is 5 or larger, we add 1 to the digit in the second decimal place. Otherwise, we simplify drop the third digit (plus any others).

For example, consider these numbers:

- 67.37$\boxed{4}$ becomes 67.37 correct to 2 dec. places as third digit is less than 5.
- 67.37$\boxed{5}$ becomes 67.38 correct to 2 dec. places as third digit is 5.
- 67.37$\boxed{6}$ becomes 67.38 correct to 2 dec. places as third digit is greater than 5.

A case to be careful with is rounding off 2.7996 correct to 3 decimal places:

Consider 2.799$\boxed{6}$ = 2.800

Here the fourth digit must be considered. It is greater than 5, so the third digit increases by 1, but it is already 9. It becomes 10, which is written as a 0, plus 1 is carried over to the second digit etc.

Also, remember that any other digits beyond those being considered are disregarded completely.

For instance, when correcting 0.914 949 73 to 2 decimal places, we have to consider only 0.914 as these 3 digits are the only ones that need checking.

Then $0.914\underbrace{949\,73}_{\text{Ignore}} = 0.91$ (2 dec. places)

For Example

1 Write correct to 2 decimal places:

a 47.6284 **b** 7.0831

c 194.7291 **d** 0.995

2 Round off to nearest tenth:

a 47.214 **b** 9.58

3 Round off to nearest cent:

a \$1.476 **b** \$2.1481

1 a 47.62$\boxed{8}$4 = 47.63 (to 2 dec. places)

(third digit is greater or equal to 5)

b 7.08$\boxed{3}$1 = 7.08 (to 2 dec. places)

(third digit is less than 5)

c $194.72\boxed{9}1 = 194.73$ (to 2 dec. places) (third digit is greater than 5)

d $0.99\boxed{5} = 1.00$ (to 2 dec. places) (third digit is 5)

2 a $47.214 = 47.2$ to nearest tenth (i.e. one dec. place)

b $9.58 = 9.6$ (to nearest tenth) (second digit is greater than 5)

3 a \$1.476 = \$1.48 (to 2 dec. places)

b \$2.148 = \$2.15 (to 2 dec. places)

Four Operations Involving Decimals

Addition and Subtraction Involving Decimals

It is helpful to:

- Write questions down the page
- Align decimal points under each other
- Use zeros: insert zeros (remember 0.21 = 0.2100) to overcome ragged ends.

For Example

1 $4 + 2.76$

2 $3.8 + 0.714 + 2$

3 $14 + 2.1 + 0.763 + 0.041$

4 $24 - 2.1$

5 $3.6 - 0.415$

6 $2 - 1.06 + 2.41$

7 Complete the patterns:

a 4.2, 4.5, 4.8, —, —

b 6.04, 5.96, 5.88, —, —

1
```
 4.00 +
 2.76
 ----
 6.76
```

2
```
 3.800
 0.714
 2.000
 -----
 6.514
```

3
```
 14.000
  2.100
  0.763
  0.041
 ------
 16.904
```

4
```
 24.0 -
  2.1
 ----
 21.9
```

5
```
 3.600 -
 0.415
 -----
 3.185
```

6 $2 - 1.06 + 2.41$:
```
 2.00 -        ∴ 0.94 +
 1.06            2.41
 ----            ----
 0.94            3.35
```
Answer = 3.35

7 a As 4.2, 4.5, 4.8 is adding 0.3
∴ Missing values are 5.1 and 5.4

b As 6.04, 5.96, 5.88 is subtracting 0.08
∴ Missing values are 5.8, 5.72

Multiplication Involving Decimals

To multiply we:

- First ignore decimal points and multiply the numbers
- Then insert the decimal point in the answer to have the same number of decimal places as the question.

For Example

Find the answers:

1 2.1×3 2 6.02×4

3 3.1×0.6 4 1.07×0.02

5 $(0.4)^2$ 6 $\sqrt{0.49}$

7 $(0.4 \times 0.2)^2$ 8 3.47×200

1 $21 \times 3 = 63$, and 1 decimal place

$\therefore 2.1 \times 3 = 6.3$ [2.1 × 3 has 1 decimal place]

2 $602 \times 4 = 2408$, and 2 decimal places

$\therefore 24.08$

3 $31 \times 6 = 186$, and 2 decimal places

$\therefore 1.86$ [3.1 × 0.6 together have 2 decimal places]

4 $107 \times 2 = 214$, and 4 decimal places

$\therefore 0.0214$ [We need to insert an additional zero to make 4 dec. places.]

5 $(0.4)^2 = 0.4 \times 0.4$

$\therefore 4 \times 4 = 16$ and 2 decimal places

$\therefore 0.16$

6 As $0.7 \times 0.7 = 0.49$

$\therefore \sqrt{0.49} = 0.7$

7 $(0.4 \times 0.2)^2 = (0.08)^2$

$= 0.0064$

8 $3.47 \times 200 = 694.00$

$= 694$

Special Cases: Multiplying and Dividing by Powers of Ten (10, 100, 1000 etc.)

- When multiplying by a power of ten, the decimal point moves to the right.
- When dividing by a power of ten, the decimal point moves to the left.
- The number of places it moves equals the number of zeros in the power of ten.

For Example

1 4.27×10 2 3.9147×1000

3 4.8×10^2 4 0.076×10

5 $31.47 \div 10$ 6 $476 \div 100$

7 $12 \div 1000$ 8 $3.047 \div 10^3$

1 $4.27 \times 10 = 42.7$

2 $3.9147 \times 1000 = 3914.7$

3 $4.8 \times 10^2 = 4.80 \times 100$

$= 480$ [Remember 4.8 = 4.80.]

4 $0.076 \times 10 = 0.76$

5 $31.47 \div 10 = 3.147$

6 $476 \div 100 = 4.76$ [Remember 476 = 476.0. The point is always after the whole number.]

7 $12 \div 1000 = 0012 \div 1000$

$= 0.012$ [Adding zeros at the front of 12 helps to insert 3 decimal places.]

8 $3.047 \div 10^3 = 0003.047 \div 1000$

$= 0.003\,047$

Division Involving Decimals

Division of a decimal by a whole number is done as if the decimal point does not exist.

When dividing by a decimal we change the question so that we divide by a whole number. This **might** mean that we multiply both components of the question by a power of ten.

For Example

1 $1.4 \div 2$ 2 $3.55 \div 5$

3 $14.2 \div 4$ 4 $3.2 \div 0.4$

5 $16 \div 0.4$ 6 $3 \div 0.001$

7 $0.0452 \div 0.05$

1 $1.4 \div 2 = 0.7$ $\quad 2\overline{)1.4}$ = 0.7

2 $3.55 \div 5 = 0.71$ $\quad 5\overline{)3.55}$ (quotient 0.71)

3 $14.2 \div 4 = 3.55$ $\quad 4\overline{)14.{}^{2}2{}^{2}0}$ (quotient 3. 5 5)

Remember we can add extra zeros.

4 $3.2 \div 0.4 = 32 \div 4$

(by multiplying both by 10)

$= 8$

5 $16.0 \div 0.4 = 160 \div 4$

(by multiplying both by 10)

$= 40$

6 $3.000 \div 0.001 = 3000 \div 1$

(by multiplying both by 1000)

$= 3000$

7 $0.0452 \div 0.05 = 4.52 \div 5$

(by multiplying both by 100)

$\therefore 5\overline{)4.52{}^{2}0}$ (quotient 0.90 4)

$= 0.904$

Remember it is the divisor that we need as a whole number.

Decimals and Order of Operations

Remember the rules from Chapter 1:

- Grouping symbols first, then
- Strong signs (÷, ×) before weak signs (+, −).

For Example

Calculate the following:

1 $3.2 + 2.1 \times 0.4$

2 $\frac{1.6}{0.4} + (0.5)^2$

3 Find the average of 4.7, 3.6, 2.8 and 4.5

1 $3.2 + 2.1 \times 0.4 = 3.2 + 0.84$

$= 4.04$

2 $\frac{1.6}{0.4} + (0.5)^2 = \frac{16}{4} + 0.5 \times 0.5$

$= 4 + 0.25$

$= 4.25$

3 Average $= \frac{4.7 + 3.6 + 2.8 + 4.5}{4}$

$= \frac{15.6}{4}$

$= 3.9$

$\therefore$ The average is 3.9.

Applications of Decimals

For Example

1 A town's maximum daily temperature during one week is recorded as follows: 28°, 27.4°, 21.5°, 22°, 28.2°, 24.3°, 27.8°. Find the average maximum temperature for the week.

2 Daniel purchased 40 litres of petrol at 172.9 cents per litre. Find the total cost of the petrol.

3 It is noticed that $1 Australian has a value of $0.72 United States. How much United States money is received after exchanging $300 Australian?

4 Find the sum of 42.1, 3.06 and 11.04.

5 Jenny needs 0.4 m lengths of ribbon to complete her quilt. If she buys 6 m of the ribbon, how many lengths will she have?

6 Brett's 10 m driveway is to be bordered by a hedge on one side. If the plants are to be placed 0.5 m apart, how many plants will be required?

7 Two towns are 4.8 cm apart on a map. If the scale on the map is 1 cm represents 100 km, find the distance the two towns are apart.

8 How long will it take Mitchell to save \$40 if he saves \$2.50 each week?

1 Total = 28 + 27.4 + 21.5 + 22 + 28.2 + 24.3 + 27.8

= 179.2

Average $= \dfrac{\text{Total}}{\text{No. of terms}}$

$= \dfrac{179.2}{7}$

= 25.6

[28.0 + 27.4 28.2 21.5 22.0 24.3 27.8 = 179.2]

Average maximum temperature is 25.6°.

2 172.9 cents = \$1.729 [Add 2 dec. places.]

∴ Total cost = \$1.729 × 40

or = \$17.29 × 4

= \$69.16

Petrol cost is \$69.16.

3 Money received = \$0.72 × 300

= \$216

\$216 United States received.

4 Sum = 42.1 + 3.06 + 11.04

= 56.2

Sum is 56.2.

[42.10 + 3.06 11.04 = 56.20]

5 No. of lengths = 6 ÷ 0.4

= 60 ÷ 4

(by multiplying both by 10)

= 15

15 lengths of ribbon may be used.

6 No. of plants = 10 ÷ 0.5 + 1

[+1 because must put a plant at the beginning of the 10 m.]

= (100 ÷ 5) + 1

= 20 + 1

= 21

So 21 plants are needed.

7 Distance = 4.8 × 100

= 480

480 kilometres is the actual distance.

8 Weeks = 40 ÷ 2.5

= 400 ÷ 25

∴ It takes 16 weeks.

[25)400 = 16; 25; 150; 150; 0]

Go to p. 255 for quick answers, or to pp. 280–282 for worked solutions.

1 Write the value of the 3 in: p. 56

a 0.317 b 28.031 c 104.763

2 Rewrite in expanded notation (form): p. 56

a 3.21 b 76.047

3 Write the number of decimal places in: p. 56

a 3.476 b 21.5048

4 Write in expanded form: p. 56

a 4.25 b 8.017 c 0.1022

5 Rewrite as decimals: p. 56

a $4 + \frac{1}{10} + \frac{7}{100}$ b $6 + \frac{3}{100} + \frac{7}{1000}$

c $\frac{7}{10} + \frac{3}{100}$ d $74 + \frac{11}{100} + \frac{3}{1000}$

6 Express as fractions in their simplest form: p. 56

a 0.3 b 0.5 c 0.8

d 0.45 e 0.42 f 0.125

g 0.56 h 6.4 i 1.41

j 4.25 k 10.35 l 1.002

7 Change these fractions to decimals: pp. 56–57

a $\frac{21}{100}$ b $\frac{9}{10}$ c $\frac{21}{10}$

d $\frac{3}{100}$ e $\frac{3}{1000}$ f $\frac{3}{5}$

g $\frac{7}{20}$ h $\frac{7}{25}$ i $\frac{7}{50}$

j $\frac{17}{200}$

8 Convert to decimals: pp. 56–57

a $\frac{1}{3}$ b $\frac{7}{8}$ c $\frac{3}{11}$

d $\frac{8}{9}$ e $\frac{4}{27}$

9 Arrange in ascending order: p. 57–58

a 0.3, 0.14, 0.07 b 1.2, 1.27, 1.217

c 0.8, $\frac{37}{100}$, 0.23 d $4\frac{1}{2}$, 4.27, 4.3

10 Insert the correct symbol, < or >, between the two numbers: pp. 57–58

a 7.2   7.3   **b** 0.48   0.46

c $2\frac{1}{2}$ 2.54   **d** 0.07   0.17

11 Round off correct to 2 decimal places: pp. 58–59

a 4.7084   **b** 20.8649

c 11.0594   **d** 12.795 13

12 Round off the following to the nearest thousandth: pp. 58–59

a 4.247 61   **b** 200.415 29

13 Round off to the nearest tenth: pp. 58–59

a 11.076   **b** 0.42

14 Round off to the nearest cent: pp. 58–59

a $4.762 94   **b** $21.439 78

15 Evaluate: pp. 59–60

a $3.06 + 0.89$ **b** $3.7 + 1.205 + 21.62$

c $4 - 2.4$ **d** $4.467 - 2.8$

e $1.2 + 3.1 - 0.6$ **f** $1.4 - 0.72 + 3.14$

g $13 - (1.2 + 2.61)$ **h** $4.1 - 2 - 0.7$

16 Continue the patterns: pp. 59–60

a 1.2, 1.6, 2, ..., ...   **b** 3.7, 3.2, 2.7, ..., ...

c 4.07, 4.11, 4.15, ..., ...   **d** 7.4, 6.5, 5.6, ..., ...

17 Evaluate: p. 60

a 4.76×10 **b** 31.476×100

c 21.4×100 **d** 0.076×10

18 Evaluate: pp. 60–61

a $12.61 \div 10$ **b** $1764 \div 100$

c $9.47 \div 100$ **d** $0.0021 \div 10$

19 Evaluate: pp. 59–60

a 3.4×2 **b** 1.47×3

c 2.8×0.4 **d** 4×0.03

e 2.01×0.6 **f** $(0.5)^2$

g $(1.2 \times 0.1)^2$ **h** 4.3×200

20 Evaluate: pp. 60–61

a $1.4 \div 2$ **b** $1.4 \div 0.2$

c $48 \div 1.2$ **d** $3.05 \div 0.05$

e $4 \div 0.02$ **f** $0.18 \div 0.9$

21 Simplify: p. 61

a $1.2 + 3 \times 0.6$ **b** $1.5 \div 5 + 2 \times 1.2$

c $\dfrac{3.4 + 0.6}{0.2}$ **d** $\dfrac{1.2 \times 0.4}{1.2 - 0.4}$

22 Find the: pp. 59–61

a Sum of 3.2 and 4.01 **b** Difference between 3.8 and 2.07

c Product of 7.6 and 0.2 **d** Quotient of 8.4 and 0.4.

23 **a** Find the average of 12.2, 4.7, 3.9 and 11.4. pp. 59–61

b By how much does 3.7 exceed 2.08?

24 Electricity costs 16.69 cents per unit. What will be the cost of 479 units of electricity? pp. 61–62

25 In a gymnastic championship, Gabrielle received the following scores: pp. 61–62

8.9, 9.2, 9.4, 9.2, 9.0, 8.7, 8.4, 9.0, 9.1, 9.0

If the highest and lowest scores are ignored, what was her total score?

26 The cost of four pineapples is $3.40. Find the cost of each pineapple. pp. 61–62

27 The product of two numbers is 25.2. If one of the numbers is 6, find the other number. pp. 61–62

28 The sum of five numbers is 25. If we know that four of the numbers are 3.7, 2.9, 11 and 4.9, find the missing number. pp. 61–62

29 At the start of a trip Karen noted that the odometer in her car displayed 47 214.8 (km). If she travelled a distance of 475.3 kilometres, calculate the number now showing on her odometer. pp. 61–62

30 How long does it take Craig to earn: pp. 61–62

a $40, if he earns $4 per hour?

b $130, if he earns $6.50 per hour?

c $300, if he earns $7.50 per hour?

31 A car travels 98.4 kilometres on 8 litres of petrol: pp. 61–62

a How far has the car travelled on each litre of petrol?

b If the petrol cost 171.9 cents/litre, what will be the cost of the trip?

32 The table shows some exchange rates for the Australian dollar for three countries: pp. 61–62

Country	$1 Australian buys
United States dollars	72.2 cents
Japanese yen	79.41 yen
New Zealand dollars	$1.23

From the table, what would I receive if I exchanged $200 Australian for:

a United States dollars?

b Japanese yen?

c New Zealand dollars?

Go to p. 255 for quick answers, or to pp. 280–282 for worked solutions.

YOUR CHECKLIST

For a complete understanding of this topic you must be able to:

✓	Write decimal fractions in expanded notation		p. 56
✓	Know the place value of a digit in a number		p. 56
✓	Recognise the number of decimal places used		p. 56
✓	Convert decimals to fractions		p. 56
✓	Convert fractions to decimals		pp. 56–57
✓	Understand the difference between terminating and recurring (or repeating) fractions		p. 57
✓	Compare and order decimals		pp. 57–58
✓	Round off (or correct) decimals to a certain number of decimal places		pp. 58–59
✓	Add and subtract decimals		p. 59
✓	Multiply decimals		pp. 59–60
✓	Multiply and divide decimals by powers of ten		p. 60
✓	Divide decimals by whole numbers and decimals		pp. 60–61
✓	Simplify expressions involving decimals using order of operations rules		p. 61
✓	Solve a variety of real-life problems involving decimals.		pp. 61–62

Now you are ready to do the tests!

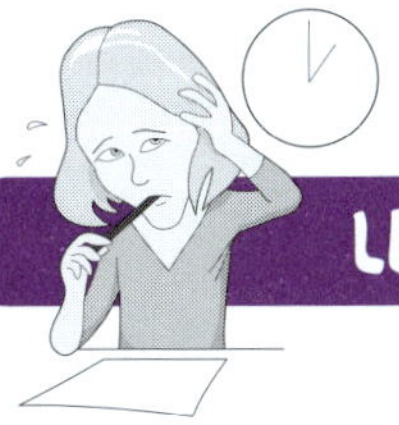

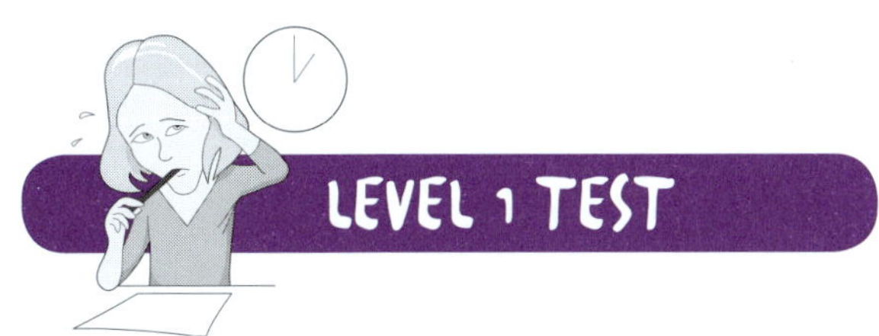

LEVEL 1 TEST

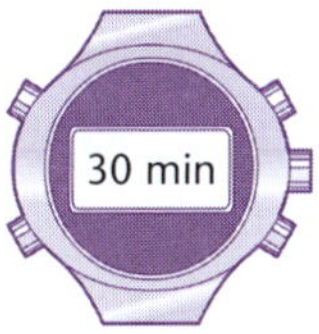

(30 marks)

1 Insert $>$ or $<$ to make the statement true:
4.76 4.577 (1 mark)

2 State the value of 9 in 31.097. (1 mark)

3 Convert $\frac{31}{1000}$ to a decimal. (1 mark)

4 Express 0.04 as a fraction. (1 mark)

5 Round off 47.628 correct to one decimal place. (1 mark)

6 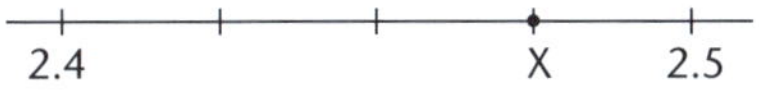

Write down the value of X. (1 mark)

7 Rewrite $\frac{21}{100} + \frac{73}{1000}$ as a decimal. (1 mark)

8 What fraction of the shape has been shaded? Write your answer as a decimal.

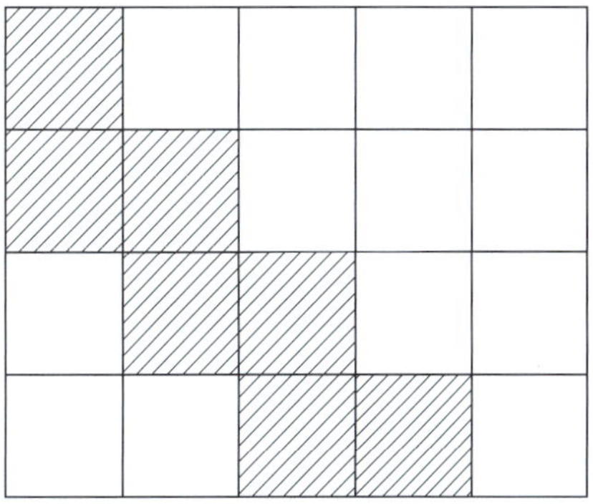

(1 mark)

9 Evaluate:
a $3.62 + 0.413$ **b** $7 - 1.4$ **c** $1 - 0.6 - 0.2$ (3 marks)

10 Complete $3.4 + ____ = 7.6$. (1 mark)

11 Evaluate:
a 3.41×5 **b** 6.047×10 **c** 4.07×0.3 (3 marks)

12 Evaluate:
a $5.2 \div 2$ **b** $18.43 \div 100$ **c** $1.44 \div 1.2$ (3 marks)

13 Phil bought 7 pencils at 24c each. Find the cost of the pencils. (1 mark)

14 From a 24.6 m length of rope a piece 3.55 m long is removed. What length of rope remains? (1 mark)

15 Evaluate $1 - 0.2 \times 0.3$. (2 marks)

16 Find the average of 1.4, 0.8, 3.6 and 0.6. (2 marks)

17 A tank contains water to a depth of 3.4 m. During the day the water drops 0.12 m. What is the new depth?

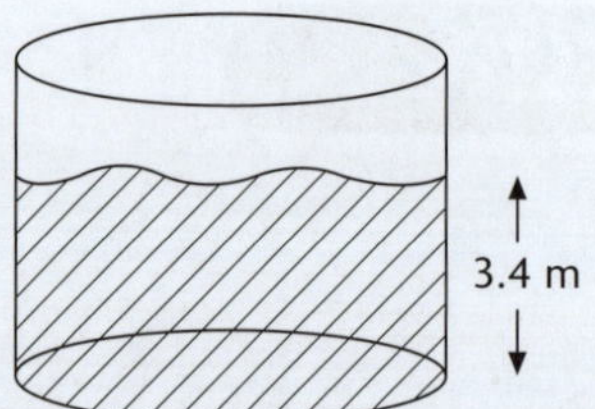

(2 marks)

18 Pete is paid $0.07 for each pamphlet he delivers. How much will he be paid for a street consisting of 126 houses? (2 marks)

19 How many 0.5 m lengths of ribbon can be cut from a roll of 20 metres? (2 marks)

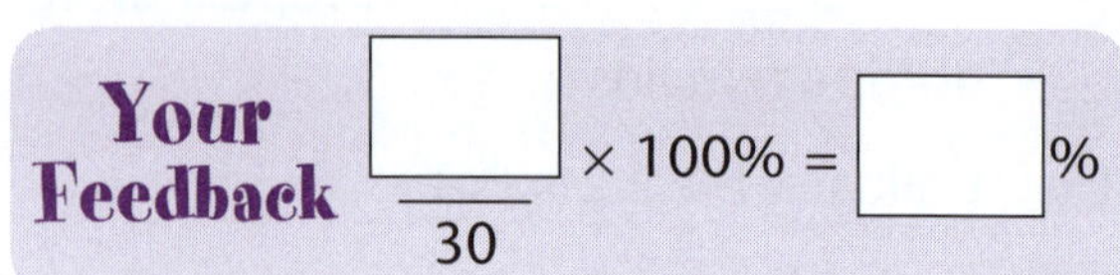

☞ Quick answers on page 261
☞ Worked solutions on page 317

(30 marks)

1 Rewrite in ascending order: 4.71, 4.701, 4.17, 4.7 (1 mark)

2 Convert $\frac{7}{8}$ to a decimal. (1 mark)

3 Rewrite 0.48 as a fraction. (1 mark)

4 Correct 8.0376 to the nearest hundredth. (1 mark)

5 Find the missing number: 2.01, 2.4, 2.79, ____, 3.57 (1 mark)

6 Evaluate:
a $6.015 + 2.1 + 3.06$ b $3.05 - 0.14$ (2 marks)

7 Complete: $4.05 - \_\_\_\_ = 1.09$. (1 mark)

8 Evaluate:
a 6.02×0.3 b 31.486×100 (2 marks)

9 Find the value of:
a $3.6 \div 0.4$ b $64.76 \div 1000$ c $\dfrac{25.51}{0.5}$ (3 marks)

10 Complete $8.5 \div \_\_\_\_ = 1.7$. (1 mark)

11 The sum of a number and 3.2 is 6.47. What is the number? (2 marks)

12 Sharon, Bob, Mary and Kev went out for lunch. They decided to split the bill of $104.80 in four equal ways. How much did Bob pay? (2 marks)

13 Find the middle of 1.6 and 2.8. (2 marks)

14 The contents of a 1.25 L bottle are poured equally into 5 glasses. How much is in each glass? (2 marks)

15 Evaluate 3.2×0.5 and plot your answer on the number line:

1 2

(2 marks)

16 Find the value of:
a $(0.5)^2 + 2.5$ b $3.2 \times 0.2 + \dfrac{1.5}{5}$ (4 marks)

17 Laura needs to save $68 in 8 weeks. How much should she save each week? (2 marks)

☞ Quick answers on page 262
☞ Worked solutions on page 318

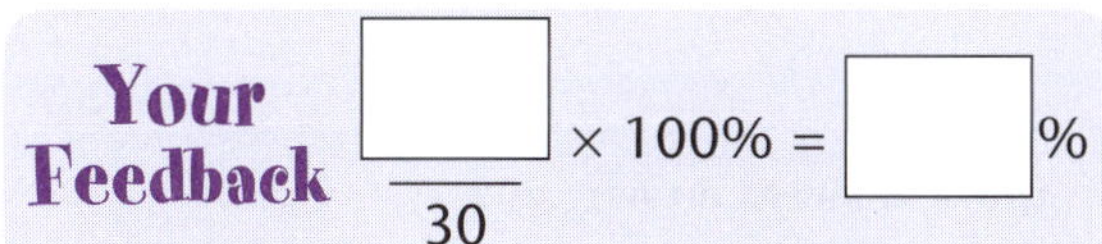

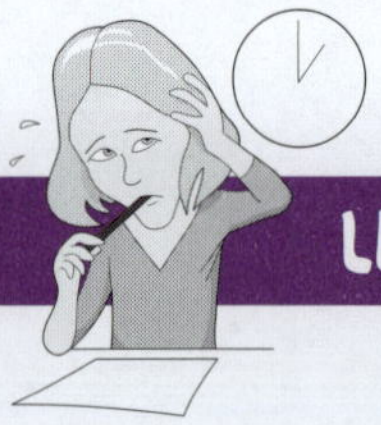

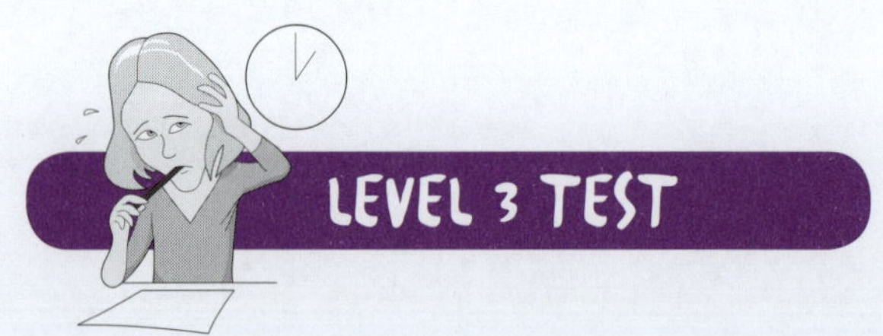

LEVEL 3 TEST

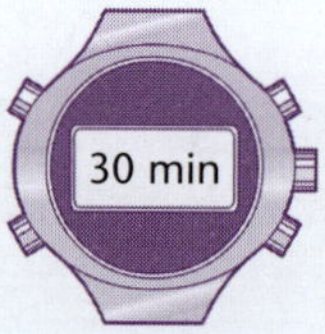

(30 marks)

1 Rewrite in ascending order: 3.47, 3.407, 3.471, 3.074 (1 mark)

2 Convert $\frac{7}{11}$ to a decimal. (1 mark)

3 Express 8.055 as a simplified mixed numeral. (1 mark)

4 Express $\frac{317}{500}$ as a decimal, correct to two decimal places. (1 mark)

5 Evaluate:

a $13.04 + 2.9 + 5$ **b** $6 - 2.031$ (2 marks)

6 Complete: ________ $- 1.76 = 2.3$ (1 mark)

7 Evaluate:

a $(1.2)^2$ **b** 7.476×10^3 **c** $0.6 \times 0.03 \times 0.1$ (3 marks)

8 Find the value:

a $2.05 \div 0.05$ **b** $\dfrac{10.44}{0.4}$ **c** $\dfrac{0.047}{100}$ (3 marks)

9 Find the length of AB in units:

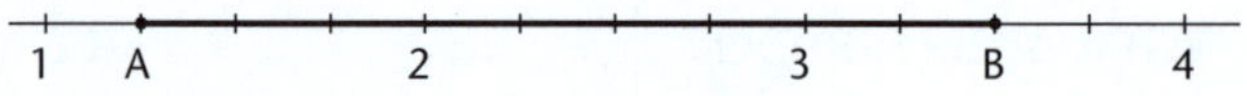

(1 mark)

10 Evaluate $\sqrt{0.16}$. (1 mark)

11 If $3.4 \times 0.9 = 3.06$, what is the value of $30.6 \div 0.9$? (1 mark)

12 Find the sum of 0.4 and the product of 4 and 0.5. (2 marks)

13 Evaluate $\dfrac{3.6}{0.4} + \dfrac{0.32}{0.8}$. (2 marks)

14 If 6 squash courts are hired for $99, find the cost of 1 court. (2 marks)

15 Find the product of 3.47 and 2.8. (2 marks)

16 I have 1.6 kg of mince to make rissoles. If each rissole takes 0.08 kg of mince, how many rissoles can I make? (2 marks)

17 Petrol costs 169.9c per litre. Find the cost of 32 litres. (2 marks)

18 The average temperature over four days was 13.7°. If the temperatures over the first three days were 14.8°, 12.7° and 14.7°, what was the temperature on the fourth day? (2 marks)

Your Feedback $\dfrac{\square}{30} \times 100\% = \square\%$

☞ **Quick answers on page 262**
☞ **Worked solutions on page 319**

4 PERCENTAGES

- Conversions
- Finding a Percentage of a Quantity
- Increasing/Decreasing Quantities by Percentages
- Expressing One Quantity as a Percentage of Another
- Unitary Method and Percentages
- Discounts

KEYWORDS

Commission **Discount**

Conversion **Percentage**

Convert

A percentage is a special fraction that has a denominator of 100. (Note that the symbol, %, is made up of two zeros.)

Conversions

Converting Fractions to Percentages

We multiply the fraction by 100% (i.e. by 1).

For Example

Convert to percentages:

1 $\frac{3}{4}$ **2** $\frac{2}{3}$

3 $\frac{7}{8}$ **4** $\frac{41}{1000}$

5 $\frac{3}{40}$ **6** $2\frac{1}{2}$

1 $\frac{3}{4} = \frac{3}{4} \times 100\%$
$= 75\%$

2 $\frac{2}{3} = \frac{2}{3} \times 100\%$
$= 66\frac{2}{3}\%$ (or $66.\dot{6}\%$)

3 $\frac{7}{8} = \frac{7}{8} \times 100\%$
$= \frac{700}{8}\%$
$= 87\frac{1}{2}\%$

4 $\frac{41}{1000} = \frac{41}{1000} \times 100\%$
$= 4.1\%$

5 $\frac{3}{40} = \frac{3}{40} \times 100\%$
$= 7\frac{1}{2}\%$

6 $2\frac{1}{2} = 2\frac{1}{2} \times 100\%$
$= 250\%$

Converting Decimals to Percentages

We multiply the decimal by 100%.

For Example

Convert to percentages:

1 0.7 **2** 0.05

3 0.27 **4** 0.407

5 3.41

1 $0.7 = 0.7 \times 100\%$
$= 70\%$

2 $0.05 = 0.05 \times 100\%$
$= 5\%$

3 $0.27 = 0.27 \times 100\%$
$= 27\%$

4 $0.407 = 0.407 \times 100\%$
$= 40.7\%$

5 $3.41 = 3.41 \times 100\%$
$= 341\%$

Converting Percentages to Fractions

Remember that writing % is another way of writing a fraction with a denominator of 100.

For Example

Convert to fractions in their simplest form:

1 7% **2** 70%

3 $4\frac{1}{2}\%$ **4** $7\frac{1}{4}\%$

5 120% **6** 600%

1 $7\% = \frac{7}{100}$

2 $70\% = \frac{70}{100}$
$= \frac{7}{10}$

3 $4\frac{1}{2}\% = \dfrac{4\frac{1}{2}}{100}$

$= \dfrac{\frac{9}{2}}{100}$ [Write the numerator as an improper fraction.]

$= \dfrac{9}{200}$ [Multiply the numerator and denominator by 2.]

4 $7\frac{1}{4}\% = \dfrac{7\frac{1}{4}}{100}$

$= \dfrac{\frac{29}{4}}{100}$ [Write the numerator as an improper fraction.]

$= \dfrac{29}{400}$ [Multiply the numerator and denominator by 4.]

5 $120\% = \frac{120}{100}$

$= \frac{6}{5}$

$= 1\frac{1}{5}$

6 $600\% = \frac{600}{100}$

$= 6$

Converting Percentages to Decimals

We divide the percentage by 100. This is equivalent to putting the percentage over 100:

e.g. $70\% = \dfrac{70}{100} = 70 \div 100 = 0.7$

For Example

Convert to decimals:

1	40%	2	4%
3	$18\frac{1}{2}\%$	4	$9\frac{3}{4}\%$
5	$66\frac{2}{3}\%$	6	247%

1 $40\% = 40 \div 100$
$= 0.4$

2 $4\% = 4 \div 100$
$= 0.04$

3 $18\frac{1}{2}\% = 18\frac{1}{2} \div 100$
$= 0.185$ [The $\frac{1}{2}$ puts a 5 on the end.]

4 $9\frac{3}{4}\% = 9\frac{3}{4} \div 100$
$= 0.0975$ [The $\frac{3}{4}$ puts a 75 on the end.]

5 $66\frac{2}{3}\% = 66\frac{2}{3} \div 100$
$= 0.666$
$= 0.\dot{6}$

6 $247\% = 247 \div 100$
$= 2.47$

The Top Ten Conversions

Students should be familiar with these conversions:

Fractions	Decimals	Percentages
1	1	100%
$\frac{1}{2}$	0.5	50%
$\frac{1}{3}$	$0.\dot{3}$	$33\frac{1}{3}\%$
$\frac{2}{3}$	$0.\dot{6}$	$66\frac{2}{3}\%$
$\frac{1}{4}$	0.25	25%
$\frac{3}{4}$	0.75	75%
$\frac{1}{8}$	0.125	12.5%
$\frac{3}{8}$	0.375	37.5%
$\frac{5}{8}$	0.625	62.5%
$\frac{7}{8}$	0.875	87.5%

Finding the Percentage of a Quantity

Remember that 'of' means multiplication.

For Example

Find:

1	16% of 300	2	8% of 76
3	$4\frac{1}{2}\%$ of 350	4	32% of \$15
5	12.5% of 72 days		
6	$33\frac{1}{3}\%$ of \$15.60		
7	$\frac{3}{4}\%$ of \$70	8	145% of \$60

Alternative sets of solutions are offered for these examples.

The first set of solutions converts the percentage to a decimal as the first step in the solution (i.e. 75% = 0.75).

The second set of solutions uses the conversion of the percentage to a fraction (over 100) (i.e. 75% = $\frac{75}{100}$).

Solutions 1

1 16% of 300

$0.16 \times 300 = 48$

2 8% of 76

$0.08 \times 76 = 6.08$

3 $4\frac{1}{2}$% of 350

$0.045 \times 350 = 15.75$

4 32% of $15

$0.32 \times 15 = 4.8$

$\therefore$ $4.80

5 12.5% of 72 days

$0.125 \times 72 = 9$

$\therefore$ 9 days

6 $33\frac{1}{3}$% of $15.60

$\frac{1}{3} \times 1560 = 520$

$\therefore$ $5.20

7 $\frac{3}{4}$% of $70

$0.0075 \times 70 = 0.525$

$\therefore$ $0.53 (to nearest cent)

8 145% of $60

$1.45 \times 60 = 87$

$\therefore$ $87

Solutions 2

1 16% of 300

$= \frac{16}{100} \times 300$

$= 48$

[Calculator = 16 ÷ 100 × 300
OR using fractions:
$\frac{16}{\cancel{100}_1} \times \frac{\cancel{300}^3}{1} = 48$]

2 8% of 76

$= \frac{8}{100} \times 76$ [$8 \div 100 \times 76$]

$= 6.08$

3 $4\frac{1}{2}$% of 350

$= \frac{4.5}{100} \times 350$

$= 15.75$

4 32% of $15

$= \frac{32}{100} \times 15$

= $4.80

5 12.5% of 72 days

$= \frac{12.5}{100} \times 72$

= 9 days

6 $33\frac{1}{3}$% of $15.60 [Note: $\frac{1}{3} = 33\frac{1}{3}\%$]

$= \frac{1}{3} \times 15.6$

= $5.20

7 $\frac{3}{4}$% of $70

$= \frac{0.75}{100} \times 70$

$= 0.525$

= $0.53 (to nearest cent)

8 145% of $60

$= \frac{145}{100} \times 60$

= $87

Increasing/Decreasing Quantities by Percentages

- To increase an amount by 20%, we find 120% (100 + 20) of the amount.
- To decrease an amount by 20%, we find 80% (100 – 20) of the amount.

For Example

1 Increase $400 by 20%.

2 Increase $75 by 15%.

3 Decrease 200 litres by 12%.

4 Decrease 400 by 17%.

5 Increase \$200 by 30% and then decrease the amount by 30%.

6 \$600 is to be increased by 12% each month for the next three months. What will be the final amount?

1 Find 120% of \$400
i.e. $120\% \times 400$
$= 1.2 \times 400$ [100 + 20]
$= 480$
∴ \$480

2 Find 115% of \$75
i.e. $115\% \times 75$
$= 1.15 \times 75$ [100 + 15]
$= 86.25$
∴ \$86.25

3 Find 88% of 200 L
i.e. 0.88×200 [100 − 12]
$= 176$
∴ 176 L

4 Find 83% of 400
i.e. 0.83×400 [100 − 17]
$= 332$
i.e. 332

5 ↑by 30%: 130%, ↓ by 30%: 70%
∴ $130\% \times 200 \times 70\%$
$= 1.3 \times 200 \times 0.7$
$= 182$
∴ \$182

[Or $130\% \times 200$
$= 1.3 \times 200$
$= 260$
Then 70% of 260
$= 0.7 \times 260$
$= 182$.]

6 112% of 112% of 112% of \$600
$1.12 \times 1.12 \times 1.12 \times 600$
$= 842.9568$
= \$842.96 (to nearest cent)

Alternatively, we could have found the percentage of the amount and then added (for increasing) or subtracted (for decreasing).

For Example

1 Increase \$270 by 15%.

2 Decrease \$760 by 18%.

1 15% of \$270
$= 0.15 \times 270$
$= 40.5$
i.e. \$40.50
∴ New amount = \$270 + \$40.50
= \$310.50

2 18% of \$760
0.18×760
$= 136.8$
i.e. \$136.80
∴ New amount = \$760 − \$136.80
= \$623.20

Expressing One Quantity as a Percentage of Another

We express the first quantity as a fraction of the second using consistent units and then convert this fraction to a percentage (by multiplying by 100%).

For Example

1 What percentage is:

a 15 of 20?

b 75c of \$1?

2 Express 30 seconds as a percentage of 2 minutes.

3 Rewrite 25 out of 40 as a percentage.

4 What percentage of each figure has been shaded?

a

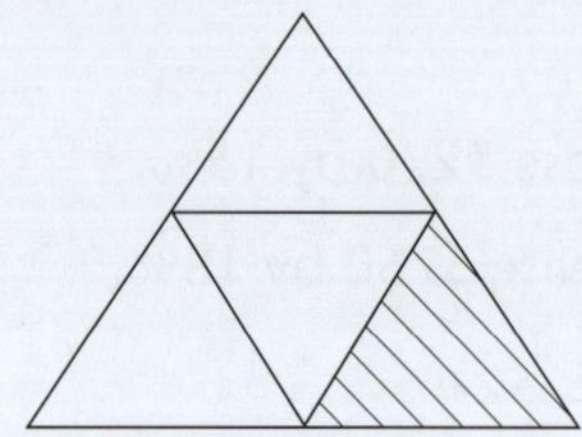

b

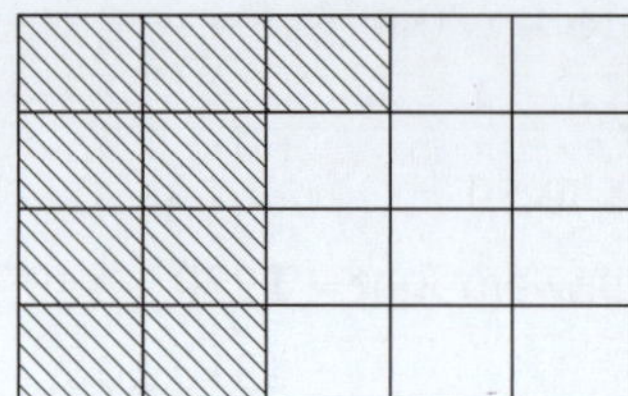

1 a $\frac{15}{20} \times 100\% = 75\%$ [Fraction is $\frac{15}{20}$]

b $1 → 100c [We must have the same units.]

$\therefore \frac{75}{100} \times 100\% = 75\%$

2 2 minutes → 120 seconds

$\therefore \frac{30}{120} \times 100\% = 25\%$

3 $\frac{25}{40} \times 100\% = 62.5\%$

4 a $\frac{1}{4} \times 100\% = 25\%$

b 9 shaded out of 20

$\therefore \frac{9}{20} \times 100 = 45\%$

Unitary Method and Percentages

We can solve some problems by dividing to find 1%, then multiplying to find the whole amount (100%).

For Example

1 If 7% of an amount is 56, find the amount.

2 Seventy-two per cent of a number is 360. What is the number?

3 15% of an amount is $180. Find half of the amount.

1 7% of an amount = 56

$\therefore$ 1% of the amount = 56 ÷ 7

= 8

100% of the amount = 8 × 100

= 800

$\therefore$ The amount is 800.

2 72% of a number = 360

1% of the number = 360 ÷ 72

= 5

100% of the number = 5 × 100

= 500

$\therefore$ The number is 500.

3 15% of an amount = 180

1% of the amount = 180 ÷ 15

= 12

[When convenient, we can drop to say 5% (by dividing by 3) and then 'build', rather than dropping to 1%.]

50% of an amount = 12 × 50

= 600

$\therefore$ The amount is $600.

Discounts

Businesses often express reductions as a percentage discount.

For Example

1 How much is saved if a CD player, valued at $240, is discounted by 20%?

2 What will be paid for a pair of jeans, marked at $80, but discounted by 30%?

3 A 'Seniors Card' entitles Bridget to a 15% discount off the price of a smorgasbord meal that costs $13. What will Bridget pay for the meal?

4 A fridge is discounted by 20% and now costs $960. What was its original price?

5 The price of a dress is cut by 40%. If this is a saving of $96, what will the dress now cost?

6 Which is the better deal on a television marked at $1200:

a A discount of 20%?

b A discount of 10% and then a further discount of 10%?

1 Savings = 20% of 240

= 0.2 × 240

= 48

∴ $48 is saved.

2 Amount of discount = 30% of 80

= 24

∴ new price = 80 – 24

= 56

∴ The new price is $56.

Or alternatively,

discounted price = 70% of 80

= 0.7 × 80

= 56 [100 – 30]

∴ The new price is $56.

[This second method is quicker.]

3 New discounted price = 85% of 13

= 0.85 × 13

[100 – 15]

= 11.05

∴ Meal will cost $11.05.

4 If already discounted,

∴ 80% of original price = 960

1% of original price = 960 ÷ 80

[100 – 20]

= 12

100% of original price = 12 × 100

∴ The original price was $1200.

5 40% of original price = 96

1% of original price = 96 ÷ 40

= 2.4

∴ 60% of original price = 2.4 × 60

[after 40% discount]

= 144

∴ The dress (with 40% discount) will now cost $144.

6 **a** Discount of 20%

∴ 80% of $1200

= 0.8 × 1200

= 960

∴ The television will cost $960.

b Discount of 10%, then discount of 10%

∴ 90% of 1200

= 0.9 × 1200

= 1080

Now, 90% of 1080

= 0.9 × 1080

= 972

[Note: A discount of 20% is NOT the same as two consecutive discounts of 10%.]

∴ The television will cost $972.

∴ The better deal is **a**.

PRACTISE, PRACTISE

Go to p. 256 for quick answers, or to pp. 282–285 for worked solutions.

1 Express each fraction as a percentage: p. 72

a $\frac{2}{5}$ **b** $\frac{9}{20}$ **c** $\frac{13}{25}$

d $\frac{7}{40}$ **e** $\frac{41}{100}$ **f** $\frac{1}{3}$

g $\frac{5}{6}$ **h** $3\frac{1}{4}$

2 Convert the following to percentages, correct to one decimal place: p. 72

a $\frac{2}{7}$ **b** $\frac{5}{11}$ **c** $2\frac{4}{15}$

3 Rewrite the following decimals as percentages: p. 72

a 0.3 **b** 0.03 **c** 0.003

d 0.19 **e** 0.576 **f** 1.12

g 4.7 **h** 0.075

4 Express as fractions, whole numbers or mixed numerals: pp. 72–73

a 60% **b** 3% **c** 95%

d 127% **e** 300% **f** $5\frac{1}{2}$%

g $12\frac{1}{4}$% **h** $3\frac{4}{5}$%

5 Convert these percentages to decimals: p. 73

a 16% **b** 7% **c** 60%

d 7.6% **e** 12.5% **f** 176%

g 104.6% **h** $\frac{1}{4}$% **i** $7\frac{3}{4}$%

j $33\frac{1}{3}$%

6 Find: pp. 73–74

a 27% of 600 **b** 16% of $140 **c** 60% of $42

d 7% of 6 km **e** $4\frac{1}{2}$% of 8 L **f** $12\frac{1}{2}$% of 4 hours

g $\frac{1}{4}$% of 6 kg **h** 145% of $300 **i** 35.2% of $100 000

j $15\frac{7}{8}$% of 4000

7 Increase the following by the given percentage: pp. 74–75

a $20 by 10% **b** 700 by 18% **c** 45 L by 30%

d 3 kg by 8%

8 Decrease the following by the given quantity: pp. 74–75

a 260 by 20% **b** $65 by 5% **c** 3000 by $12\frac{1}{2}$%

d $60 by 42%

9 One hundred dollars is increased by 30% and the result decreased by 30%. What is the result? pp. 74–75

10 What percentage is: pp. 75–76

a 12 of 25?
b 20 cents of $2?
c 4 cm of 5 m?
d 250 mL of 2 L?
e 20 seconds of 4 minutes?
f 45c of $9?

11 Express the first amount as a percentage of the second amount: pp. 75–76

a 4 cm, 40 cm
b 3.5 g, 2 kg
c 400 cm, 2 m
d 3 minutes, 3 seconds

12 a If 20% of an amount is 700, find the amount. p. 76
b If 9% of an amount is 45, find the amount.

13 If we know that 15% of an object weighs 525 g, how much will the whole object weigh? p. 76

14 25% OFF EVERYTHING

A store had a sale offering 25% off all its merchandise. Find the discounted price of a product that was originally marked: pp. 76–77

a $95
b $38

15 A plumber received a discount of $12\frac{1}{2}$% on a shovel that was priced at $22: pp. 76–77

a How much will the plumber save?
b What will the shovel cost the plumber?

16 The price of a television set was reduced from $800 to $700 as part of a discount sale. What was the percentage discount? pp. 76–77

17 A dress marked at $84 was reduced in price by 35%. What was the new price for the dress? pp. 76–77

18 As part of its 'Runout Sale', Honest Joe's Car Yard dropped its prices by 15%. How much will be saved on a car marked at $16 999? pp. 76–77

19 Adam's basketball team played twenty games during the season. If the team won twelve of the games, what percentage of games did they lose if there were no draws? pp. 75–76

20 A group of fifty students in Year 7 were surveyed regarding the number of televisions in their homes. The results are recorded below: pp. 75–76

Number of televisions	Number of students
0	1
1	14
2	18
3	7
4	6
5	4

What percentage of the students had:

a Two televisions at home?
b More than four televisions at home?

21 In a bag of 40 jelly beans, Kathy counts 8 black, 7 red, 7 green, 4 pink, 6 white and the rest yellow: pp. 75–76

a What percentage of the jelly beans are yellow?

b If Kathy decides to eat all the black jelly beans first, what percentage of the remaining jelly beans are yellow?

22 Jacylyn is presently being paid $602 per week. If she was to receive a pay rise of 4%, what will be her new weekly pay? p. 76

23 The Cessnock to Coonabarabran car rally attracted a large field of eighty cars. By the time the cars arrived at their destination, only sixteen cars remained. What percentage of the original number failed to finish? pp. 75–76

24 The enrolment of Year 7 at a particular school is 198. If this represents 22% of the school's total enrolment, find the number of students at the school. p. 76

25 Ben, Ken and Len combined their savings to purchase the complete set of 1994 All-Stars basketball cards. Ben's contribution of $68 represents 40% of the total cost of the cards: p. 76

a Find the cost of the complete set.

b Ken contributed 20% of the total cost. How much was his contribution?

26 Glendon Brook had a population of 240 in 1980. By 1995 it had increased to 270. Find the population increase as a percentage. pp. 75–76

27 During the big drought of 1994, Trevor reduced the number of sheep on his property from 4500 to 1400: pp. 75–76

a What was the percentage decrease, to two decimal places?

b By the end of 1995 the better season enabled Trevor to increase his number of sheep by 30% from their lowest level in 1994. How many sheep did he have by the end of 1995?

Go to p. 256 for quick answers, or to pp. 282–285 for worked solutions.

YOUR CHECKLIST

For a complete understanding of this topic you must be able to:

✓	Convert fractions to percentages		p. 72
✓	Convert decimals to percentages		p. 72
✓	Convert percentages to fractions		pp. 72–73
✓	Convert percentages to decimals		p. 73
✓	Find a percentage of a quantity		pp. 73–74
✓	Increase a quantity by a percentage		pp. 74–75
✓	Decrease a quantity by a percentage		pp. 74–75
✓	Express one quantity as a percentage of another		pp. 75–76
✓	Use the unitary method approach		p. 76
✓	Find discounts.		pp. 76–77

Now you are ready to do the tests!

(25 marks)

1 Convert to a percentage:

a $\frac{3}{5}$ **b** $1\frac{1}{2}$ **c** 0.45

d 0.8 **e** 0.06 **f** 0.355 (6 marks)

2 Convert to a fraction:

a 27% **b** 8% **c** 40% (3 marks)

3 Convert to a decimal:

a 4% **b** $7\frac{1}{2}$% **c** 215% (3 marks)

4 What percentage is:

a 12 of 48? **b** \$3.20 of \$4? **c** 12 of 25? (3 marks)

5 Find:

a 15% of \$20 **b** 8% of \$600 **c** 12% of \$20 (3 marks)

6 Decrease \$40 by 10%. (1 mark)

7 Express 15 minutes as a percentage of an hour. (1 mark)

8 If 25% of an amount is \$30, find the amount. (1 mark)

9 Athena has \$200 and gives 25% of it to her sister Lucy. How much money has Athena left? (2 marks)

10 Increase \$450 by 20%. (2 marks)

☞ **Quick answers on page 262**
☞ **Worked solutions on page 320**

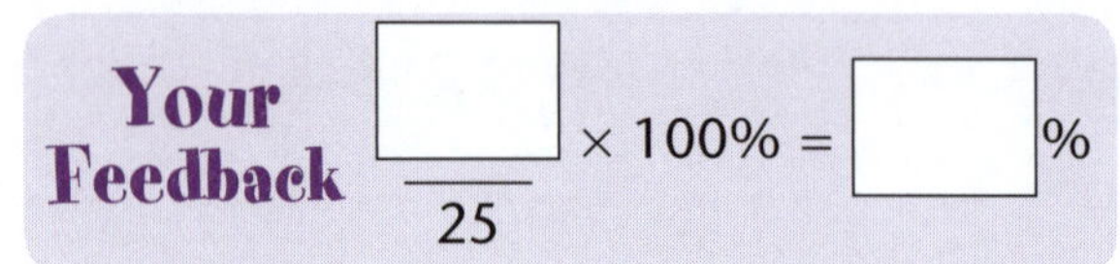

LEVEL 2 TEST

(30 marks)

1 Rearrange in ascending order:

a 41%, $\frac{2}{5}$, 0.42 b 37%, $\frac{3}{8}$, 0.4 (2 marks)

2 Find:

a 4% of \$290 b 16% of \$360

c $8\frac{1}{2}$% of \$7000 d $5\frac{1}{4}$% of 4 metres (4 marks)

3 What percentage is:

a 30c of \$3? b 5 mL of 2 L?

c 6 minutes of 2 hours? d 35 mg of 2 g? (4 marks)

4 Increase:

a \$25 by 10% b \$650 by 8% (4 marks)

5 Decrease:

a \$650 by 8% b \$4000 by $9\frac{1}{2}$% (4 marks)

6 Find the discount on the following items if they originally cost:

a \$45 with a discount of 15%

b \$2480 with a discount of 20%. (4 marks)

7 Michelle received a discount of $12\frac{1}{2}$% on a dress that was priced at \$160.

a How much did Michelle save?

b What will Michelle pay for the dress? (2 marks)

8 Andrea shared a pizza with her brother Savvas. If Andrea ate 40% of the pizza, what fraction did Savvas eat? (2 marks)

9 If 5% of an amount is 24, find the amount. (2 marks)

10 Sophie won \$2400 in a lottery prize. She decided to put half of it in the bank and used 60% of the remainder to buy a bike. How much did the bike cost? (2 marks)

☞ **Quick answers on page 262**
☞ **Worked solutions on page 320**

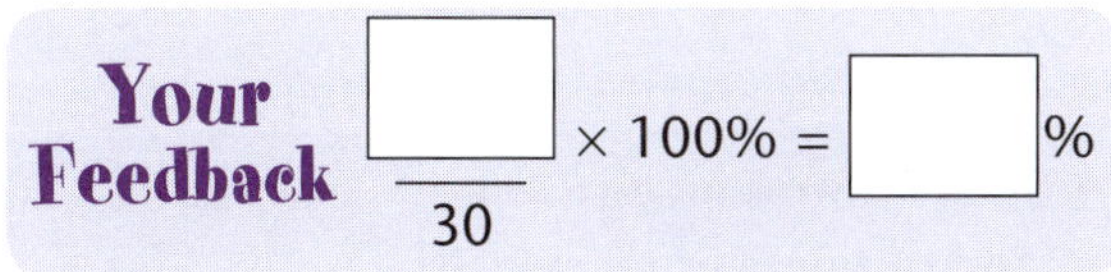

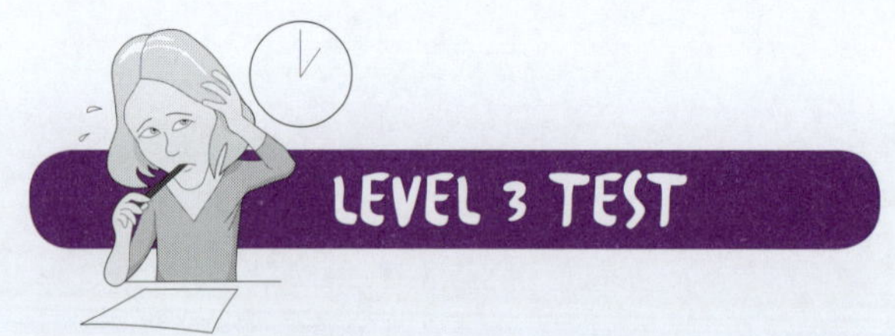

(25 marks)

1 Rewrite as a percentage:

a $\frac{2}{3}$ b $\frac{7}{8}$ (2 marks)

2 Convert the following to decimals:

a $8\frac{1}{2}\%$ b 3.4% c $12\frac{3}{4}\%$ (3 marks)

3 Find:

a $12\frac{1}{2}\%$ of \$70 b $\frac{4}{5}\%$ of \$50 (2 marks)

4 Increase 120 by 10% and then decrease the result by 10%. (2 marks)

5 Anastasia saves 60% of her income and spends the rest. What fraction of her income does Anastasia spend? (2 marks)

6 What percentage is:

a 50 centimetres of 2 metres?

b 125 millilitres of 1 litre? (2 marks)

7 Olivia earns \$800 per week. She spends 21% on rent, 13% on food and 6% on petrol. She saves the rest. How much does Olivia save per week? (2 marks)

8 A shop offers '18% OFF EVERYTHING'. How much did Dimitra pay for a television originally marked at \$1200? (2 marks)

9 Jamie paid \$23.40 for an item that was marked at \$25.74:

a How much discount did Jamie receive?

b What is the discount expressed as a percentage of the original price? (2 marks)

10 A washing machine is dicounted by 15% and now costs \$1275. What was the original price? (2 marks)

11 Manaia received a discount of $22\frac{1}{2}\%$ on an item marked \$205.50. How much did she pay for the item? (2 marks)

12 Decrease \$369.60 by $33\frac{1}{3}\%$. (2 marks)

Your Feedback $\frac{\square}{25} \times 100\% = \square\%$

☞ **Quick answers on page 262**
☞ **Worked solutions on page 321**

INTEGERS 5

- The Integer Number Line
- Addition of Directed Numbers
- Subtraction of Directed Numbers
- Multiplication and Division of Directed Numbers
- Algebra and Directed Numbers
- Substitution in Algebraic Expressions
- Some Directed Number Problems
- Directed Numbers and the Calculator
- Ordered Pairs and the Number Plane
- Directed Number Operations Involving Fractions and Decimals

KEYWORDS

Directed number
Integer
Negative
Number plane
Opposite
Operation
Ordered pairs
Positive
Substitution

The Integer Number Line

The number line helps us see relationships between numbers. It extends both sides of zero:

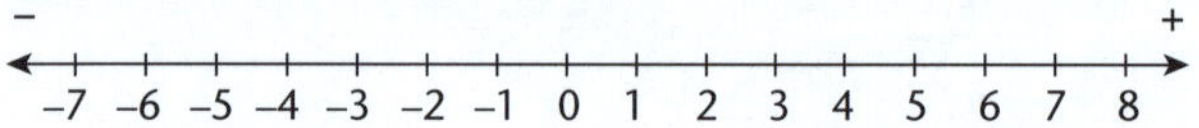

Negative direction	Positive direction
In this direction: ←	In this direction: →
numbers become smaller.	numbers become larger.

Any number to the *left* of another number is smaller than that number.

For example: $-3 < -2$
$-1 < 0$
$-99 < -98$

Or, any number to the *right* of another number is larger than that number.

For example: $-2 > -3$
$0 > -1$
$-98 > -99$

For Example

1 Insert the correct sign to make these statements true:

a $-6 \square -4$

b $-3 \square 0$

c $-2 \square -4$

2 Indicate whether these statements are true or false:

a $-6 < -7$

b $-3 > -1$

c $0 > -4$

3 Arrange in ascending order:
–1, –8, –3, 4, 1, 0

1 **a** $-6 < -4$ [–6 is left of –4 on number line]

b $-3 < 0$ [–3 is left of 0]

c $-2 > -4$ [–2 is to right of –4 on number line]

2 **a** $-6 < -7$ False [–6 is to right of –7 $\therefore -6 > -7$]

b $-3 > -1$ False [–3 is to left of –1 $\therefore -3 < -1$]

c $0 > -4$ True [0 is to right of –4]

3 –8, –3, –1, 0, 1, 4 [Ascending means going up ↑]

Addition of Directed Numbers

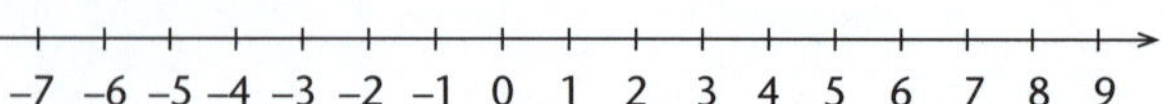

The number line can be used to add directed numbers. Consider the question $-6 + 8$:

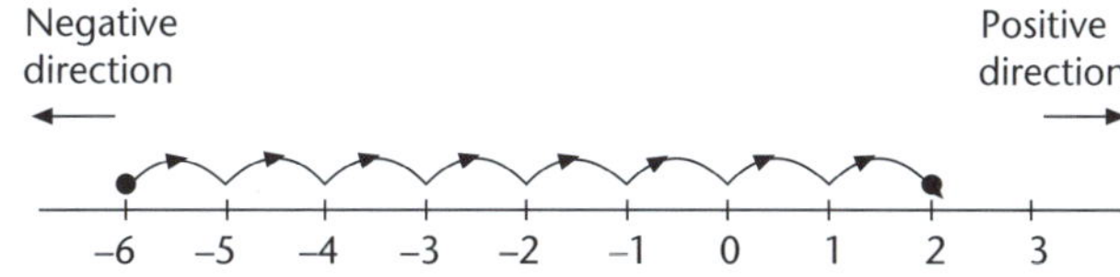

Start at –6 and move 8 in a positive direction. Finish point is 2.

∴ $-6 + 8 = 2$

Now consider $-2 + (-4)$.

Start at –2 and move 4 in negative direction (–4). Finish point is –6.

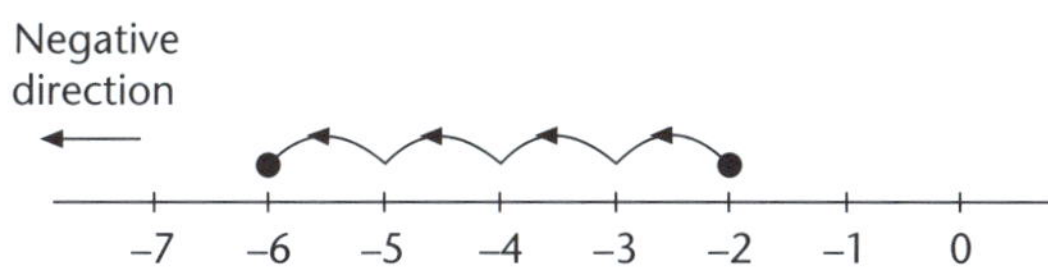

∴ $-2 + (-4) = -6$

Now consider $-4 + 5 + (-3)$.

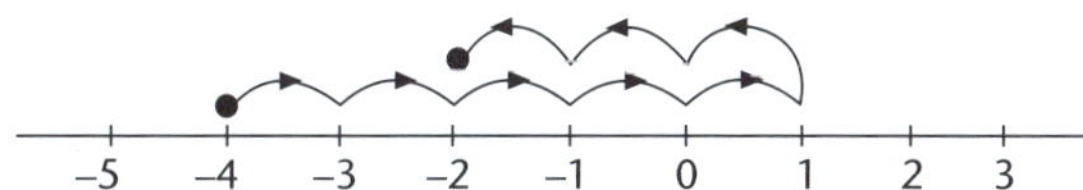

Start at –4, move 5 in a positive direction (at 1 now) and then move 3 in a negative direction. Finish point is –2.

∴ $-4 + 5 + (-3) = -2$

For Example

1 Work out the value of:

a $-3 + (-4)$

b $-2 + 7 + (-4)$

c $2 + (-4) + (-4) + 6$

2 Complete these statements to make them true:

a $-4 + \square = 2$

b $\square + (-5) = -7$

3 a Find the total of –6, 4 and –2.

b Calculate the sum of –9 and 7.

A number line is drawn for each solution in Question 1. In reality you would visualise the number line in your head or use the same number line for all questions.

1 a $-3 + (-4)$

$= -7$

b $-2 + 7 + (-4)$

$= 1$

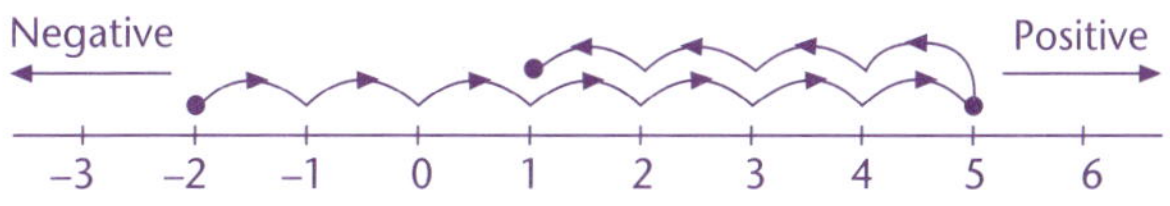

c $2 + (-4) + (-4) + 6$

$= 0$

2 a $-4 + \square = 2$

∴ $\square = 6$

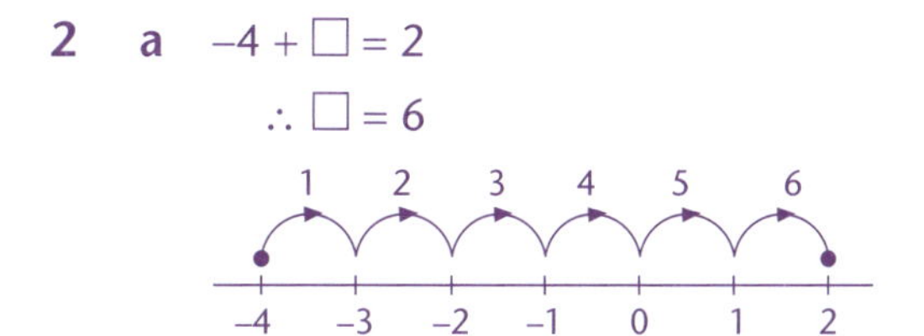

Start at –4, end at 2. Instruction must have been to add 6.

b $\square + (-5) = -7$

∴ $\square = -2$

Where do we start from? Finish is –7 and we moved –5 to get there. Undo this by going back 5 positive. Start must have been at –2.

3 a $(-6) + 4 + (-2)$

$= -4$ [Total means add.]

b $-9 + 7$

$= -2$ [Sum means add.]

Subtraction of Directed Numbers

We do not subtract—we add the opposite of the directed number.

Note: The opposite of –8 is 8.
The opposite of 7 is –7.

So to evaluate (–8) – 4, the question is changed to an addition question:

i.e. $(-8) - 4$
$= (-8) + (-4)$
$= -12$

- The first number is unchanged.
- Change subtraction sign to addition sign.
- Change 4 to opposite (–4).
- Add.

Also $(-6) - (-4)$
$= (-6) + 4$
$= -2$

- First number is unchanged.
- Change subtraction to addition.
- Change (–4) to opposite: 4.
- Add.

For Example

Work out the values of the following:

1 $-6 - 7$

2 $6 - 14$

3 $5 - (-5)$

4 $-3 - (-7)$

5 $5 - 8 - 6$

6 $-3 - (-4) - (-2)$

7 $-6 + 4 - (-3)$

8 $5 + (-3) - (-4)$

1 $-6 - 7$
$= -6 + (-7)$
$= -13$

2 $6 - 14$
$= 6 + (-14)$
$= -8$

3 $5 - (-5)$
$= 5 + 5$
$= 10$

4 $-3 - (-7)$
$= -3 + 7$
$= 4$

5 $5 - 8 - 6$
$= 5 + (-8) + (-6)$
$= -9$

Each subtraction becomes an addition of the opposite.

6 $-3 - (-4) - (-2)$
$= -3 + 4 + 2$
$= 3$

7 $-6 + 4 - (-3)$
$= -6 + 4 + 3$
$= 1$

– (–3) becomes +3. Additions do not change.

8 $5 + (-3) - (-4)$
$= 5 + (-3) + 4$
$= 6$

– (–4) becomes +4. Addition does not change.

Multiplication of Directed Numbers

When multiplying two directed numbers:

- If the signs are the **same**, the answer is **positive**.
- If the signs are **different**, the answer is **negative**.

This means that, if both numbers are negative, or if both are positive, the answer will be **positive**. If the numbers have different signs, one is positive and one is negative, then the answer will be **negative**.

For Example

Evaluate:

1 -6×4 2 $6 \times (-4)$

3 $(-6) \times (-4)$ **4** $-6 \times -3 \times -2$

5 $4 \times (-9) \times (-5)$ **6** $(-7)^2$

1 $-6 \times 4 = -24$

2 $6 \times -4 = -24$

[$6 \times 4 = 24$ However, signs different ∴ answer negative.]

3 $(-6) \times (-4) = 24$ [Same signs ∴ answer positive.]

4 $-6 \times -3 \times -2$ [Work from left to right.]
$= (-6 \times -3) \times -2$
$= 18 \times -2$
$= -36$

5 $4 \times (-9) \times (-5)$ [Work from left to right.]
$= [4 \times (-9)] \times (-5)$
$= (-36) \times (-5)$
$= 180$

6 $(-7)^2$
$= (-7) \times (-7)$
$= 49$

Division of Directed Numbers

The same rules as for multiplication apply to division.

That is, if the signs are the same, the answer is **positive**, and, if the signs are different, the answer is **negative**.

For Example

Calculate the value of:

1 $-32 \div -4$ **2** $42 \div -7$

3 $-63 \div 9$ **4** $\dfrac{-56}{8}$

5 $\dfrac{72}{-8}$ **6** $\dfrac{-35}{-7}$

1 $-32 \div -4$
$= 8$
[$32 \div 4 = 8$. Signs the same ∴ answer positive.]

2 $42 \div -7$
$= -6$
[Different signs ∴ answer negative.]

3 $-63 \div 9$
$= -7$

4 $\dfrac{-56}{8}$
$= -56 \div 8$
$= -7$

5 $\dfrac{72}{-8}$
$= 72 \div -8$
$= -9$

6 $\dfrac{-35}{-7}$
$= -35 \div -7$
$= 5$

Order of Operations

The rules for order of operations as set out in Chapter 1 apply for all integers, not just positive numbers.

For Example

Work out the value of these expressions:

1 $6 - 3 \times 4$

2 $-6 + 4 \times 5$

3 $15 - 3 \times (-2)$

4 $-8 \times 2 + 6 \times (-3)$

5 $12 + 3 \times (-6) + (-4)$

6 $(72 - 96) \div (-8)$

7 $(-3 + 8) \times (-8 + 3)$

8 $(-18) - (-24) \div (-6)$

1 $6 - 3 \times 4$
$= 6 - 12$
$= -6$

[When there are no grouping symbols, strong signs are done first ×, ÷]

2 $-6 + 4 \times 5$
$= -6 + 20$
$= 14$

3 $15 - 3 \times (-2)$
$= 15 - (-6)$
$= 15 + 6$
$= 21$

4 $-8 \times 2 + 6 \times (-3)$
$= -16 + (-18)$
$= -34$

5 $12 + 3 \times (-6) + (-4)$
$= 12 + (-18) + (-4)$
$= -10$

6 $(72 - 96) \div (-8)$
$= -24 \div (-8)$
$= 3$

7 $(-3 + 8) \times (-8 + 3)$
$= 5 \times (-5)$
$= -25$

8 $(-18) - (-24) \div (-6)$
$= (-18) - (-24 \div -6)$
$= (-18) - 4$
$= (-18) + (-4)$
$= -22$

Algebra and Directed Numbers

Addition and Subtraction

Terms that share a common pronumeral are called 'like terms'. For example a, $3a$, $7a$ are like terms, while $4x$, $-3y$, $2x^2$ are unlike terms.

For Example

Simplify these algebraic expressions:

1 $6a - 11a$

2 $-8y - 4y$

3 $-9t + 7t$

4 $-2a - 4a + 11a$

5 $16 + 3a - 11a$

6 $4y - 3x + y - 2x$

7 $2x + 3 - 5x + 2$

8 $3w - 2u - w + 8u$

1 $6a - 11a$
$= -5a$
[$6 - 11 = -5$]

2 $-8y - 4y$
$= -12y$
[$-8 - 4$
$= -8 + (-4) = -12$]

3 $-9t + 7t$
$= -2t$
[$-9 + 7 = -2$]

4 $-2a - 4a + 11a$
$= 5a$
[$-2 - 4 + 11$
$= -2 + (-4) + 11 = 5$]

5 $16 + 3a - 11a$
$= 16 + 3a + (-11a)$
$= 16 + (-8a)$
$= 16 - 8a$

6 $4y - 3x + y - 2x$
$= 4y + y - 3x - 2x$
$= 5y - 5x$

[
- Move like terms so that they are together.
- Move the sign in front with the term.
- $-3 - 2$
$= -3 + (-2) = -5$
]

7 $2x + 3 - 5x + 2$
$= 2x - 5x + 3 + 2$
$= -3x + 5$
[$2 - 5 = -3$]

8 $3w - 2u - w + 8u$
$= 3w - w - 2u + 8u$
$= 2w + 6u$
[$-2 + 8 = 6$]

Multiplication and Division

Like terms and unlike terms can be multiplied and divided.

For Example

Simplify:

1 $-8 \times 4y$ 2 $-6a \times 4a$

3 $9t \times (-3s)$ 4 $(-8y)^2$

5 $\frac{-8y}{4}$ 6 $-24y^2 \div -6y$

7 $\frac{32a}{-4a}$ 8 $\frac{-6a}{12}$

9 $8a \times (-4a) \times (-3a)$ 10 $-6a^3 \times 8a^4$

1 $-8 \times 4y$ $[-8 \times 4 = -32]$
$= -32y$

2 $-6a \times 4a$ $\left[\begin{array}{c}-6 \times 4 = -24 \\ a \times a = a^2\end{array}\right]$
$= -24a^2$

3 $9t \times (-3s)$ $\left[\begin{array}{c}9 \times -3 = -27 \\ t \times s = ts\end{array}\right]$
$= -27ts$

4 $(-8y)^2$
$= -8y \times -8y$ $\left[\begin{array}{c}-8 \times -8 = 64 \\ y \times y = y^2\end{array}\right]$
$= 64y^2$

5 $\frac{-8y}{4}$
$= -8y \div 4$ $[-8 \div 4 = -2]$
$= -2y$

6 $-24y^2 \div -6y$ $\left[\begin{array}{c}-24 \div -6 = 4 \\ y^2 \div y = y\end{array}\right]$
$= 4y$

7 $\frac{32a}{-4a} = 32a \div -4a$ $\left[\begin{array}{c}32 \div -4 = -8 \\ a \div a = 1\end{array}\right]$
$= -8$

8 $\frac{-6a}{12} = \frac{-a}{2}$ $\left[\frac{6}{12} = \frac{1}{2}\right]$

9 $8a \times (-4a) \times (-3a)$ $[a^2 \times a = a^3]$
$= -32a^2 \times (-3a)$ [See Chapter 7 — Algebra]
$= 96a^3$

10 $-6a^3 \times 8a^4$ $\left[\begin{array}{c}-6 \times 8 = -48 \\ a^3 \times a^4 = a^7\end{array}\right]$
$= -48a^7$

Substitution in Algebraic Expressions

Numbers replace pronumerals and the expression is evaluated.

For Example

If $a = -6$, $b = 4$, $c = -3$, evaluate the following expressions:

1 $a + b$

2 $a - b$

3 $b - a$

4 $4 - a$

5 $4 - a - c$

1 $a + b = -6 + 4$
$= -2$

2 $a - b = -6 - 4$
$= -6 + (-4)$
$= -10$

3 $b - a = 4 - (-6)$
$= 4 + 6$
$= 10$

4 $4 - a = 4 - (-6)$
$= 4 + 6$
$= 10$

5 $4 - a - c = 4 - (-6) - (-3)$
$= 4 + 6 + 3$
$= 13$

Some Directed Number Problems

The rules developed for evaluating expressions involving directed numbers can be applied to problems.

For Example

1 If the temperature at Fredbo is –6°C and:

a it rises 4°C

b it falls by 4°C,

what are the new temperatures?

2 The temperature at midnight is 2°C. By 3 a.m. it has fallen by 6°C, but at 6 a.m. it has risen 4°C from 3 a.m. Find the temperature at 6 a.m.

1 **a** $-6 + 4 = -2$

New temperature is –2°C.

b $-6 - 4 = -6 + (-4)$

$= -10$

New temperature is –10°C.

2 $2 - 6 + 4 = 2 + (-6) + 4$

$= -4 + 4 = 0$

The temperature at 6 a.m. is 0°C.

Directed Numbers and the Calculator

You are able to answer directed number questions on your calculator. You should familiarise yourself with the appropriate buttons and methods for your brand of calculator. Check the manual.

Ordered Pairs and the Number Plane

Each point on the number plane is represented by two numbers, one from the *x* axis and one from the *y* axis. The *x* number comes first. Look at point A. It is the ordered pair (2, 3). B is (3, –2), C is (–2, –3) and D is (–3, 2):

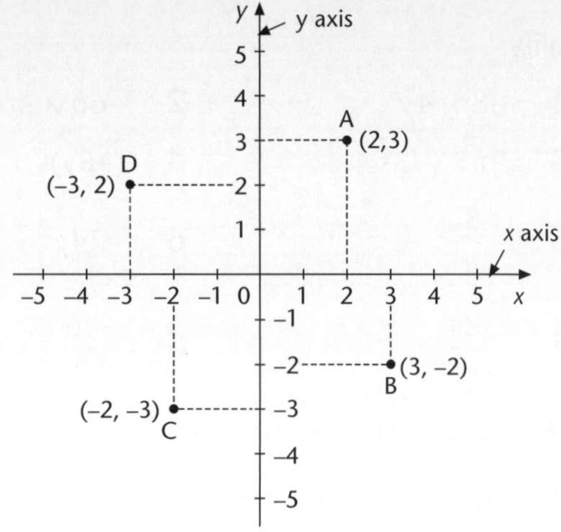

These sets of ordered pairs are also called **co-ordinates**. The point of intersection of the two axes is called the **origin**. The origin has co-ordinates (0, 0). Any point on the *x* axis has a *y* co-ordinate of zero, i.e. (–, 0). Any point on the *y* axis has an *x* co-ordinate of zero, i.e. (0, –).

For Example

1 Write down the co-ordinates of the points marked A, B, C, D and E:

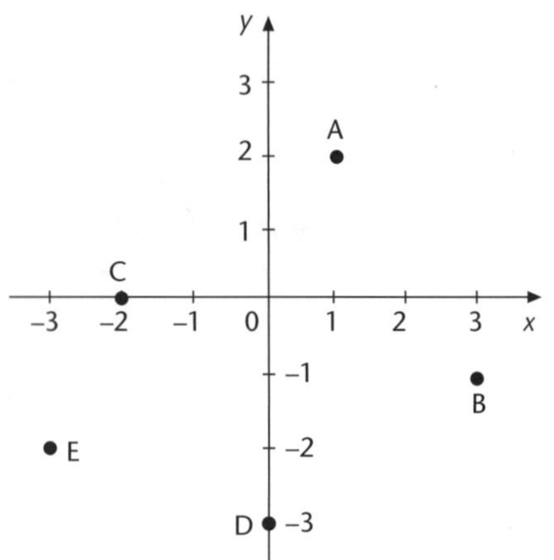

2 On a number plane, with a scale between –4 and 4 on both the x and y axes, plot the following ordered pairs: P(–3, 3), Q(2, –4), R(0, –2), S(4, 0), T(–4, –2).

3 Plot the points A(–3, –1), B(2, –1), C(3, 2) and D(–2, 2) on the number plane. Join A to B to C to D to A with straight lines. Name the shape that has been drawn.

1 A(1, 2), B(3, –1), C(–2, 0), D(0, –3), E(–3, –2).

2

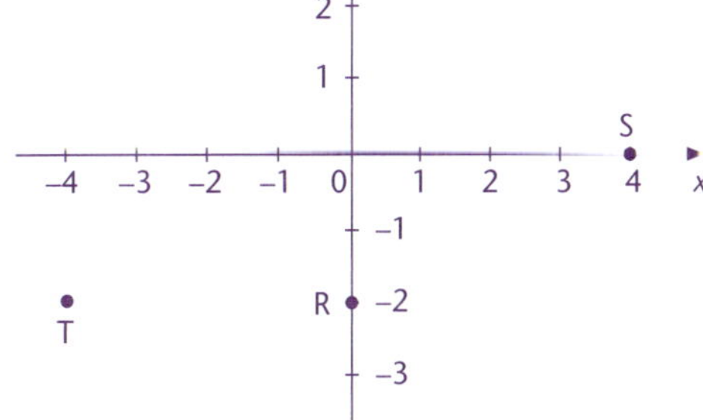

3

The shape is a parallelogram.

Directed Number Operations Involving Fractions and Decimals

The rules developed for evaluating directed numbers can be applied to questions involving fractions and decimals. Use Chapters 2 and 3 for further assistance.

For Example

Calculate:

1 $\frac{3}{5} - \frac{4}{5}$

2 $\frac{-2}{3} + \frac{1}{2}$

3 $\frac{-3}{4} - \frac{1}{4}$

4 $1.4 - 2.4$

5 $\frac{1}{2}a - \frac{3}{4}a$

6 $\frac{-1}{2}y - \frac{1}{2}y$

1 $\frac{3}{5} - \frac{4}{5} = \frac{-1}{5}$ $\quad [3 - 4 = -1]$

2 $\frac{-2}{3} + \frac{1}{2} = \frac{-4}{6} + \frac{3}{6}$ $\quad [-4 + 3 = -1]$

$= \frac{-1}{6}$

3 $\frac{-3}{4} - \frac{1}{4} = \frac{-4}{4} = -1$ $\quad [-3 - 1 = -4]$

4 $1.4 - 2.4 = -1$ $\quad [1.4 + (-2.4)]$

5 $\frac{1}{2}a - \frac{3}{4}a$

$= -\frac{1}{4}a$ $\quad \left[\frac{2}{4} - \frac{3}{4} = \frac{-1}{4}\right]$

6 $\frac{-1}{2}y - \frac{1}{2}y$ $\quad \left[\frac{-1}{2} - \frac{1}{2} = -1\right]$

$= -y$

PRACTISE, PRACTISE

Go to p. 256 for quick answers, or to pp. 285–289 for worked solutions.

1 a I have \$40 in my bank account. If I withdraw \$105, how much am I overdrawn? p. 86

b The temperature at 1 a.m. is 2°C. If it falls by 2°C every hour, what is the temperature at 4 a.m.?

c The temperature at Murrurundi was –8°C. If it rose by 6°C, find the new temperature.

2 Insert the correct sign in the following statements (< or >): p. 86

a $-5 \square -4$	b $-3 \square -4$	c $0 \square 5$
d $0 \square -5$	e $55 \square -55$	f $-99 \square -100$

3 Indicate whether these statements are true or false: p. 86

a $-3 < -5$	b $-4 > -3$	c $5 > -5$
d $-6 = 6$	e $-30 < -31$	f $-45 > -44$

4 Arrange in ascending order: p. 86

a –9, 2, –1, –4
b 0, –5, 2, –4, 6
c –7, 14, 5, –14, 8, 0
d –3, –5, –7, –1, 1, 3
e 0, –10, 5, –12, 1, –4, –6
f –112, –114, –115, –113, –131, –141, –151

5 Arrange in descending order: p. 86

a –6, –5, –2, 0
b 0, 3, –3, 2, –2
c –4, –2, 0, –5, 5
d –10, 2, –8, 6, –4, –5
e –9, –10, –8, –11, –12, –7, –5, –6
f –112, –115, –114, –124, –123, –126

6 Complete the following additions: pp. 86–87

a $-8 + 5$	b $-6 + 8$	c $(-3) + (-9)$
d $6 + (-4)$	e $(-5) + (-8)$	f $9 + (-9)$
g $-8 + 8$	h $-6 + -4$	i $4 + (-12)$
j $-18 + 12$	k $-30 + (-12)$	l $-15 + 16$
m $-12 + 14$	n $16 + (-17)$	o $-31 + 42$

7 Evaluate: pp. 86–87

a $-8 + 3 + 2$
b $(-6) + (-4) + 7$
c $-9 + 14 + (-6)$
d $(-9) + (-5) + (-8)$
e $-14 + 6 + 8$
f $7 + (-9) + 2$
g $-54 + 17 + 19$
h $-20 + (-18) + 38$
i $(-8) + (-6) + (-8) + 14 + (-4)$
j $16 + (-9) + 17 + (-20) + (-14) + 10$

8 Complete these statements to make them true: pp. 86–87

a $-5 + \square = 9$	b $7 + \square = 2$	c $4 + \square = -3$
d $-6 + \square = 0$	e $-8 + \square = -4$	f $-3 + \square = 7$
g $5 + \square = -9$	h $\square + (-8) = -2$	i $\square + -3 = -8$

j $-5 + \square = -3 + 7$ k $-6 + 2 = 2 + \square$ l $4 + (-6) = 5 + \square$
m $\square + -2 = -6 + 4$ n $\square + 3 = -4 + 7 + (-3)$

9 Find the sum of: pp. 86–87
a -8 and 2 b -6 and -4 c 4, 7 and -7
d -4, -6 and -8 e 9, -4, 4, -9

10 Calculate the totals of these numbers: pp. 86–87
a $-7, -4, 3, 5$ b $-8, -2, -6, -4$
c $9, -9, 4, -4, 8, -8$ d $-112, 99, -4, 5, -6$
e $114, -53, -114, 71, 53, -71$

11 Complete the following subtractions: p. 88
a $-4 - 6$ b $11 - 13$ c $6 - 14$
d $-9 - 5$ e $5 - 13$ f $6 - (-4)$
g $5 - (-4)$ h $-5 - (-7)$ i $-8 - (-3)$
j $7 - 3$ k $5 - 17$ l $-4 - 18$
m $31 - 45$ n $-18 - 12$ o $21 - (-19)$

12 Evaluate: p. 88
a $-6 - 3 - 4$ b $7 - 12 - 5$ c $5 - 8 - 4$
d $-8 - 3 - 9$ e $5 - (-5) - 6$ f $(-6) - (-4) - 5$
g $3 + (-9) - (-4)$ h $-11 - (-4) + 3$ i $-6 - 4 + 8$
j $13 - (-8) + 9$ k $7 - 15 - 9 + (-3)$ l $-4 - 3 + 11 - (-4)$

13 a By how much does 5 exceed -2? p. 88
b Find the difference between 3 and -4.
c How much bigger than -5 is -3?

14 a Find the sum of -7 and -9 pp. 86–88
b By how much does 4 exceed -16?

15 Complete the following multiplications: pp. 88–89
a -9×5 b $(-8) \times (-7)$ c $4 \times (-8)$
d -3×7 e $(-8) \times (-5)$ f 4×-6
g $9 \times (-3)$ h 6×-6 i $(-9) \times (-4)$
j $7 \times (-1)$ k $(-3) \times (-3)$ l $5 \times (-9)$
m $(-7)^2$ n $(-8)^2$ o $15 \times (-8)$

16 Evaluate: pp. 88–89
a $(-3) \times (-5) \times 7$ b $(-2) \times 4 \times (-6)$
c $8 \times (-3) \times 2$ d $(-5) \times 3 \times (-2)$
e $(-2)^3$ f $(-9) \times (-3) \times (-1)$
g $5 \times (-3)^2$ h $(-5)^2 \times 3$
i $(-2)^2 \times (-3)^2$ j $(-3) \times 5 \times (-4) \times 2$

17 Find the products of: pp. 88–89
a -8 and 6 b 4 and -12
c -9 and -7 d -10 and -20

18 Perform the following divisions: p. 89

a $-27 \div 3$ **b** $42 \div -6$ **c** $\frac{49}{-7}$

d $\frac{-35}{5}$ **e** $\frac{-54}{-9}$ **f** $\frac{36}{-9}$

g $56 \div -8$ **h** $-72 \div -9$ **i** $(-81) \div (-9)$

j $60 \div (-15)$

19 Find the quotient of: p. 89

a 40 and –8 **b** –30 and –6

c 84 and –12 **d** –24 and –4

20 Evaluate: pp. 86–89

a $-8 + 3$ **b** -8×3 **c** $-8 - 3$

d $(-8) \times (-3)$ **e** $-8 - (-3)$ **f** $-8 + (-3)$

g $(-12)^2$ **h** -12×2 **i** $-12 + 2$

j $-12 - 2$ **k** $-15 - 15$ **l** $-15 + 15$

m $(-3) \times 15$ **n** $15 - 18$ **o** $-32 \div -8$

21 Calculate the values of the following expressions: pp. 89–90

a $4 - 3 \times 5$ **b** $-8 + 2 \times 5$

c $10 - (-5) \times 3$ **d** $20 - (-6) \times (-3)$

e $-4 \times 5 + 3 \times 5$ **f** $5 \times 4 - 7 \times 4$

g $(5 - 8) \times (3 - 7)$ **h** $(-6 - 4) \times (-7 - 3)$

i $4 + 3 \times (-4) + 8$ **j** $(27 - 12) \div (-5)$

k $15 - 27 \div (-9)$ **l** $(-8 + 5) \times (-3 + 7)$

m $(13 - 37) \div 4$ **n** $3 - 36 \div 4$

22 Simplify these algebraic expressions: pp. 90–91

a $a - 7a$ **b** $-9t + 4t$ **c** $-8y - 6y$

d $5y - 11y$ **e** $3w - (-3w)$ **f** $-5t + 13t$

g $9y - 17y$ **h** $-9n^2 + 6n^2$ **i** $5m^2 - 6m^2$

j $-12z + 13z$ **k** $-a + 2a$ **l** $-a - 4a$

m $2y - 3y - 5y$ **n** $-8a + 4a + 4a$ **o** $3t - 8t + 5t$

p $4x - 11x - 7x$ **q** $-a - a - a$ **r** $y^2 - 2y^2 - 7y^2$

s $a + 2a - 3a - a$ **t** $7b - 11b + 5b - 3b$

23 Simplify: pp. 90–91

a $14 + 3a - 7$ **b** $7y + 3b - 4y$

c $7a + 4b - 5a$ **d** $-9y + z - 4y$

e $5t + 6s - 11t + 2s$ **f** $9a - 7c - 4a + 5c$

g $a + a + a - 4b - 2b$ **h** $t + 3u - 7t + 2u$

i $4y - 8 - 6y + 12$ **j** $a - b - 2a + 4b - a + 2b$

24 Simplify: pp. 90–91

a $-7 \times a$ **b** $-6 \times 2a$ **c** $8 \times -3y$

d $-9 \times -4b$ **e** $(-4t)^2$ **f** $6a \times (-7a)$

g $-24y \div 6y$ **h** $15a \div -5$ **i** $\dfrac{-36y}{9}$

j $\dfrac{-42a}{-7a}$ **k** $5a \times (-7b)$ **l** $(-9a) \times (-8c)$

25 Simplify: pp. 90–91

a $6a + 4 \times 3a$ **b** $12y - 3 \times 4y$

c $2a \times 3a - 4a \times 5a$ **d** $24y - 18y^2 \div 6y$

e $(5y - 8y) \times (3t - 8t)$

26 If $x = -4$, $y = 5$, $z = -2$, evaluate: p. 91

a $x + y$ **b** $x - y$

c $y - x$ **d** xy

e $\dfrac{x}{z}$ **f** $-5x$

g $\dfrac{-35}{y}$ **h** xz

i $x + y + z$ **j** $x + y - z$

k $x - y + z$ **l** $x - (y + z)$

m x^2 **n** $-3x^2$

o $6 - x$ **p** $-5 + x$

q $xy + x^2$ **r** $x(y + x)$

s $xy - x^2$ **t** $x(y - x)$

u $x^2 - y^2$ **v** $(x - y)(x + y)$

w $3x - 2y$ **x** $3x - 2y + 4z$

y $(x + y) \times (z - y) \div (y + 2)$

27 Write down the ordered pairs for the points marked A, B, C, D, E, F, G, H, I, J, K, L, M, N, P, Q, R, S, T, U, V, W: pp. 92–93

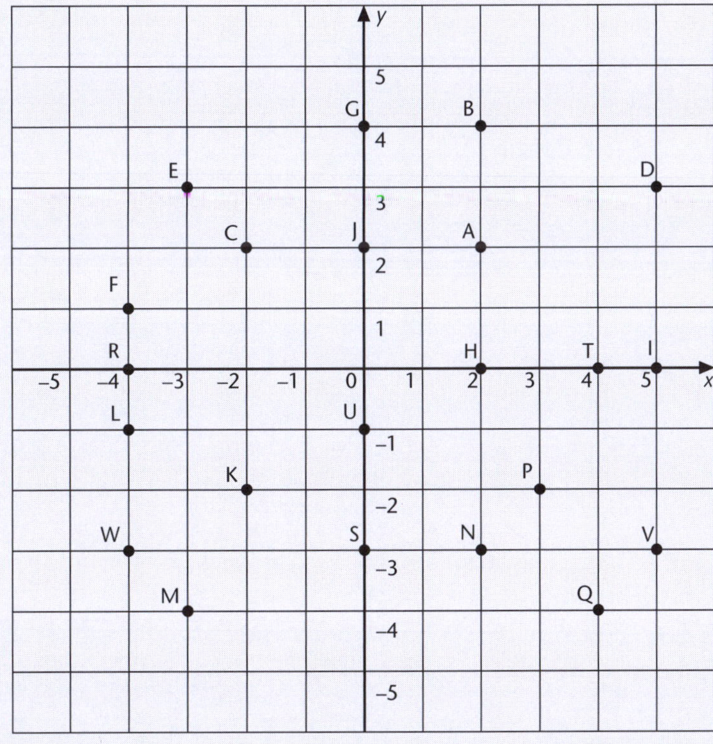

28 Plot the following ordered pairs and join them with straight lines. A full stop (.) indicates that those two points are not to be joined: pp. 92–93

a A(−4, −2), B(2, −2), C(4, 1), D(−2, 1), A.
E(2, 0), F(3, 2), G(2, 4), H(1, 2), E.
M(−4, 2), J(−2, 0), K(0, 2), L(−2, 4), M.
N(0, −3), P(1, −4), Q(1, −1), R(0, −2), N.
Name the shapes ABCD, EFGH, MJKL and NPQR.

b (−3, −2), (0, 1), (−1, 1), (0, 3), (1, 1), (0, 1), (3, −2), (−3, −2).
(−1, 1), (−1, 3), ($-\frac{1}{2}$, 2).
(1, 1), (1, 3), ($\frac{1}{2}$, 2).
($2\frac{1}{2}$, $-1\frac{1}{2}$), (3, −1), (4, −1), (5, −1), (4, $-1\frac{1}{2}$), (3, −2).

29 The temperature at noon was 8°C. If it fell 2°C every hour for the next 5 hours, find the temperature at 5 p.m. p. 92

30 By first plotting the ordered pairs P(−5, −1) and Q(3, 5), write down the ordered pair for the mid-point of the line joining P and Q. pp. 92–93

31 Evaluate: p. 93

a $\frac{1}{2} - 1$	**b** $\frac{3}{5} - \frac{4}{5}$	**c** $1\frac{1}{2} - 2\frac{1}{4}$	**d** $-\frac{3}{5} + \frac{1}{5}$
e $0.6 - 0.8$	**f** $4 - 6.8$	**g** $-0.8 + 0.6$	**h** $\frac{1}{3}a - \frac{3}{4}a$

Go to p. 256 for quick answers, or to pp. 285–289 for worked solutions.

YOUR CHECKLIST

For a complete understanding of this topic you must be able to:

✓	Understand the idea of positive and negative directions		p. 86
✓	Arrange directed numbers by size		p. 86
✓	Add integers		pp. 86–87
✓	Find the opposite of an integer		p. 88
✓	Subtract integers		p. 88
✓	Multiply and divide integers		pp. 88–89
✓	Apply the order of operation rules to integers		pp. 89–90
✓	Apply knowledge of directed number to simplifying algebraic expressions		pp. 90–91
✓	Substitute integers into algebraic expressions		p. 91
✓	Use directed numbers in problem-solving		p. 92
✓	Write down an ordered pair for any point on the number plane		pp. 92–93
✓	Plot the position of an ordered pair on the number plane		pp. 92–93
✓	Apply directed numbers to fractions and decimals.		p. 93

Now you are ready to do the tests!

LEVEL 1 TEST

(30 marks)

1 Insert the correct sign (< or >) to make these statements true:

a $-5 \square 0$ b $-3 \square -4$ (2 marks)

2 Complete the following calculations:

a $-3 + 7$ b $(-9) + (-3)$ c $-6 + 8 + (-4)$

d $-7 - 11$ e $-3 - (-8)$ (5 marks)

3 Find the sum of 11, −3 and −5. (1 mark)

4 Complete these statements to make them true:

a $-3 + \square = -5$ b $\square - 3 = -8$ (2 mark)

5 Evaluate:

a $-8 + 2$ b $2 - 6$ c $(-5) \times (-3)$ d $12 \div (-4)$ (4 marks)

6 Simplify:

a $3t - 5t$ b $-8a + 4a$ c $-2m - 3m$

d $-3a \times 4$ e $15a \div (-3a)$ (5 marks)

7 If $m = -7$ and $n = 5$, evaluate the expressions:

a $m + n$ b $n - m$ c $2 - m$

d $m - n$ e $3 + m - n$ (5 marks)

8 The temperature at Vladivostok at 8 a.m. was −18°C. If the temperature rose 4°C and 3°C over the next two hours, calculate the temperature at 10 a.m. (2 marks)

9 Write down the ordered pairs for the points:

a A b B (2 marks)

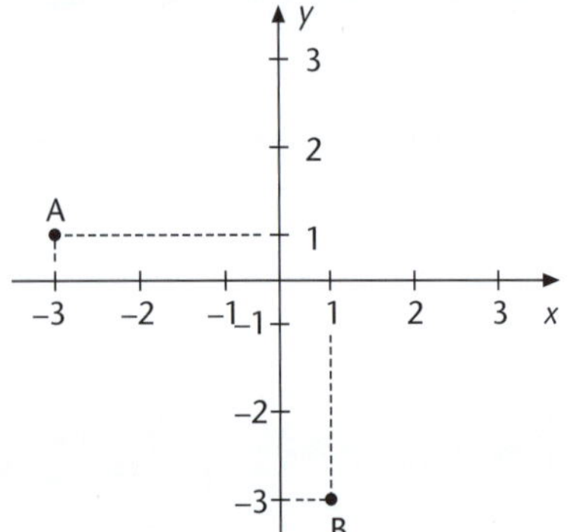

10 Evaluate:

a $-\frac{1}{2} + \frac{3}{4}$ b $1.2 - 3.5$ (2 marks)

Your Feedback $\frac{\square}{30} \times 100\% = \square\%$

☞ Quick answers on page 262
☞ Worked solutions on page 322

(40 marks)

1 Arrange in ascending order: $-1, -8, 0$ (1 mark)

2 Evaluate:

a $11 + (-8)$ b $-13 + 5$

c $-3 - 11$ d $11 - 17$

e $5 - (-3) - 7$ f $(-9) + 5 + (-8)$ (6 marks)

3 Complete the statements to make them true:

a $9 - \square = -13$ b $(-8) - \square = 4$ (2 marks)

4 Total the integers -18, -3, 11 and -5. (1 mark)

5 Evaluate:

a $(-4) \times (-6)$ b $40 \div (-8)$

c $(-100) \div 10$ d $(-12) \div (-6)$

e $(-3) \times 2 \times (-2)$ f $(-6)^2$ (6 marks)

6 Simplify the expressions:

a $t - 7t$ b $-8t + 2t$

c $7b - b$ d $-3a + 3a$

e $-2t + 3t - t$ f $6a + 12 - 8a + 4$ (6 marks)

7 Evaluate:

a $-1\frac{1}{2} + \frac{1}{4}$ b $(5 - 9) - (-3 + 7)$

c $-12 + 3 - 2$ d $1 - 7 + 2$ (4 marks)

8 If $a = -7$, $b = -4$ evaluate the expressions:

a $b + a$ b $3 - a$

c $b - a + 2$ d ab

e a^2 f $2b - a$ (6 marks)

9 Evaluate:

a $\frac{2}{3} - \frac{4}{5}$ b $0.7 + (-1.5)$ c $-6 + 2.3$ (3 marks)

10 The temperature at Ketchican measured at 8 a.m., 9 a.m. and 10 a.m. was -11°C, -8°C and -5°C. Calculate the average temperature. (2 marks)

11 Evaluate:

a $3 - 4 \times 2$ b $4 \times 3 - 14$ c $(-12) \div 6 - 4$ (3 marks)

☞ Quick answers on page 262
☞ Worked solutions on page 322

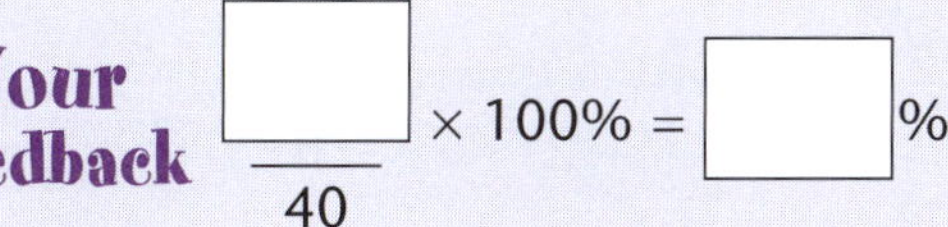

LEVEL 3 TEST

(40 marks)

1 Arrange in descending order: –7, –1, –10, 5 (1 mark)

2

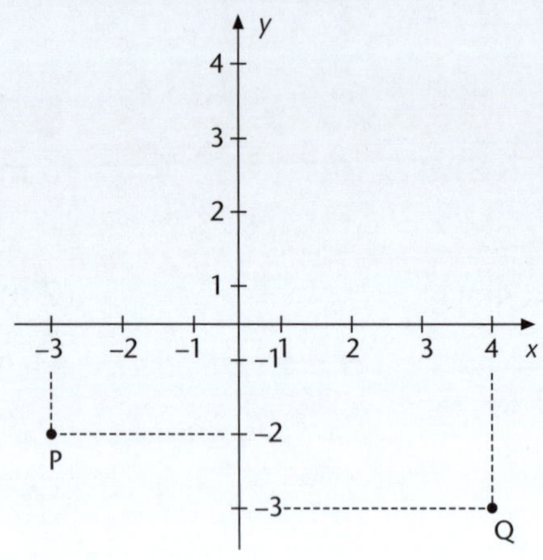

a Write down the ordered pairs for P and Q.

b Hence, or otherwise, deduce the mid-point of the interval joining P and Q.

(4 marks)

3 Evaluate:

a $-19 - 21$ b $5 - 7 - 13$

c $-16 + 13 + (-2)$ d $-5 - 3 - (-8)$ (4 marks)

4 Find the sum of –9 and –7. (1 mark)

5 Find the difference of 3 and –10. (1 mark)

6 Evaluate:

a $-9 - 4$ b $-9 - (-5) + 3$ c $-7 - 3 + 10$ (3 marks)

7 Complete the statements to make them true:

a $\triangle + (-8) = -40$ b $\square - (-9) = 6$ (2 marks)

8 Evaluate the expressions:

a $10 - 16 \div (-4)$ b $(-16 \times 2) \div (-8)$ (2 marks)

9 Complete this statement to make it true: $5 - \square - (-3) = 15$ (2 marks)

10 Simplify the expressions:

a $-5m - 3m$ b $-8y + 2y$ c $-20a - a$

d $(-4y)^2$ e $10a - 4a \times 3$ f $(-15y) \div (-3y) \times (-2)$ (6 marks)

11 If $x = -9$, $y = 8$ evaluate the expressions:

a $4 - x - y$ b $x - y$ c $y - x$

d $x + y$ e $7 - x$ (5 marks)

12 Evaluate:

a $-\frac{2}{3} + 2$ b $-\frac{3}{4} - \frac{5}{8}$ c $1 - (-\frac{4}{5}) + \frac{1}{10}$ (3 marks)

13 Given $a = -2$, evaluate $-5a + 3$. (1 mark)

14 Simplify:

a $4x - 5x + 2$ b $7a - 3 + a$ c $12 - 2a + 3a - 1$ (3 marks)

15 If $x = -4$, $y = -3$, evaluate:

a $2x^2 - 3y^2$ b $\dfrac{x + 2y}{5}$ (2 marks)

☞ **Quick answers on page 263**
☞ **Worked solutions on page 323**

Your Feedback $\dfrac{\square}{40} \times 100\% = \square\%$

RATIOS 6

 Ratio

 Using Ratios to Solve Problems

KEYWORDS

Equivalent **Unitary**

Ratio

Ratio

Ratio is the comparison of two or more quantities expressed in the same units. The ratio of a to b is written as $a:b$ (or sometimes $\frac{a}{b}$).

Order is important so that $a:b$ is different to $b:a$. The value of the ratio $a:b$ is $\frac{a}{b}$, of $b:a$ is $\frac{b}{a}$.

Examples of ratios are 2:5, 5:2, 3:4, 4:3, 7:2. The value of the ratio 2:5 is $\frac{2}{5}$, while the value of 5:2 is $\frac{5}{2}$.

Equivalent Ratios

Ratios are simplified in the same way as fractions—we divide (or multiply) both components of the ratio by the same number (except 0).

For Example

Simplify:

1	6:4	**2**	4:6
3	30:50	**4**	16:64
5	120:10	**6**	40:20:10
7	$\frac{3}{4}:\frac{1}{4}$	**8**	$\frac{2}{3}:1$
9	$\frac{1}{2}:\frac{1}{4}$	**10**	$1\frac{1}{2}:2$
11	$\frac{4}{5}:1\frac{1}{2}$	**12**	0.2:0.8
13	0.5:1.5:4		

1 $6:4 = \frac{6}{2}:\frac{4}{2}$ (dividing both by 2)
$= 3:2$

2 $4:6 = \frac{4}{2}:\frac{6}{2}$
$= 2:3$

3 30:50 = 3:5 (divide both parts by 10)

4 16:64 = 1:4 (divide by 16)

5 12Ø:1Ø = 12:1
(cross off zeros—divide by 10)

6 4Ø:2Ø:1Ø = 4:2:1 (divide all by 10)

7 $\frac{3}{4}:\frac{1}{4} = \frac{3}{4} \times 4:\frac{1}{4} \times 4$ (multiplying both by 4)
$= 3:1$

8 $\frac{2}{3}:1 = \frac{2}{3} \times 3:1 \times 3$
$= 2:3$

9 $\frac{1}{2}:\frac{1}{4} = \frac{1}{2} \times 4:\frac{1}{4} \times 4$
$= 2:1$

or $\frac{1}{2}:\frac{1}{4} = \frac{2}{4}:\frac{1}{4}$
$= 2:1$ (multiplying both by 4)

10 $1\frac{1}{2}:2 = 1\frac{1}{2} \times 2:2 \times 2$
$= 3:4$

or $1\frac{1}{2}:2 = \frac{3}{2}:2$
$= \frac{3}{2}:\frac{4}{2}$
$= 3:4$ (multiplying by 2)

11 $\frac{4}{5}:1\frac{1}{2} = \frac{4}{5} \times 10:1\frac{1}{2} \times 10$
$= \frac{8}{10} \times 10:\frac{15}{10} \times 10$
$= 8:15$

Multiply both parts by 10 as this is the LCD of 5 and 2.

12 $0.2:0.8 = 0.2 \times 10:0.8 \times 10$
$= 2:8$
$= 1:4$ (dividing by 2)

13 0.5:1.5:4 = 5:15:40 (multiplying by 10)
= 1:3:8 (dividing by 5)

Sometimes we find the value of the pronumeral to complete the simplification.

For Example

Find the value of pronumerals:

1 $x:4 = 15:20$ **2** $4:28 = 2:x$

1 $15:20 = 3:4$
$\therefore x = 3$

2 $4:28 = 2:14$
$\therefore x = 14$

Often we have to change one part of the ratio so that both parts have the same units. We then simplify the resulting ratio.

For Example

Simplify:

1 20 cents:80 cents

2 40 cents:\$2

3 \$1.20:\$6

4 6 days:3 weeks

5 4 cm:1 m

6 2 L:150 mL

7 8 secs:2 min

8 4 hours:1 day

9 $1\frac{1}{2}$ mL:1 L

10 400 kg:2 t

11 \$20:\$40:\$80

12 7.2 L:21.6 L

13 $3x:12x$

14 $21ab:14ac$

15 $\frac{1}{p}:p$

16 $16ab:4ab^2$

1 20 cents:80 cents = 20:80
= 1:4 (divide by 20)

2 40 cents:\$2 = 40 cents:200 cents
= 40:200
= 1:5 (divide by 40)

3 \$1.20:\$6 = 1.2:6
= 6:30 (multiply by 5)
[or 120c:600c = 1:5]
= 1:5 (divide by 6)

4 6 days:3 weeks = 6 days:21 days
= 6:21
= 2:7

5 4 cm:1 m = 4 cm:100 cm
= 4:100
= 1:25

6 2 L:150 mL = 2000 mL:150 mL
= 200~~0~~:15~~0~~
= 200:15
= 40:3 (dividing by 5)

7 8 secs:2 min = 8 secs:120 secs
= 8:120
= 1:15

8 4 hours:1 day = 4 hours:24 hours
= 4:24
= 1:6

9 $1\frac{1}{2}$ mL:1 L = $1\frac{1}{2}$ mL:1000 mL
= $1\frac{1}{2}$:1000
= 3:2000 (multiply by 2)

10 400 kg:2 t = 400 kg:2000 kg
= 400:2000
= 4:20
= 1:5

[t is the symbol for metric tonne = 1000 kg]

11 $\$20:\$40:\$80 = 20:40:80$
$= 1:2:4$

12 $7.2\text{ L}:21.6\text{ L} = 7.2:21.6$
$= 72:216$ (multiply by 10)
$= 1:3$ (divide by 72)

13 $3x:12x = \frac{3x}{x}:\frac{12x}{x}$ (divide both parts by x)
$= 3:12$
$= 1:4$

14 $21ab:14ac = \frac{21ab}{a}:\frac{14ac}{a}$ (divide by a)
$= 21b:14c$
$= 3b:2c$ (divide by 7)

15 $\frac{1}{p}:p = \frac{1}{p} \times p:p \times p$
$= 1:p^2$

16 $16ab:4ab^2 = \frac{\overset{4}{\cancel{16}}\overset{1}{\cancel{a}}\overset{1}{\cancel{b}}}{\underset{1}{\cancel{4}}\underset{1}{\cancel{a}}\underset{1}{\cancel{b}}}:\frac{\overset{1}{\cancel{4}}\overset{1}{\cancel{a}}\overset{1}{\cancel{b}}b}{\underset{1}{\cancel{4}}\underset{1}{\cancel{a}}\underset{1}{\cancel{b}}}$
(divide by $4ab$)
$= 4:b$

Finding a Ratio of Two or More Quantities

From the information in a problem, a ratio is formed and simplified if necessary.

For Example

1 A netball team won 8 games in a season and lost 4 games. What is the ratio of games won to games lost?

2 Sylvio mixes 6 buckets of gravel, 4 buckets of sand and 2 buckets of cement into a batch of concrete. Determine the ratio of gravel to sand to cement that was used.

3 A street in a new subdivision has 40 blocks of land, of which 25 have been built on. What is the ratio of occupied to vacant blocks?

4 The results of an election were announced:

Candidate	No. of votes
Christine Howard	16 000
Noel Horton	12 000
Peter Chapman	20 000

What is the ratio of the votes for:

- **a** Christine Howard to Noel Horton?
- **b** Peter Chapman to the total votes?

5 In a class survey the hair colour of boys and girls was recorded:

Colour	Boys	Girls
Blonde	3	4
Brown	5	7
Black	2	4
Red	2	1

- **a** Write down the number of students:
  - **i** who are boys
  - **ii** who are girls
  - **iii** in the class.
- **b** What is the ratio of:
  - **i** Blonde boys to blonde girls?
  - **ii** Blonde girls to black-haired girls?
  - **iii** Brown-haired to black-haired to red-haired students?
  - **iv** Non-blonde students to the total number of students?

6 Find the ratio of:

a Shaded squares to unshaded squares:

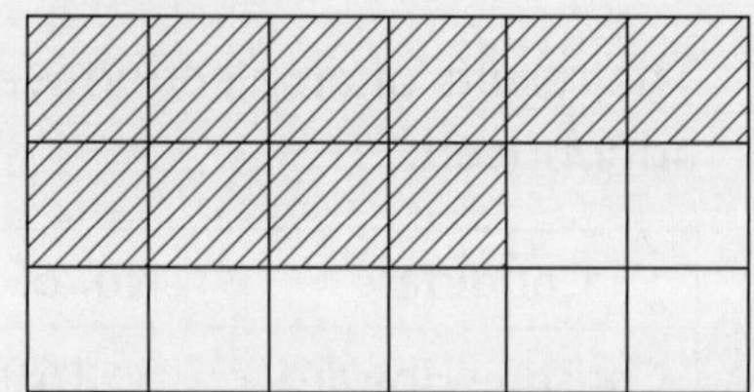

b Shaded squares to total number of squares.

7

Walkers	Bus travellers	Cyclists

10 20 30 40 50 60 70 80 90 100%

Students in Year 7 were surveyed to determine their method of travelling to school each day and the results recorded in the above graph. Find the ratio of:

a Bus travellers to walkers

b Bus travellers to the total number of students surveyed.

8

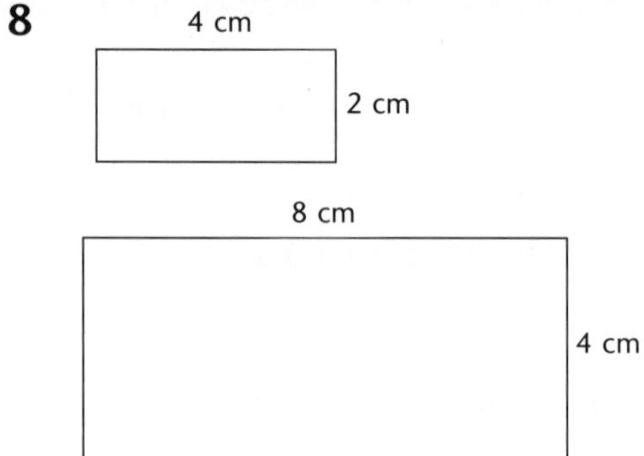

What is the ratio of the:

a Lengths of the two rectangles?

b Perimeters of the two rectangles?

c Areas of the two rectangles?

1 8:4 = 2:1

2 6:4:2 = 3:2:1

3 Vacant = 40 – 25
= 15
∴ 25:15 = 5:3

4 a 16 000:12 000 = 16:12
= 4:3

b Total = 16 000 + 12 000 + 20 000
= 48 000

∴ Ratio = $20\not{0}\not{0}\not{0}:48\not{0}\not{0}\not{0}$
= 20:48
= 5:12

5 a i Boys = 3 + 5 + 2 + 2
= 12
∴ There were 12 boys.

ii Girls = 4 + 7 + 4 + 1
= 16
∴ There were 16 girls.

iii Total = 12 + 16
= 28
∴ There were 28 students.

b i 3:4

ii 4:4 = 1:1

iii 12:6:3 = 4:2:1

iv (28 – 7):28 = 21:28
= 3:4

6 a Shaded = 10, unshaded = 8
∴ 10:8 = 5:4

b Shaded = 10, total = 18
∴ 10:18 = 5:9

7 a Bus travellers = 50, walkers = 30
Ratio = 50%:30%
= 50:30
= 5:3

b Bus travellers = 50, total students = 100
Ratio = 50%:100%
= 50:100
= 1:2

8 a 4 cm:8 cm = 4:8 = 1:2

b Perimeter of small rectangle = 2(4 + 2)
= 2(6) = 12

∴ The perimeter is 12 cm.

Perimeter of large rectangle = 2(8 + 4)
= 2(12)
= 24

∴ The perimeter is 24 cm.

∴ Ratio of perimeters = 12 cm:24 cm
= 12:24
= 1:2

c Area of small rectangle = 4×2
= 8

∴ The area is 8 cm^2.

Area of large rectangle = 8×4
= 32

∴ The area is 32 cm^2.

∴ Ratio of areas = 8 cm^2:32 cm^2
= 8:32
= 1:4

Using Ratios to Solve Problems

The Unitary Method

The **unitary method** can be used in ratio problems. It involves finding the value of one item (or one *unit*) and then multiplying this value to find the required answer.

1 The ratio of cats to dogs is 4:5. If there are 30 dogs, how many cats?

2 On a bus trip the ratio of males to females is 4:7. If there were 28 females on the bus, how many males?

3 Three sums of money are in the ratio of 2:3:5. If the smallest amount is $4.80, find the largest amount.

4 It is known that a metal alloy is made of copper, tin and zinc in the ratio 14:17:10. In a certain batch, 85 kg of tin is used. How much copper and zinc will be required to make the alloy?

5 Joe's will stated that his estate should be divided such that for each $2 that Carol receives, Keith should receive $5. If Keith gained $74 000 from Joe's estate, how much will Carol receive?

6 The ratio of the ages of three brothers is 10:8:5. If the youngest is 15 years old, how old is the eldest?

1 Cats to dogs = 4:5 [Order is important.]
Now, 30 dogs:
∴ 5 parts = 30 [We can use 'parts' or 'shares'.]
1 part = 6
4 parts = 4×6
= 24
∴ There are 24 cats. [*Check*: 24:30 = 4:5]

2 Males to females = 4:7
Now, 28 females:
∴ 7 parts = 28
1 part = 4
4 parts = 16 [*Check*: 16:28 = 4:7]
∴ There are 16 males.

3 Sums of money = 2:3:5
∴ smallest amount = 2 parts
∴ 2 parts = $4.80
1 part = $2.40
∴ 5 parts = $12
∴ $12 is largest amount.

4 Copper, tin, zinc = 14:17:10

$\therefore$ 17 parts = 85

1 part = 5

$\therefore$ 14 parts = 70

and 10 parts = 50

$\therefore$ 70 kg copper and 50 kg zinc are required.

5 Carol:Keith = \$2:\$5

= 2:5

$\therefore$ Keith's share is 5 parts
(Carol's share is 2 parts)

$\therefore$ 5 parts = \$74 000

1 part = \$14 800 (74 000 ÷ 5)

2 parts = \$29 600 (14 800 × 2)

$\therefore$ Carol's share of the estate is \$29 600.

6 Brothers' ages = 10:8:5 (5 is the youngest)

$\therefore$ 5 parts = 15

1 part = 3

10 parts = 30

$\therefore$ The eldest brother is 30 years old.

Dividing Quantities in a Given Ratio

To divide quantities in a given ratio, find the total of the parts and then use the unitary method.

For Example

1 Divide \$250 in the ratio of 7:3.

2 Share \$72 in the ratio of 4:3:2.

3 If \$200 is to be split in the ratio of 3:2, find the smaller amount?

4 A 2 L container is used to make up an orange fruit juice drink. 500 mL of concentrate is poured into the container, and then water is used to fill the remainder of the container:

a What is the ratio of concentrate to water in the fruit juice drink?

b Two-hundred and forty millilitres of the fruit juice drink is poured into a glass. How much of the drink is water?

5

The ratio of two supplementary angles is 4:5. What is the size of each angle?

6 Two sisters, Margaret and Robyn, contribute \$8000 and \$4000 respectively to buy a car. Three years later they decide to sell the car for \$7200 and split the money in the same ratio as their investment. What was Margaret's share?

7 The ratio of a father's age to that of his son is 9:2. If the sum of their ages is 44, how old is the father?

8 In an orchard the ratio of orange trees to lemon trees is 3:1, while the ratio of lemon trees to mandarin trees is 2:5:

a What is the ratio of orange trees to lemon trees to mandarin trees?

b If there are 390 citrus trees, how many of each variety are in the orchard?

1 Total parts = 7 + 3

= 10

$\therefore$ 10 parts = \$250

$\therefore \frac{7}{10} \times \$250 = \$175$

$\frac{3}{10} \times \$250 = \$75$

$\therefore$ The two amounts are \$175 and \$75.

2 Total parts = 4 + 3 + 2
= 9
∴ 9 parts = \$72
∴ $\frac{4}{9} \times \$72 = \$32$
$\frac{3}{9} \times \$72 = \$24$
$\frac{2}{9} \times \$72 = \$16$
∴ \$32, \$24, \$16

3 5 parts = \$200
∴ $\frac{2}{5} \times \$200 = \$80$
∴ The smaller amount is \$80.

4 **a** Water = 2000 − 500
= 1500
∴ 1500 mL water
∴ Concentrate:water = 500 mL:1500 mL
= 1:3
∴ The ratio of concentrate to water is 1:3.

b Total parts = 1 + 3
= 4
∴ 4 parts = 240 mL
∴ $\frac{3}{4} \times 240 = 180$
∴ 180 mL of the drink is water.

5 Supplementary angles add to 180°.
As 4 + 5 = 9, there are 9 parts:
∴ 9 parts = 180°
$\frac{4}{9} \times 180 = 80$
$\frac{5}{9} \times 180 = 100$
∴ The angles are 80° and 100°.

6 Ratio of investments = \$8000:\$4000
= 2:1
∴ Total parts = 2 + 1
= 3
∴ $\frac{2}{3} \times \$7200 = \$4800$
∴ Margaret's share of the car is worth \$4800.

7 Total parts = 9 + 2
= 11
∴ $\frac{9}{11} \times 44 = 36$
∴ The father is 36 years old.

8 **a** Orange:lemon = 3:1 = 6:2,
lemon:mandarin = 2:5
∴ 6:2:5 is the ratio of orange to lemon to mandarin.

b Total parts = 6 + 2 + 5
= 13
∴ 13 parts = 390
$\frac{6}{13} \times 390 = 180$
$\frac{2}{13} \times 390 = 60$
$\frac{5}{13} \times 390 = 150$
∴ There are 180 orange trees, 60 lemon trees and 150 mandarin trees in the orchard.

PRACTISE, PRACTISE

Go to p. 257 for quick answers, or to pp. 289–291 for worked solutions.

1 Express the following in simplest form: p. 105

a 25:15 **b** 30:70 **c** 9:18:27
d 14:49 **e** 100:100 000 **f** 75:125

2 Find the value of the pronumeral: p. 106

a $6:2 = x:1$ **b** $15:45 = 1:x$ **c** $12:8 = 3:x$
d $7:5 = x:35$ **e** $13:3 = x:9$

3 Simplify the following: pp. 105–107

a 3.5:1.5 **b** 2:1.4 **c** 8.4:4.2
d 3.25:4.25 **e** $\frac{3}{4}:\frac{1}{2}$ **f** $\frac{2}{3}:\frac{5}{6}$
g $\frac{3}{5}:1\frac{1}{2}$ **h** $2\frac{1}{4}:2\frac{1}{2}$ **i** $\frac{4}{5}:2.2$

4 Simplify the following ratios by first changing to the same units: pp. 105–107

a \$1.50:40c **b** \$2.40:\$1.60
c 8 hours:2 days **d** 1 mm:1 cm:1 m
e 3 weeks:6 days **f** $1\frac{1}{2}$ mins:45 secs
g 1 mL:1 kL **h** 3 decades:2 centuries

5 Express in simplest form: pp. 105–107

a $12ab:6a$ **b** $3xy:15xy$
c $2a^2:(2a)^2$ **d** $\frac{3}{p}:3p$

6 A motorbike was bought for \$700 and after two years was sold for \$900. Find the: pp. 107–109

a Profit
b The ratio of the:
i Cost price to selling price **ii** Profit to cost price.

7 A school has a teaching staff of 56. If 31 are male, find the: pp. 107–109

a Number of female teachers
b Ratio of the:
i Male teachers to female teachers
ii Total teaching staff to female teachers.

8 On a particular day, there are 10 hours of sunlight. What is the ratio of the number of hours of sunlight to hours in the day? pp. 107–109

9 A bag contains 6 black, 4 white and 10 red marbles. What is the ratio of: pp. 107–109

a Black to red marbles?
b Red to non-red marbles?

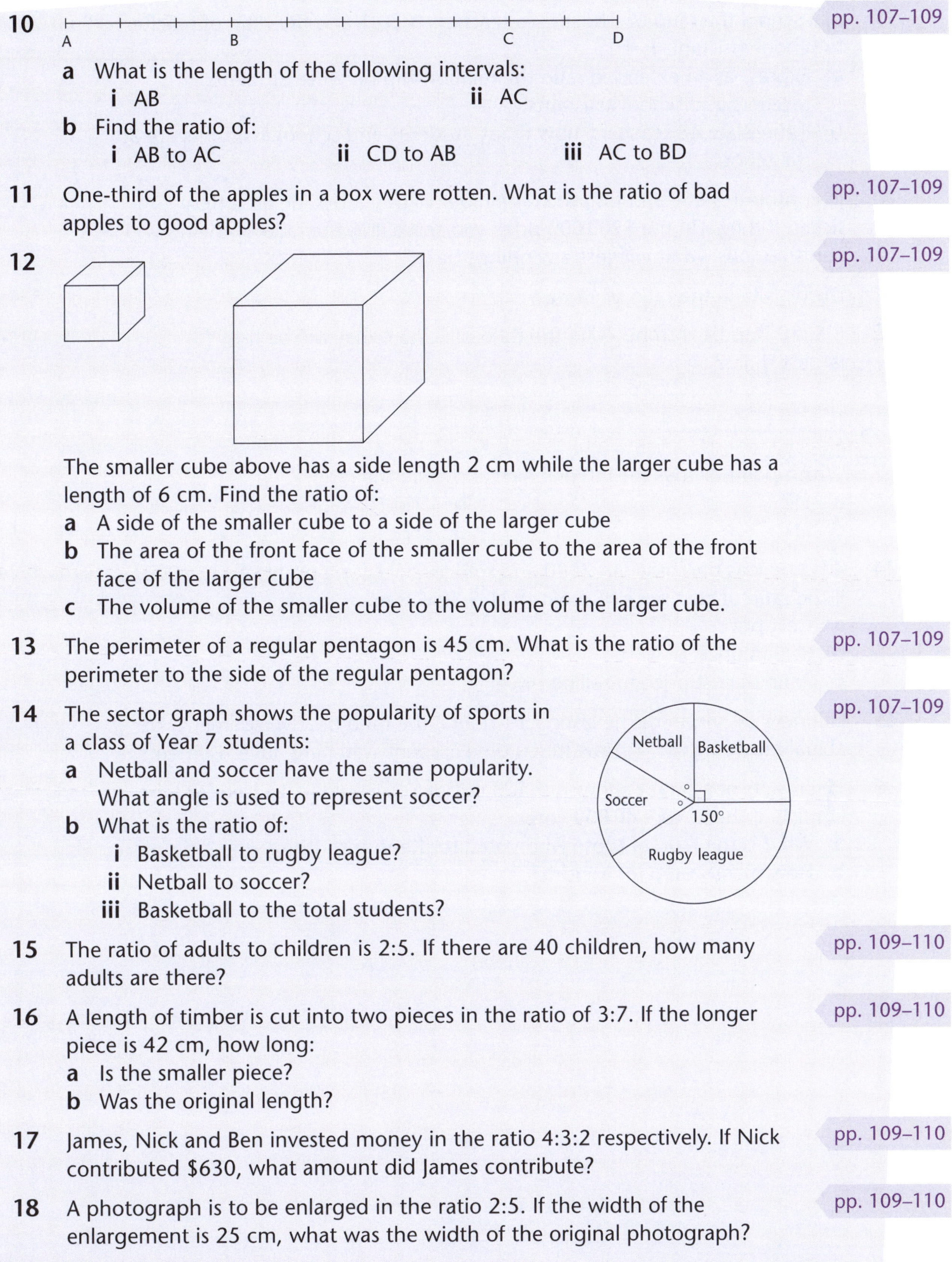

10 pp. 107–109

a What is the length of the following intervals:

  i AB **ii** AC

b Find the ratio of:

  i AB to AC **ii** CD to AB **iii** AC to BD

11 One-third of the apples in a box were rotten. What is the ratio of bad apples to good apples? pp. 107–109

12 pp. 107–109

The smaller cube above has a side length 2 cm while the larger cube has a length of 6 cm. Find the ratio of:

a A side of the smaller cube to a side of the larger cube

b The area of the front face of the smaller cube to the area of the front face of the larger cube

c The volume of the smaller cube to the volume of the larger cube.

13 The perimeter of a regular pentagon is 45 cm. What is the ratio of the perimeter to the side of the regular pentagon? pp. 107–109

14 The sector graph shows the popularity of sports in a class of Year 7 students: pp. 107–109

a Netball and soccer have the same popularity. What angle is used to represent soccer?

b What is the ratio of:

  i Basketball to rugby league?

  ii Netball to soccer?

  iii Basketball to the total students?

15 The ratio of adults to children is 2:5. If there are 40 children, how many adults are there? pp. 109–110

16 A length of timber is cut into two pieces in the ratio of 3:7. If the longer piece is 42 cm, how long: pp. 109–110

a Is the smaller piece?

b Was the original length?

17 James, Nick and Ben invested money in the ratio 4:3:2 respectively. If Nick contributed $630, what amount did James contribute? pp. 109–110

18 A photograph is to be enlarged in the ratio 2:5. If the width of the enlargement is 25 cm, what was the width of the original photograph? pp. 109–110

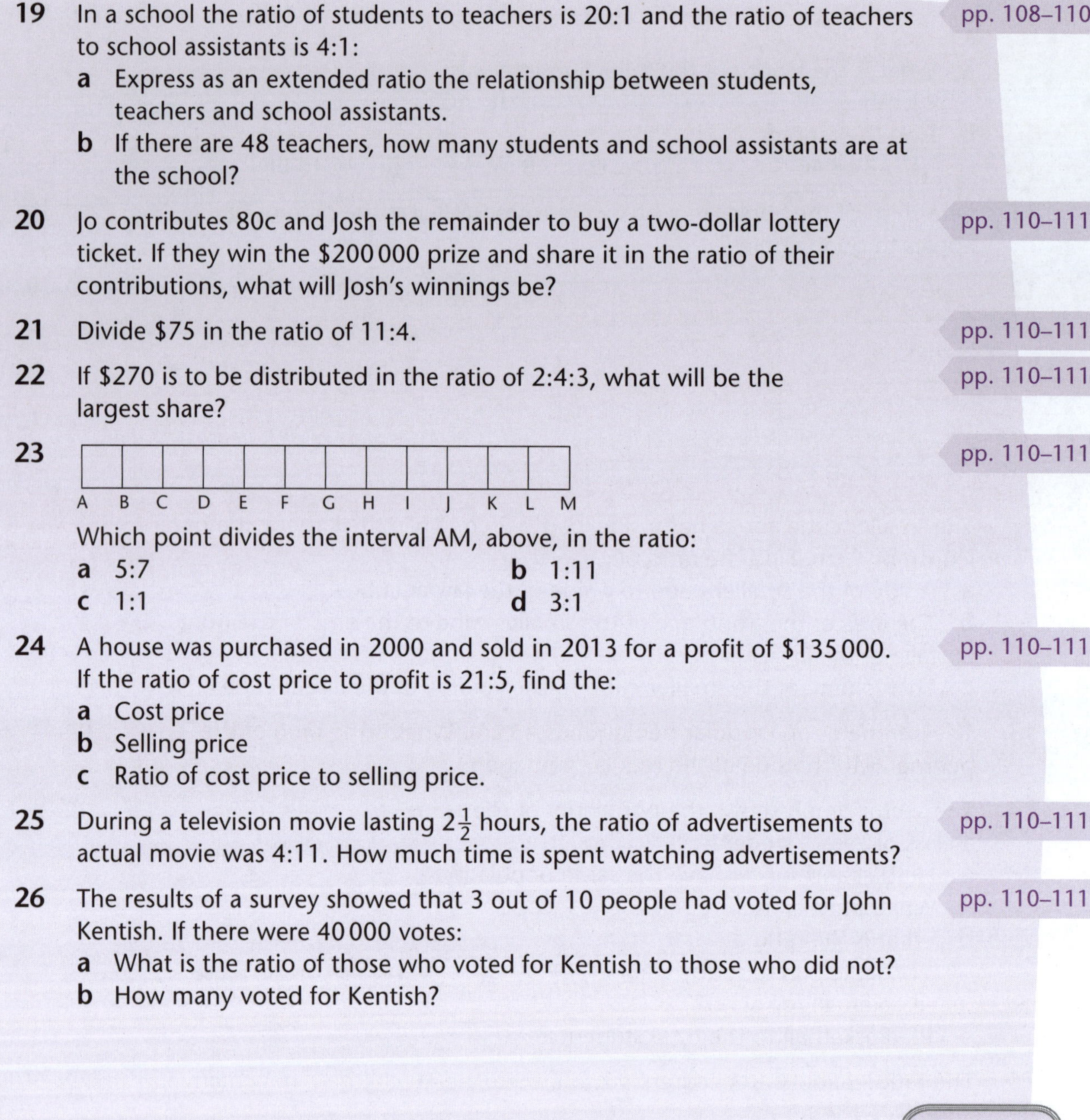

19 In a school the ratio of students to teachers is 20:1 and the ratio of teachers to school assistants is 4:1: pp. 108–110

a Express as an extended ratio the relationship between students, teachers and school assistants.

b If there are 48 teachers, how many students and school assistants are at the school?

20 Jo contributes 80c and Josh the remainder to buy a two-dollar lottery ticket. If they win the $200 000 prize and share it in the ratio of their contributions, what will Josh's winnings be? pp. 110–111

21 Divide $75 in the ratio of 11:4. pp. 110–111

22 If $270 is to be distributed in the ratio of 2:4:3, what will be the largest share? pp. 110–111

23 pp. 110–111

A B C D E F G H I J K L M

Which point divides the interval AM, above, in the ratio:

a 5:7 **b** 1:11

c 1:1 **d** 3:1

24 A house was purchased in 2000 and sold in 2013 for a profit of $135 000. If the ratio of cost price to profit is 21:5, find the: pp. 110–111

a Cost price

b Selling price

c Ratio of cost price to selling price.

25 During a television movie lasting $2\frac{1}{2}$ hours, the ratio of advertisements to actual movie was 4:11. How much time is spent watching advertisements? pp. 110–111

26 The results of a survey showed that 3 out of 10 people had voted for John Kentish. If there were 40 000 votes: pp. 110–111

a What is the ratio of those who voted for Kentish to those who did not?

b How many voted for Kentish?

Go to p. 257 for quick answers, or to pp. 289–291 for worked solutions.

YOUR CHECKLIST

For a complete understanding of this topic you must be able to:

✓	Define the term ratio		p. 105
✓	Simplify ratios		p. 105
✓	Find the value of a pronumeral in equivalent ratios		pp. 105–106
✓	Find a ratio by comparing two or more quantities		pp. 106–109
✓	Use ratios to solve problems		pp. 109–110
✓	Divide quantities in a given ratio.		pp. 110–111

Now you are ready to do the tests!

LEVEL 1 TEST

(25 marks)

1 Simplify:

a 3:12 b 45:35 c \$2:20c (3 marks)

2 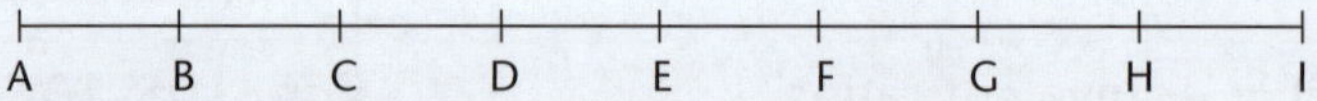

Find the ratio of:

a AB:EF b CE:AI c BG:AF (3 marks)

3 A class contains 16 boys and 14 girls. Find the ratio of:

a Boys to girls b Girls to total number of students. (2 marks)

4 An orchard has 120 citrus trees, comprising orange and mandarin trees. If there are 80 orange trees, find the ratio of:

a Orange trees to mandarin trees b Total trees to orange trees. (2 marks)

5 The ratio of cordial to water is 1:3. If there is 270 mL of water, how much cordial is in the mixture? (3 marks)

6 A small town has a population of 464. If the ratio of adults to children in the town is 5:3, find the number of adults in the town. (3 marks)

7 A 30-metre cable is cut in the ratio 2:3. Find the longer length. (2 marks)

8 Find the ratio of:

a Shaded squares to unshaded squares

b Unshaded squares to all squares.

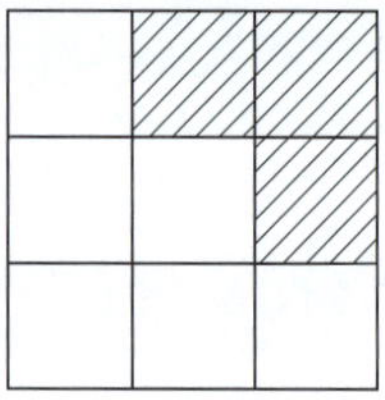

(2 marks)

9 The ratio of boys to girls in a class is 2:3. If there are 30 students in the class, how many boys are there? (2 marks)

10 The ratio of boys to girls in a school is 5:4. There are 550 boys at the school. How many students are at the school? (3 marks)

☞ Quick answers on page 263
☞ Worked solutions on page 324

Your Feedback $\frac{\square}{25} \times 100\% = \square\%$

(25 marks)

1 Simplify:

a $3.20:$4   b 15:12:6   c 1 metre:10 kilometres (3 marks)

2 Find the value of x:

a $3:x = 12:16$ b $5:15 = x:30$ (2 marks)

3 Of 72 blocks of land in a new subdivision, 24 are already sold. Find the ratio of total blocks to unsold blocks. (3 marks)

4 At a school dance the ratio of teachers to students is 1:35. If there are 280 students, how many teachers are there? (2 marks)

5 Divide 540 g in the ratio of 7:2. (3 marks)

6 A piece of timber is cut in the ratio of 2:3:4. If the longest piece is 1.2 m, how long is the shortest piece? (2 marks)

7 A farmer planted 480 ha of land with wheat, oats and barley, in the ratio of 3:5:4. Find the area planted with oats. (2 marks)

8 In a bag the ratio of red to black jellybeans is 4:5. If there are 45 jellybeans in the bag, how many of them are black? (2 marks)

9

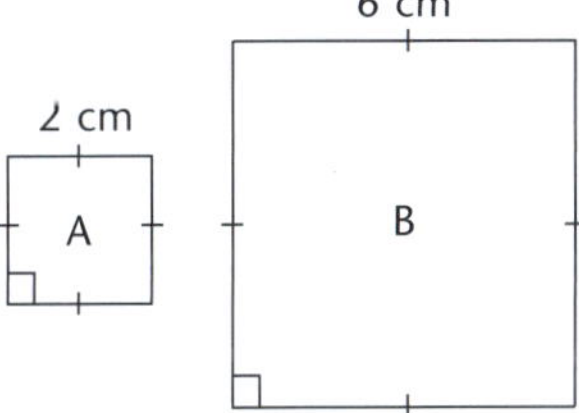

What is the ratio of the:

a Length of square A to the length of square B?

b Perimeter of square A to the perimeter of square B?

c Area of square A to the area of square B? (3 marks)

10 An alloy is made up of three elements A, B and C in the ratio 5:4:3 respectively. In a particular batch, 21 kg of element C is used. How much of element A is required to make up the alloy? (3 marks)

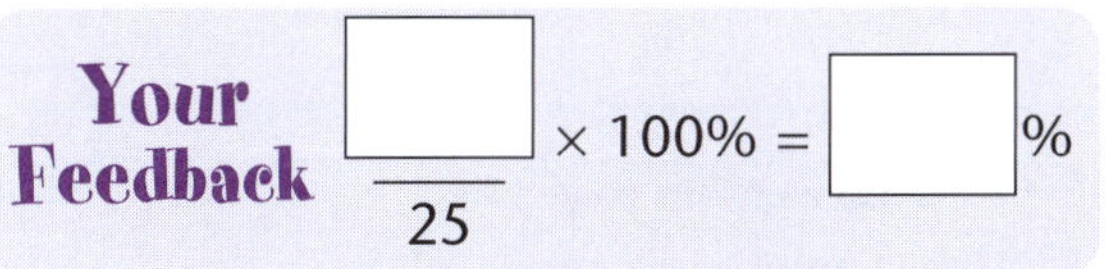

☞ Quick answers on page 263
☞ Worked solutions on page 324

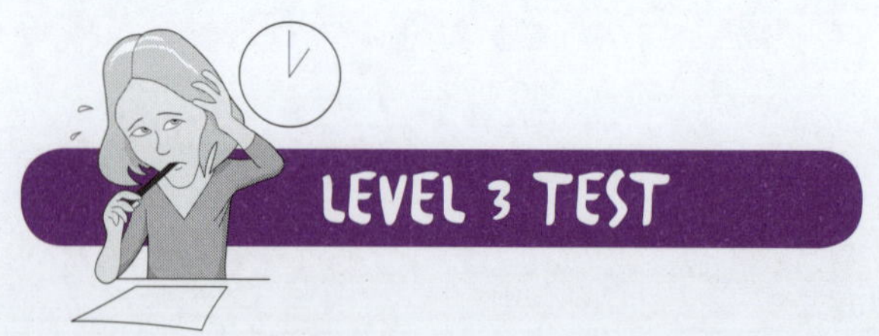

LEVEL 3 TEST

(25 marks)

1 Simplify:

a $\frac{3}{4}:1$ b 5000 m^2:2 ha c $\frac{2}{3}:\frac{3}{4}$ (3 marks)

2 Find the value of x:

a $x:4 = 48:64$ b $25:35 = 10:x$ (2 marks)

3 A triangle has two angles of 60° and 40°. Find the ratio of the largest angle to the smallest angle in the triangle. (1 mark)

4 A DVD recorder is priced at $650, but is discounted by 15%. Express the discount as a ratio of the original price. (1 mark)

5 Divide $360 in the ratio of 2:3:5. (2 marks)

6 The perimeter of a rectangle is 80 cm and its sides are in the ratio of 2:3. Find the length of each side. (2 marks)

7 An alloy contains copper and iron in the ratio of 3:4. If the quantity of iron is 5.6 kg, find the amount of copper required to make the alloy. (2 marks)

8 Divide $8.70 in the ratio 2:1. How much is the smaller share? (2 marks)

9 What is the ratio of the:

a Perimeters of the rectangles?

b Areas of the rectangles? (2 marks)

3 cm
2 cm
4.5 cm
3 cm

10 The ratio of a mother's age to that of her daughter is 10:3. If the daughter is 12 years old, what is the sum of their ages? (2 marks)

11 Two friends Christopher and Maraiah contribute $2.50 and $7.50 respectively to buy a lottery ticket. The lottery ticket won $25 000. They decided to split the money in the same ratio as their investment. What was Christopher's share? (2 marks)

12 A survey is conducted with a group of students to find their mode of transport to school. The ratio of train travellers to bus travellers was 3:2, while the ratio of bus travellers to car travellers was 5:2.

a Find the ratio of train travellers to bus travellers to car travellers.

b If there were 48 car travellers, how many students went by train? (4 marks)

Your Feedback $\frac{\square}{25} \times 100\% = \square\%$

☞ **Quick answers on page 263**
☞ **Worked solutions on page 325**

7 ALGEBRA

- Number Patterns and Pronumerals
- Operations on Pronumerals
- Multiplication of Pronumerals
- Substitution in Algebraic Expressions
- Like Terms
- Indices (Extension)
- Division of Pronumerals

KEYWORDS

Algebra	**Patterns**
Expand	**Power**
Evaluate	**Pronumerals**
Expression	**Relationship**
Formula	**Simplify**
Indices	**Substitution**

Number Patterns and Pronumerals

The study of number patterns is very important in the understanding of pronumerals.

1 By considering the pattern below, made of matchsticks, complete the table.

Number of squares	1	2	3	4	5	6	7	8	9
Number of matchsticks									

2 In your own words, write a rule that relates the number of matchsticks needed to build the pattern to the number of squares.

3 If N stands for the number of matchsticks used to create S squares in the pattern, write the rule in Question 2 using pronumerals.

4 Using the rule, find the number of matchsticks needed to create 30 squares.

5 Using the rule, find the number of squares created if 240 matchsticks are used.

1

Number of squares	1	2	3	4	5	6	7	8	9
Number of matchsticks	4	8	12	16	20	24	28	32	36

2 Number of matchsticks

= 4 × number of squares in the pattern

3 $N = 4 \times S$

where N = number of matchsticks

and S = number of squares.

4 Using the rule found in Question 3, substitute $S = 30$ and find N.

$$N = 4 \times S$$
$$= 4 \times 30$$
$$= 120$$

Therefore, 120 matchsticks are needed to build a pattern with 30 squares.

5 Using the rule found in Question 3, substitute $N = 240$ and find S.

$$N = 4 \times S$$
$$240 = 4 \times S$$
$$S = 60$$

Therefore, 60 squares can be built with 240 matchsticks.

Note:

- N and S are called **pronumerals**.
- Pronumerals stand for numerals. In the above example, the pronumeral N stands for the number of matchsticks and S stands for the number of squares.
- In algebra, a rule is called a **formula**. For the above example, the formula is $N = 4 \times S$, where N stands for the number of matchsticks used in the pattern and S stands for the number of squares.

For Example

1 Find a formula relating x and y in the following pattern:

x	1	2	3	4	5
y	3	5	7	9	11

1 It is observed that 'to obtain the number in the second row you need to multiply the number in the first row by 2 and then add 1', as follows:

x	1	2	3	4	5
y	3	5	7	9	11
	$3 = 1 \times 2 + 1$	$5 = 2 \times 2 + 1$	$7 = 3 \times 2 + 1$	$9 = 4 \times 2 + 1$	$11 = 5 \times 2 + 1$

Therefore, the rule is $y = 2 \times x + 1$.

For Example

1 Complete the table using the formula $N = 3 \times t - 1$:

t	1	2	3	4	5
N					

1 When $t = 1$, $N = 3 \times 1 - 1 = 2$
$t = 2$, $N = 3 \times 2 - 1 = 5$
$t = 3$, $N = 3 \times 3 - 1 = 8$
$t = 4$, $N = 3 \times 4 - 1 = 11$
$t = 5$, $N = 3 \times 5 - 1 = 14$

Therefore, the completed table is as follows:

t	1	2	3	4	5
N	2	5	8	11	14

Definition

A pronumeral represents a number and may take any numerical value. For example, $x = 2$ and $y = 5$.

Operations on Pronumerals

The four operations of arithmetic (+, −, × and ÷) have the same meaning in algebra as they have in arithmetic:

- $a + b$ means the **sum** of the numbers represented by the pronumerals a and b. The actual value of $a + b$ can be found only if the values of a and b are known. If $a = 7$ and $b = 3$ then $a + b = 7 + 3 = 10$.
- $a - b$ means the **difference** of the numbers represented by the pronumerals a and b. The value of $a - b$ can be found only if the values of a and b are known. If $a = 10$ and $b = 6$ then $a - b = 10 - 6 = 4$.
- ab means the **product** (i.e. $ab = a \times b$) of the numbers represented by the pronumerals a and b. The value of ab can be found only if the values of a and b are known. If $a = 2$ and $b = 6$ then $ab = 2 \times 6 = 12$.
- $\frac{a}{b}$ means the quotient (i.e. $\frac{a}{b} = a \div b$) of the numbers represented by the pronumerals a and b. The value of $\frac{a}{b}$ can be found only if the values of a and b are known. If $a = 12$ and $b = 4$ then $\frac{a}{b} = a \div b = \frac{12}{4} = 12 \div 4 = 3$.

Note: ab is the shorthand way of writing $a \times b$ (that is, $a \times b = ab$). $\frac{a}{b}$ means $a \div b$ (that is, $\frac{a}{b} = a \div b$). ab is called the simplified form. $a \times b$ is called the expanded form.

Multiplication of Pronumerals

When multiplying pronumerals, the multiplication sign, ×, is often omitted.

For example, $a \times b$ is written as ab
$4 \times a = 4a$
$a \times b \times c = abc$

Simplify the following algebraic expressions:

1 $3 \times b$ **2** $a \times 5$
3 $2 \times a \times b$ **4** $5a \times 2b$
5 $a \times a$ **6** $b \times 3a \times 4$
7 $3 \times a \times 2a$

1 $3 \times b = 3b$ [Omit the multiplication sign. Note the number is always written first.]

2 $a \times 5 = 5a$ [Always write the number first.]

3 $2 \times a \times b = 2ab$ [This expression could be written as $2ba$ because $ab = ba$.]

4 $5a \times 2b = 10ab$ [$5 \times 2 = 10$, $a \times b = ab$. Multiply the numbers first and then write the letters.]

5 $a \times a = a^2$ [$a \times a$ is not written as aa but as a^2.]

6 $b \times 3a \times 4 = 12ba$
[$b \times a = ba$, $3 \times 4 = 12$
Remember this expression could be written as $12ab$ since $ba = ab$.]

7 $3 \times a \times 2a = 6a^2$
[Remember: multiply numbers first; $a \times a = a^2$ and not aa.]

Write the following expressions in expanded form:

1 $4b$ **2** $3abc$ **3** $3b^2$
4 ab^2c **5** $(2a)^2$

1 $4b = 4 \times b$
2 $3abc = 3 \times a \times b \times c$
3 $3b^2 = 3 \times b \times b$
4 $ab^2c = a \times b \times b \times c$
5 $(2a)^2 = 2a \times 2a = 2 \times a \times 2 \times a$

Substitution in Algebraic Expressions

In substitution we replace the pronumeral with its numerical value (which must be given) and find the value of an arithmetic expression.

If $a = 2$, $b = 4$ and $c = 5$ evaluate:

1 ab **2** $a + b$
3 $\frac{b}{2}$ **4** $c - a$
5 $2b - c$ **6** abc
7 $3a^2$ **8** $\frac{2b + 4}{a}$
9 $c(b - a)$

1 $ab = a \times b$
$= 2 \times 4$
$= 8$
[$ab = a \times b$.]
[Remember that $a = 2$ and $b = 4$.]

2 $a + b = 2 + 4$
$= 6$

3 $\frac{b}{2} = \frac{4}{2}$
$= 4 \div 2$
$= 2$

[Note $\frac{b}{2} = b \div 2$]

4 $c - a = 5 - 2$
$= 3$

5 $2b - c = 2 \times b - c$
$= 2 \times 4 - 5$
$= 8 - 5$
$= 3$

[Note order of operations. Do multiplication first.]

6 $abc = a \times b \times c$
$= 2 \times 4 \times 5$
$= 40$

7 $3a^2 = 3 \times a \times a$
$= 3 \times 2 \times 2$
$= 12$

[$3a^2 \neq (3a)^2$ but $3a^2 = 3 \times a \times a$]

8 $\frac{2b + 4}{a} = \frac{2 \times b + 4}{a}$
$= \frac{2 \times 4 + 4}{2}$
$= \frac{8 + 4}{2}$
$= \frac{12}{2}$
$= 6$

[$\frac{12}{2}$ means $12 \div 2$]

9 $c(b - a) = 5 \times (4 - 2)$
$= 5 \times 2$
$= 10$

Substitution into Simple Formulae

A formula combines at least two pronumerals. The value of one of the pronumerals can be found by substituting numbers for the other pronumerals.

For Example

If $T = 4n - 1$, find the value of T when $n = 5$.

$T = 4n - 1$
$= 4 \times n - 1$
$= 4 \times 5 - 1$
$= 20 - 1$
$= 19$

[Replace n by 5.]

For Example

1 If $P = 2(n + b)$, and $n = 28$, $b = 12$, find P.

1 $P = 2(n + b)$
$= 2(28 + 12)$
$= 2(40)$
$= 2 \times 40$
$= 80$

[$2(40) = 2 \times 40$]

For Example

1 If $K = 4a^2$, find the value of K when $a = 3$.

1 $K = 4a^2$
$= 4 \times a \times a$
$= 4 \times 3 \times 3$
$= 36$

For Example

1 If $x = ut + \frac{1}{2}at^2$, and $u = 12$, $t = 3$ and $a = 8$, find x.

1 $x = ut + \frac{1}{2}at^2$
$= 12 \times 3 + \frac{1}{2} \times 8 \times 3^2$
$= 12 \times 3 + \frac{1}{2} \times 8 \times 9$
$= 36 + 36$
$= 72$

Like Terms

Like terms have the same pronumeral or pronumeral parts. For example, x, $-5x$, $4x$ and $20x$ are like terms.

Like Terms of an Expression

An expression such as $2x + 3y + 3t$ has 3 terms. Similarly, $5x + 3x^2$ has 2 terms. Like terms of an expression are those that have the same pronumeral parts. For example, the expression $5x + 3y + 2x + 4x$ has 4 terms. $5x$, $2x$ and $4x$ are like terms of the expression. The following are examples of groups of like terms: $\{3y, y, -2y\}$, $\{ab, ba, 3ab\}$, $\{2x^2, -7x^2, 3x^2\}$. Similarly, $5x$, $3y$, $-2t$ are unlike terms since their pronumeral parts are all different.

Collecting Like Terms

When adding or subtracting pronumerals, only like terms can be added or subtracted.

For Example

Simplify the following:

1 $3a + 2a + a$

2 $6a + 2b + 2a + 7b$

3 $7t - t$

4 $4ab - 2ba$

5 $5x^2 + 3x - 2x^2 + x$

6 $4t + k - 3t + 3k$

1 $3a + 2a + a = 6a$ [$a = 1a$; $(3 + 2 + 1)a$]

2 $6a + 2b + 2a + 7b = 8a + 9b$

[This expression could be rearranged so that like terms are grouped together.]

$$6a + 2b + 2a + 7b = (6a + 2a) + (2b + 7b) = 8a + 9b$$

3 $7t - t = 7t - 1t = 6t$ [$t = 1t$]

4 $4ab - 2ba = 2ab$ [$ab = ba$, therefore $4ab$ and $2ba$ are like terms.]

5 $5x^2 + 3x - 2x^2 + x = (5x^2 - 2x^2) + (3x + x) = 3x^2 + 4x$

[Grouping like terms together; x^2 and x are unlike terms.]

6 $4t + k - 3t + 3k = (4t - 3t) + (k + 3k)$

[Grouping like terms.]

$= t + 4k$

Indices (Extension)

Index Notation

$a \times a$ is written as a^2 (a is squared)
$a \times a \times a$ is written as a^3 (a is cubed)
$a \times a \times a \times a$ is written as a^4
(it is a to the power of 4)

Note $a^m = \underbrace{a \times a \times a \times \ldots \times a}_{m \text{ times}}$ (it is called the mth power of a)

In the expression a^m the m is called the **index** or **power**. The plural of index is **indices**.

For Example

1 Evaluate:

a 2^3 b 12^2 c 3^4

2 Write the following expressions in index form:

a $5 \times 5 \times 5 \times 5$

b $b \times b \times b \times a \times a$

c $3 \times a \times a \times a \times b \times b \times b \times b \times b$

1 a $2^3 = 2 \times 2 \times 2 = 8$

b $12^2 = 12 \times 12 = 144$

c $3^4 = 3 \times 3 \times 3 \times 3 = 81$

2 **a** $5 \times 5 \times 5 \times 5 = 5^4$

b $b \times b \times b \times a \times a = b^3a^2$ or a^2b^3

c $3 \times a \times a \times a \times b \times b \times b \times b \times b = 3a^3b^5$

Laws of Indices

When multiplying, add indices. When dividing, subtract indices:

- $a^m \times a^n = a^{m+n}$
- $a^m \div a^n = a^{m-n}$
- $(a^m)^n = a^{mn}$
- $(ab)^m = a^mb^m$
- $a^0 = 1$

For Example

Simplify:

1 $b^5 \times b^2$ **2** $a^8 \div a^2$

3 $(x^4)^3$ **4** $(ab)^5$

5 $(2x)^3$ **6** y^0

7 $5a^2 \times 4a$ **8** $\dfrac{a^8 \times a^4}{a^3}$

9 $12a^4b^5 \div 4a^2b^2$

1 $b^5 \times b^2 = b^{5+2}$
$= b^7$

2 $a^8 \div a^2 = a^{8-2}$
$= a^6$

3 $(x^4)^3 = x^{4 \times 3}$
$= x^{12}$

4 $(ab)^5 = a^5b^5$

5 $(2x)^3 = 2^3 \times x^3$
$= 8 \times x^3$
$= 8x^3$

6 $y^0 = 1$

7 $5a^2 \times 4a = (5 \times 4)a^{2+1}$
$= 20a^3$
[Multiply numbers first; $a = a^1$.]

8 $\dfrac{a^8 \times a^4}{a^3} = \dfrac{a^{12}}{a^3}$ [$\dfrac{a^{12}}{a^3}$ means $a^{12} \div a^3 = a^9$]
$= a^{12-3}$
$= a^9$

9 $12a^4b^5 \div 4a^2b^2 = 3a^{4-2}b^{5-2}$
$= 3a^2b^3$
[Divide numbers first.]

Division of Pronumerals

When dividing pronumerals, we can cancel components such as numbers and like terms.

For Example

Simplify:

1 $18b \div 3$ **2** $12a \div 4a$

3 $15xy^2t \div 5xy$ **4** $15m^2n \div 10mn^2$

5 $25x^7y^4 \div 5x^3y^2$

1 $18b \div 3 = \dfrac{18b}{3}$ [Write in fractional form. Note $a \div b = \dfrac{a}{b}$]

$= \dfrac{\cancel{18}_6 \times b}{\cancel{3}_1}$ [Write it in expanded form.]

$= \dfrac{6b}{1}$ [Cancel common factors.]

$= 6b$ [$\dfrac{6b}{1}$ is written as $6b$.]

2 $12a \div 4a = \dfrac{12a}{4a}$ [Write in fractional form.]

$= \dfrac{\cancel{12}_3 \times \cancel{a}_1}{\cancel{4}_1 \times \cancel{a}_1}$ [Write in expanded form.]

$= \dfrac{3}{1}$ [Cancel common factor.]

$= 3$

3 $15xy^2t \div 5xy = \dfrac{15xy^2t}{5xy}$

$$= \frac{\cancel{15}_3 \times \cancel{x}_1 \times \cancel{y}_1 \times y \times t}{\cancel{5}_1 \times \cancel{x}_1 \times \cancel{y}_1}$$

$$= \frac{3yt}{1}$$

$$= 3yt$$

4 $15m^2n \div 10mn^2 = \dfrac{15m^2n}{10mn^2}$

$$= \frac{\cancel{15}_3 \times \cancel{m}_1 \times m \times \cancel{n}_1}{\cancel{10}_2 \times \cancel{m}_1 \times \cancel{n}_1 \times n}$$

$$= \frac{3m}{2n}$$

5 $25x^7y^4 \div 5x^3y^2 = \dfrac{25}{5}x^{7-3}y^{4-2}$

[Use the rules of indices.]

$$= 5x^4y^2$$

PRACTISE, PRACTISE

Go to p. 257 for quick answers, or to pp. 291–295 for worked solutions.

1 **a** By considering the pattern below, built with matchsticks, complete the table: pp. 120–121

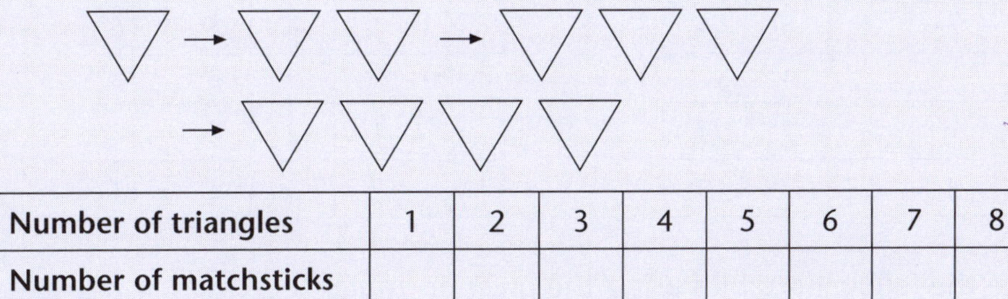

Number of triangles	1	2	3	4	5	6	7	8
Number of matchsticks								

b Write in your own words the relationship between the number of matchsticks and number of triangles in the above pattern.

c If T stands for the number of matchsticks used to create t triangles in the pattern, write the rule in (b) using pronumerals.

d Using the rule, find the number of matchsticks needed to build a pattern with:

i 10 triangles **ii** 50 triangles **iii** 100 triangles

e Using the rule, and given 285 matchsticks, how many triangles can be constructed in the pattern?

2 Study the following pattern built with matchsticks: pp. 120–121

a Complete the table below:

Number of squares	1	2	3	4	5	6	7	8
Number of matchsticks								

b Write in your own words a statement relating the number of matchsticks to the number of squares in the above number pattern.

c If y stands for the number of squares created in the pattern when x matchsticks are used, write the formula that relates x and y in the above pattern.

d Using the formula find the number of matchsticks needed to construct a pattern with:

i 15 squares **ii** 52 squares **iii** 400 squares

e Using the formula found in (**c**), find the number of squares formed using:

i 61 matchsticks **ii** 313 matchsticks **iii** 901 matchsticks

3 Study the patterns and find a formula relating x and y in the following: pp. 120–121

a

x	1	2	3	4
y	2	4	6	8

b

x	1	2	3	4	5
y	2	3	4	5	6

c

x	3	4	5	6
y	1	2	3	4

d

x	1	2	3	4
y	1	4	9	16

e

x	1	2	3	4
y	3	5	7	9

f

x	1	2	3	4
y	2	5	8	11

4 Complete the following tables using the given formula: pp. 120–121

a $y = x + 3$

x	1	2	3	4	5
y					

b $T = t - 1$

t	1	2	3	4	5
y					

c $n = 6 - m$

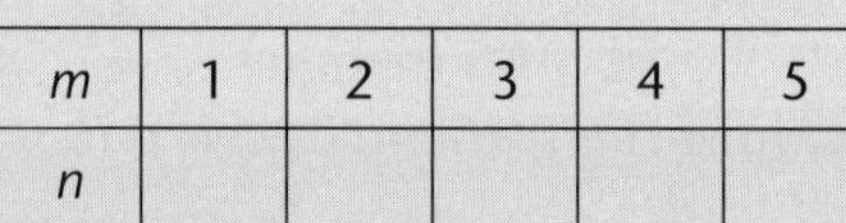

m	1	2	3	4	5
n					

d $Q = 2 \times P$

P	1	2	3	4	5
Q					

e $y = 3 \times x - 1$

x	1	2	3	4	5
y					

f $w = 12 - 3 \times v$

v	1	2	3	4	5
w					

5 Write these expressions in simplest algebraic form: p. 122

a $a \times b$ **b** $3 \times y$ **c** $t \times 2$

d $4 \times t \times s$ **e** $7 \times y \times y$

6 Write the following expressions in expanded form: p. 122

a xy **b** $3yb$ **c** $4b^2$

d $2mnq$ **e** $2m^2nq^2$ **f** $(3b)^2$

7 If $a = 4$, $b = 2$ and $c = 5$, evaluate: pp. 122–123

a ab **b** $4a$ **c** $a + b$

d $a - b + c$ **e** c^2 **f** $2a - c$

g abc **h** $c^2 - a^2$ **i** $a(c - 3)$

j $ac - 3a$ **k** $\frac{4bc}{2a}$ **l** $3b^2$

m $(3b)^2$ **n** $\frac{4c^2}{25a}$

8 If $y = \frac{3}{4}$, $b = \frac{1}{3}$ evaluate: pp. 122–123

a by **b** $b + y$
c $4y$ **d** $6b$
e y^2 **f** $2y + 3b$
g $9b^2$ **h** $y - 2b$

9 If $A = nb$, find A when $n = 8$ and $b = 4$ p. 123

10 If $P = 2(n + b)$, find P given that $n = 10$ and $b = 3$ p. 123

11 If $T = 4n - 2$, find T when $n = 5$ p. 123

12 If $N = 4(n + 1)$, find the value of N when $n = 8$ p. 123

13 If $V = u + at$, find V when $u = 32$, $a = 2$ and $t = 6$ p. 123

14 If $E = mc^2$, find E when $m = 3$ and $c = 5$ p. 123

15 If $D = \dfrac{M}{V}$, find D when $M = 24$ and $V = 5$ p. 123

16 If $Q = \dfrac{10A}{P + 12}$, find Q when $A = 6$ and $P = 3$ p. 123

17 Simplify: p. 122

a $m \times m$ **b** $n \times 5$ **c** $4 \times a \times a$
d $p \times q$ **e** $y + y$ **f** $4x \div 2$
g $5a \times b$ **h** $p \times 6$ **i** $4b \times 3a$
j $x \times y \times z$ **k** $3a \times 2a$ **l** $2y \times 5xy$
m $c \times 2a \times 4$ **n** $4a \times a \times a$ **o** $a \times 4 \times b \times a$
p $(5p)^2$ **q** $16x \div 4x$ **r** $18ab \div 3a$

18 Simplify: pp. 124–126

a $a^3 \times a^8$ **b** $a^{12} \div a^2$ **c** $(b^3)^5$
d $(ab)^1$ **e** $y \times y^4 \times y^2$ **f** $3xy \times 2y^2$
g $4a^{15} \div 2a^5$ **h** $\dfrac{a^4 \times a^6}{a^5}$ **i** $(6a)^2$
j $(3a^4)^2$ **k** $(x^3y^2)^3$ **l** $12a^4b^5 \div 6a^6b^2$
m $15x^2y \div 10xy^2$

19 Evaluate: pp. 124–126

a 4^3 **b** 10^2 **c** 2^5
d 5^0 **e** 6^3 **f** $2^2 \times 2^3$
g $(2^3)^2$

20 Simplify the following by collecting like terms: pp. 123–124

a $8t + 10t - 3t$ **b** $10a - a$ **c** $5x + 2y + 3x - 2y$
d $4x + 7x - 3x$ **e** $5a - 2 + 3a$ **f** $4a + 2 + a + 7$
g $5ab + 2ba - 3ab$ **h** $5x^2 + 3x - 2x^2 + x$ **i** $3a^2 + 2a + 4a - a^2$
j $4y + 5 - y - 1$ **k** $5x + 8 - 3x - 2$ **l** $12p - 8p - p$
m $x + 9y + 3x - 4y$ **n** $4p + 8 - p + 2$ **o** $4x + 10 + x - 3$

21 Simplify: pp. 123–126

a $a + a + a + a$	**b** $4 \times a$	**c** $a \times a \times a \times a$
d $\frac{4ab}{2a}$	**e** $a \times b \times a \times b \times a$	**f** $7x + 3y$
g $8x - 3x - x$	**h** $(2x)^3$	**i** $15x^2y \div 10xy^3$
j $12xy \div 8x$	**k** $4a + 3 - 2a + 10$	**l** $2a \times 4b \times a$
m $2a \times 3a \times 4a$	**n** $8b - b$	**o** $(4x^3)^2$
p $5x^2 - x - x^2$	**q** $2x^2 + 5x - 3x + 3x^2$	**r** $2b \div 3b$
s $6 \times a + b$		

22 Simplify by collecting like terms: pp. 123–126

a $8x + 2x - 4x - 6x$	**b** $5t - 7t + 4t$	**c** $-12x + 4x$
d $-a - a$	**e** $2x - 5y + x + 3y$	**f** $4x^2 + 7x - 2x^2 - 10x$
g $10x - 4 - 3x + 2$	**h** $2p + 5 - 4p - 3$	**i** $2x - 7 + 3x + 4$
j $2x - 4 + 5x - 7$	**k** $x + 9y - 4x - y$	**l** $4p - 8 - 6p + 3$
m $-x + 5 - 2x - 7$	**n** $x - 9y + x - y$	**o** $2x^2 - x - x^2 - x$
p $a - a + a - 2a$	**q** $x^2 - 5x - x^2 + 2x$	**r** $x - 8y + x - 2y$
s $2 - 7t - 6 + 6t$		

Go to p. 257 for quick answers, or to pp. 291–295 for worked solutions.

YOUR CHECKLIST

For a complete understanding of this topic you must be able to:

✓	Build and describe patterns in words and using pronumerals		p. 120
✓	Determine rules algebraically from a table of values		pp. 120–121
✓	Simplify pronumerals that involve multiplication and division		pp. 122–126
✓	Substitute in algebraic expressions given the values of pronumerals		pp. 122–123
✓	Recognise like terms and simplify algebraic expressions that involve addition and subtraction		pp. 123–124
✓	Use rules of index notation.		pp. 124–125

Now you are ready to do the tests!

(45 marks)

1 For the following pattern made of matchsticks:

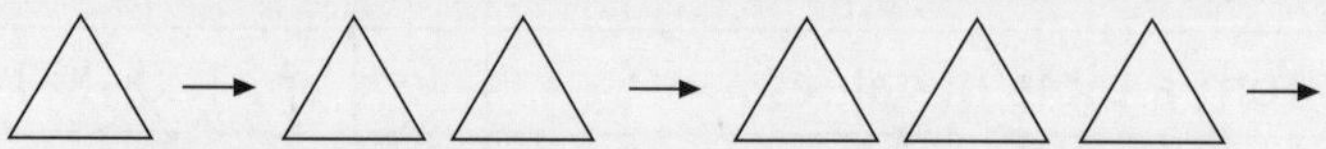

a Draw the next one in the pattern.

b Using the above pattern, complete the table:

Number of triangles	1	2	3	4	5	6
Number of matchsticks						

c Complete the statement, relating the number of matchsticks to the number of triangles, about the above pattern:

Number of matches used = … × Number of triangles formed

d Using the pattern above, find the number of matchsticks needed to form 15 triangles.

e Find the number of triangles formed in the above pattern if 30 matchsticks were used. (5 marks)

2 Write the next number in each of the following number patterns:

a 3, 4, 5, … b 5, 7, 9, …

c 5, 10, 20, … d 30, 27, 24, … (4 marks)

3 Complete the following tables using the given rule:

a $y = x + 2$

x	1	2	3	4
y				

b $T = 3 \times t$

t	1	2	3	4
T				

(4 marks)

4 Study the patterns below and complete the rules relating x and y:

a

x	3	4	5	6
y	5	6	7	8

$y = x + \ldots$

b

x	2	3	4	5
y	1	2	3	4

$y = x - \ldots$ (2 marks)

5 Simplify:

a $5 \times a$ b $b \times 2$ c $3 \times x \times 4$ d $c \times c$ (4 marks)

6 Write the following in expanded form:

a $4k$ b ab c $5xy$ d y^2 (4 marks)

7 Simplify:

a $15y \div 5$ b $4b \div 2$ (2 marks)

8 If $t = 5$, find the value of:

a $t + 3$ **b** $t - 2$ **c** $2t$ **d** $3t - 10$ (4 marks)

9 If $a = 4$ and $b = 2$, evaluate:

a ab **b** $\frac{a}{b}$ **c** $3ab$ **d** a^2 (4 marks)

10 If $T = a + b - 2$ find T when $a = 3$, $b = 5$. (1 mark)

11 Simplify $5 \times a \times a$. (1 mark)

12 Simplify the following:

a $5b + 3b + 2b$ **b** $3y + 2y - y$

c $2a + 3b + 5a$ **d** $2t + 6n + 3t - 2n$ (4 marks)

13 Write the following expressions in index form:

a $2 \times 2 \times 2$ **b** $a \times a \times a \times a$ **c** $3 \times a \times a \times a$ (3 marks)

14 If $x = 3$ find the value of $(x + 1)^2$. (1 mark)

15 Simplify $5a \div a$. (2 marks)

☞ **Quick answers on page 263**
☞ **Worked solutions on page 326**

Your Feedback $\frac{\square}{45} \times 100\% = \square\%$

(40 marks)

1 By considering the pattern below, built with matchsticks:

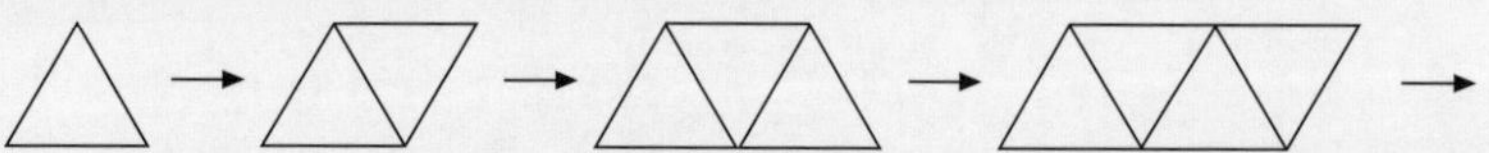

a Draw the next pattern.

b Complete the table below:

Number of triangles	1	2	3	4	5	6	7
Number of matchsticks							

c Write in your own words the relationship between the number of matchsticks and the number of triangles in the above pattern.

d If T stands for the number of triangles in the pattern and M stands for the number of matchsticks used to create that pattern, write the rule in (c) using pronumerals.

e Using the formula found in part (d) find the:

i Number of matchsticks needed to build a pattern with 20 triangles

ii Number of triangles in a pattern where 257 matchsticks are used. (6 marks)

2 Write the next two numbers in each of the following number patterns:

a 15, 12, 9, ..., ...

b 1, 4, 9, 16, ..., ...

c 1, 1, 2, 3, 5, 8, ..., ...

d 144, 72, 36, ..., ... (4 marks)

3 Complete the following tables using the given relationship between x and y:

a $y = 2x - 1$

x	1	2	3	4
y				

b $y = 15 - 2x$

x	2	3	4	5
y				

(2 marks)

4 Study the tables below and complete the rules relating x and y:

a

x	1	2	3	4
y	5	8	11	14

b

x	2	3	4	5
y	1	3	5	7

(2 marks)

5 Simplify:

a $5 \times 2a$

b $a \times 2b \times 3c$

c $4a \times a$

d $12a \div a$

e $15x \div 3x$

f $5a \times 2a \times b$

g $(3b)^2$

h $5a - a$

i $7a + 3b + 5a + b$

j $5a^2 + 3a + 2a^2 - 2a$ (10 marks)

6 Simplify:

a $2 \times a \times 3 \times a \times a$ b $a^5 \times a^2$

c $a^8 \div a^2$ d $(a^5)^2$ (4 marks)

7 a If $m = 5$ and $n = 4$ find the value of $mn - 3$.

b If $a = 3$ find the value of $2a^2$.

c If $x = 4$ find the value of $10 - 2x$.

d If $A = \frac{5B - 3}{C}$ find the value of A when $B = 3$ and $C = 4$. (4 marks)

8 If $2x = 8$ find the value of $x - 2$. (1 mark)

9 If $a = 1$ find the value of y given $y = 2a^2 - a + 2$. (1 mark)

10 Simplify:

a $15x^2y \div 5xy$ b $8xy + 5x - 2yx + x$

c $7xy^2 \times 3x^4y^3$ d $(2a^4)^3$

e $a \times 2 \times b \times 3$ f $2 \times b \times 3 \times b \times a$ (6 marks)

☞ **Quick answers on page 264**

☞ **Worked solutions on page 326**

Your Feedback $\frac{\square}{40} \times 100\% = \square\%$

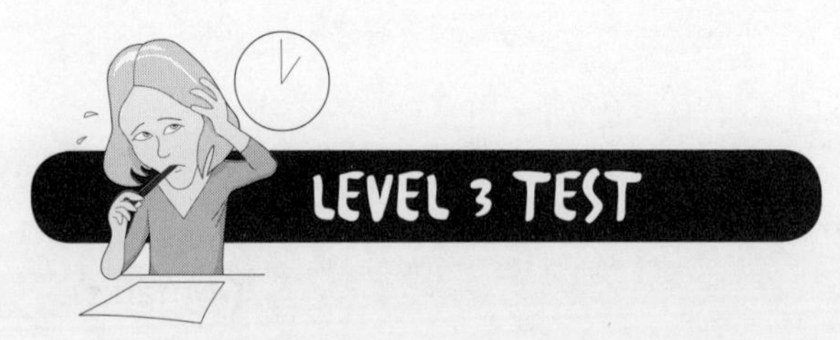

(40 marks)

1 Consider the following pattern of squares built with matchsticks:

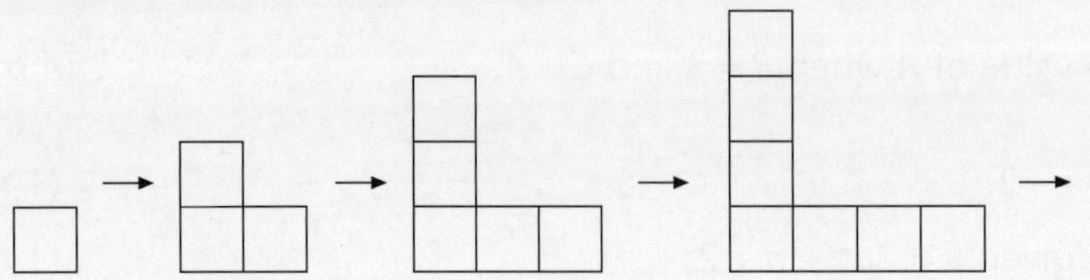

a How many matchsticks are needed to make the next pattern?

b Complete the table for the above pattern:

Number of squares	1	3	5	7	9	11
Number of matchsticks						

c Describe the relationship between the number of matchsticks and the number of squares used in the pattern in your own words.

d If m stands for the number of matchsticks used to create T squares in the above pattern, write using pronumerals the formula that relates m and T.

e How many matchsticks are needed to create a pattern with 57 squares?

f Find the number of squares created if 601 matchsticks are used in the pattern. (6 marks)

2 Write down the next three numbers in each of the following number patterns and describe the rule in your own words:

a 2, 4, 6, 8, …, …, …

b 31, 27, 23, 19, …, …, …

c 1, 1, 2, 3, 5, 8, …, …, … (3 marks)

3 Complete the following tables using the given rule:

a $y = 2x^2 - 1$

x	1	2	3	4
y				

b $y = 3x - 2$

x	1		21	
y		13		148

(4 marks)

4 Study the tables below and write down the formula relating the pronumerals used:

a

x	1	2	3	4
y	−2	1	6	13

b

x	1	2	3	4
y	4	1	−4	−11

(4 marks)

5 Simplify:

a $6xy + 5x - 4yx - x$

b $5tm^2 \times 3t^2m^3$

c $8x^2y \div 4xy$

d $\frac{25a^2 \times a^4}{5a^3}$

e $\frac{4x^3}{6x^2}$

f $(2a)^3$

g $(3a^4b^2)^2$

h $8a^4b^2 \div 2a^3b$ (8 marks)

6 a Given that $P = (220 - A) \times 4 \div 5$, find P given $A = 30$.

b If $m = 3$ find the value of $2m^2 - m + 4$.

c If $2x - 1 = 7$ find the value of $20 - 3x$. (3 marks)

7 Simplify:

a $\frac{5t^2 \times 4t^3}{(2t)^3}$

b $\frac{8x^2 - 2x \times 3x}{4x^2}$ (4 marks)

8 If $c = \frac{nx}{n + 12}$ find the:

a Value of c when $n = 3$ and $x = 30$.

b Value of x when $n = 6$ and $c = 12$. (4 marks)

9 a If $a = 2$, find the value of $3a^2$.

b If $x = 5$, evaluate $x(x^2 - 2)$. (2 marks)

10 a Simplify $(5a)^3 - (5a)^0$.

b Simplify $8a^2 \div 4ab$. (2 marks)

☞ **Quick answers on page 264**

☞ **Worked solutions on page 327**

Your Feedback $\frac{\square}{40} \times 100\% = \square\%$

MEASUREMENT 8

- Units of Length
- Limits of Measurement
- Perimeter
- Area
- Volume
- Capacity

KEYWORDS

Accuracy	**Kilometre**
Area	**Measurement**
Capacity	**Metre**
Centimetre	**Millimetre**
Composite figures	**Perimeter**
Hectare	**Prism**
	Volume

Many questions in this chapter use fractions and decimals. Students can review Chapters 2 and 3 for assistance.

Units of Length

In Australia, the basic unit of length is the metre (abbreviated m). Other lengths are multiples of the metre, or divisions of it:

1000 mm = 1 m
100 cm = 1 m
∴ 10 mm = 1 cm
Also 1000 m = 1 km

For Example

1 **a** Write down the length of each interval to the nearest millimetre:

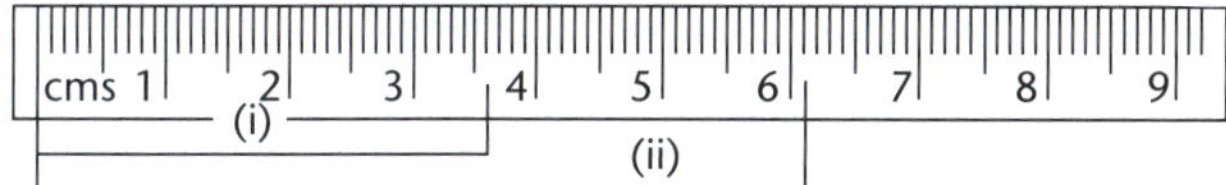

b Write down the length of each interval to the nearest centimetre:

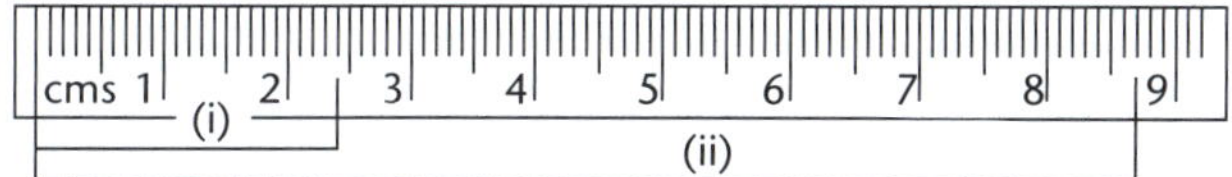

2 **a** Convert the following to the given units:

i 4 m = ____ cm
ii 3 km = ____ m
iii 7000 mm = ____ m
iv 270 cm = ____ m
v 64 mm = ____ cm
vi 3 400 000 mm = ____ km
vii 800 mm = ____ m
viii 0.35 km = ____ m

b Complete the following (using sensible units):

i Length of pen = 15 ____
ii Height of door = 2 ____
iii Distance from Singleton to Walgett = 511 ____
iv Length of a fingernail = 11 __

c If a 20c coin rolls 85 mm in one revolution:

i How far, in metres, will it travel in 25 revolutions?
ii How many revolutions of the coin is required for it to travel 30.6 m?

1 **a** **i** 36 mm **ii** 61 mm
b **i** 2 cm **ii** 9 cm

2 **a** **i** $4 \times 100 = 400$ i.e. 400 cm
ii $3 \times 1000 = 3000$ i.e. 3000 m
iii $7000 \div 1000 = 7$ i.e. 7 m
iv $270 \div 100 = 2.7$ i.e. 2.7 m
v $64 \div 10 = 6.4$ i.e. 6.4 cm
vi $3\,400\,000 \div 1\,000\,000 = 3.4$ i.e. 3.4 km

vii $800 \div 1000 = 0.8$ i.e. 0.8 m

viii $0.35 \times 1000 = 350$ i.e. 350 m

b **i** 15 cm **ii** 2 m

iii 511 km **iv** 11 mm

c **i**

$$85 \text{ mm} = 0.085 \text{ m}$$
$$\therefore \text{Distance} = 0.085 \text{ m} \times 25$$
$$= 2.125 \text{ m}$$

ii Each revolution = 0.085 m

$\therefore$ No. of revolutions

$$= 30.6 \div 0.085$$
$$= 360$$

$\therefore$ 360 revolutions are needed.

Limits of Measurement

A variety of measuring instruments are used, depending on the length of the object to be measured; for example, a ruler, tape measure or trundle wheel. However, when we measure we are really only approximating. If we say a length is 8 cm, the exact measure could be between 7.5 cm and 8.5 cm. This is called the **limit of measurement**.

For Example

Find the limit of measurement of the following:

1. A length of hair is measured as 17 cm.
2. A piece of timber is measured as 4.2 m.

1 16.5 cm to 17.5 cm

2 4.15 m to 4.25 m

Perimeter

The perimeter of a shape is the distance around the shape.

For Example

1 Measure the perimeter of these shapes with your ruler, leaving your answer in centimetres (cm):

a

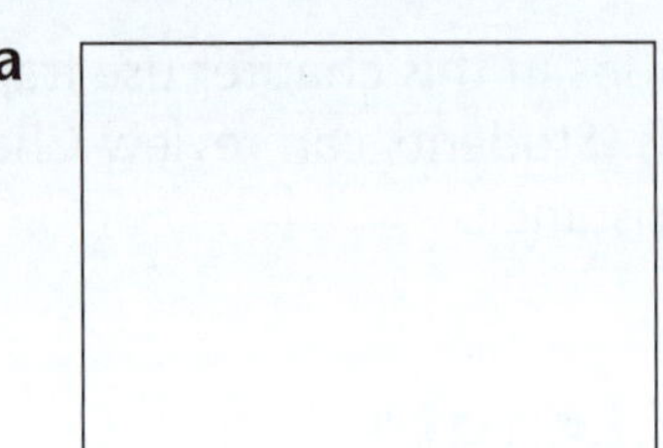

b

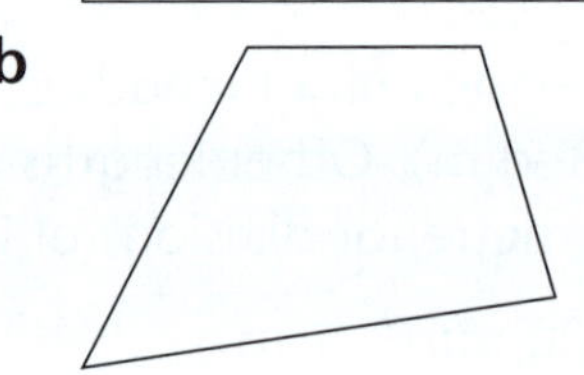

2 Find the perimeter of the following:

a

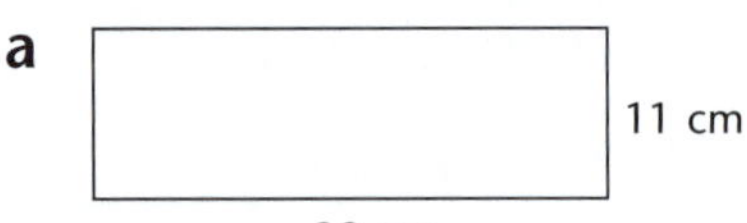

b

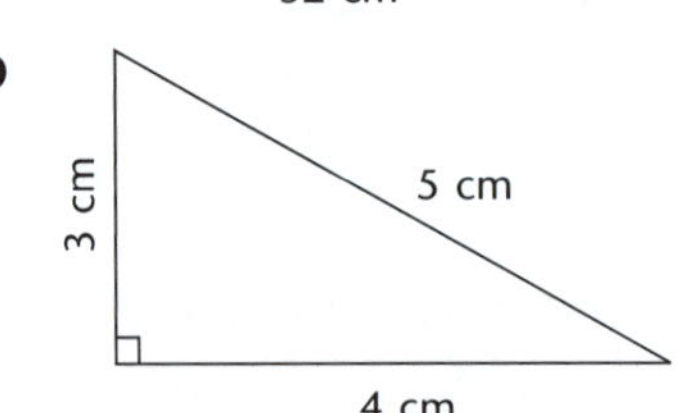

c A regular hexagon with side lengths of 14 cm.

d

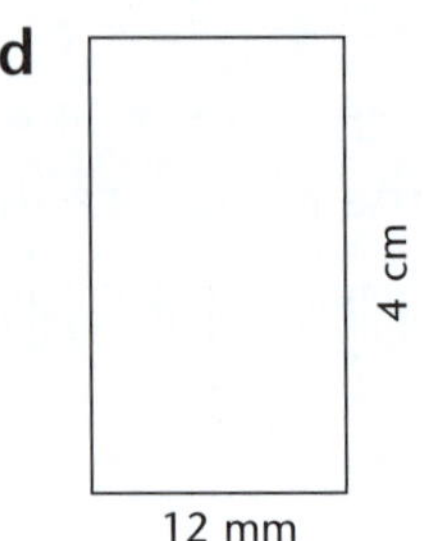

e

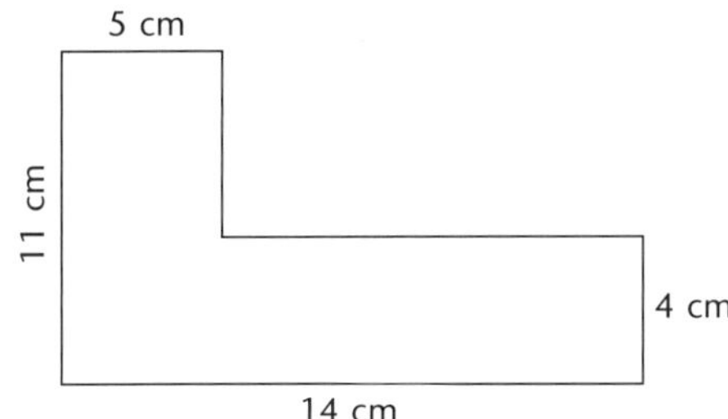

f

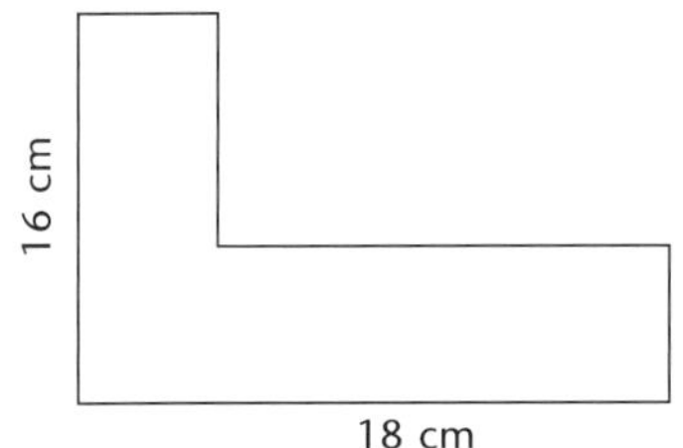

3 **a** If the perimeter of a square was 116 cm, find the length of each side.

b A farmer wishes to fence a paddock that measures 125 metres by 220 metres:

- **i** Find the perimeter of the paddock.
- **ii** How much will it cost the farmer at 95 cents per metre to fence the paddock?

1 **a** $P = 4 + 3 + 4 + 3$

$= 14$ ∴ Perimeter is 14 cm.

b $P = 2.4 + 1.6 + 1.7 + 3.3$

$= 9.0$ ∴ Perimeter is 9.0 cm.

2 **a** $P = 32 + 11 + 32 + 11$

$= 2(32 + 11)$

$= 2(43)$

$= 86$ ∴ Perimeter is 86 cm.

b $P = 3 + 4 + 5$

$= 12$ ∴ Perimeter is 12 cm.

c $P = 6 \times 14$ [A hexagon has 6 sides.]

$= 84$

∴ Perimeter is 84 cm.

d $P = 12 + 40 + 12 + 40$ [4 cm = 40 mm]

$= 2(12 + 40)$

$= 2(52)$

$= 104$ ∴ Perimeter is 104 mm.

e

5 cm, 11 cm, 7 cm, 9 cm, 4 cm, 14 cm

$P = 14 + 4 + 9 + 7 + 5 + 11$

$= 50$ ∴ Perimeter is 50 cm.

f

A, D, E, F, C, B, 16 cm, 18 cm

From A to B via C is the same distance as A to B via D, E, F:

$P = 2(16 + 18)$

$= 2(34)$

$= 68$ ∴ Perimeter is 68 cm.

3 **a** Perimeter = 116 cm

∴ Side = 116 ÷ 4

= 29

∴ Length of side is 29 cm.

b **i** $P = 2(125 + 220)$

$= 2(345)$

$= 690$ ∴ Perimeter is 690 m.

ii Cost = 690 × \$0.95

[Change 95c to \$0.95]

= \$655.50

∴ Cost to farmer is \$655.50.

Area

Area is a measure of the space contained within a plane shape.

For Example

1 Find the area of the shaded region:

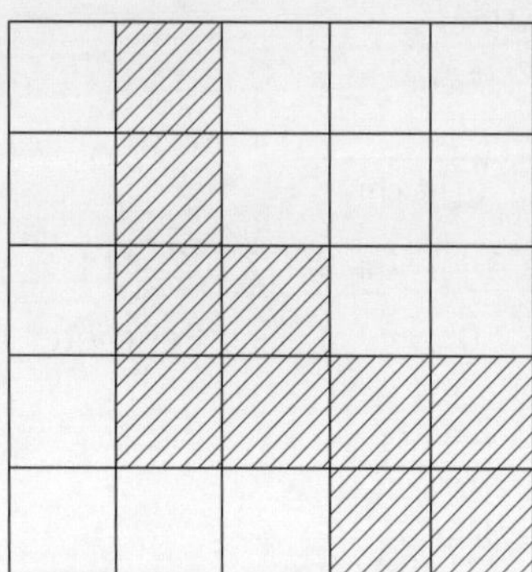

Each square on the diagram measures 1 unit by 1 unit, called a square unit (i.e. unit^2).

2 Find the approximate area of the shaded region:

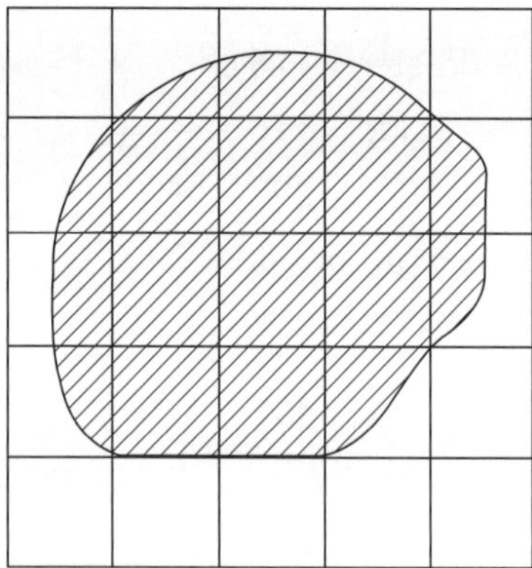

1 By counting, Area = 10 square units

$= 10 \text{ units}^2$

2 By counting, Area = 8 squares plus 9 'half squares'

$= 8 + 4\frac{1}{2}$

$= 12\frac{1}{2}$

$\therefore$ Area is approximately $12\frac{1}{2}$ units^2.

Area of Rectangles and Squares

For a rectangle, Area = length × breadth

i.e. $A = lb$

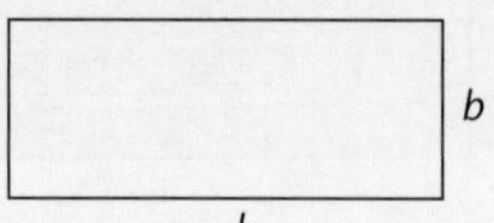

For a square, Area = side × side

$= \text{side}^2$

$\therefore A = s^2$

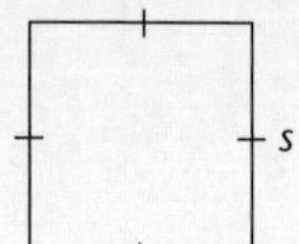

For Example

Find the area:

1

16 cm

5 cm

2

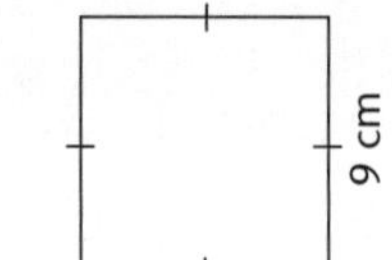

3

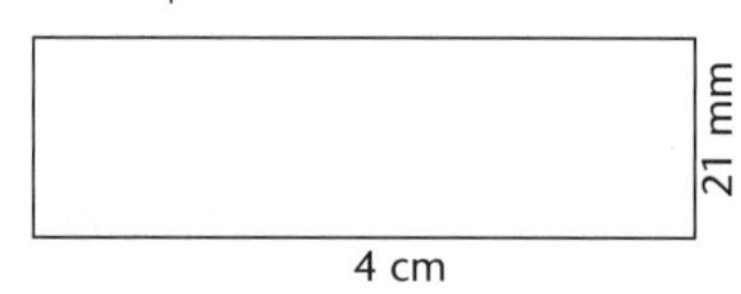

Remember that if the units in the question are in cm, then the area units are cm^2.

1 $A = lb$

$= 16 \times 5$

$= 80$

$\therefore$ Area is 80 cm^2.

2 $A = s^2$

$= 9^2$

$= 81$

$\therefore$ Area is 81 cm^2.

3 4 cm = 40 mm

$\therefore A = lb$

$= 40 \times 21$

$= 840$

$\therefore$ Area is 840 mm^2.

[A good way to learn a formula is to write it down in your solution before you use it.]

Area of a Parallelogram

For a parallelogram, Area = base × height

i.e. $A = bh$

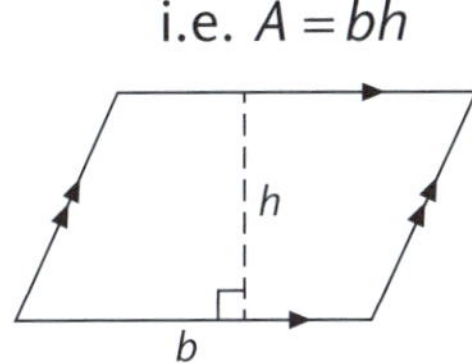

For Example

Find the area:

1

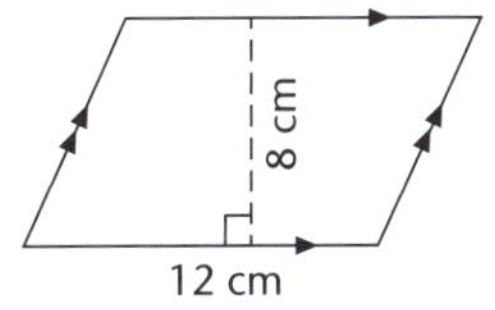

2

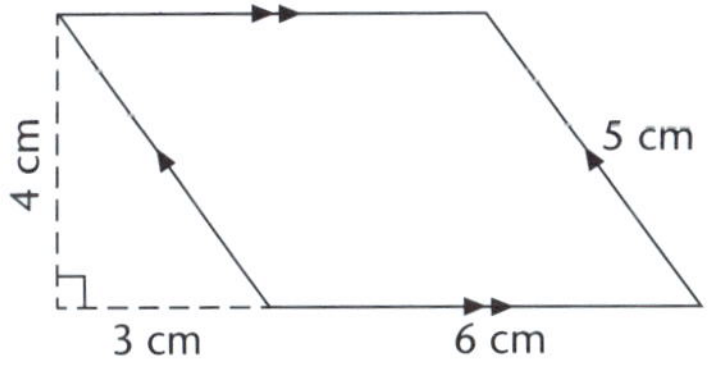

1 $A = bh$

$= 12 \times 8$

$= 96$

$\therefore$ Area is 96 cm^2.

2 $A = bh$

$= 6 \times 4$

$= 24$

$\therefore$ Area is 24 cm^2.

Area of a Triangle

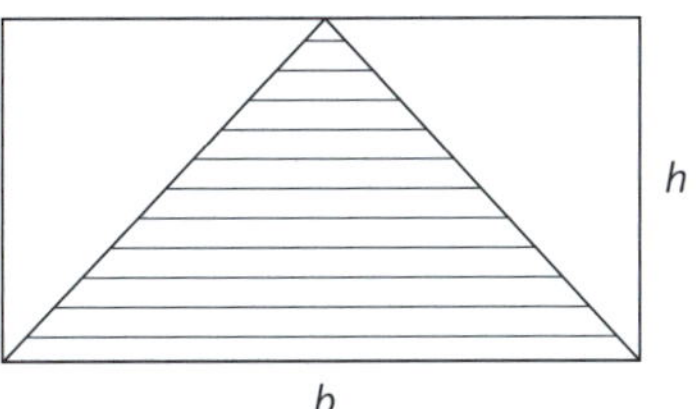

The area of the shaded triangle is **exactly** half the area of the rectangle.

$\therefore$ For a triangle, Area = $\frac{1}{2}$ × base × height

i.e. $A = \frac{1}{2}bh$ [$A = \frac{1}{2} \times b \times h$]

For Example

Find the area:

1

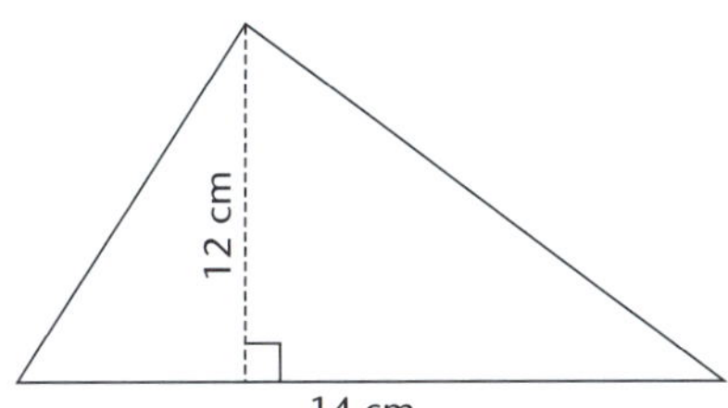

2

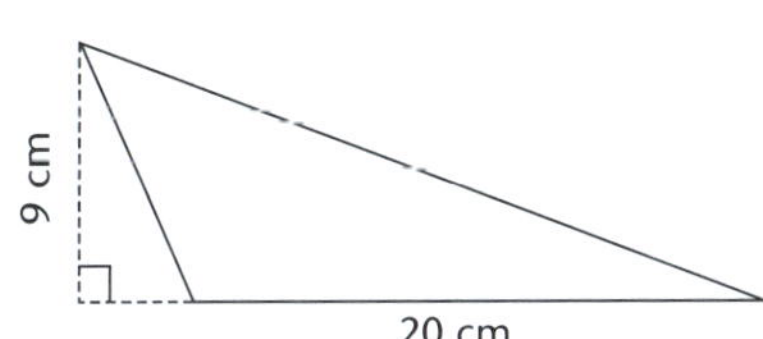

1 $A = \frac{1}{2}bh$

$= \frac{1}{2} \times 14 \times 12$

$= 7 \times 12$

$= 84$

$\therefore$ Area is 84 cm^2.

2 $A = \frac{1}{2}bh$

$= \frac{1}{2} \times 20 \times 9$

$= 10 \times 9$

$= 90$

$\therefore$ Area is 90 cm^2.

Units Used in Area Measurement

- One hectare (ha) is a square area 100 metres long by 100 metres wide:

100 m × 100 m square

$$1 \text{ hectare} = 100 \times 100 \text{ m}^2 = 10\,000 \text{ m}^2$$

[1 hectare = 10 000 m^2]

- Also:

100 cm × 100 cm square

$$1 \text{ m}^2 = 100 \times 100 \text{ cm}^2 = 10\,000 \text{ cm}^2$$

[1 m^2 = 10 000 cm^2]

For Example

Complete these statements:

1. 30 000 cm^2 = ____ m^2
2. 4.2 m^2 = ____ cm^2
3. 2 ha = ____ m^2
4. 84 000 m^2 = ____ ha

1. $30\,000 \div 10\,000 = 3 \text{ m}^2$
2. $4.2 \times 10\,000 = 42\,000 \text{ cm}^2$
3. $2 \times 10\,000 = 20\,000 \text{ m}^2$
4. $84\,000 \div 10\,000 = 8.4 \text{ ha}$

Composite Areas and Other Area Problems

Sometimes the area to be calculated is a combination of two or more regular shapes; for example, two rectangles, or a rectangle and a triangle.

Find the area of this composite shape (all angles are right angles):

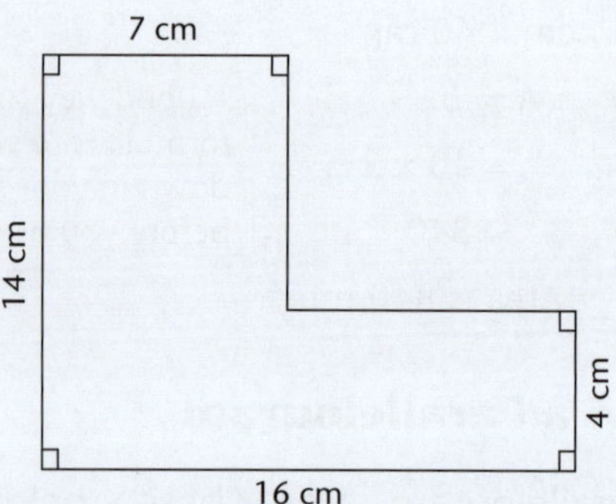

Method 1

Divide the shape into two rectangles with a vertical (or horizontal) line. Call the areas formed A_1 and A_2:

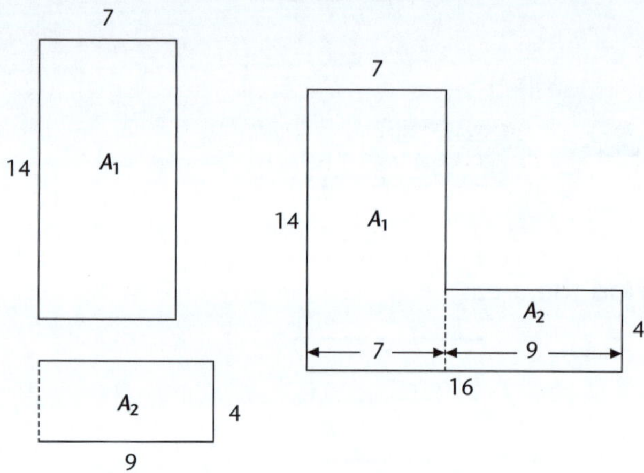

$\therefore$ Total area = $A_1 + A_2$

For A_1 we have used a vertical line (dotted):

$$A_1 = lb = 14 \times 7 = 98$$

For A_2, we must find the length before we can calculate the area. This is 16 − 7 = 9 cm:

$$A_2 = lb = 9 \times 4 = 36$$

$$\text{Total area} = 98 + 36 = 134 \text{ (i.e. } A_1 + A_2\text{)}$$

$\therefore$ Area is 134 cm^2.

Method 2

This method involves completing the full rectangle (dotted line) and then subtracting the extra piece (shaded):

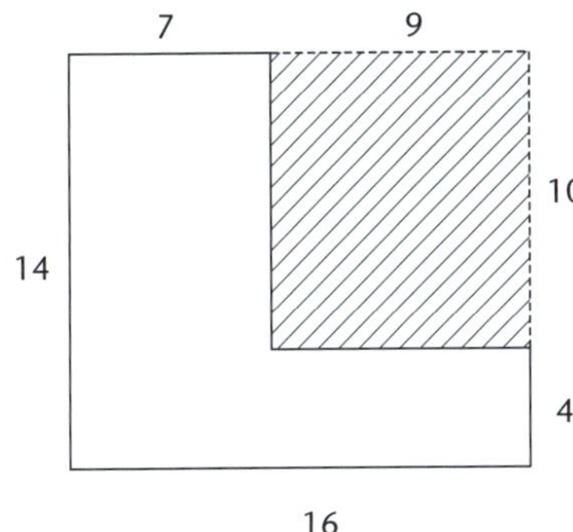

$$\begin{aligned}\text{Area large rectangle} &= lb\\ &= 16 \times 14\\ &= 224\end{aligned}$$

$$\begin{aligned}\text{Area shaded} &= lb\\ &= 9 \times 10\\ &= 90\end{aligned}$$

$$\begin{aligned}\text{Required area} &= 224 - 90\\ &= 134\end{aligned}$$

$\therefore$ Area is 134 cm^2.

For Example

1 Find the areas:

a

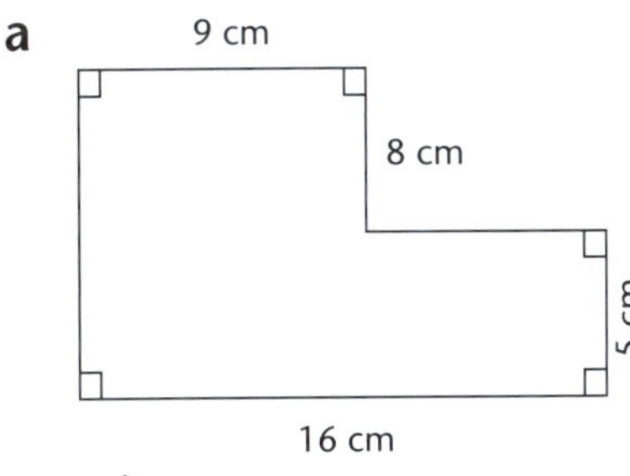

b

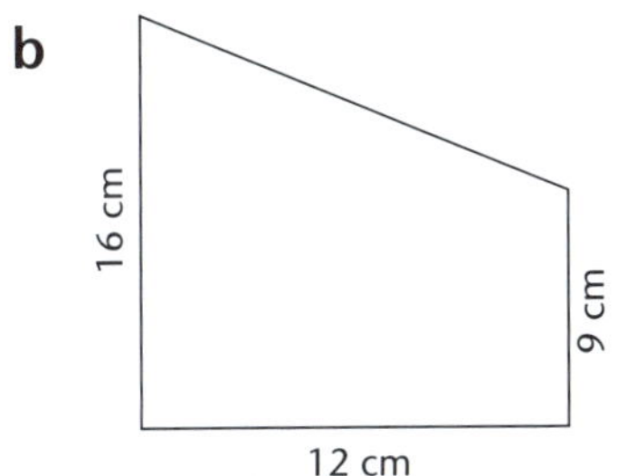

2 Find the area of the shaded regions:

a

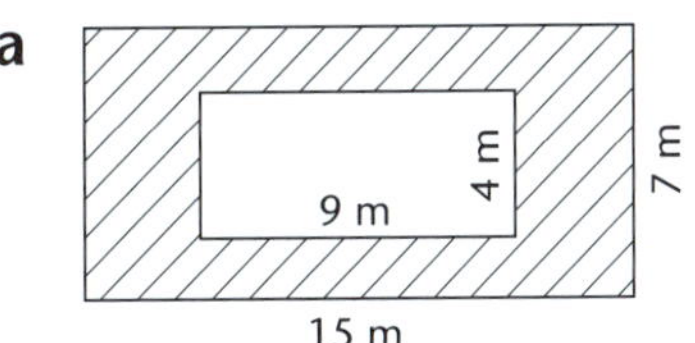

b

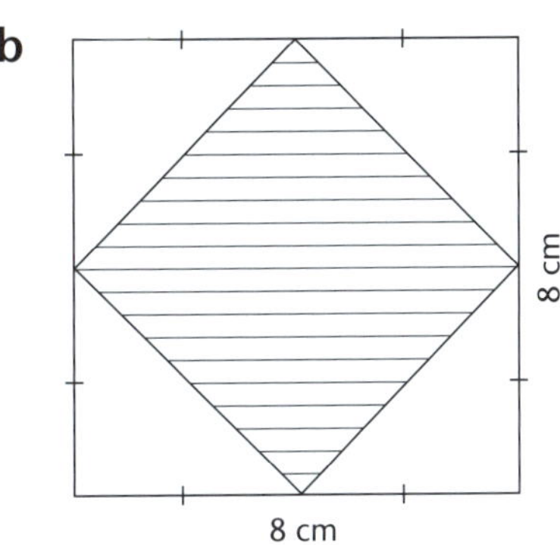

3 The diagram below shows part of the plan of a house where the size of rooms is expressed in metres. Find the:

a Area of each room

b Total cost of carpeting the lounge and dining areas if carpet costs $30/m^2.

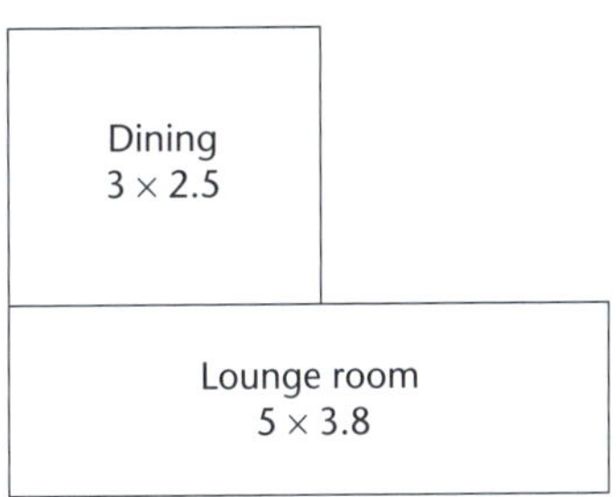

4 Kevin is to lay a 1 m wide footpath around the back of his house using rectangular pavers that are 20 cm by 10 cm:

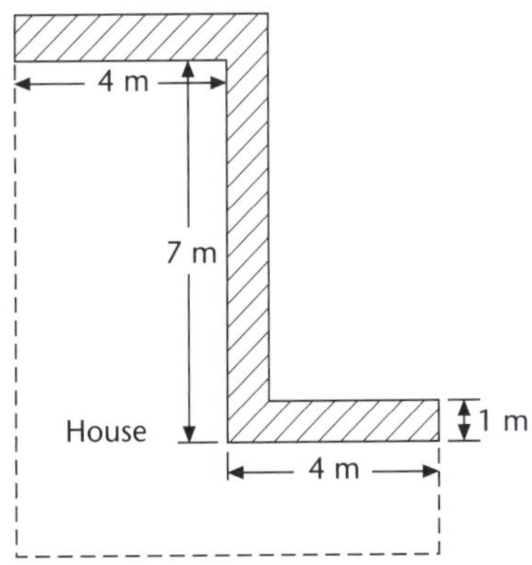

a Find the area to be paved.

b How many pavers will Kevin need?

1 **a**

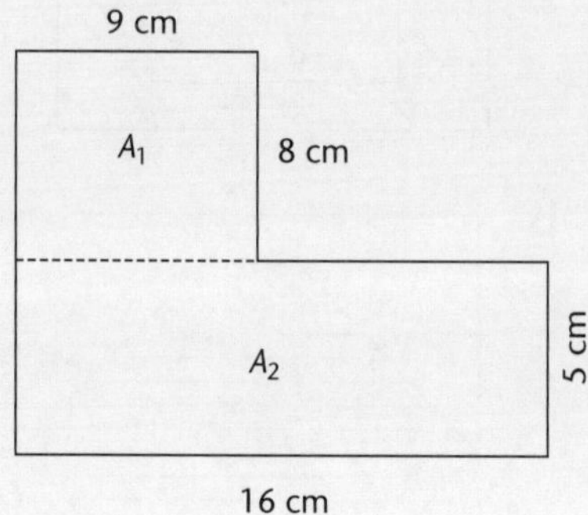

$A_1 = lb$

$= 9 \times 8$

$= 72$

$A_2 = lb$

$= 16 \times 5$

$= 80$

Total area $= A_1 + A_2$

$= 72 + 80$

$= 152$

Area is 152 cm^2.

b

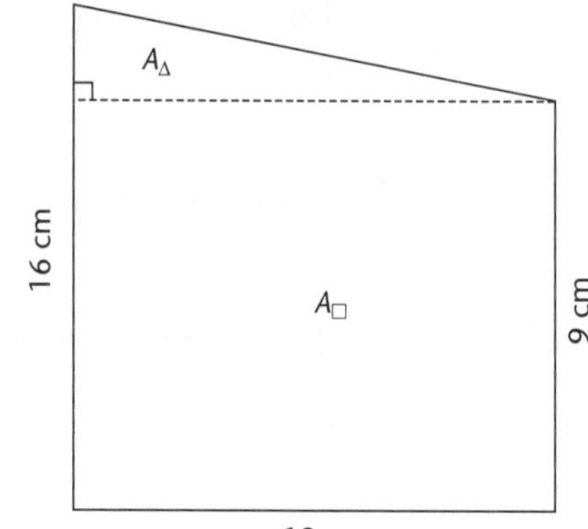

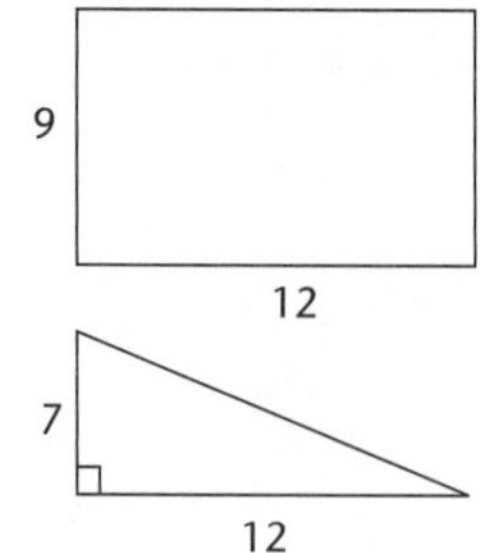

This time, with a horizontal line (dotted) we form a triangle and a rectangle:

$\text{Area}_\square = lb$

$= 12 \times 9$

$= 108$

$\text{Area}_\triangle = \frac{1}{2}bh$

$= \frac{1}{2} \times 12 \times 7$

$= 42$

[Height of $\triangle = 16 - 9 = 7$
Base $= 12$]

Total area $= A_\square + A_\triangle$

$= 108 + 42$

$= 150$

Area is 150 cm^2.

2 **a**

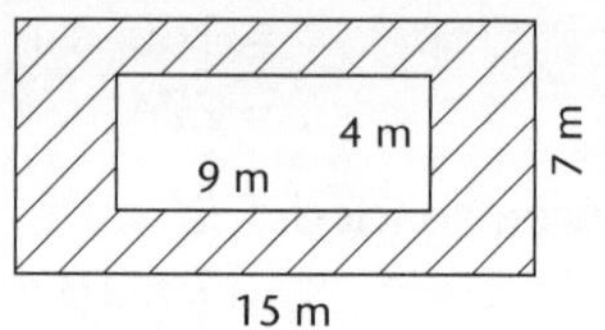

Shaded area = Area of large rectangle − Area of small rectangle

$= 15 \times 7 - 9 \times 4$

$= 105 - 36$ $[A = lb]$

$= 69$

$\therefore$ Shaded area is 69 m^2.

b

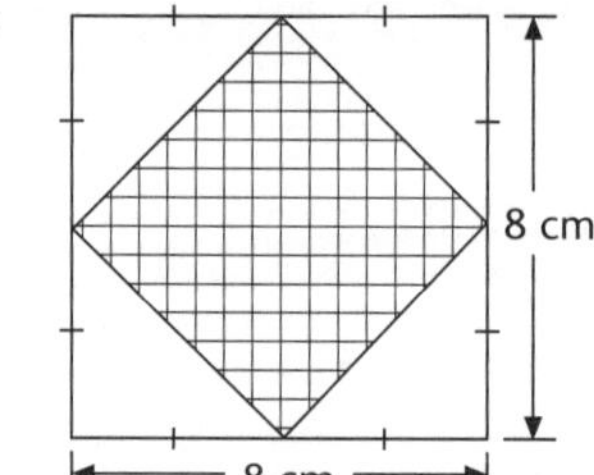

Shaded area = Area of square − Area of 4 triangles

$= 8 \times 8 - 4(\frac{1}{2} \times 4 \times 4)$

$= 64 - 4(8)$

$= 64 - 32$

$= 32$

$\therefore$ Shaded area is 32 cm^2.

3 a 

Dining room: $A = 3 \times 2.5$
$= 7.5$

$\therefore$ Area is 7.5 m^2.

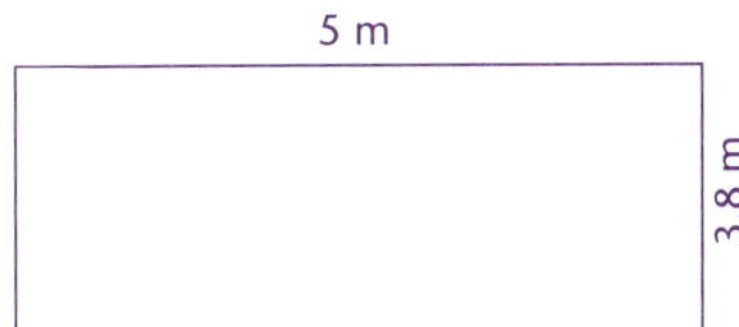

Lounge room: $A = 5 \times 3.8$
$= 19$ m^2

$\therefore$ Area is 19 m^2.

b Total area = 7.5 + 19
= 26.5 m^2

Total cost = 26.5 × \$30
= \$795

$\therefore$ Total cost is \$795.

4 a Total area = $5 \times 1 + 6 \times 1 + 4 \times 1$
= 15 [$A = lb$]

Total area is 15 m^2

Area to be paved is 15 m^2.

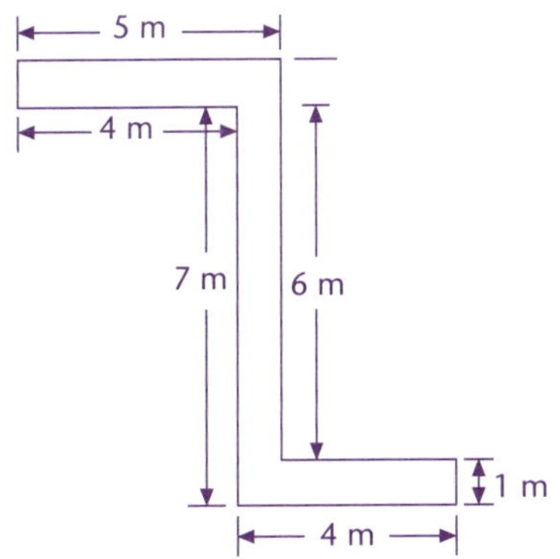

b Area of each paver = 20 cm × 10 cm
= 200 cm^2

Now, total area in cm^2 = 15 × 10 000
= 150 000 cm^2

$\therefore$ No. of pavers = 150 000 ÷ 200
= 750

$\therefore$ 750 pavers are required.

Volume

Volume is a measure of the space contained in a solid shape:

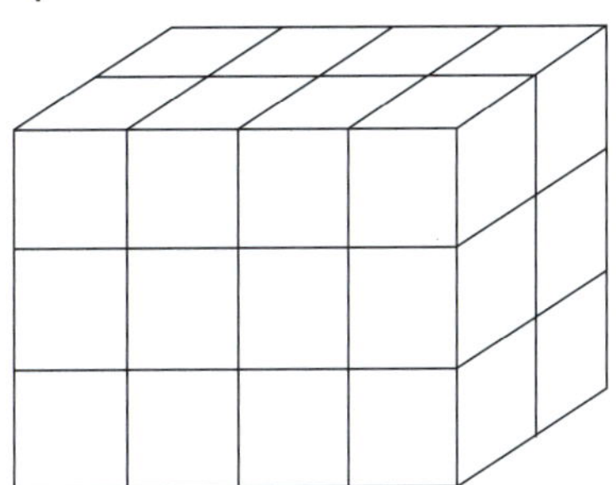

[Each cube measures 1 cm by 1 cm by 1 cm, which is called a cubic centimetre (i.e. cm^3).]

By 'counting' we can see that the volume of this solid is $4 \times 2 \times 3 = 24$ cm^3.

Volumes of Prisms

$V = Ah$, where V = Volume
A = Area of cross section
h = height

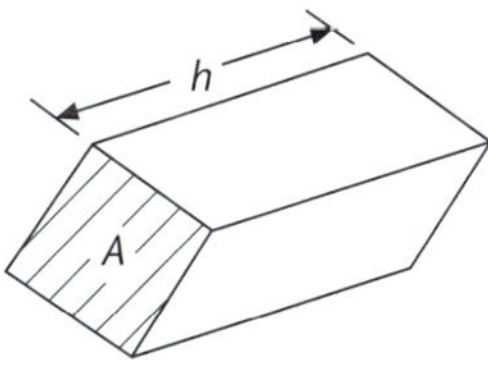

For a rectangular prism, this can be thought of as:

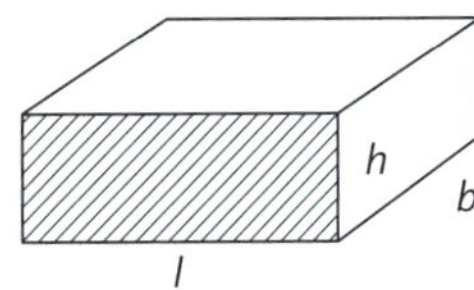

Volume = length × breadth × height

[as Area = lb]

$V = lbh$

[Remember units of volume are units^3.]

For Example

Find the volume:

1

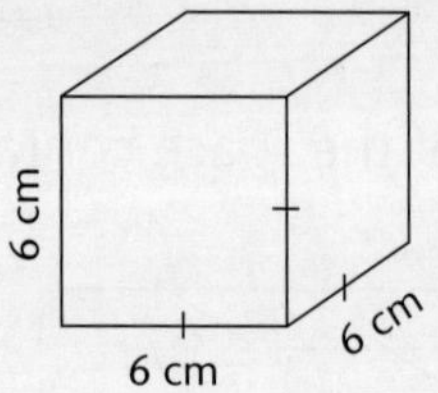

2

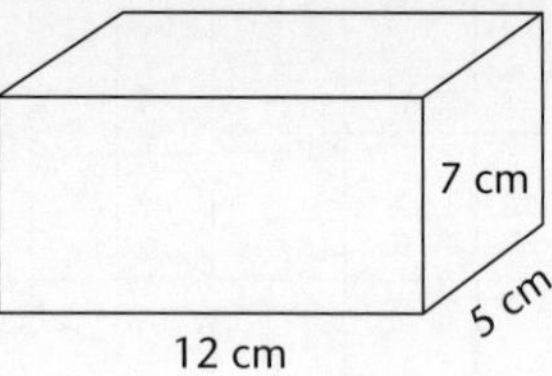

3

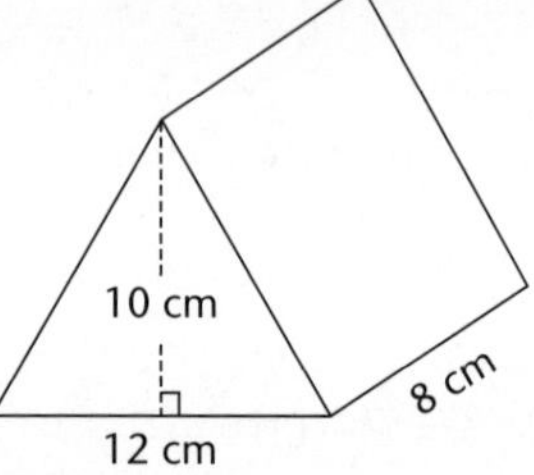

4

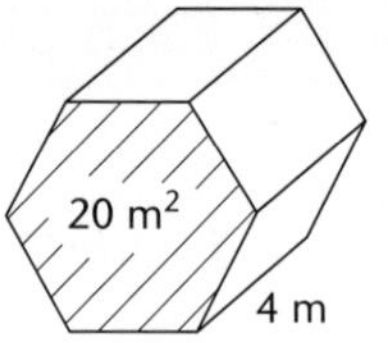

5

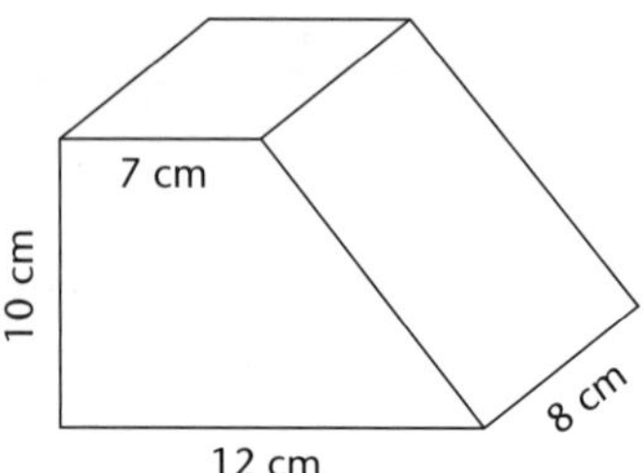

6

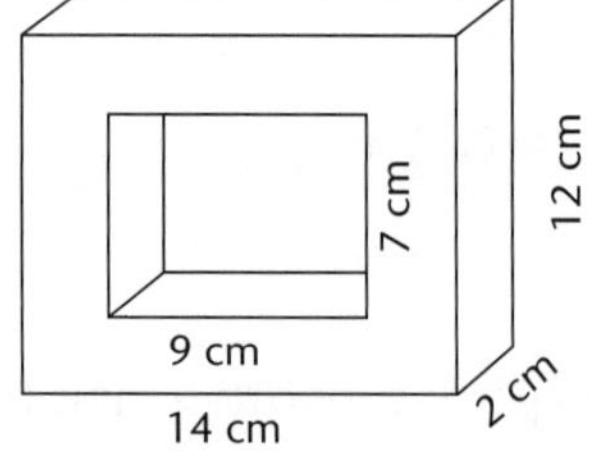

1

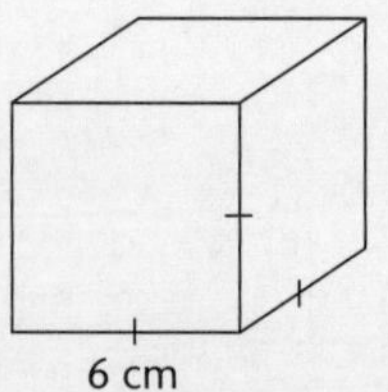

Cube (type of rectangular prism)

$V = lbh$

$\therefore V = 6 \times 6 \times 6$

$= 216$

$\therefore$ Volume is 216 cm^3.

2

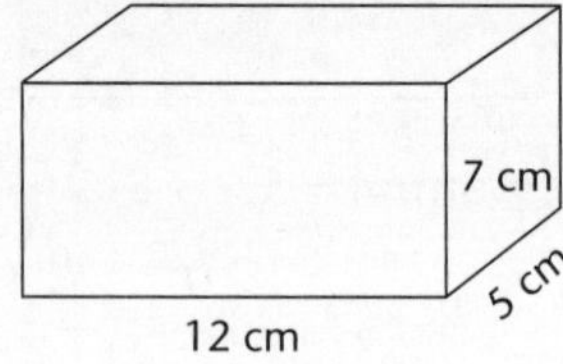

(Rectangular prism)

$V = lbh$

$\therefore V = 12 \times 7 \times 5$

$= 420$

$\therefore$ Volume is 420 cm^3.

3

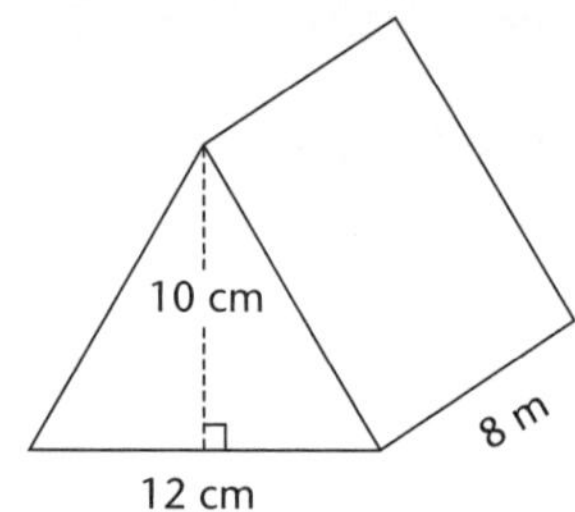

$A = \frac{1}{2}bh$

$= \frac{1}{2} \times 12 \times 10$

$= 60$

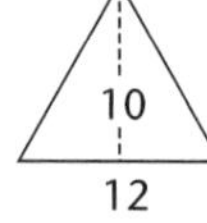

$V = Ah$

$= 60 \times 8$

$= 480$

Volume is 480 cm^3.

4

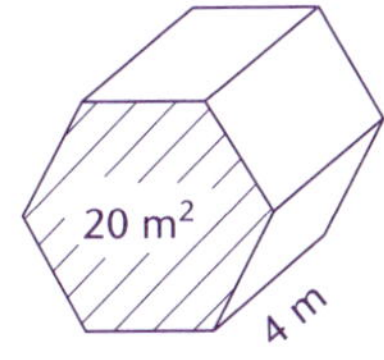

$V = Ah$

$= 20 \times 4$

$= 80$

Volume is $80\ m^3$.

5

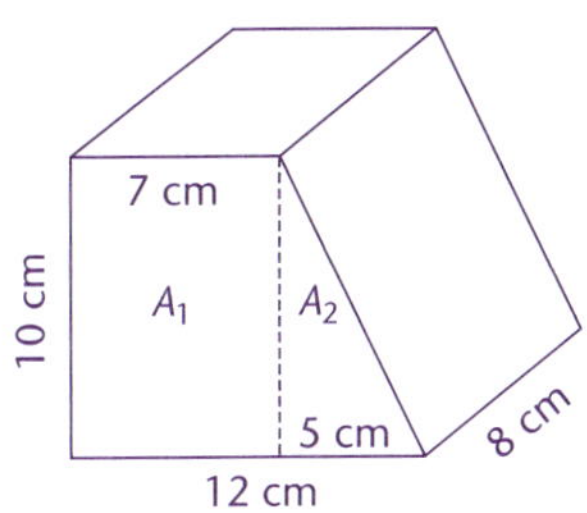

For area of cross-section:

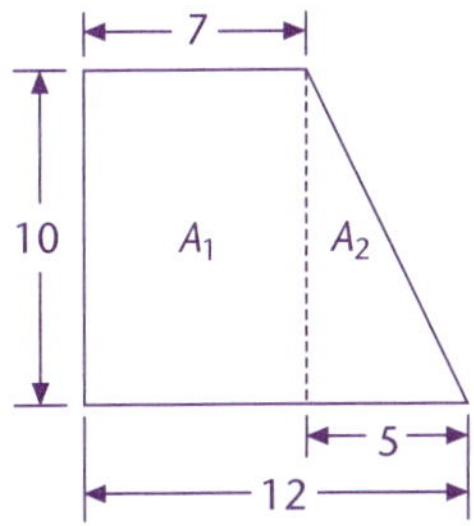

$A_1 = lb$

$A_1 = 10 \times 7$

$A_1 = 70$

$A_2 = \frac{1}{2}bh$

$A_2 = \frac{1}{2} \times 5 \times 10$

$A_2 = 25$

Cross-section $A = 70 + 25$

$= 95$

$\therefore V = Ah$

$= 95 \times 8$

$= 760$

$\therefore$ Volume is $760\ cm^3$.

6

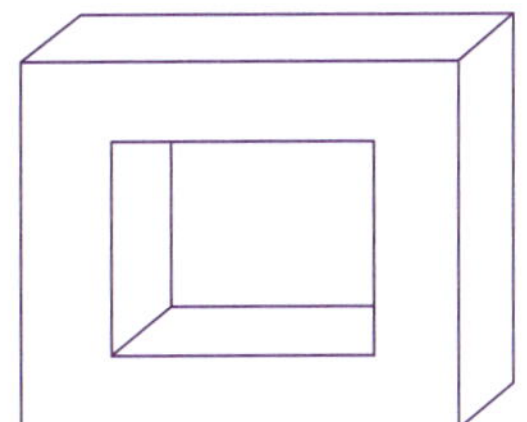

Area of cross section:

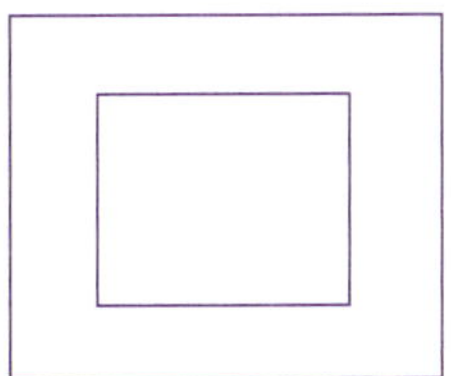

A = Area of large rect. – Area of small rect.

$= 14 \times 12 - 9 \times 7$

$= 168 - 63$

$= 105$

$\therefore V = Ah$

$= 105 \times 2$

$= 210$

Volume is $210\ cm^3$.

Units Used in Volume Measurement

Here we have two identical cubes, one with dimensions in cm, the other in mm:

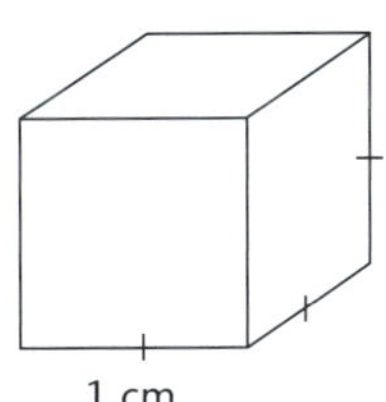

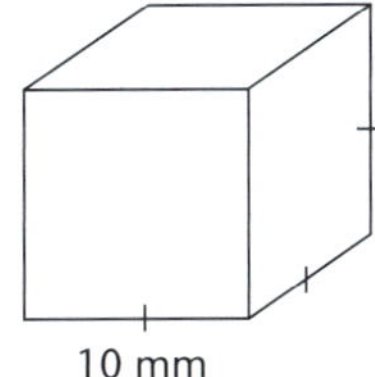

$V = 1 \times 1 \times 1$

$= 1$

Volume is $1\ cm^3$

$V = 10 \times 10 \times 10$

$= 1000$

Volume is $1000\ mm^3$

$\therefore 1\ cm^3 = 1000\ mm^3$

Complete:

1 $5\text{ cm}^3 = \_\_\_\_ \text{ mm}^3$

2 $25\,000\text{ mm}^3 = \_\_\_\_ \text{ cm}^3$

1 $5 \times 1000 = 5000$

$\therefore 5\text{ cm}^3 = 5000\text{ mm}^3$

2 $25\,000 \div 1000 = 25$

$\therefore 25\,000\text{ mm}^3 = 25\text{ cm}^3$

Capacity

Capacity is a measure of the amount of liquid within a container. The basic unit we use is the litre (L):

1000 millilitres (mL) = 1 litre (L)
1000 litres (L) = 1 kilolitre (kL)
1000 kilolitres (kL) = 1 megalitre (ML)

1 Convert:

a 3.2 litres = ____ mL

b 7600 mL = ____ L

c 4.28 kL = ____ L

d 2742 L = ____ kL

e 7 ML = ____ mL

2 Jenz opened a 2-litre bottle of cordial and poured 220 mL into each of four glasses. How much remains in the bottle (in litres)?

1 a $3.2 \times 1000 = 3200$

$\therefore$ 3200 mL

b $7600 \div 1000 = 7.6$

$\therefore$ 7.6 L

c $4.28 \times 1000 = 4280$

$\therefore$ 4280 L

d $2742 \div 1000 = 2.742$

$\therefore$ 2.742 kL

e $7 \times 1000 \times 1000 \times 1000$

$= 7\,000\,000\,000$

$\therefore$ 7 000 000 000 mL

2 Remainder in bottle $= 2000 - 4 \times 220$

$= 2000 - 880$

$= 1120$

$\therefore$ 1120 mL

$\therefore$ 1.12 L remains

Capacity and Volume

$1\text{ cm}^3 = 1\text{ mL}$

$\therefore \quad 1000\text{ cm}^3 = 1\text{ L}$

Also, $\text{cm}^3 = \text{cc}$ [cc = cubic centimetres]

$\therefore \quad 1000\text{ cc} = 1\text{ L}$

Find the capacity, in litres, of these containers:

1

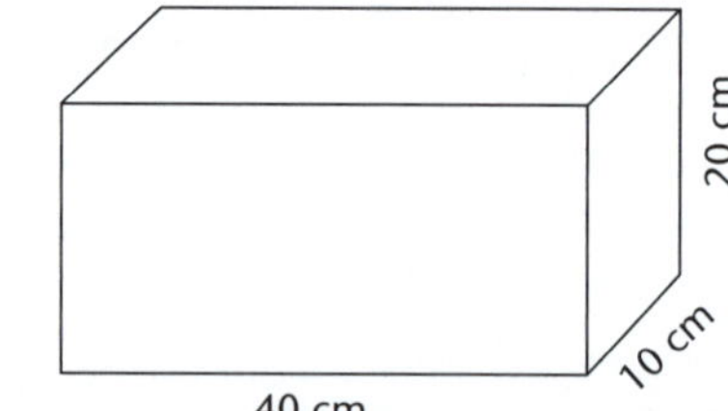

2

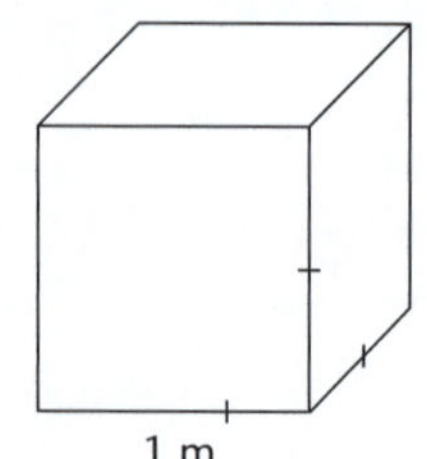

1

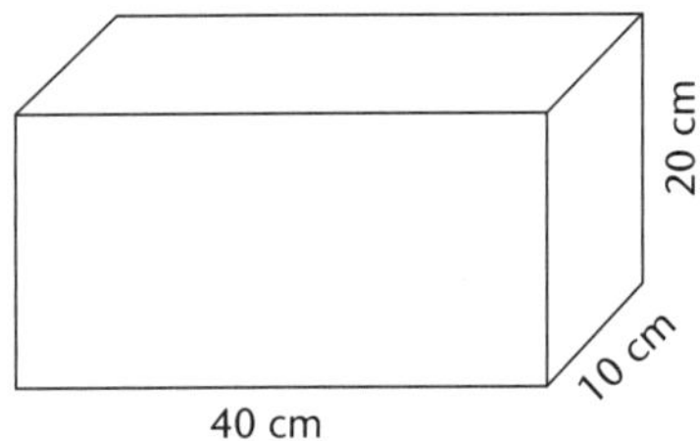

$V = lbh$

$= 40 \times 10 \times 20$

$= 8000$

$\therefore$ Volume is 8000 cm^3

$\therefore$ Capacity = 8000 ÷ 1000 = 8

Capacity is 8 litres.

2 1 m = 100 cm

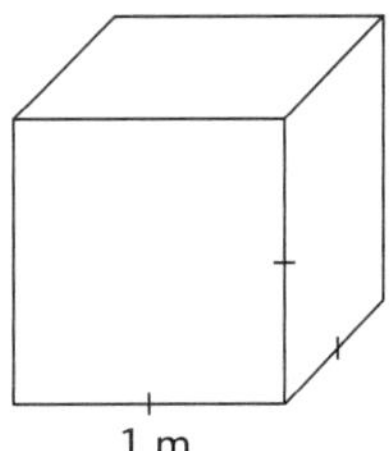

$V = lbh$

$\therefore V = 100 \times 100 \times 100$

$= 1\,000\,000$

$\therefore$ Volume is 1 000 000 cm^3

$\therefore$ Capacity = 1 000 000 ÷ 1000

= 1000 L

= 1 kL

$\therefore$ Capacity is 1 kL. [i.e. 1 m^3 = 1 kL]

Go to p. 258 for quick answers, or to pp. 295–298 for worked solutions.

1 Convert the following to the indicated units: pp. 139–140

a 2000 m = ____ km **b** 4.21 km = ____ m

c 37 cm = ____ mm **d** $3\frac{1}{2}$ cm = ____ mm

e 420 mm = ____ cm

2 Choose A, B, C or D: pp. 139–140

a The width of a pencil is closest to:
A. 1 mm B. 1 cm C. 10 cm D. 1 m

b The length of a fork is closest to:
A. 2 cm B. 10 cm C. 20 cm D. 40 cm

c The distance from Brisbane to Sydney is closest to:
A. 100 km B. 900 km C. 2000 km D. 5000 km

d The height of a ceiling is closest to:
A. 1 m B. 2.5 m C. 14 m D. 7 m

3 p. 139

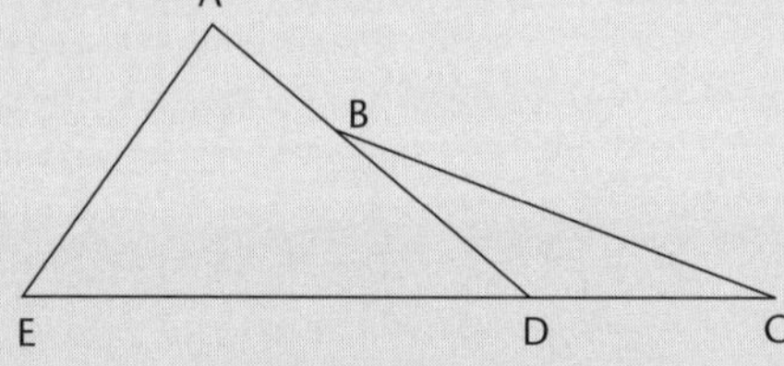

Measure the following lengths, in mm:

a AE **b** AD **c** BC

4 Find the limit of measurement of a: p. 140

a pencil measured at 12 cm

b length of rope measured at 16 m.

5 Find the perimeter of the following: pp. 140–141

a 12 cm, 30 cm **b**

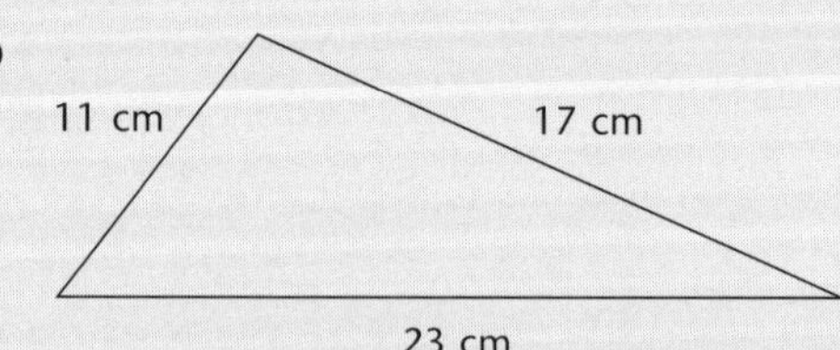

c A regular octagon with side 7 cm.

6 If the perimeter of a rectangle is known to be 96 mm and the length is 28 mm, find the width. pp. 140–141

7 Find the perimeters of these figures: pp. 140–141

a

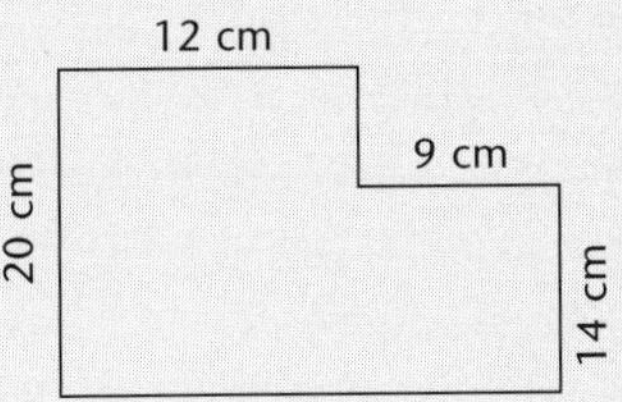

b

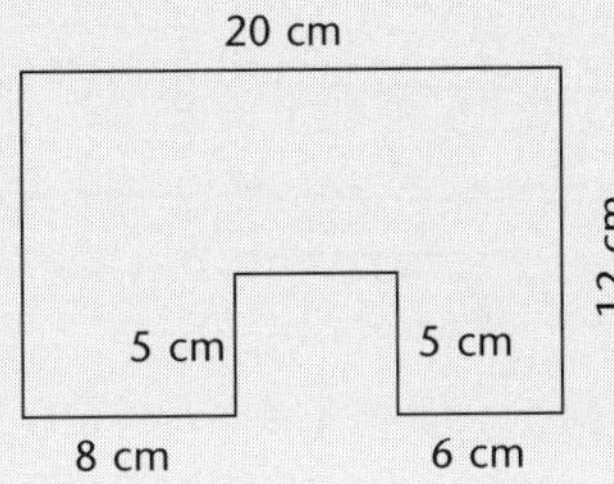

8 A lounge room is in the shape of a rectangle measuring 5.2 metres long and 3.8 metres wide. A wallpaper frieze is to be pasted along the top of each wall around the room. Each roll of the frieze costs $21.90 and covers a 5-metre length: pp. 140–141

a What is the perimeter of the room?

b How many rolls will need to be purchased to complete the job?

c What will be the total cost of the rolls?

9 Find the areas of the shaded regions in square units: pp. 141–142

a

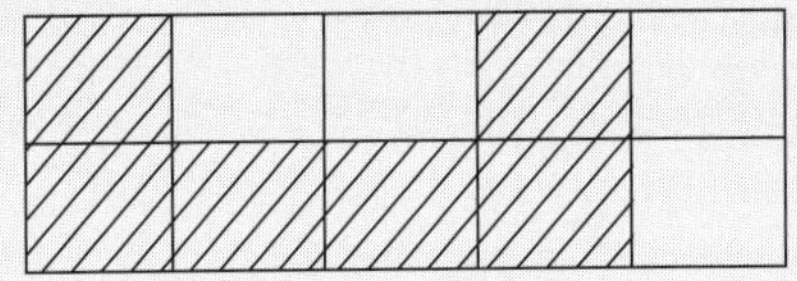

b

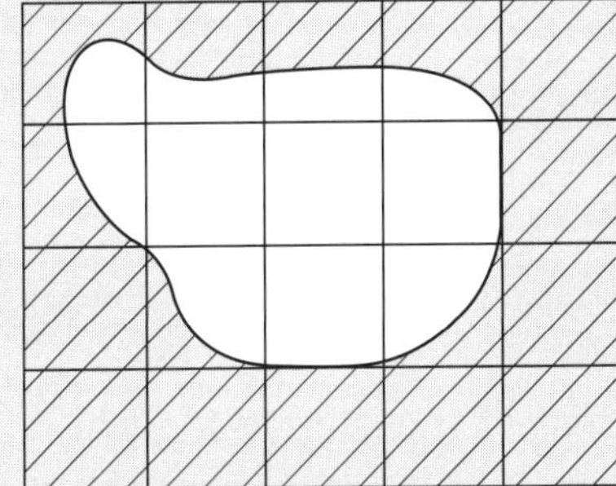

10 Find the areas of these shapes: pp. 142–143

a

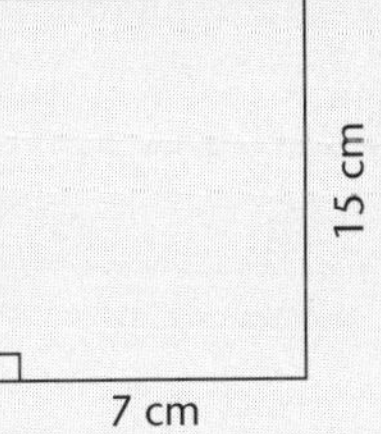

b

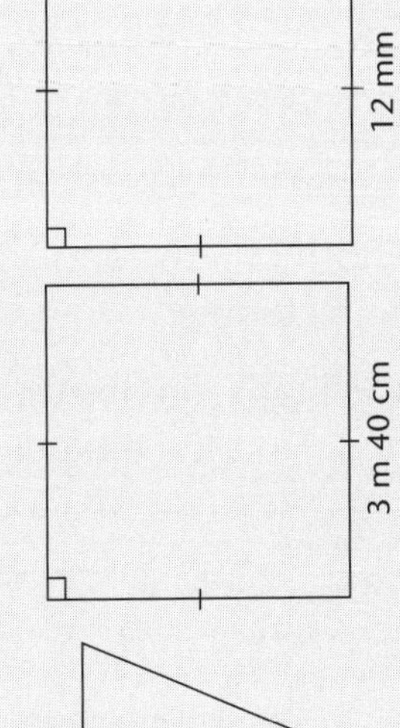

c

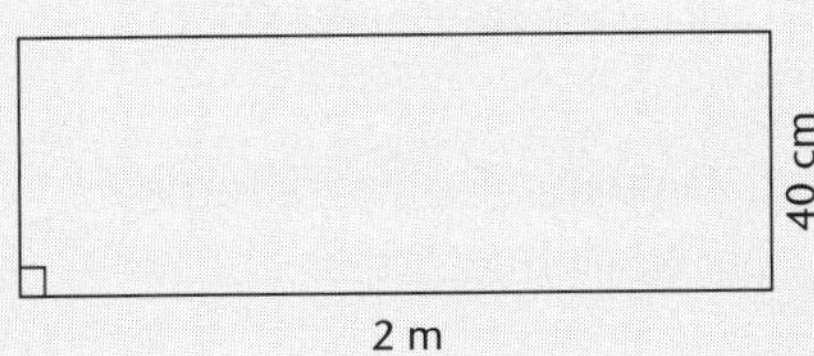

d

e

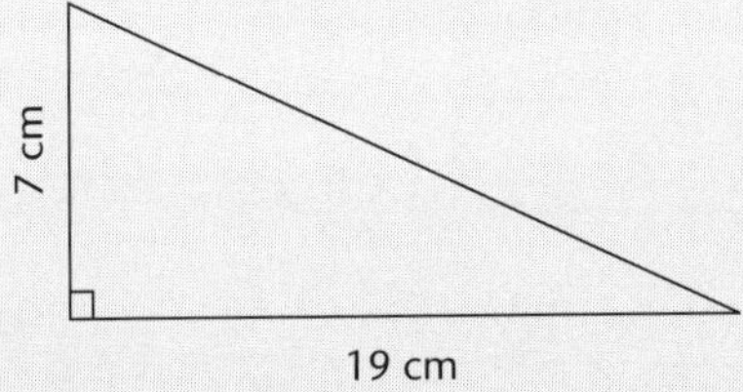

f

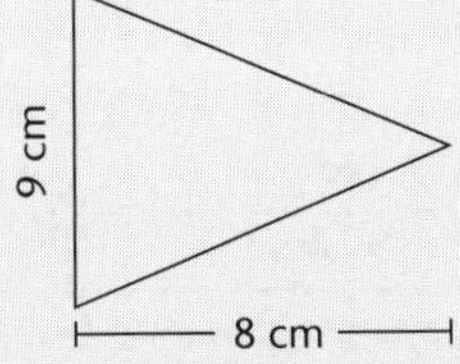

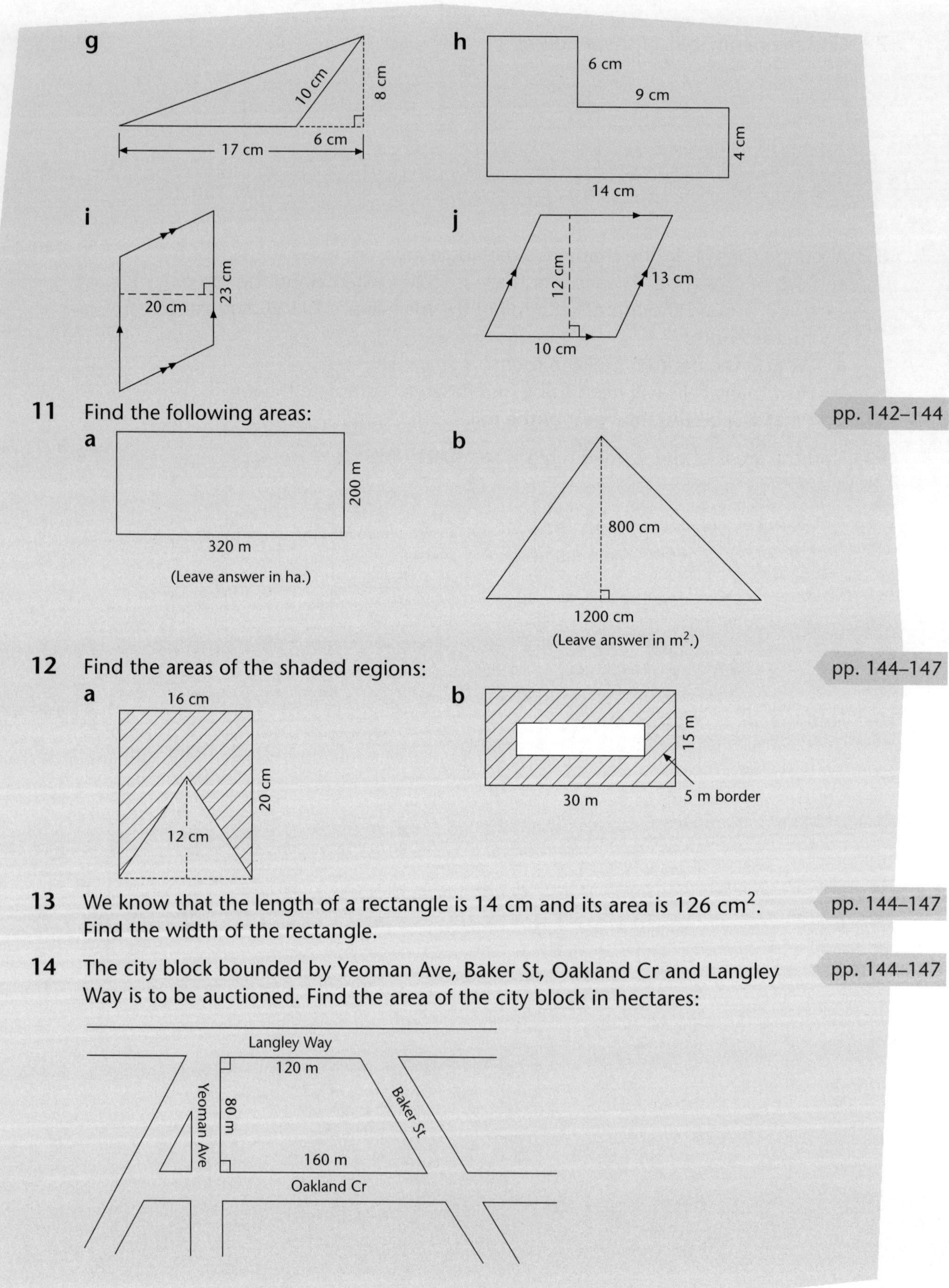

11 Find the following areas:

pp. 142–144

12 Find the areas of the shaded regions:

pp. 144–147

13 We know that the length of a rectangle is 14 cm and its area is 126 cm^2. Find the width of the rectangle.

pp. 144–147

14 The city block bounded by Yeoman Ave, Baker St, Oakland Cr and Langley Way is to be auctioned. Find the area of the city block in hectares:

pp. 144–147

15 The diagram shows the plan of a house in the shape of a rectangle: pp. 142–147

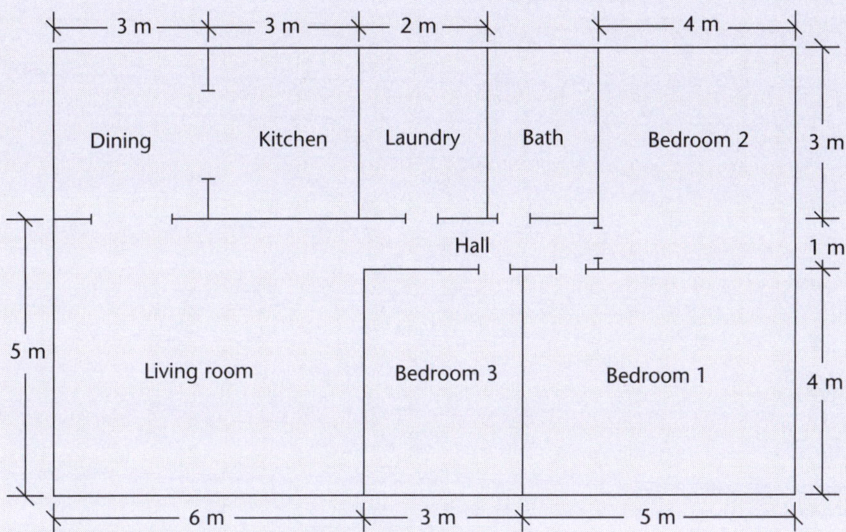

a What is the length of the living room?
b What is the width of the hall?
c Find the area of the dining room.
d Find the area of the entire house.
e The bathroom floor is to be tiled. If the tiles cost $\$30/m^2$ what will be the cost of the tiles?
f The ceilings in the three bedrooms are to be painted with a special compound. It is known that the cost of the compound is 70 cents/m^2. Find the total cost.

16 A painting measuring 90 cm by 60 cm is to be mounted in a picture frame using a border of 10 cm around the painting. pp. 142–147

Find the:
a Length and width of the picture frame.
b Perimeter of the picture frame.
c Cost of the frame moulding if it sells for $12/m.
d Total area of the framed picture.
e Cost of the glass for the picture if it sells at $\$70/m^2$.

17 The floor of a school's assembly hall measures 30 metres by 20 metres. It is to be covered with carpet squares with sides 50 centimetres: pp. 142–147
a Find the area of the floor.
b Find the area of each carpet square.
c How many carpet squares are required?

18 Convert the following units to those indicated: p. 144
a 3.2 ha = ____ m^2 **b** 16 000 m^2 = ____ ha
c 3700 cm^2 = ____ m^2 **d** 5.84 m^2 = ____ cm^2

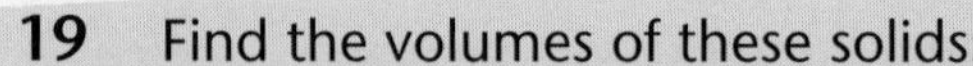

19 Find the volumes of these solids: pp. 147–149

a

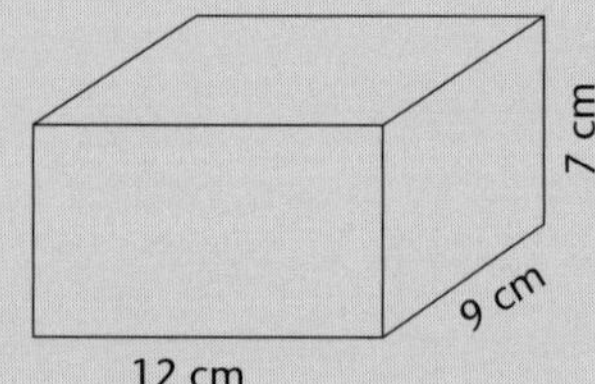

b

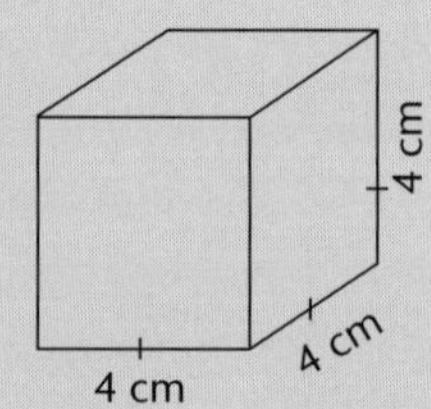

c

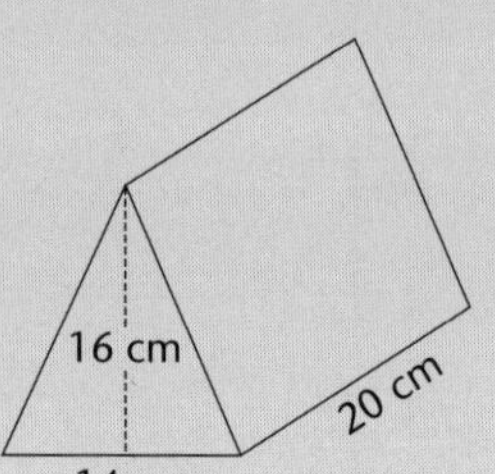

d

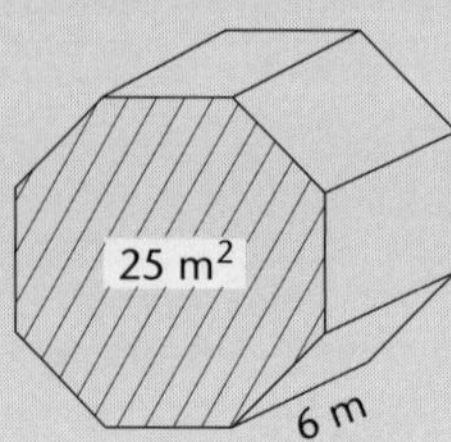

e

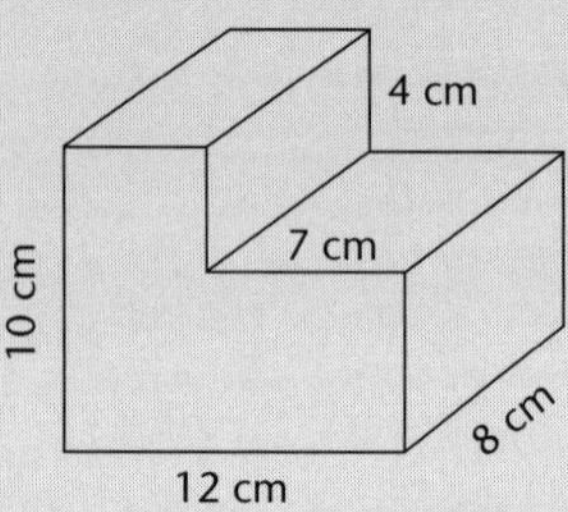

f

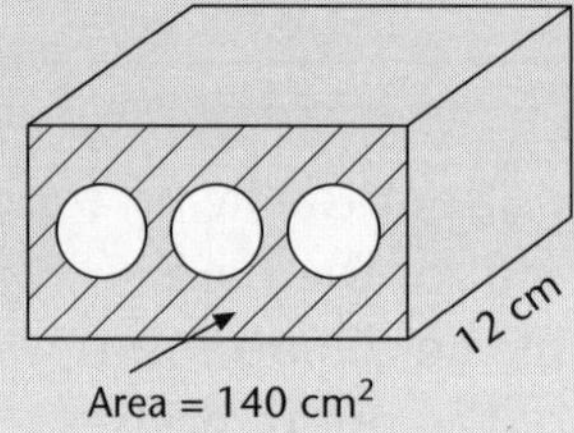

g

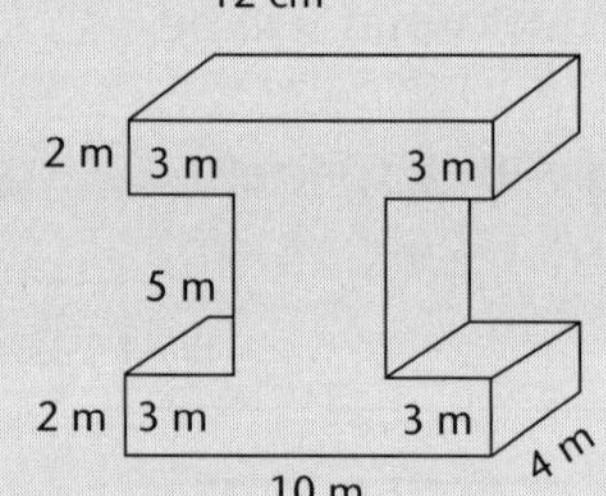

20 Convert: pp. 149–150

a $7\text{ cm}^3 = \_\_\_\_ \text{ mm}^3$ **b** $45\,000\text{ mm}^3 = \_\_\_\_ \text{ cm}^3$

21 A room measures 4 metres by 3 metres. If it is 2.5 metres in height, find the volume of the room. pp. 147–149

22 Lloyd is to construct a garage measuring 6 m by 8 m in his backyard. The concrete slab for the garage is to be 20 cm in depth. Find the volume of concrete required. pp. 147–149

23 Convert to the indicated units: p. 150

a 1250 mL = ____ L **b** 14.83 kL = ____ L **c** 3700 L = ____ kL

24 A carton of juice measures 10 cm by 20 cm by 5 cm: pp. 150–151

a What is the volume of the carton?

b Find the capacity of the carton.

Go to p. 258 for quick answers, or to pp. 295–298 for worked solutions.

YOUR CHECKLIST

For a complete understanding of this topic you must be able to:

✓	Measure lengths to the nearest millimetre/centimetre		p. 139
✓	Convert between units of length		pp. 139–140
✓	Describe the limits of accuracy of measuring instruments		p. 140
✓	Measure the perimeter of shapes		p. 140
✓	Find the perimeter of figures		pp. 140–141
✓	Estimate the area of a shape with the aid of a grid		pp. 141–142
✓	Find the area of squares, rectangles, parallelograms and triangles		pp. 142–143
✓	Convert between units of area		p. 144
✓	Find the areas of composite figures		pp. 144–147
✓	Solve problems involving areas		pp. 145–147
✓	Find the volumes of prisms		pp. 147–149
✓	Convert between units of volume		pp. 149–150
✓	Solve problems involving volumes of prisms		pp. 147–150
✓	Solve problems involving capacity.		pp. 150–151

Now you are ready to do the tests!

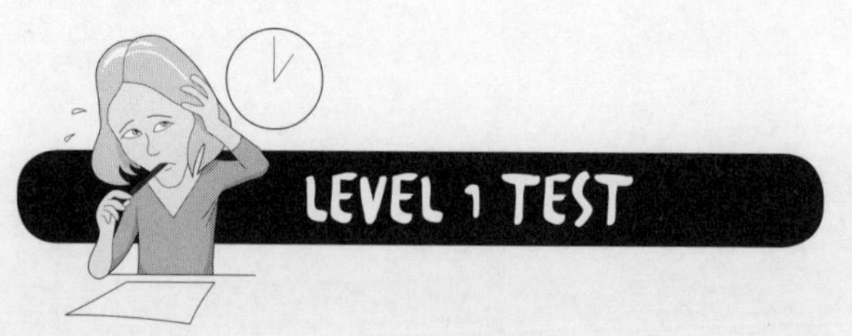

(30 marks)

1 Measure the perimeter of triangle ABC:

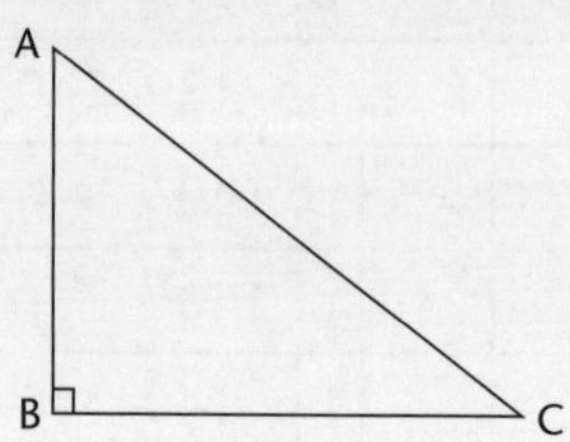

(1 mark)

2 Convert 23 000 metres to kilometres. (1 mark)

3 Find the perimeters:

a

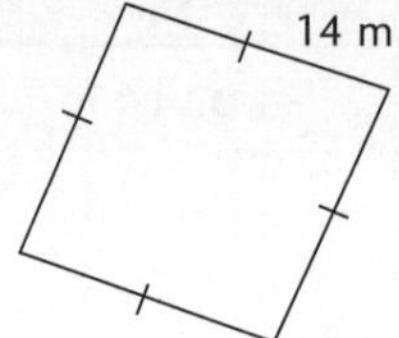

b

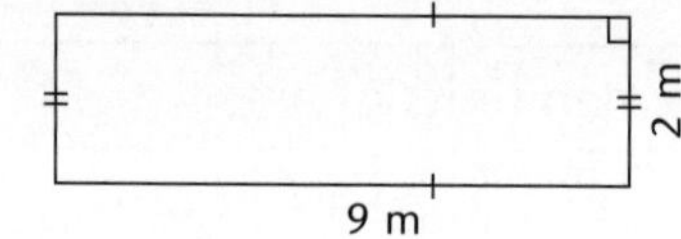

c

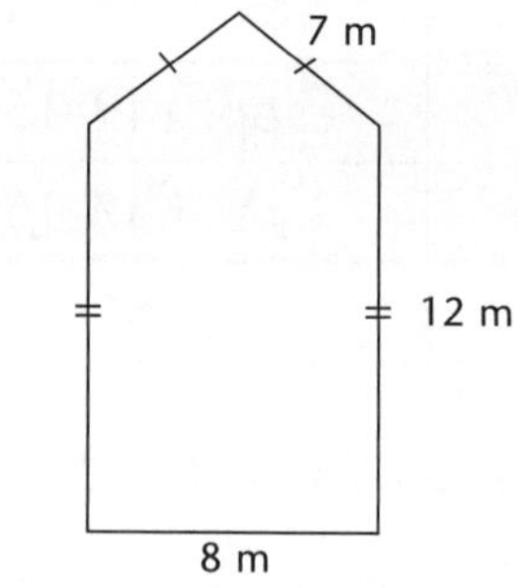

(3 marks)

4 Bob, the builder, measured a length of gyprock as 435 cm to the nearest cm. Between what two measurements would the true measurement be? (1 mark)

5 The perimeter of this isosceles triangle is 14 cm. Find the length of the missing side:

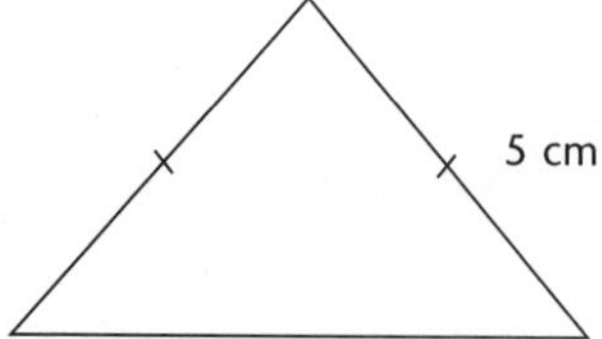

(1 mark)

6 This paddock is in the shape of a square with each side 16 m. Find the cost of fencing the paddock if it costs $2/metre:

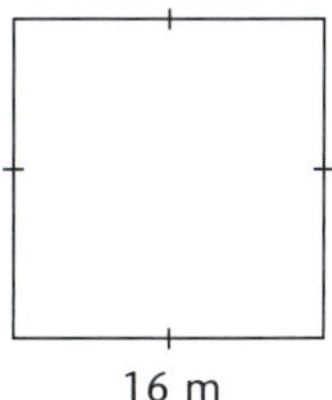

(2 marks)

7 Find the perimeter:

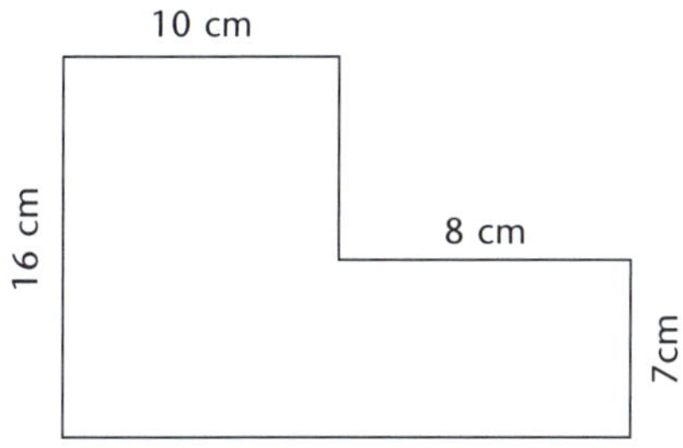

(2 marks)

8 This court is marked using special tape. Find the length of tape required:

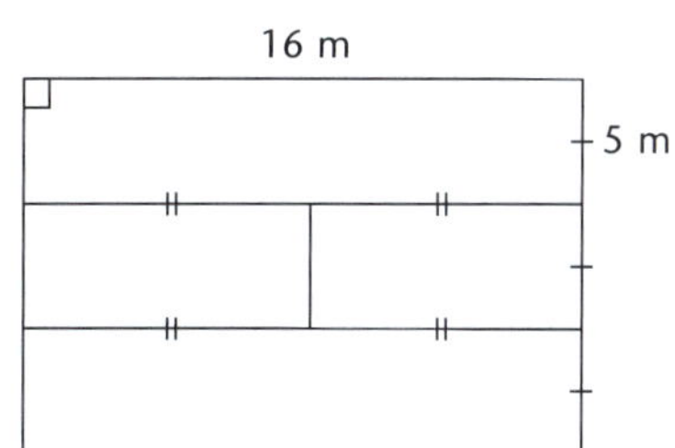

(2 marks)

9 Find the areas:

a

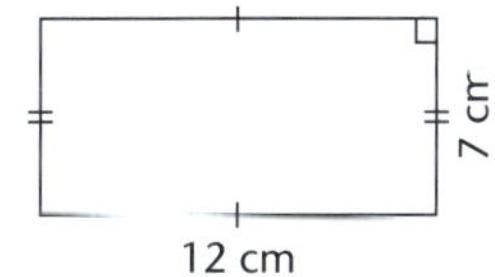

b

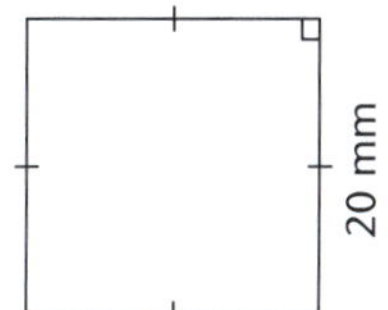

c

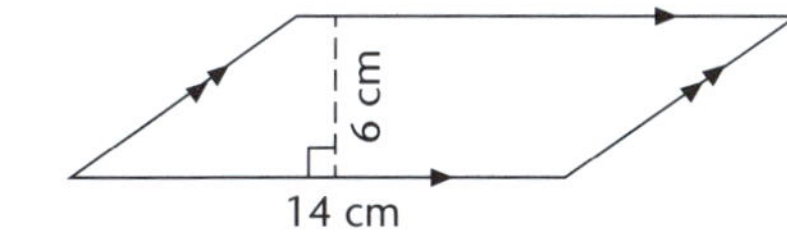

(3 marks)

10 Find the volumes:

a

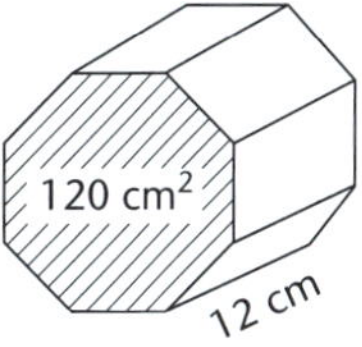

b

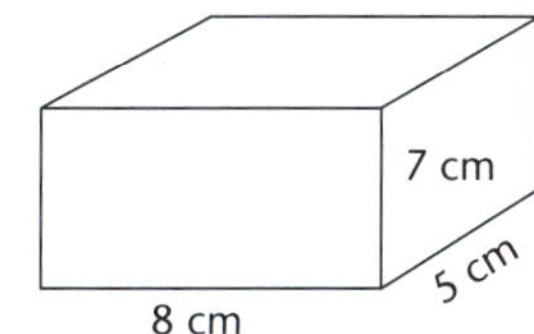

(2 marks)

11 Find the areas:

a

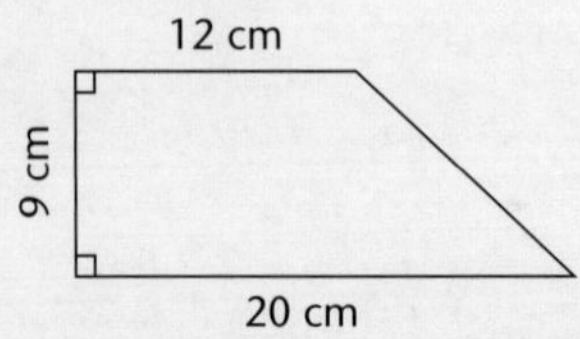

b

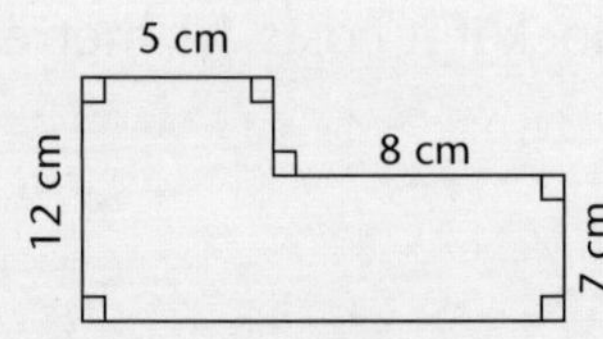

(4 marks)

12 Find the area of the shaded region:

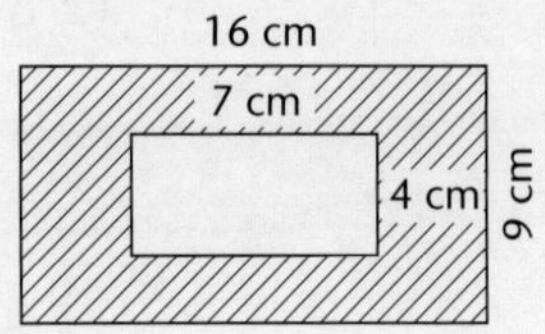

(2 marks)

13 A rectangle has an area of 48 cm^2. If the length is 10 cm, find the width:

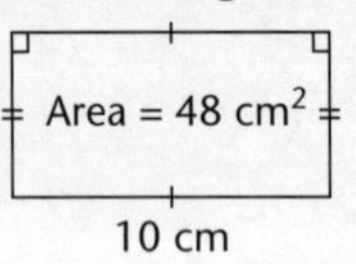

(2 marks)

14 Find the volume:

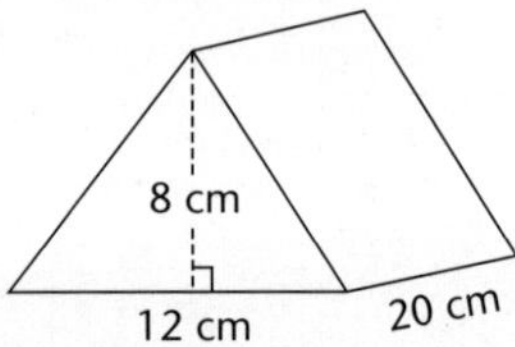

(2 marks)

15 Find the capacity, in mL:

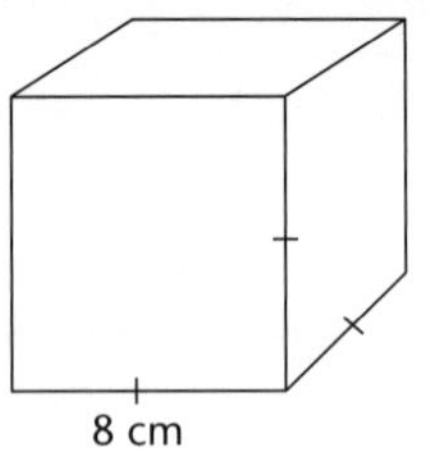

(2 marks)

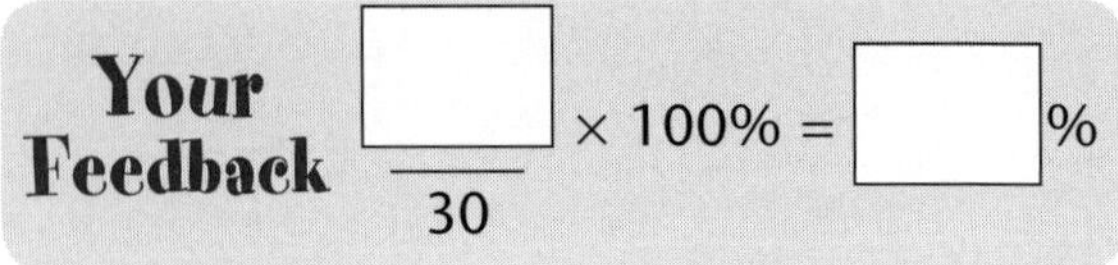

☞ **Quick answers on page 264**

☞ **Worked solutions on page 328**

LEVEL 2 TEST

(30 marks)

1 Find the length of BD: (1 mark)

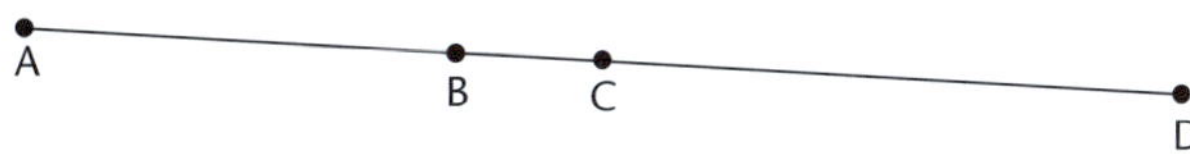

2 Convert 2400 cm to m. (1 mark)

3 Find the perimeters:

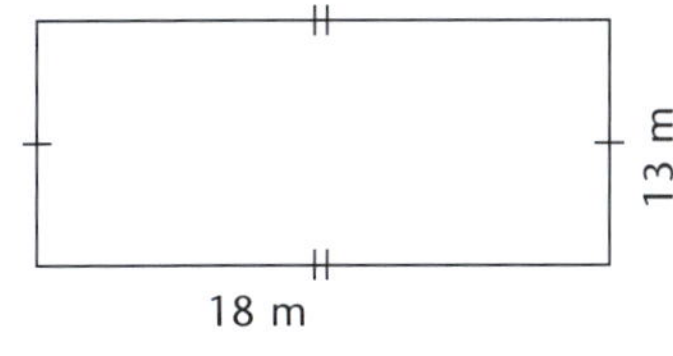

b

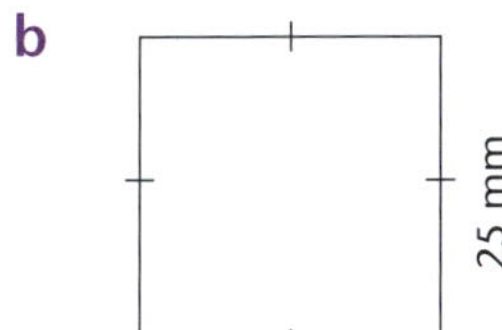

(2 marks)

4 Peter measures a piece of string to be 68 cm in length to the nearest cm. The true measurement is between what two measurements? (1 mark)

5 If the perimeter of the rectangle is 40 cm and the length is 12 cm, find the width:

Perimeter = 40 cm

(2 marks)

6 Find the perimeter:

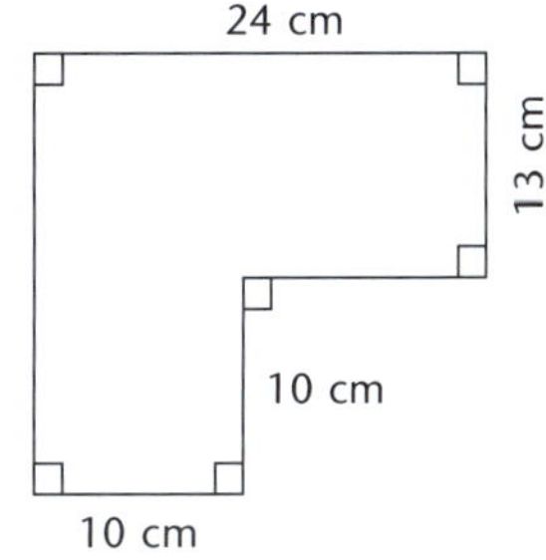

(2 marks)

7 Find the perimeter:

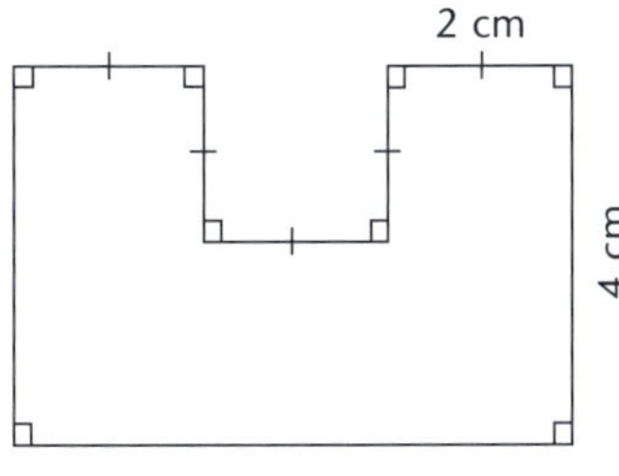

(2 marks)

8 Convert 310 000 m^2 = ____ ha. (1 mark)

9 Find the areas:

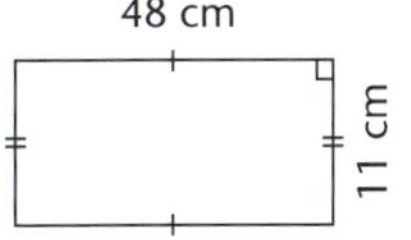

b

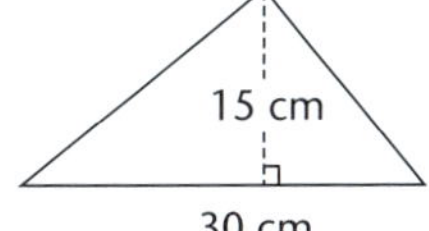

c

12.2 cm

20 cm

(3 marks)

10 Use the grid to draw a rectangle with an area of 12 cm^2:

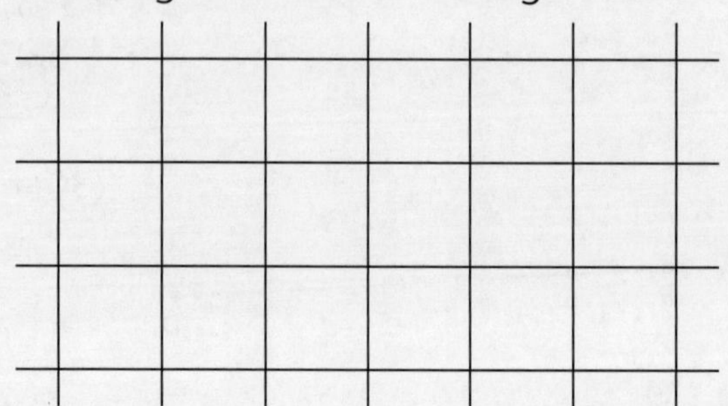

(1 mark)

11 Find the areas:

a

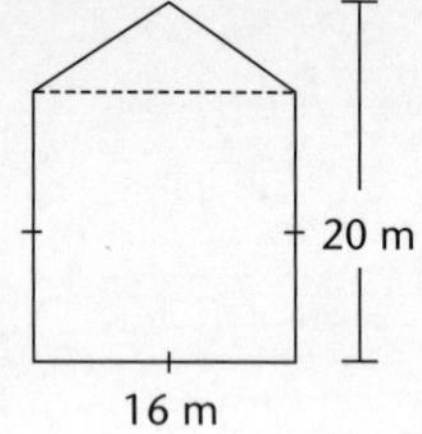

b

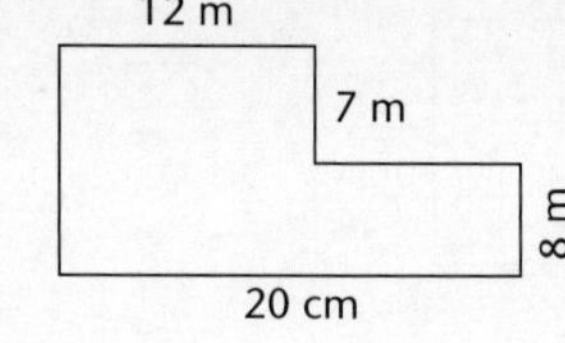

(4 marks)

12 Find the area of the shaded region:

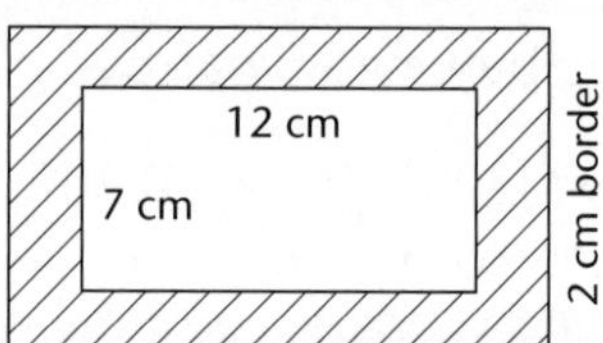

(2 marks)

13 A paddock measuring 30 m by 14 m is to be fertilised at the rate 80 g/m^2.
How much fertiliser is required? (2 marks)

14 A wall of Brae's bedroom is to be painted.
Calculate the area to be painted:

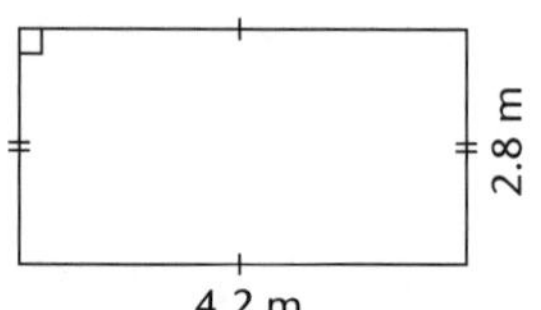

(2 marks)

15 Find the volumes:

a

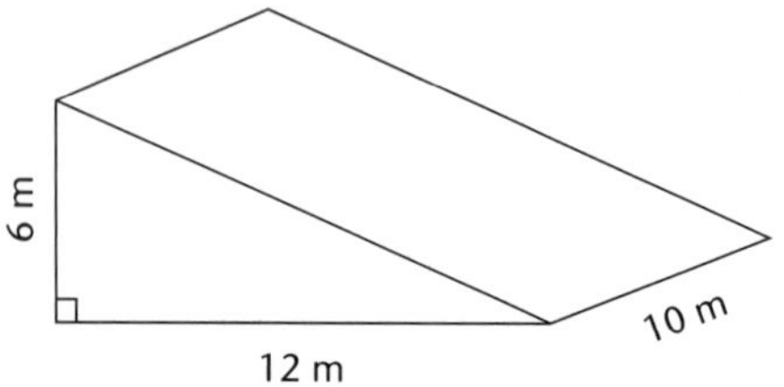

b

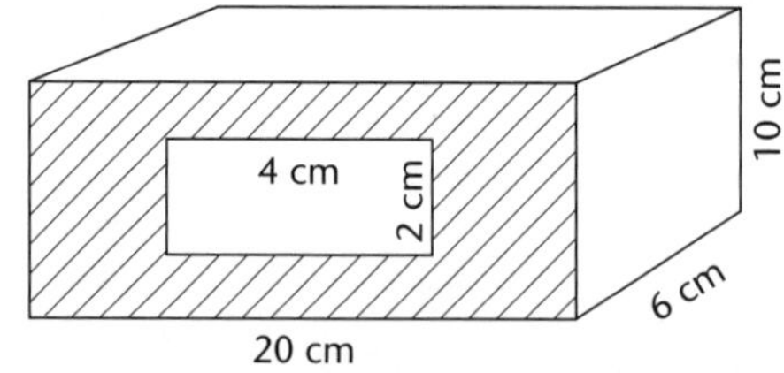

(4 marks)

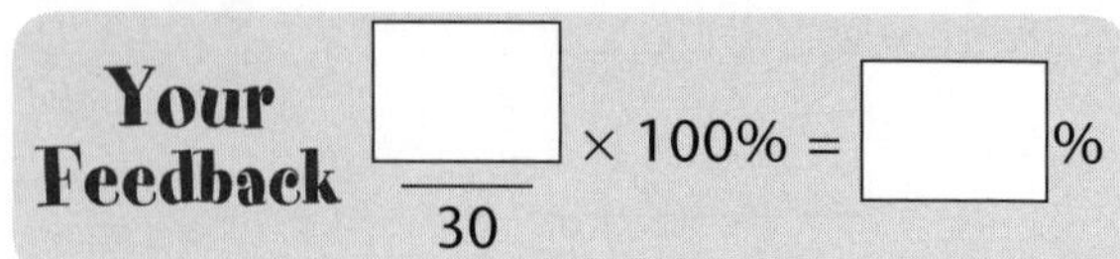

☞ Quick answers on page 264
☞ Worked solutions on page 329

(30 marks)

1 Convert 4 000 000 mm to km. (1 mark)

2 Find the perimeter, in metres:

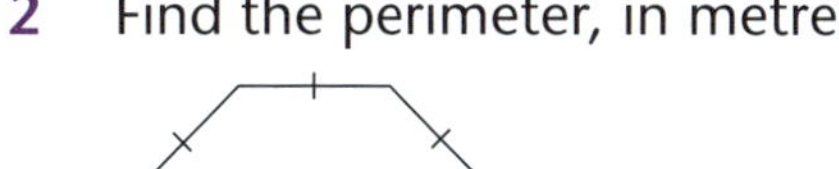

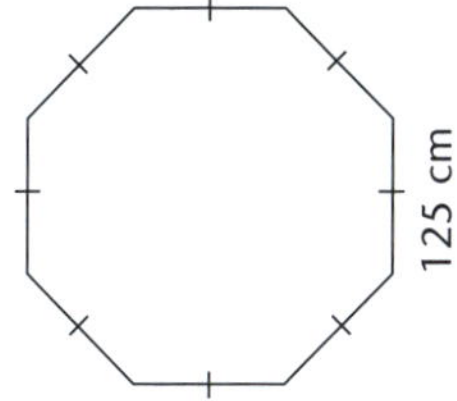

(1 mark)

3 A pool fence is to be built against a house. The length of the rectangular area to be fenced is three times the width. If the fence is 25 m in length, find the dimensions of the rectangle:

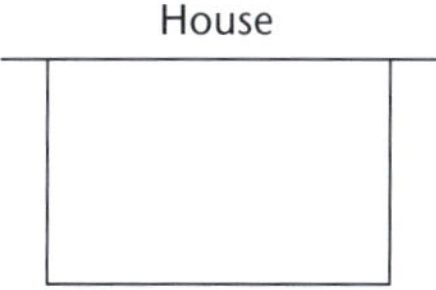

(2 marks)

4 Find the perimeter in cm:

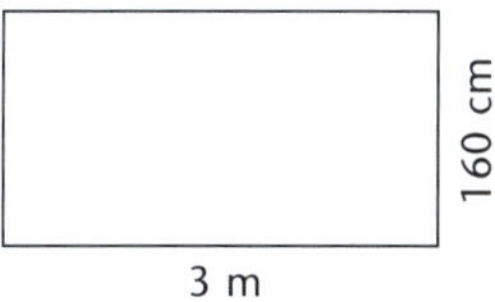

(2 marks)

5 Find the perimeter:

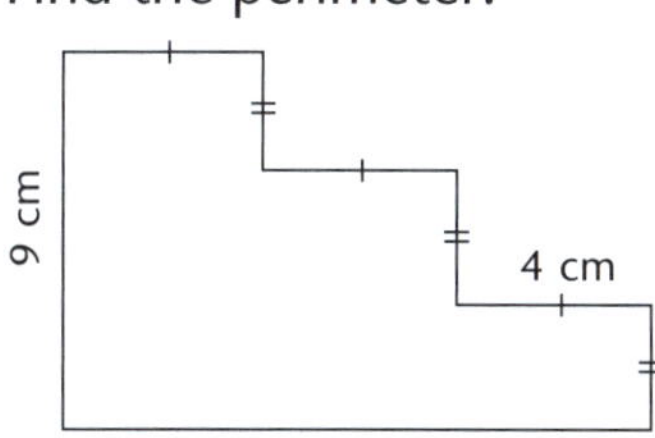

(2 marks)

6 A 12 cm × 8 cm photograph is to be mounted on cardboard and then framed.

The border is to be 3 cm wide. Find the length of frame that will be required:

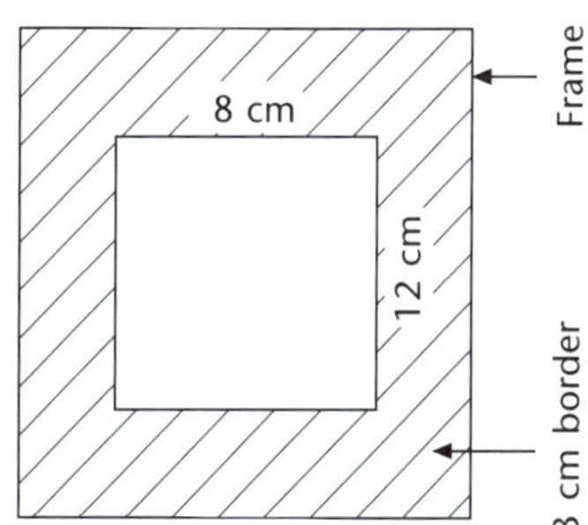

(2 marks)

7 Find the length of the base of a parallelogram if its perpendicular height is 32 cm and the area is 800 cm^2. (2 marks)

8 Use the grid to draw a triangle with an area of 8 cm^2:

(1 mark)

9 Convert 6.814 km = ____ mm. (1 mark)

10 Find the area, in m^2:

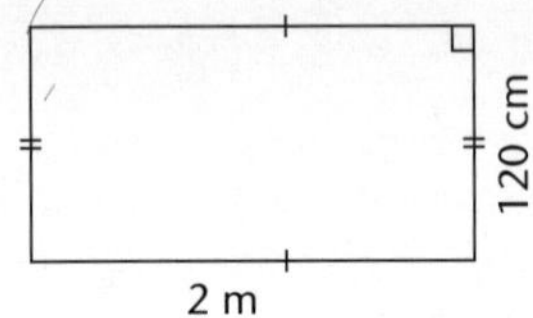

(2 marks)

11 Find the area, in hectares:

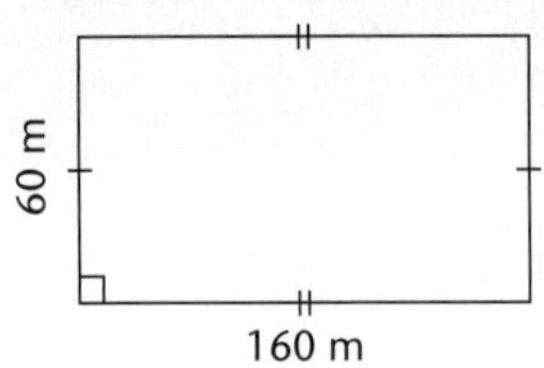

(2 marks)

12 Find the area of the shaded region:

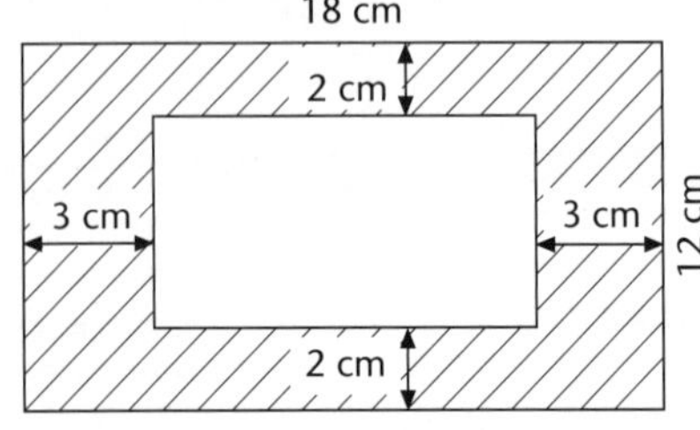

(2 marks)

13 Find the shaded area:

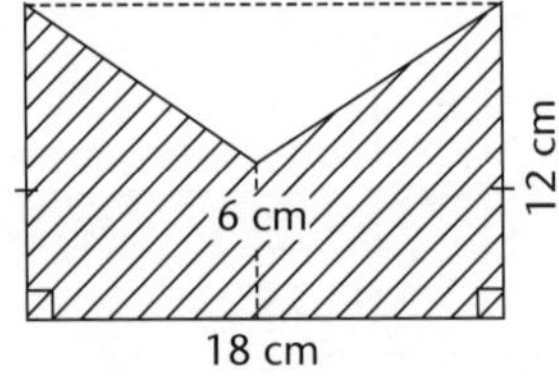

(2 marks)

14 An area is to be turfed with rolls of grass that are 2 m in length and 60 cm wide. If turf costs $3/roll, find the cost of the turf:

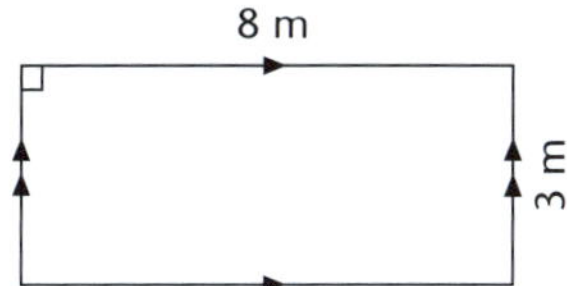

(2 marks)

15 The volume of a cube is 27 cm^3. Find the area of one of its faces. (2 marks)

16 Find the volumes:

a

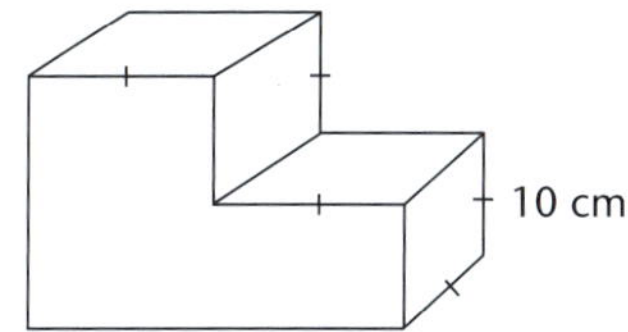

b

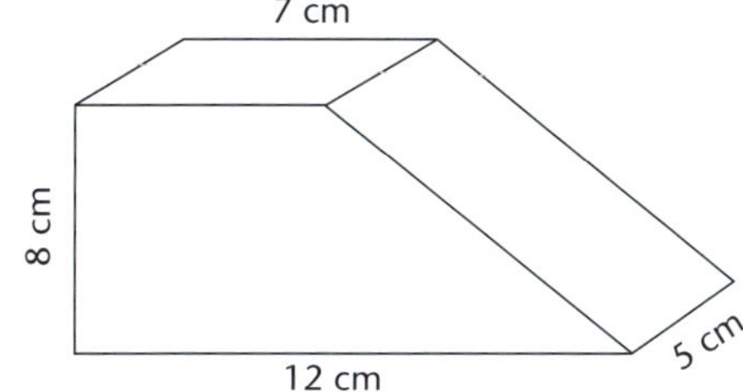

(4 marks)

☞ Quick answers on page 264
☞ Worked solutions on page 330

Your Feedback $\frac{\square}{30} \times 100\% = \square\%$

GEOMETRY

- Notation and Conventions in Geometry
- Indicators
- Polygons
- Naming Points, Intervals, Lines, Polygons
- Lines
- Symmetry
- Classification of Triangles
- Classification of Quadrilaterals
- Other Plane Shapes
- Parts of a Circle
- Classification of Solids
- Parts of Polyhedra
- Angle Terminology
- Parallel Lines
- Triangles
- Quadrilaterals

KEYWORDS

Acute	Parallel
Adjacent	Parallelogram
Alternate	Pentagon
Circle	Perpendicular
Circumference	Platonic
Co-interior	Polygon
Complementary	Polyhedron
Corresponding	Prism
Cross-section	Pyramid
Decagon	Quadrilateral
Dodecagon	Rectangle
Edges	Reflex
Ellipse	Revolution
Equilateral	Rhombus
Faces	Scalene
Hendecagon	Skew
Heptagon	Square
Hexagon	Supplementary
Isosceles	Symmetry
Kite	Trapezium
Nonagon	Vertically opposite
Obtuse	Vertices
Octagon	

Notation and Conventions in Geometry

Symbol	Meaning
∟	A right angle (90°)
△	A triangle (e.g. ΔPQR)
^ or ∠	Angle (e.g. ∠ABC or $A\hat{B}C$)
‖ or //	Parallel to (e.g. AB // PQ)
⊥	Perpendicular to (at 90° to)
°	Degrees (e.g. 27°)

Indicators

- To indicate that two intervals (sides) are equal, mark them in one of the following ways:

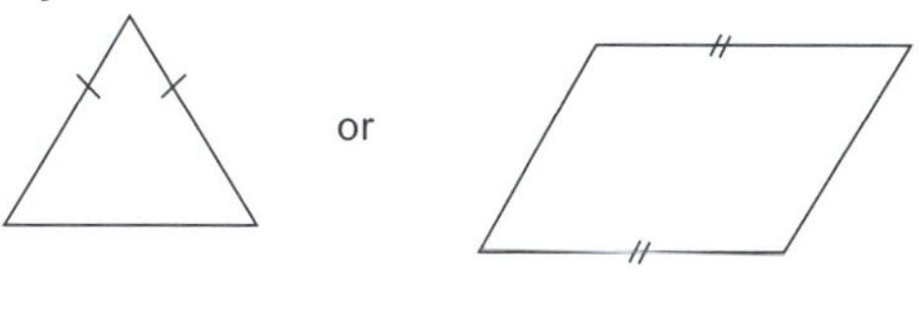

- To indicate that an angle is a right angle, mark it the following way:

- To indicate that two angles are equal, mark them in one of the following ways:

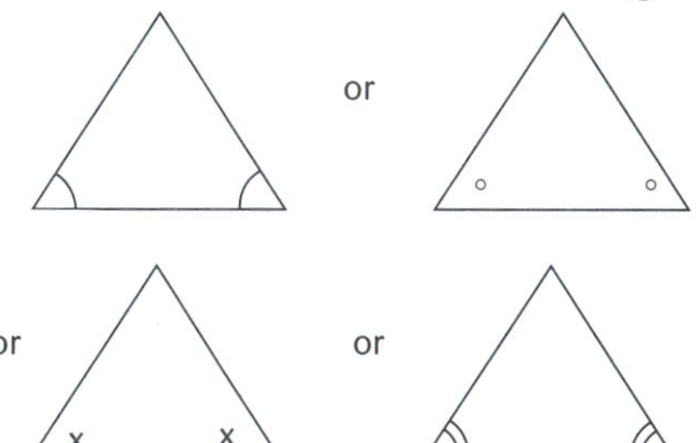

- To indicate that two lines are parallel, mark them as follows:

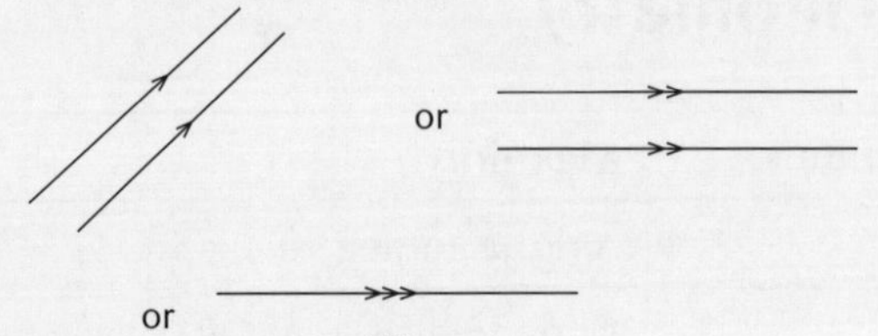

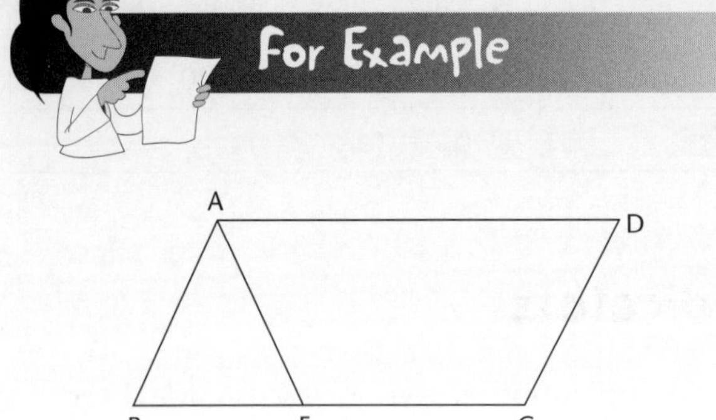

Draw the above figure in your book and mark on it the following information:

1 AB // DC

2 AB = AE

3 ∠ABE = ∠CDA

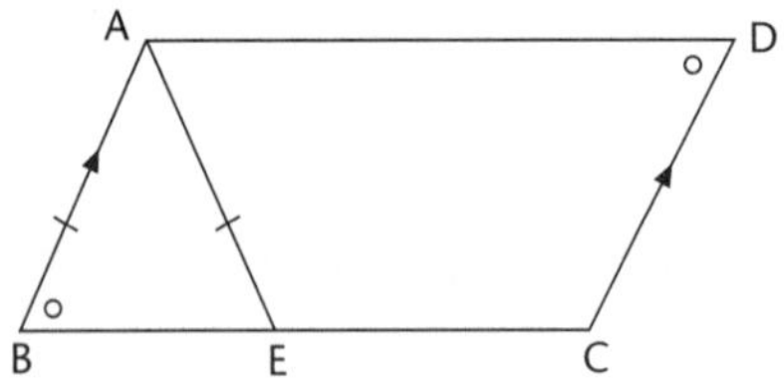

Polygons

A polygon is a closed figure with three or more straight sides.

For Example

Advise whether or not each shape is a polygon:

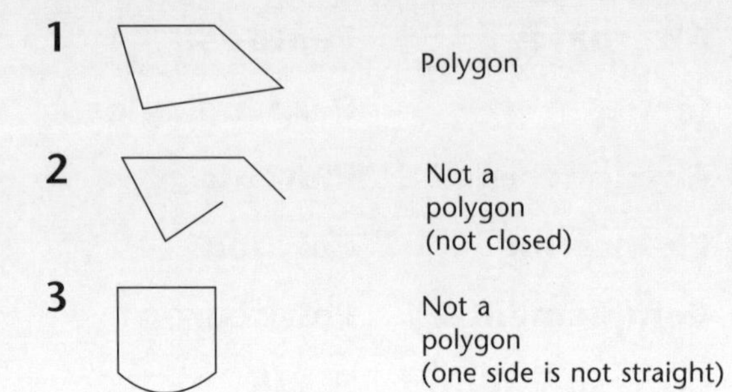

Naming Polygons

n stands for the number of sides of a polygon.

A polygon is named according to the number of sides it has:

Number of sides *n*	Name of polygon
3	Triangle
4	Quadrilateral
5	Pentagon
6	Hexagon
7	Heptagon
8	Octagon
9	Nonagon
10	Decagon
11	Hendecagon
12	Dodecagon
n	*n*-gon

Note: A polygon has the same number of angles as sides. For example, a hexagon has six angles and also six sides.

Regular Polygons

A polygon is **regular** if all its sides are equal in length and all its angles are equal.

Note:

1 The regular triangle is an **equilateral triangle**.

2 The regular quadrilateral is a **square**.

3 The regular five-sided polygon is called a **regular pentagon** etc.

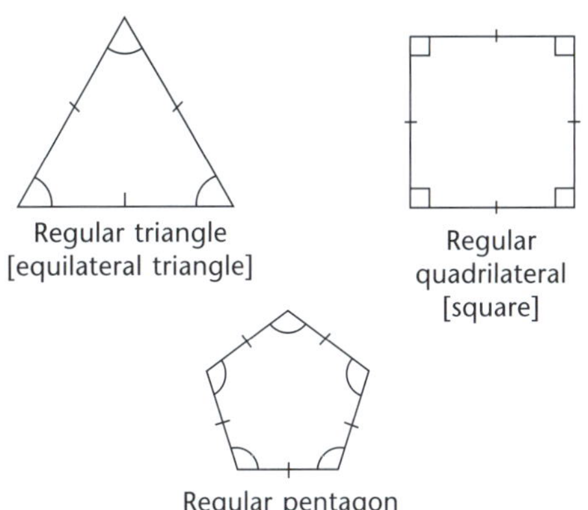

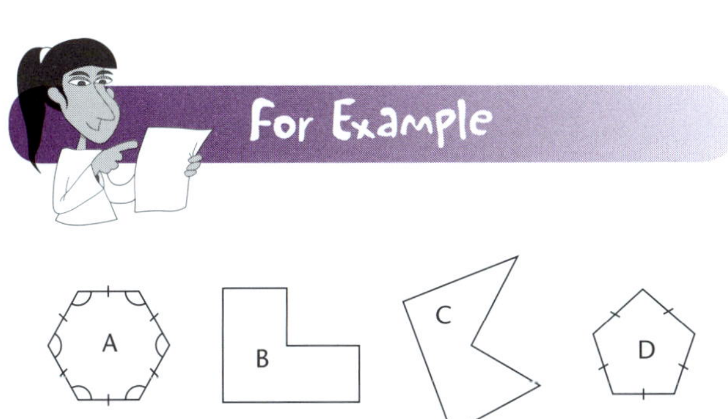

Which of the figures A, B, C or D represent:

1 Hexagons?

2 Regular hexagons?

1 A and B are hexagons, since they are polygons with six sides.

2 A is a regular hexagon, since all six sides and angles are equal.

Naming Points, Intervals, Lines, Polygons

- A **point** is named by using capital letters; for example, B.
- An **interval** is named by using capital letters; for example:

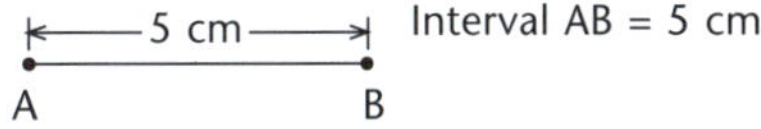

- A **line** is named by using any two points on it:

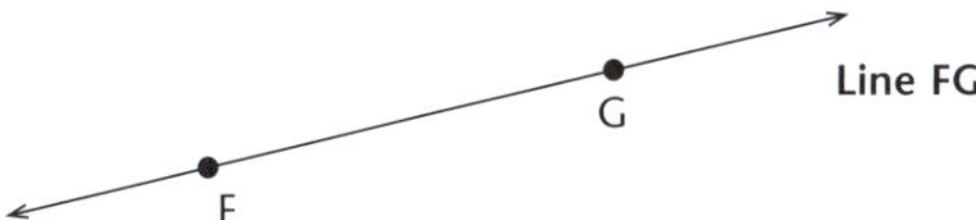

- **Polygons** are named by taking the vertices in order; for example:

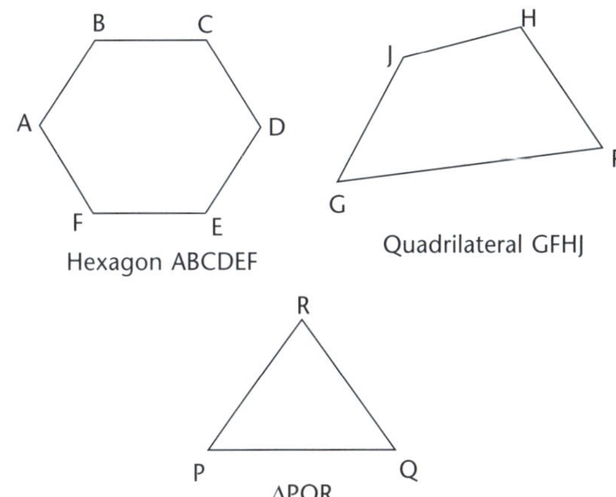

Note:

- △ stands for triangle.
- Three or more points on the one line are called **collinear points**.
- Three or more lines passing through the same point are called **concurrent lines**.

Lines

Two lines may be **parallel**, may **intersect** or may be **skew**.

- Parallel:

← These lines are parallel. (They are always the same distance apart.)

- Intersect at a point:

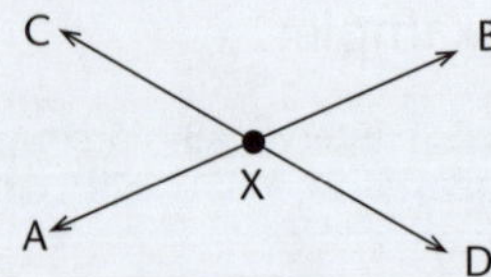

Line AB and line CD intersect at point X.

- Skew:

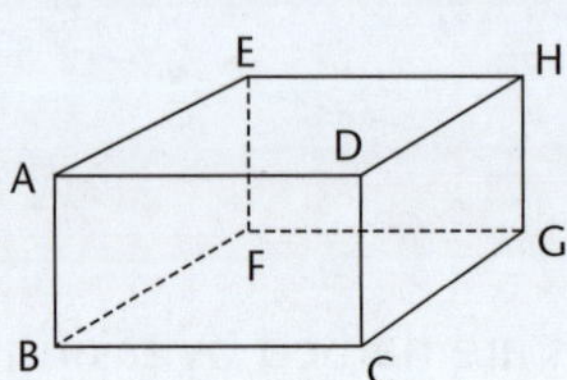

Line AB and line DH are skew. Skew lines do not intersect and are not parallel.

For Example

Name the lines in the figure that are:

1 Parallel to AD

2 Perpendicular to AD

3 Skew to AD.

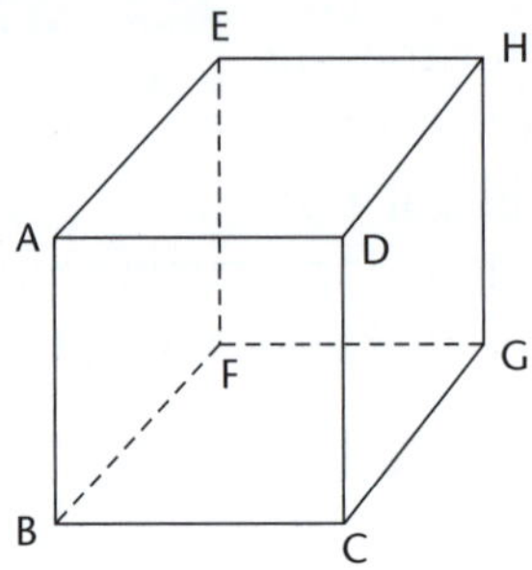

1 The lines parallel to AD are BC, EH and FG.

2 The lines perpendicular to AD are AB, DC, AE and DH.

3 The lines skew to AD are BF, EF, CG and GH.

Symmetry

A figure has **line symmetry** if one side of the figure is a mirror image of the other about a line.

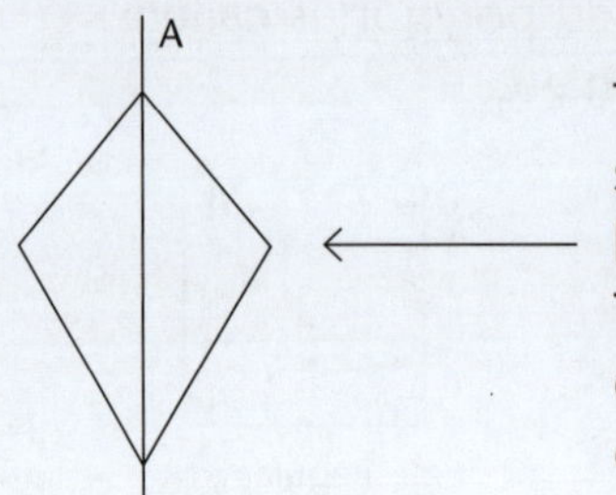

This figure is symmetrical about line AB. The line AB is called the **axis of symmetry**.

Some figures have more than one axis of symmetry and some have no axis of symmetry (i.e. they are not symmetrical).

A rectangle, for example, has two axes of symmetry:

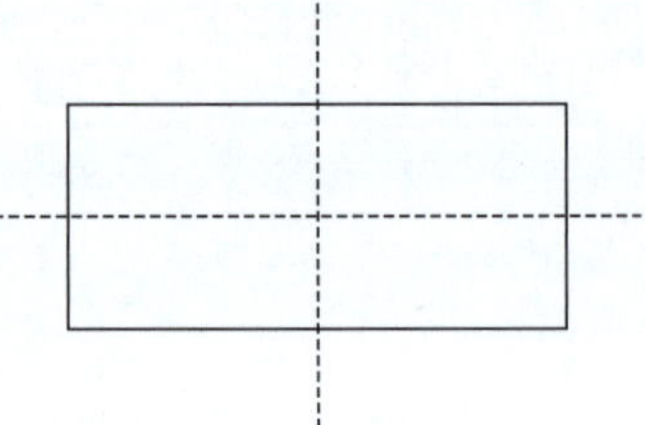

A parallelogram has no axis of symmetry:

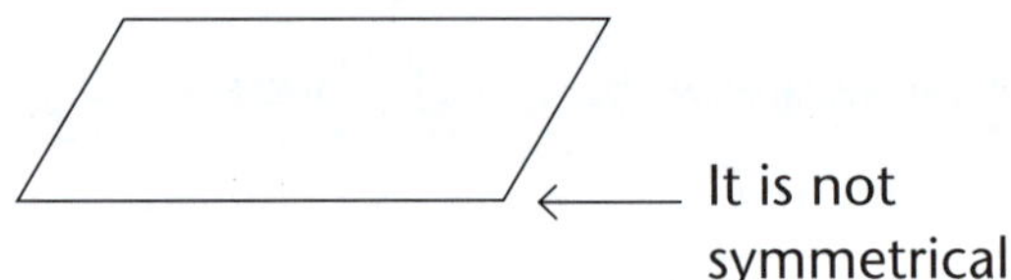

Order of Symmetry

The **order of symmetry** is the number of axes of symmetry of a figure.

An equilateral triangle has three axes of symmetry. This figure is said to have line symmetry of order three:

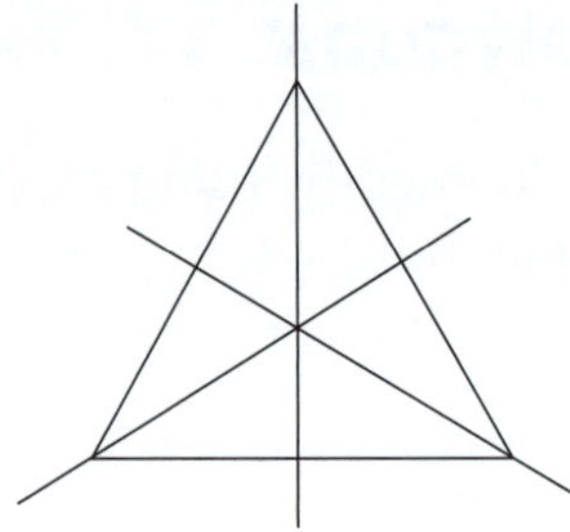

For Example

1 Draw a square and mark on it all axes of symmetry. What is the order of symmetry of a square?

1

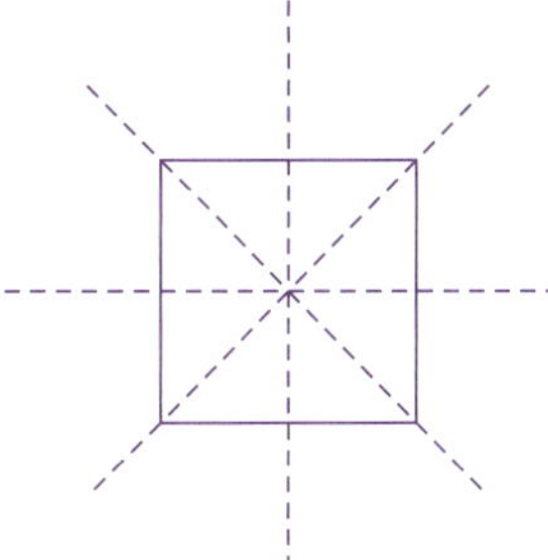

The order of symmetry = 4.

Classification of Triangles

A triangle is a polygon with three sides, and can be classified according to the lengths of its sides or the sizes of its angles.

Classification According to the Lengths of the Sides

Name	Diagram	Properties
Scalene		All sides are different lengths
Isosceles		Two sides are equal in length
Equilateral		All sides are equal in length

Classification According to the Size of the Angles

Name	Diagram	Properties
Acute-angled		All angles are acute
Obtuse-angled		One angle is obtuse
Right-angled		One angle is a right angle

Note: 'Acute' and 'obtuse' are defined in the section 'Angle Terminology'.

Note: A triangle can have the properties of two types of triangles. For example, it can be both isosceles and right-angled, as in this example:

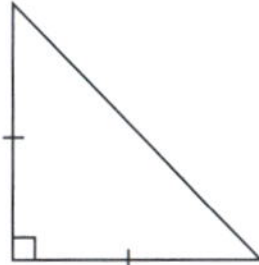

Classification of Quadrilaterals

Trapezium

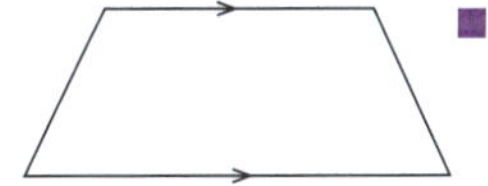

- A pair of sides are parallel

Parallelogram

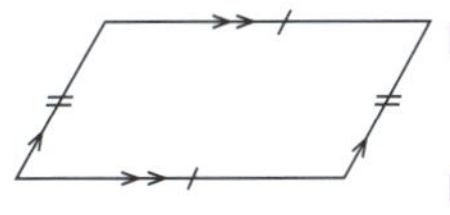

- Opposite sides are equal in length
- Opposite sides are parallel

Rhombus
(diamond shape)

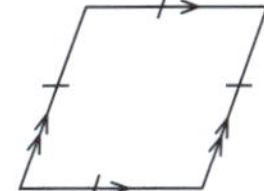

- All sides are equal in length
- Opposite sides are parallel

Rectangle

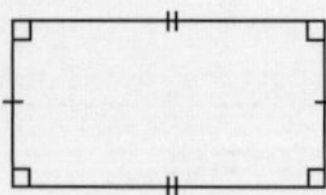

- Opposite sides are equal in length
- Opposite sides are parallel
- All angles are right angles

Square

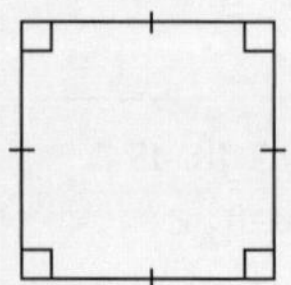

- All sides are equal in length
- Opposite sides are parallel
- All angles are right angles

Kite

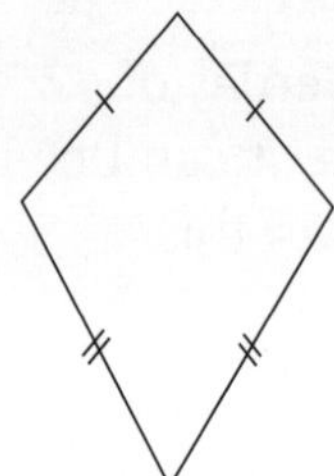

- Two pairs of adjacent sides are equal

See page 182 for more work on quadrilaterals.

Other Plane Shapes

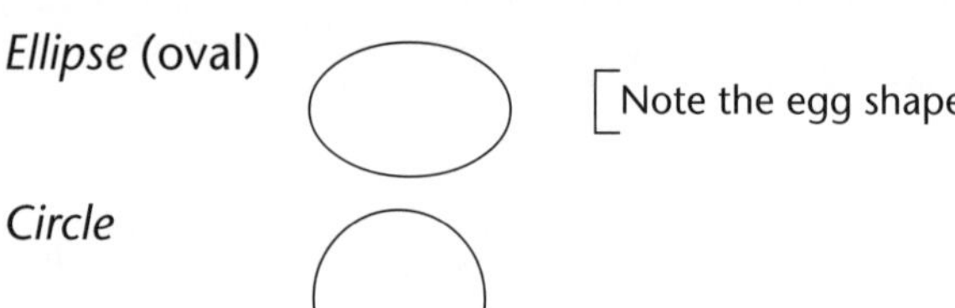

Parts of a Circle

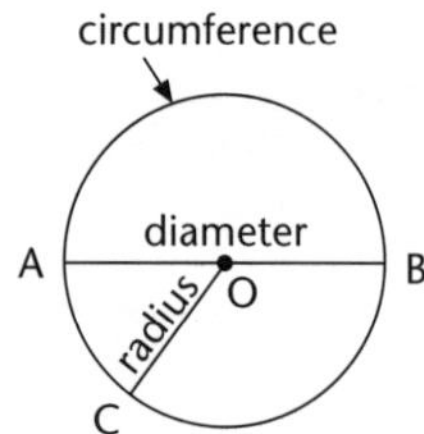

The centre of a circle is usually named O.

The circumference of a circle is its perimeter.

AB is the diameter.

OC is the radius. The radius is half the length of the diameter.

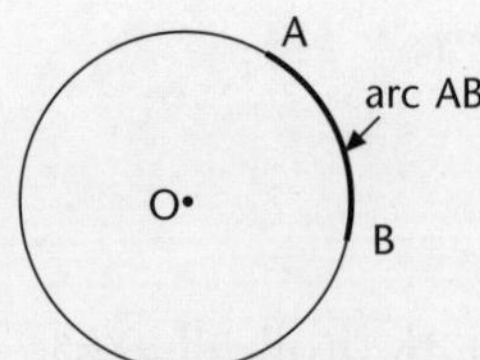

The length AB (on the circumference) is called an arc.

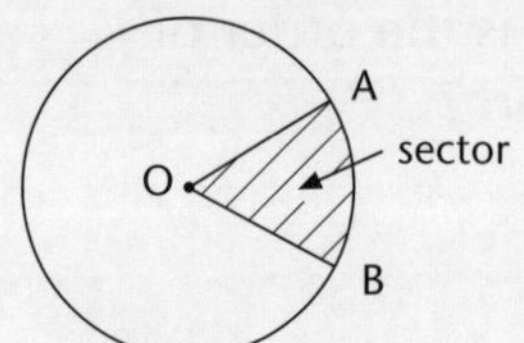

The area contained between two radii is called a sector.

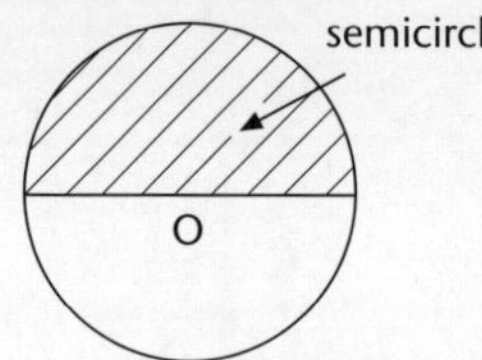

Half a circle is called a semicircle.

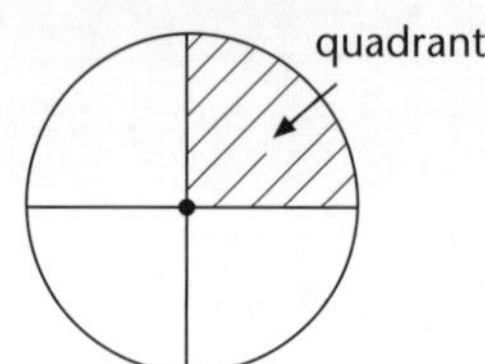

A quarter of a circle is called a quadrant.

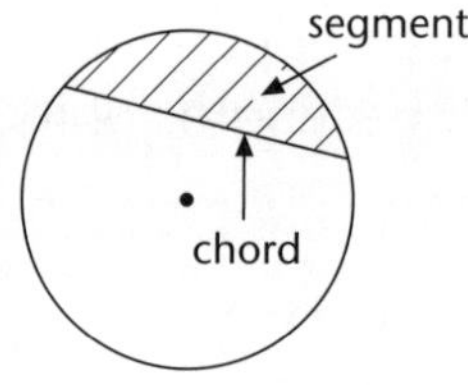

A chord joins any two points on the circumference.
A segment is the area contained between a chord and the edge of the circle.

Classification of Solids

An object that occupies space is called a solid; for example, a textbook, a ball, a cone. Solids are classified in the following way:

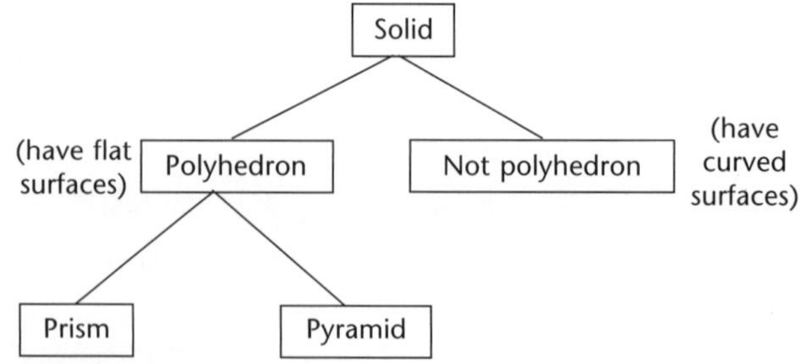

Definitions

A polyhedron is a solid made of plane surfaces with straight edges (flat surfaces).

Examples of polyhedra

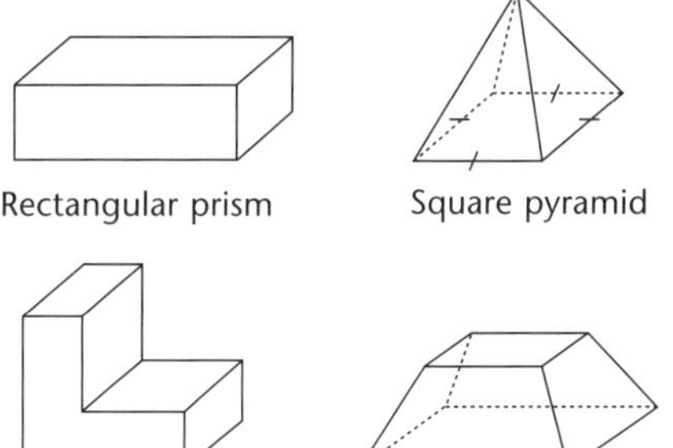

Rectangular prism

Square pyramid

Examples of non-polyhedra

Sphere

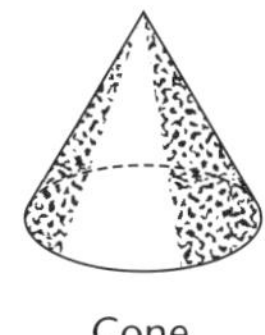

Cone

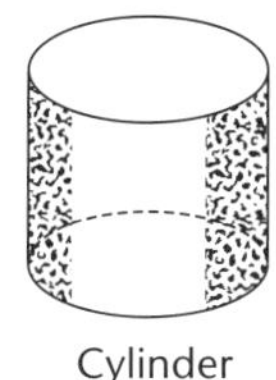

Cylinder

Note: Polyhedra is the plural for polyhedron.

Prisms

A prism is a solid that has a pair of parallel faces and, if 'sliced' parallel to these faces, the cross-sections are identical to the end faces:

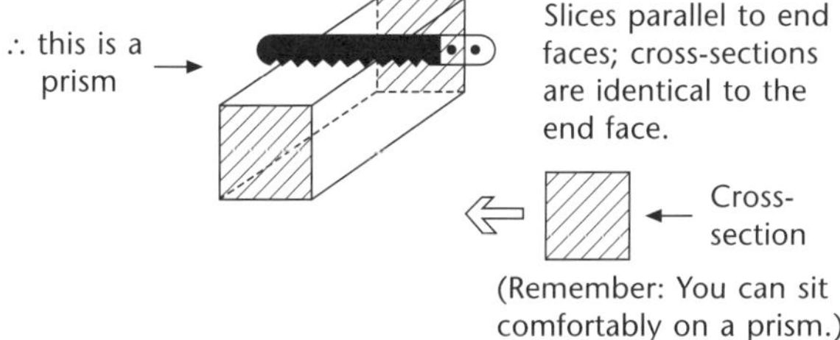

Note: A prism is named according to the shape of its cross-section.

Some familiar prisms

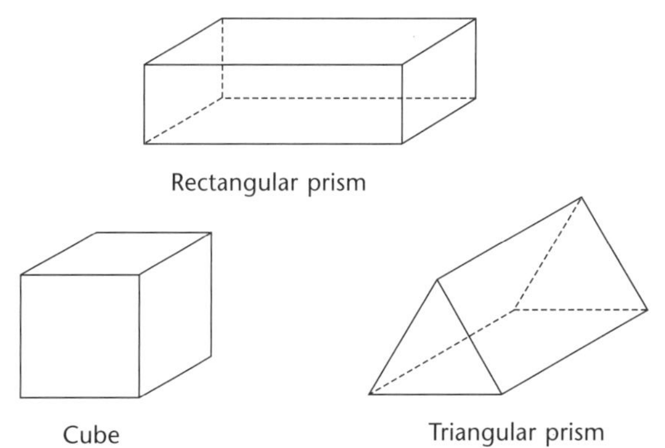

Rectangular prism

Cube

Triangular prism

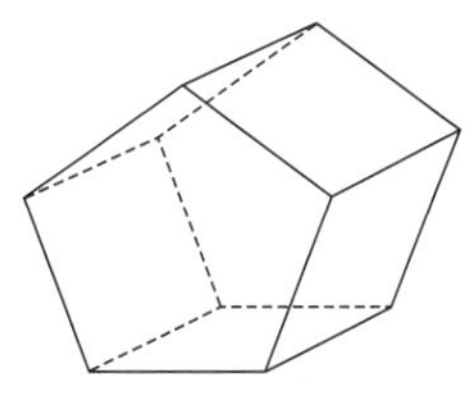

Pentagonal prism

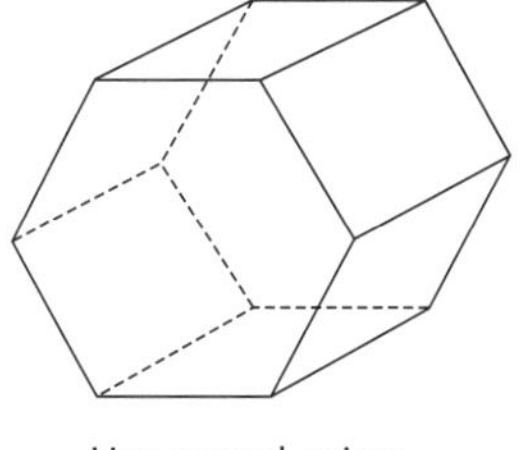

Hexagonal prism

Pyramids

A pyramid is named according to the shape of its base.

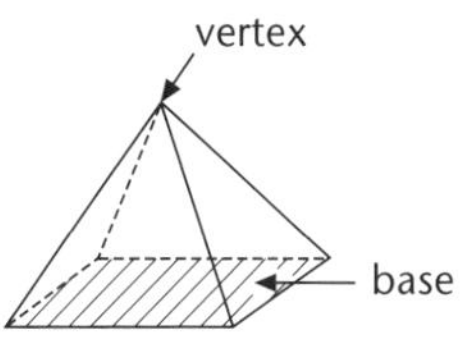

Some familiar pyramids

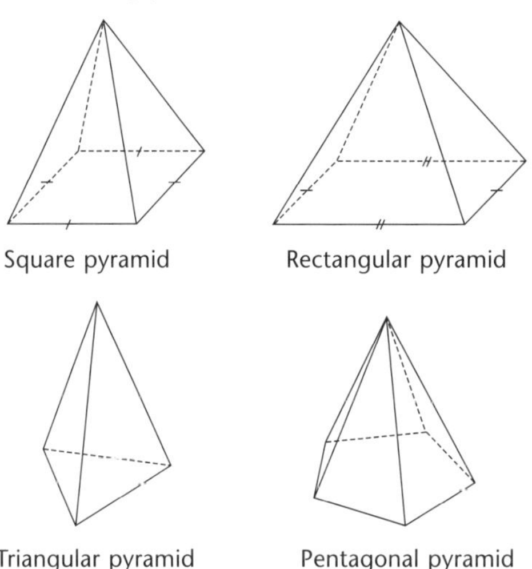

Square pyramid

Rectangular pyramid

Triangular pyramid

Pentagonal pyramid

Platonic Solids

Platonic solids are polyhedra whose faces are all congruent (identical) regular polygons. The most common example is a hexahedron (cube) whose faces are six identical squares.

Another example of a platonic solid is a regular tetrahedron, which is a triangular pyramid whose faces are four identical equilateral triangles.

The platonic solids are named after the ancient Greek philosopher and teacher Plato:

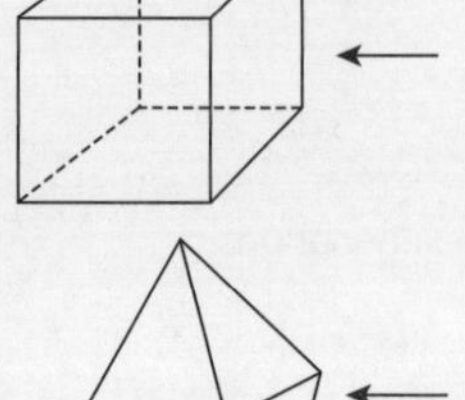

A cube is a platonic solid. It is also called a hexahedron.

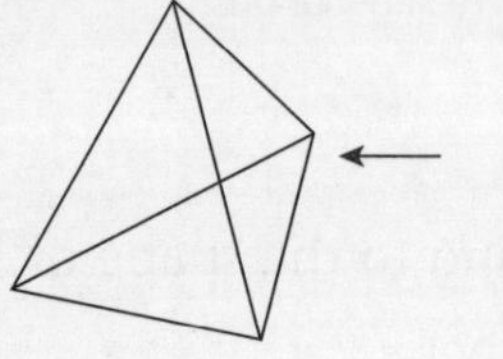

A triangular pyramid with four identical equilateral triangles as faces is called a tetrahedron.

Isometric Sketches of Solids

Solids are difficult to represent on paper, due to their three-dimensional qualities. Sketching them on isometric dot paper allows the solid to be drawn showing its true perspective.

For Example

1 Draw the following solid on isometric dot paper:

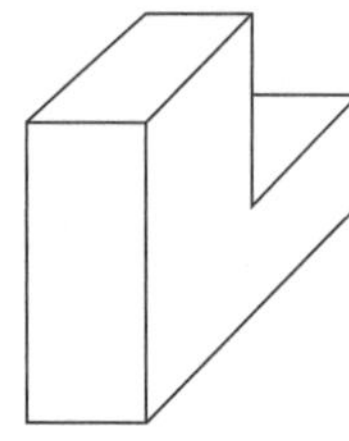

2 Draw a new solid by removing the shaded solid:

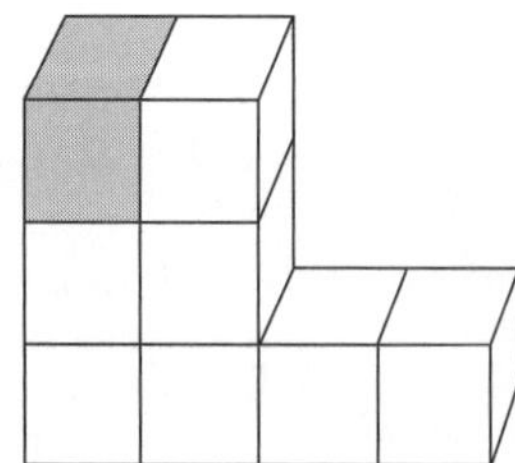

1

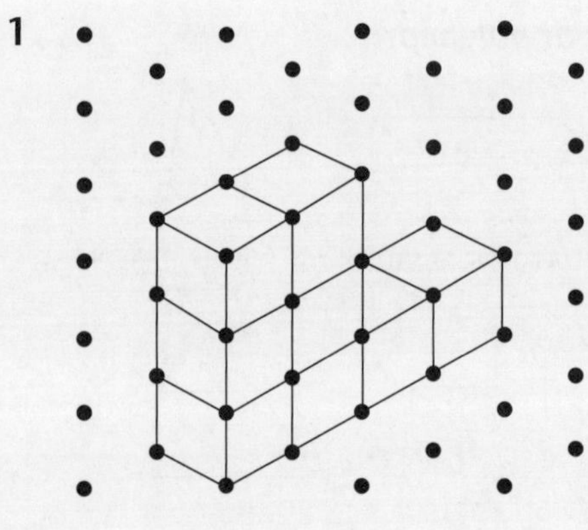

2

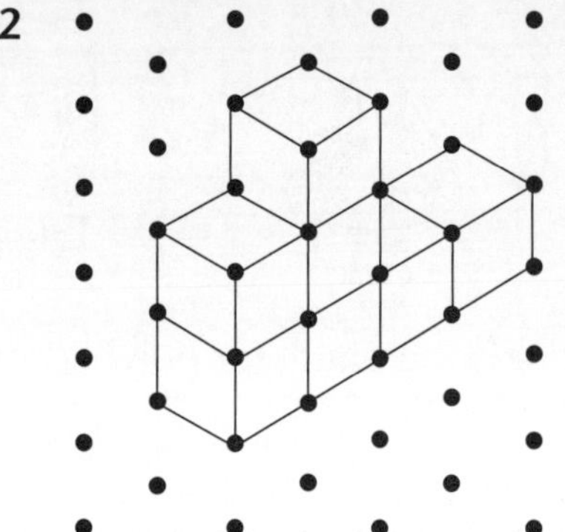

Sketching Solids from Different Views

Another method of showing the true perspective of a solid is to draw it from different views. The most common views of a solid are front, side and top views:

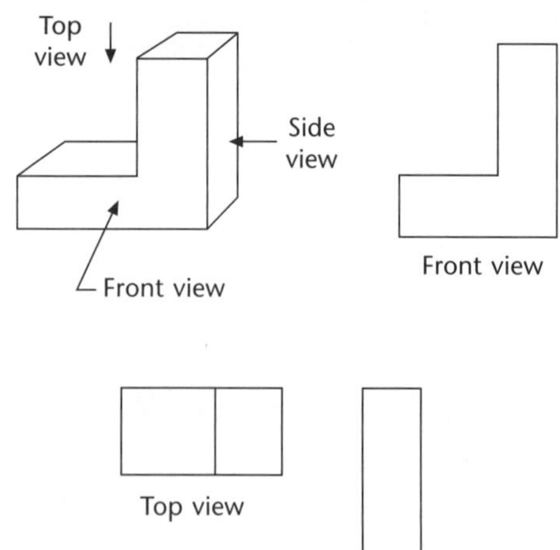

1 Consider the following solid:

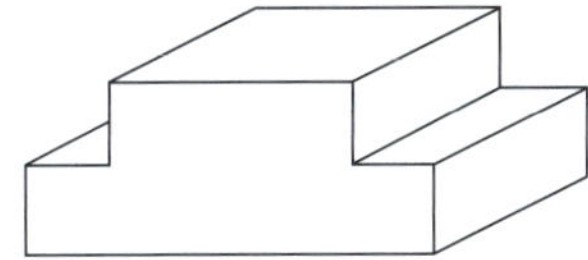

Draw the front, side and top views of the solid.

1

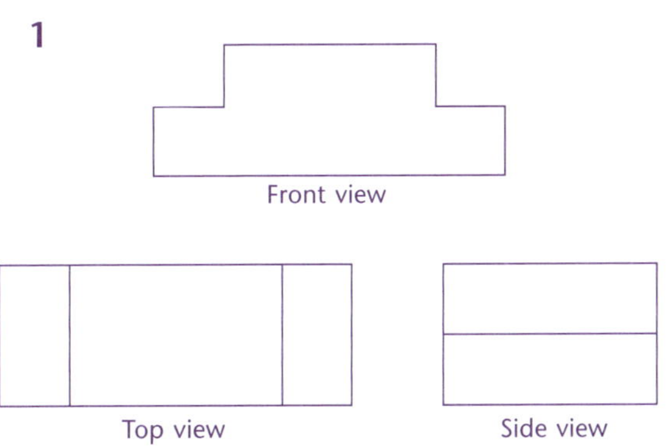

Nets of Solids

The net of a solid is a plan that, when folded, forms the solid:

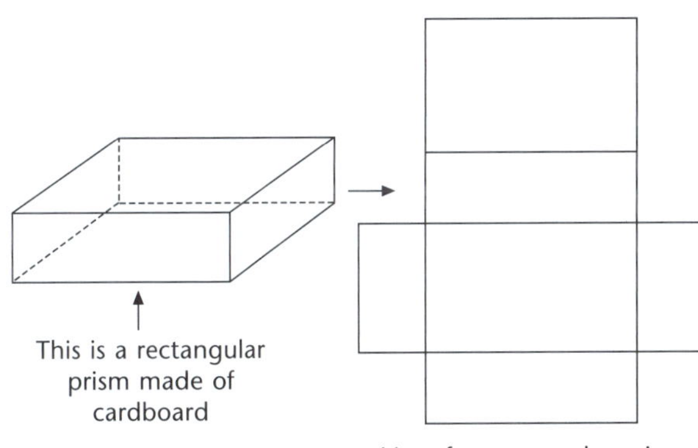

Parts of Polyhedra

- Each surface of a polyhedron is called a **face**.
- The line in which two faces meet is called an **edge**.
- The point where edges meet is called a **vertex** (corner point).

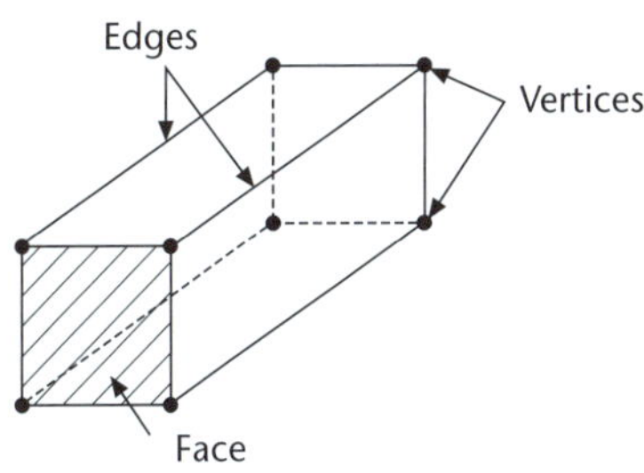

Note: Vertices is the plural for vertex.

Angle Terminology

Naming Angles

When naming angles, we start from arm → vertex → arm.

For example, the angle in the diagram below is ∠ABC or ∠CBA:

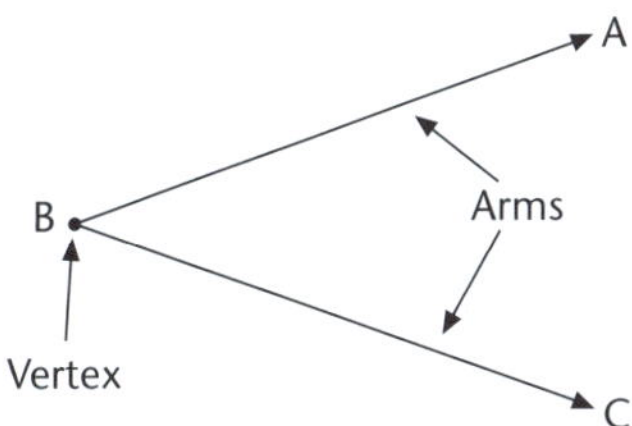

Notation

For the angle in the above diagram, we can use any of the following notations:

- ∠ABC
- $A\hat{B}C$
- $\hat{B}$ or ∠B

Types of Angles

We name angles as follows:

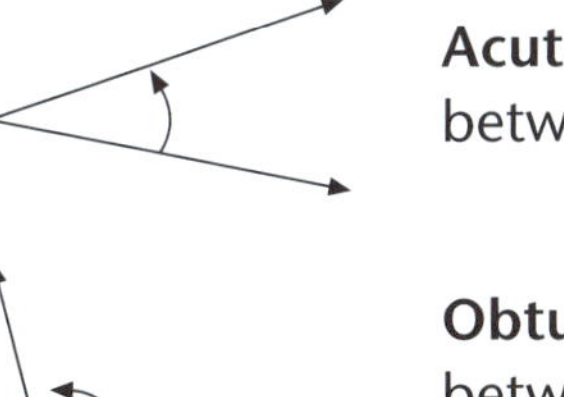

Acute angle—measures between 0° and 90°.

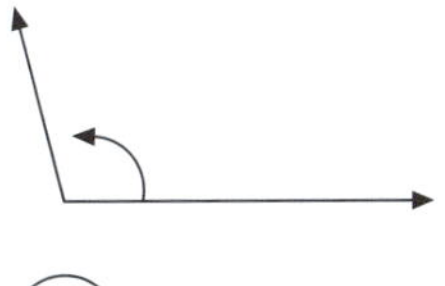

Obtuse angle—measures between 90° and 180°.

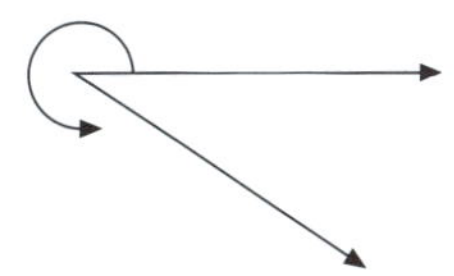

Reflex angle—measures between 180° and 360°.

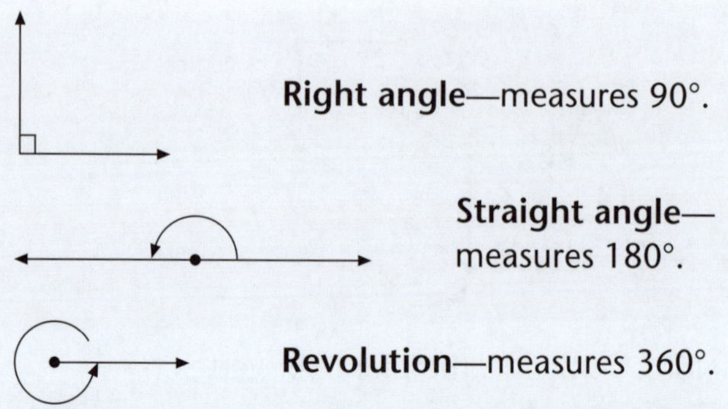

Complementary angles—two angles that add up to 90°.

Supplementary angles—two angles that add up to 180°.

Adjacent angles—two angles are adjacent if they have a common arm and common vertex. In the diagram, $\angle ABD$ is adjacent to $\angle DBC$. (BD is the common arm, and B is the common vertex.)

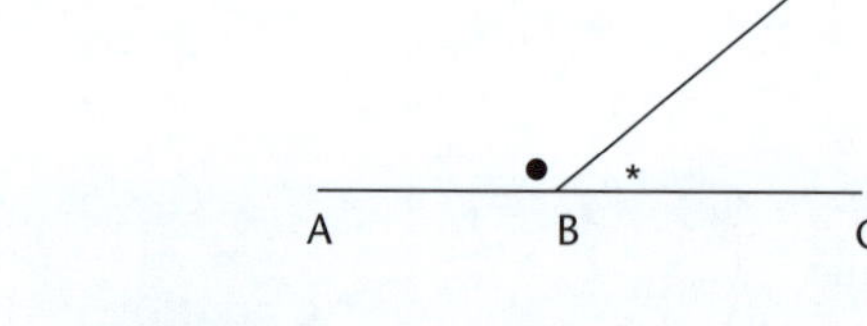

For Example

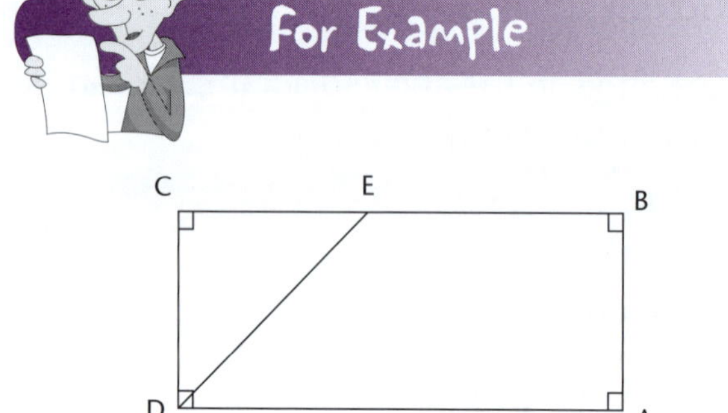

In the figure, ABCD is a rectangle.

1. From the diagram, name:
   - **a** An acute angle
   - **b** An obtuse angle
   - **c** A right angle
   - **d** A straight angle
   - **e** A pair of complementary angles
   - **f** A pair of supplementary angles.

2. From the diagram, answer these questions:
   - **a** If $\angle CDE = 40^\circ$, find the size of $\angle EDA$.
   - **b** If $\angle BED = 130^\circ$, find the size of $\angle DEC$.

3. **a** Find the complement of 70°.
   - **b** What is the supplement of 70°?

1 **a** $\angle CDE$ is an acute angle (also $\angle EDA$, $\angle CED$).

b $\angle BED$ is an obtuse angle.

c There are four right angles in the diagram, $\hat{A}$, $\hat{B}$, $\hat{C}$ and $\angle CDA$.

d $\angle BEC$ is a straight angle.

e $\angle CDE$ and $\angle EDA$ are complementary since $\angle CDE + \angle EDA = 90^\circ$.

f $\angle BED$ and $\angle DEC$ are adjacent supplementary angles since $\angle BED + \angle DEC = 180^\circ$.

2 **a** From 1e, $\angle CDE + \angle EDA = 90^\circ$
(They are complementary angles.)
If $\angle CDE = 40^\circ$, then $40^\circ + \angle EDA = 90^\circ$
$\therefore \angle EDA = 50^\circ$

b From 1f, $\angle BED + \angle DEC = 180^\circ$
(They are supplementary angles.)
If $\angle BED = 130^\circ$, then

$130^\circ + \angle DEC = 180^\circ$

$\therefore \angle DEC = 50^\circ$

3 **a** To find the **complement** of 70° we find an angle that, when added to 70°, makes 90°. Therefore, the complement of 70° = 20°.

[The complement of $x^\circ = (90 - x)^\circ$]

b To find the **supplement** of 70° we find an angle that, when added to 70°, makes 180°. Therefore, the supplement of 70° = 110°.

[The supplement of $x^\circ = (180 - x)^\circ$]

Further Angles

Angles on a straight line

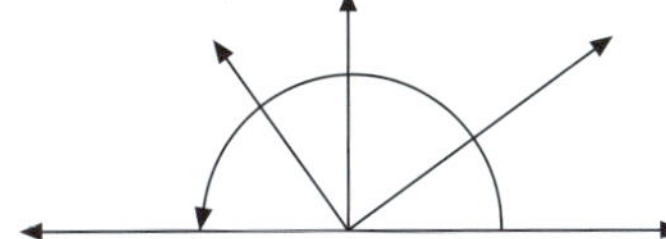

Angles on a straight line add up to 180°.

For Example

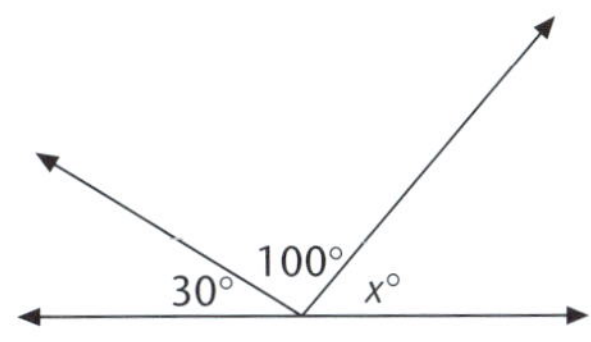

1 Find x.

1 $x + 100 + 30 = 180$

(angles on a straight line)

$\therefore x = 50$

Vertically opposite angles

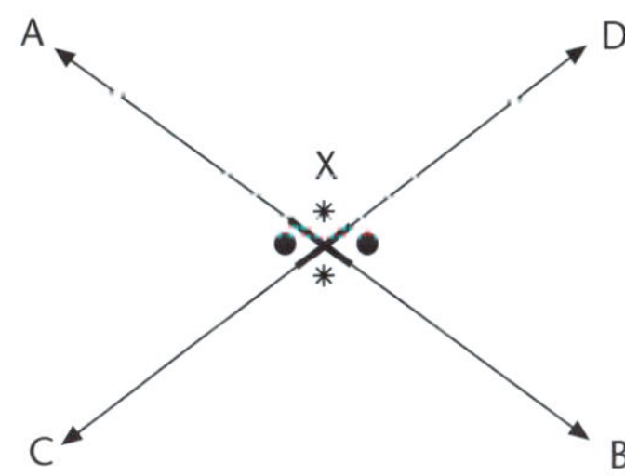

[Lines AB and CD intersect at X.]

When two lines intersect, two pairs of vertically opposite angles are formed. Vertically opposite angles are equal.

∠CXA and ∠DXB are vertically opposite. Also, ∠BXC and ∠AXD are vertically opposite.

For Example

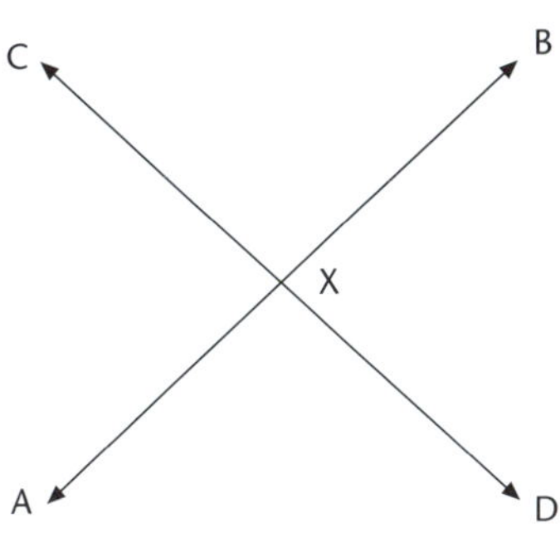

[AB intersects CD at X]

1 Name the angle vertically opposite to ∠BXD.

2 If ∠AXC = 112°, find the size of ∠BXD.

3 Does ∠CXB = ∠DXA? Why?

1 The angle vertically opposite to ∠BXD is ∠AXC.

2 ∠BXD = ∠AXC (because they are vertically opposite) ∴ ∠BXD = 112°

3 ∠CXB = ∠DXA because they are vertically opposite angles (and vertically opposite angles are equal).

For Example

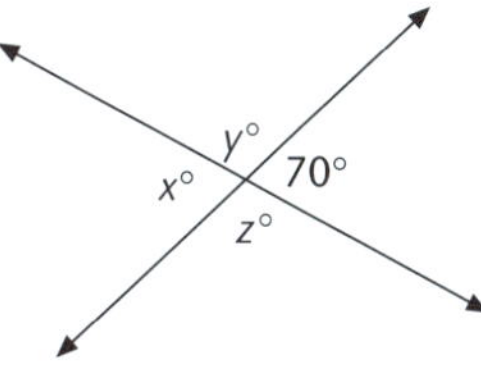

1 Find the values of x, y and z.

1 $x = 70$ (vertically opposite angles)

$y + 70 = 180$ (supplementary angles)

$\therefore y = 110, z = 110$

Angles at a point

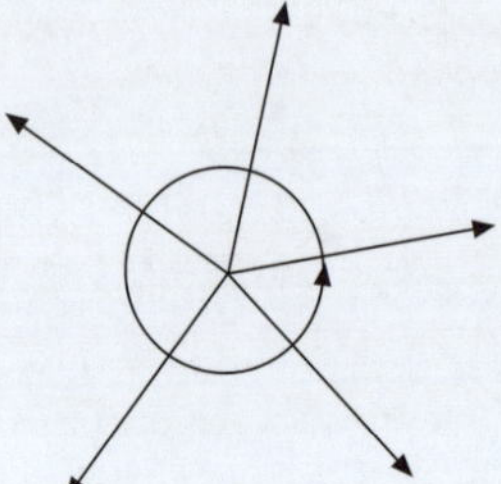

Angles at a point add up to 360° (make up one revolution).

For Example

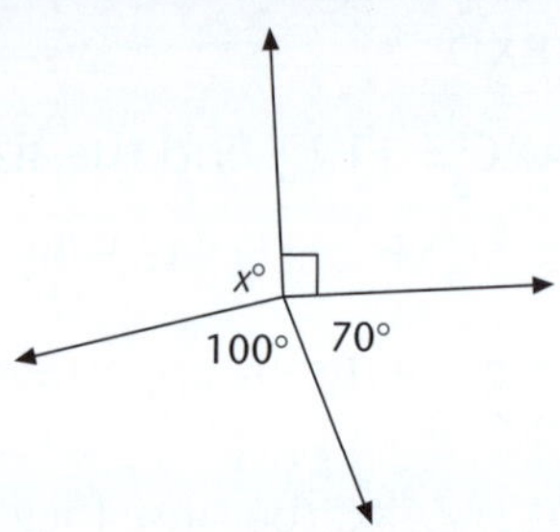

1 Find *x*.

1 $x + 100 + 70 + 90 = 360$

(angles at a point)

$\therefore\ x + 260 = 360$

$x = 100$

Parallel Lines

When a pair of parallel lines is cut by a transversal, three types of special angles are formed: F angles, Z angles and C angles.

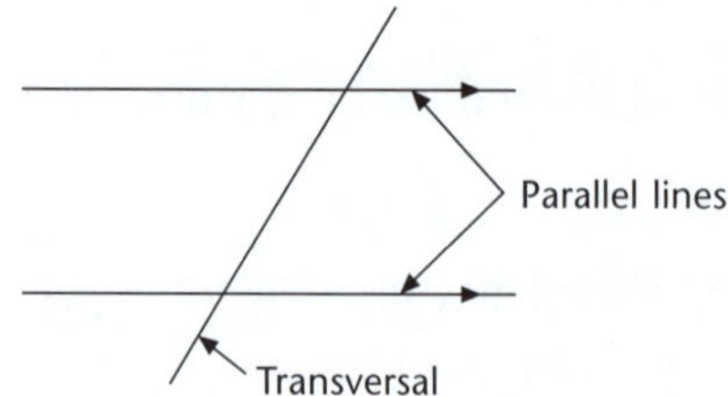

1

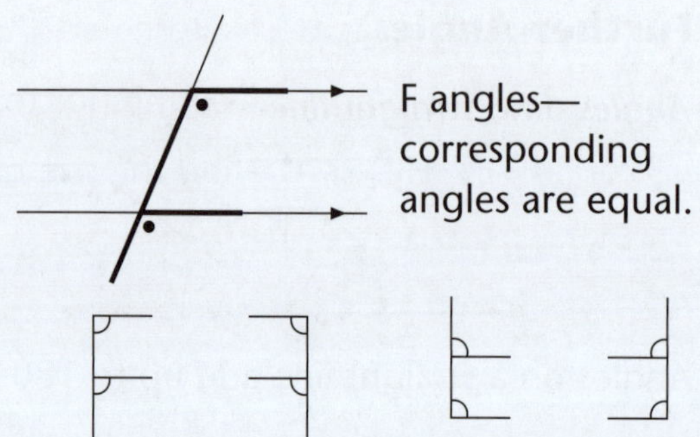

F angles—corresponding angles are equal.

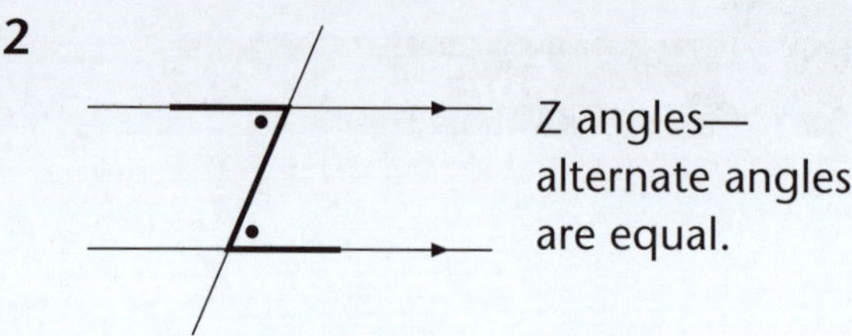

2

Z angles—alternate angles are equal.

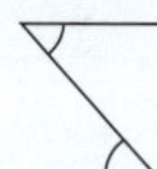

3

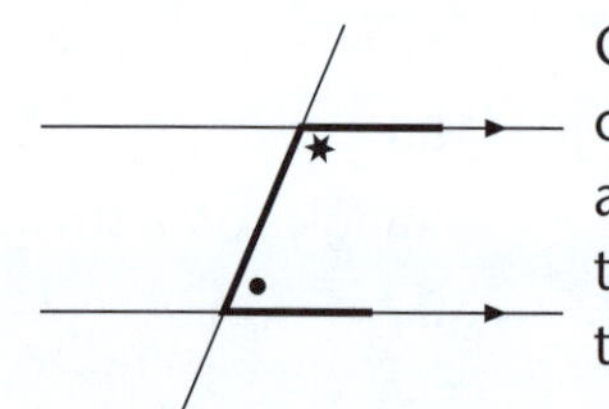

C angles—co-interior angles are supplementary; that is, they add up to 180°.

For Example

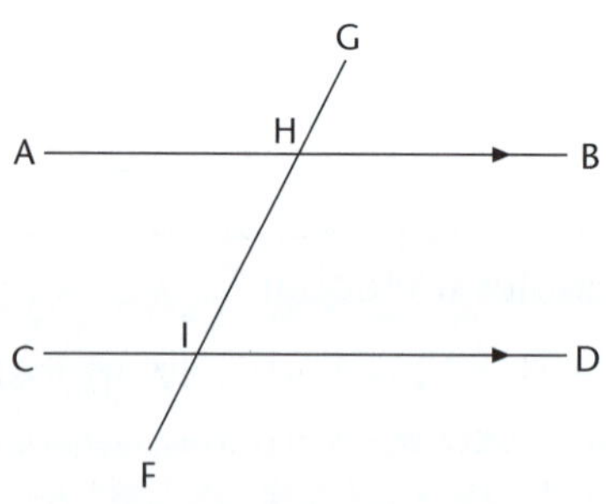

1 In the figure, AB // CD and AB and CD are cut by the transversal FG:

a Which angle is corresponding to ∠HID?

b Which angle is alternate to ∠HID?

c Which angle is co-interior to ∠HID?

d If ∠HID = 70°, find the size of ∠BHI.

2 **a** Find x.

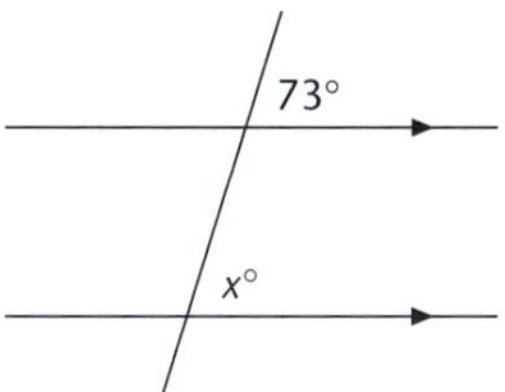

b Find y.

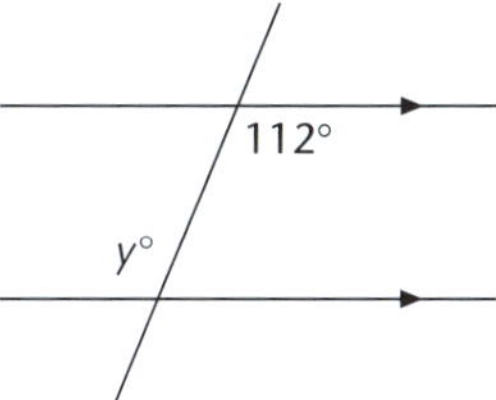

c Find x.

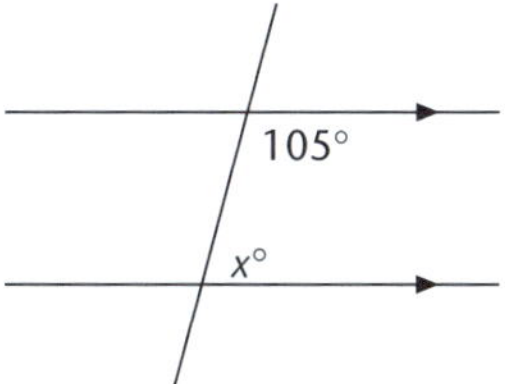

3 **a**

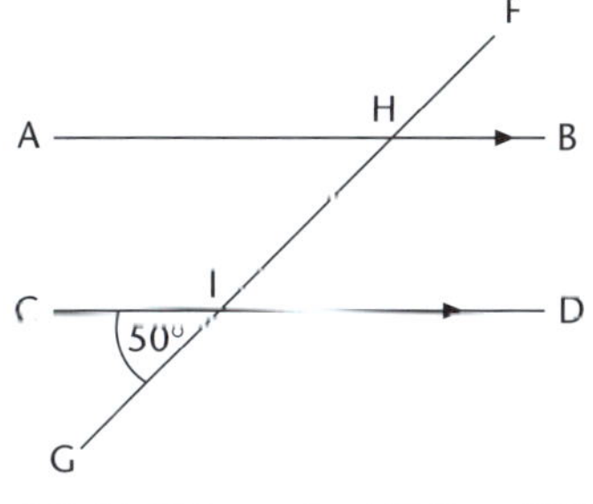

AB // CD and ∠GIC = 50°. Find the size of ∠BHI.

b

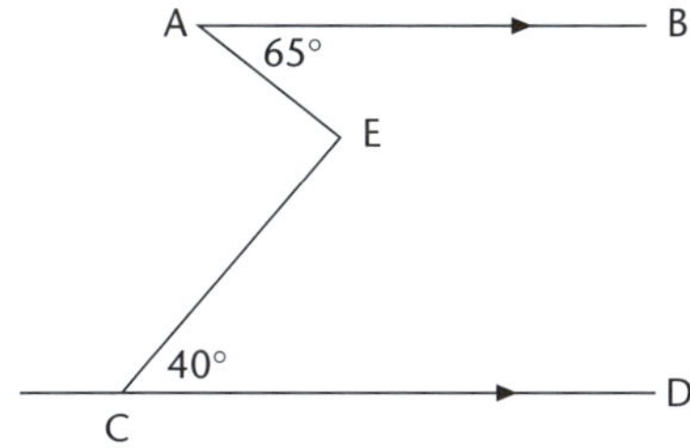

AB // CD. Find the obtuse angle ∠CEA.

1 **a** The angle corresponding to ∠HID is ∠GHB:

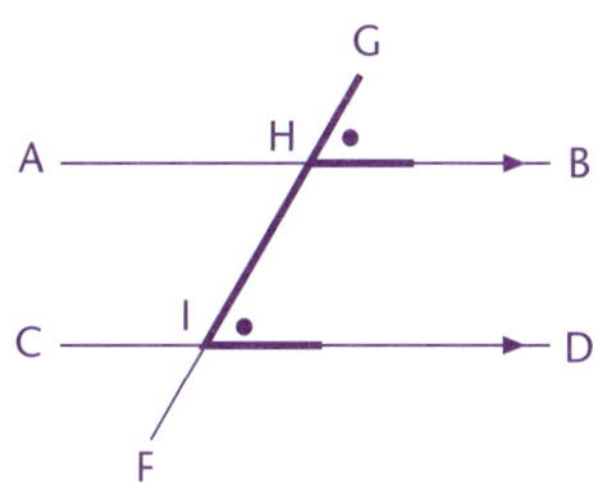

Remember that corresponding angles or alternate angles are only equal when the lines are parallel.

b The angle alternate to ∠HID is ∠IHA:

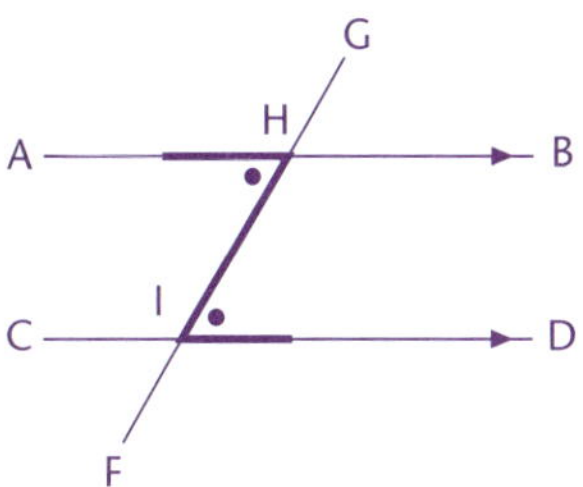

c The angle co-interior to ∠HID is ∠BHI:

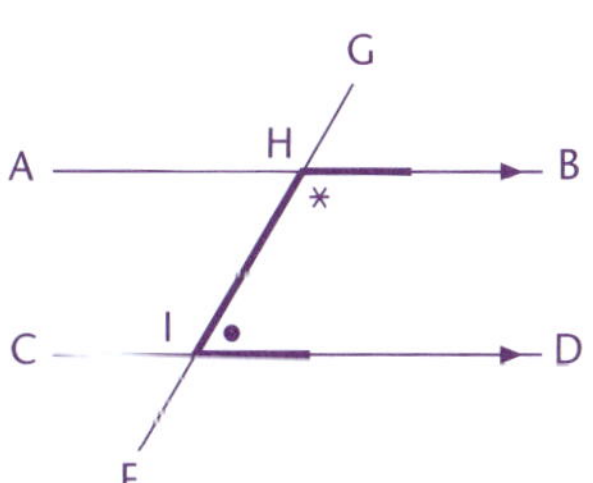

d

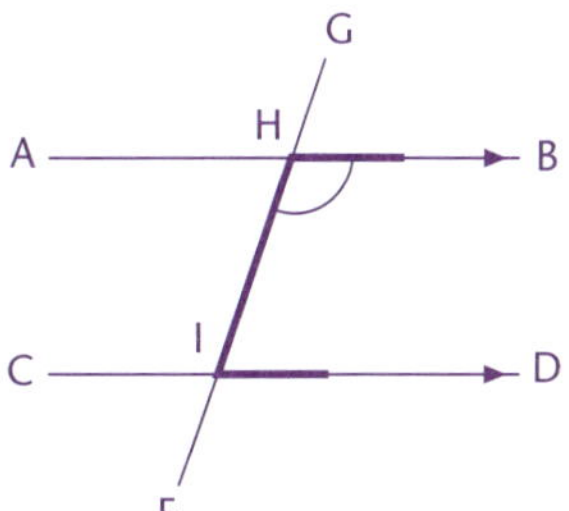

∠BHI + ∠HID = 180° (co-interior angles, AB // CD)

∴ ∠BHI + 70° = 180°

∠BHI = 110°

Remember, co-interior angles are only supplementary when the lines are parallel.

2 **a** $x = 73$ (corresponding angles in parallel lines)

b $y = 112$
(alternate angles in parallel lines)

c $x + 105 = 180$
(co-interior angles in parallel lines)
$\therefore x = 75$

3 **a**

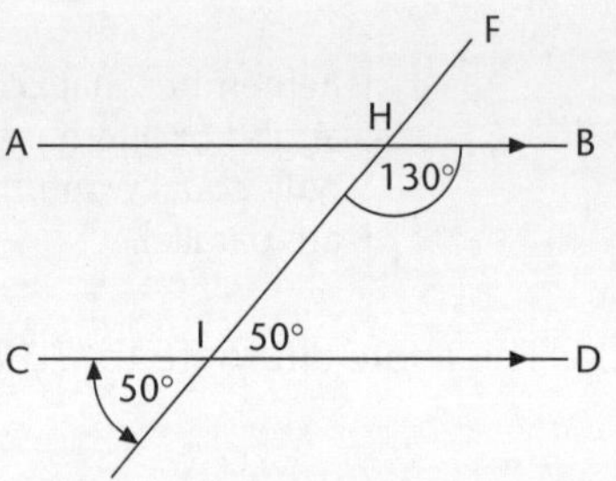

$\angle$HID = 50° (vertically opposite to $\angle$GIC)
$\angle$BHI = 130° (co-interior to $\angle$HID, and AB // CD)

b

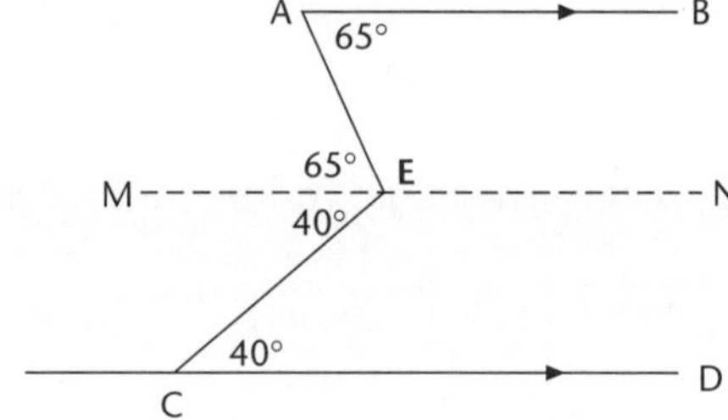

Construct a line MN through E that is parallel to AB. Since MN // AB and AB // CD, then MN // CD.

$\therefore$ $\angle$MEA = 65° (alternate to $\angle$BAE, AB // MN)

Also, $\angle$CEM = 40° (alternate to $\angle$ECD, CD // MN)

$\therefore$ $\angle$CEA = (65 + 40) ° = 105°

Conditions for Parallel Lines

Two straight lines are parallel if a transversal makes with them:

i A pair of equal corresponding angles, or

ii A pair of equal alternate angles, or

iii A pair of supplementary co-interior angles.

Also, two straight lines are parallel if they are both parallel to a third line.

For Example

1 Is AB // CD?

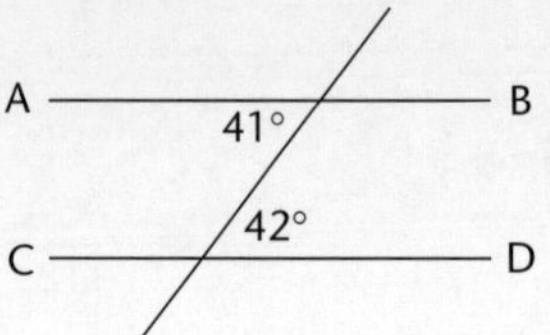

2 Are the lines XY and PQ parallel?

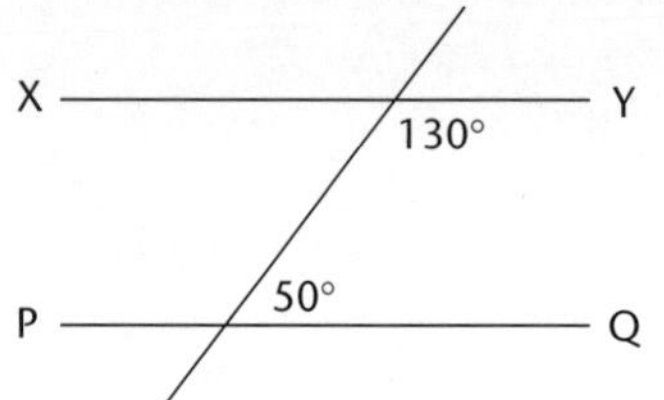

1 AB is not parallel to CD, because the pair of alternate angles are not equal.

2 XY and PQ are parallel, because the pair of co-interior angles are supplementary.

That is, 130° + 50° = 180°.

Triangles

Properties of Triangles

Angle sum of a triangle

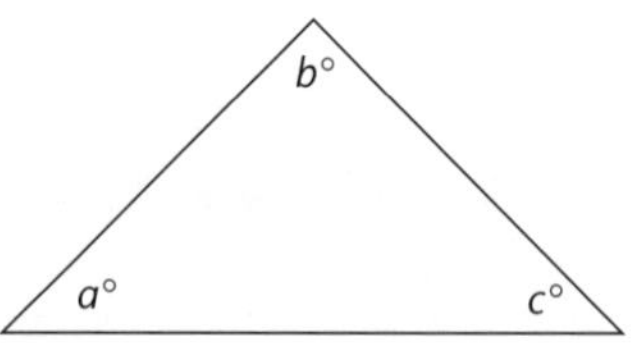

The angle sum of any triangle is 180°.

$[a + b + c = 180]$

Isosceles triangle

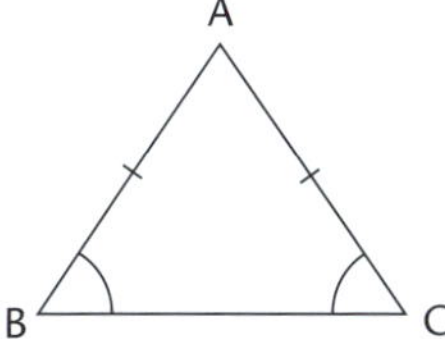

The base angles of an isosceles triangle are equal.

[∠ABC = ∠BCA]

Equilateral triangle

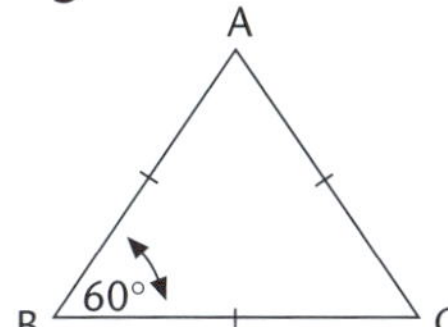

Each angle in an equilateral triangle is 60°.

[∠A = ∠B = ∠C = 60°]

For Example

1 Find x.

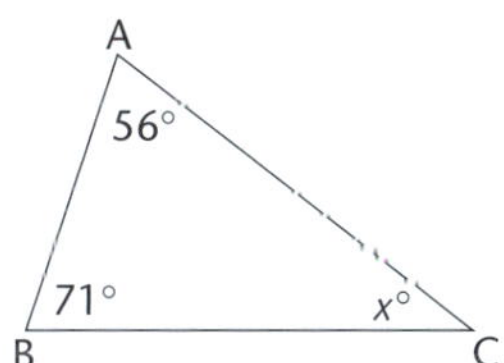

2 Find x.

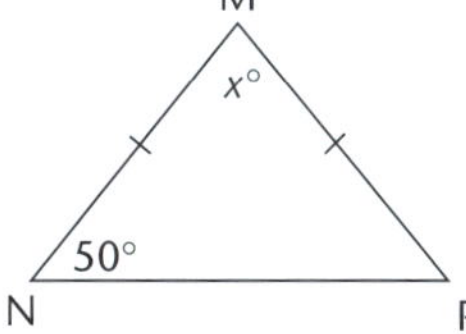

3 Find x.

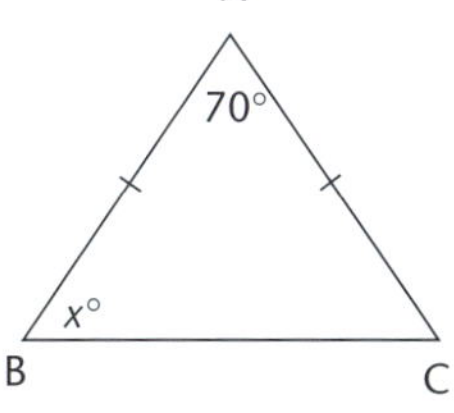

1 $x + 71 + 56 = 180$ (angle sum of △ABC)

$x + 127 = 180$

$x = 53$

2 ∠NPM = 50°

(base angles of isosceles △MNP)

$\therefore x + 50 + 50 = 180$

(angle sum of △MNP)

$\therefore x = 80$

3 ∠ABC = ∠BCA

(base angles of isosceles △ABC)

$x + x + 70 = 180$ (angle sum of △ABC)

$2x = 110$

$\therefore x = 55$

Exterior Angle of a Triangle (exterior = outside)

The exterior angle of any triangle equals the sum of the two opposite interior angles:

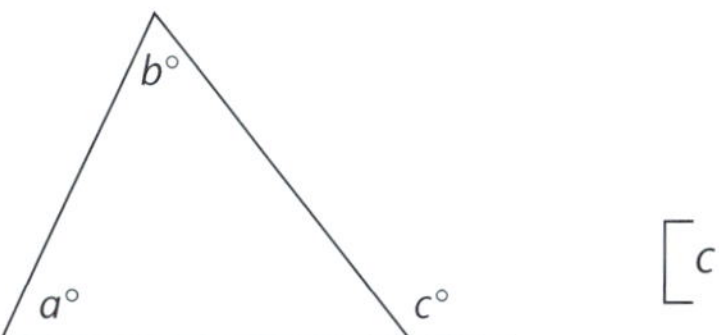

[$c = a + b$]

Note: The opposite interior angles are the two that are **not** adjacent to the exterior angle.

For Example

1 Find x.

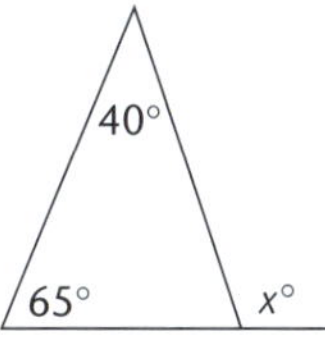

2 Find x.

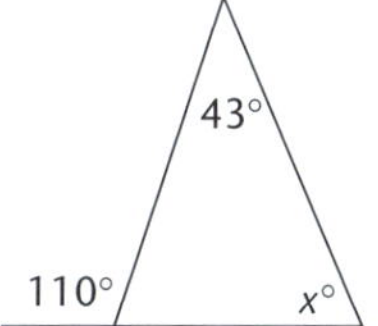

1 $x = 65 + 40$ (exterior $\angle$ of $\triangle$ result)

$\therefore x = 105$

2 $x + 43 = 110$ (exterior $\angle$ of $\triangle$ result)

$\therefore x = 110 - 43$

$\therefore x = 67$

Quadrilaterals

Angle Sum of Quadrilaterals

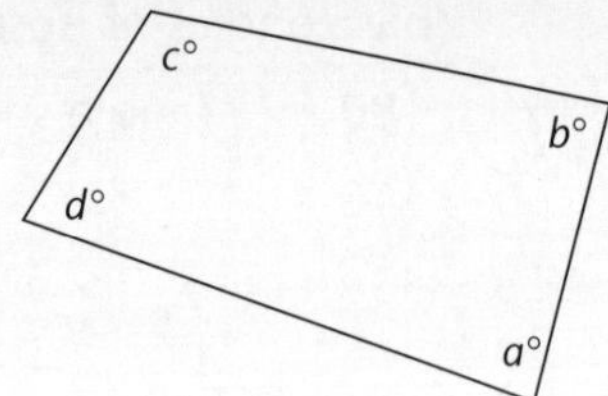

The angle sum of any quadrilateral is 360°.

That is: $a + b + c + d = 360$

Properties of Special Quadrilaterals

A **trapezium** is a quadrilateral with one pair of opposite sides parallel:

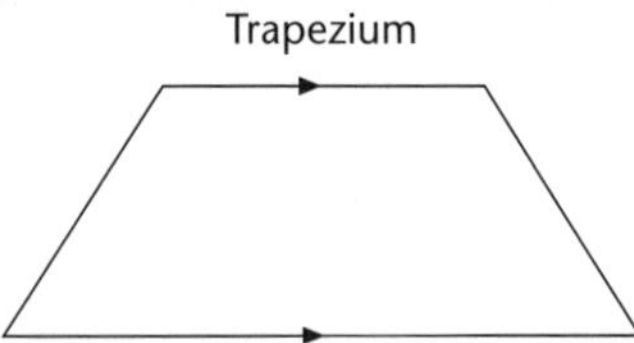

A **parallelogram** is a quadrilateral with both pairs of opposite sides parallel and equal:

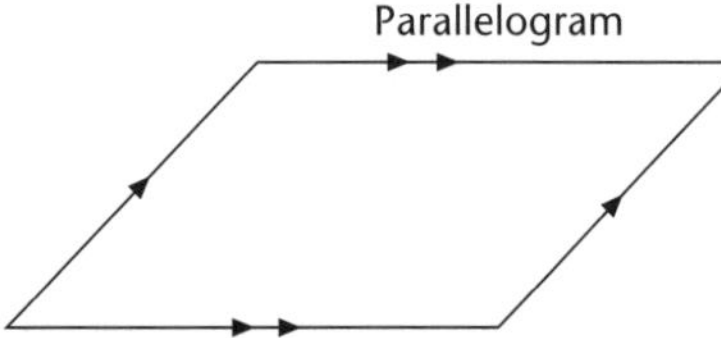

A **rhombus** is a quadrilateral with both pairs of opposite sides parallel and all sides equal:

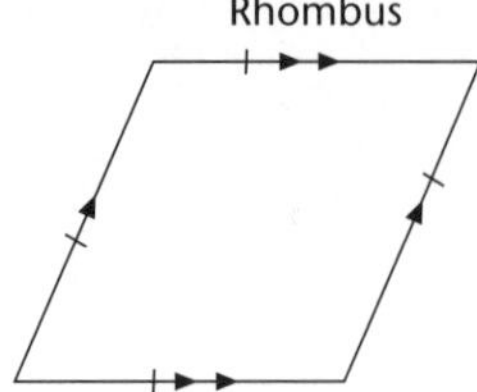

A **rectangle** is a quadrilateral with both pairs of opposite sides parallel and equal, and all four angles are right angles:

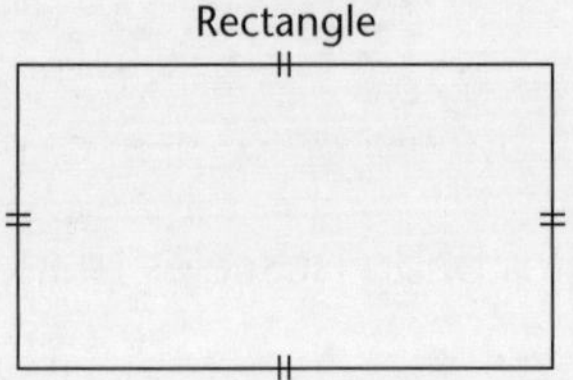

A **square** is a quadrilateral with all four sides equal and all four angles are right angles:

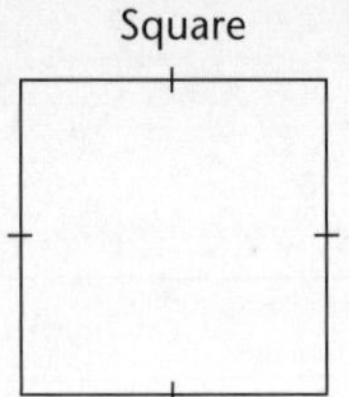

A **kite** is a quadrilateral with two pairs of adjacent sides equal:

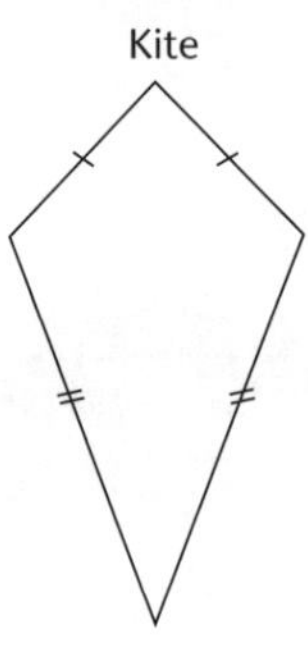

Family of Quadrilaterals

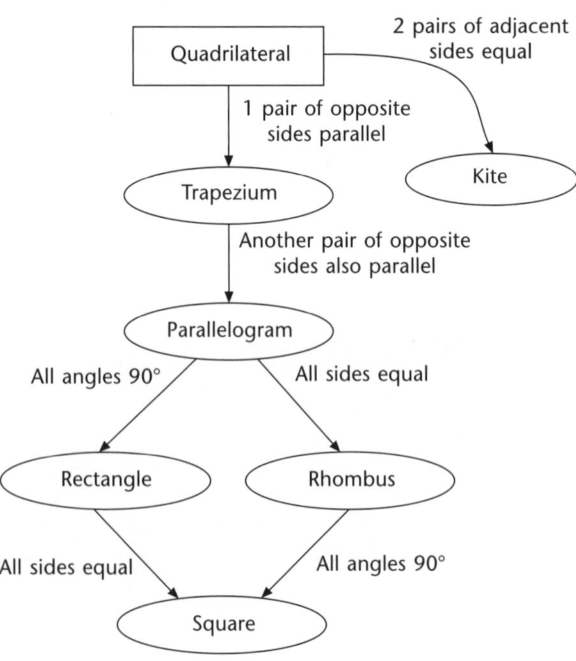

For Example

1 Find x, giving reasons for your answer:

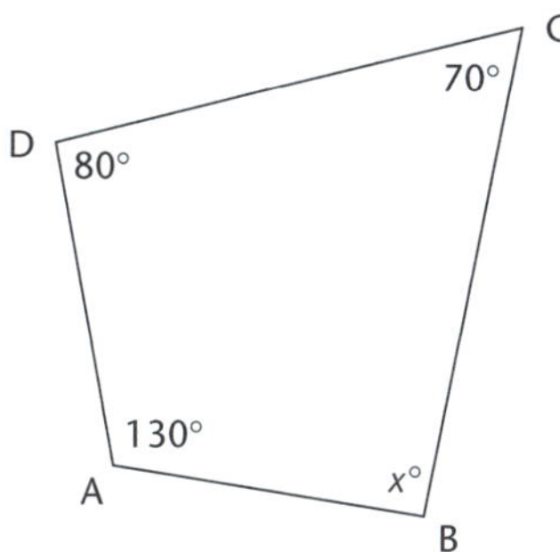

1 $x + 130 + 80 + 70 = 360$

(Angle sum of quadrilateral ABCD.)

$\therefore x + 280 = 360$

$x = 80$

For Example

1 Find x, giving reasons for your answer:

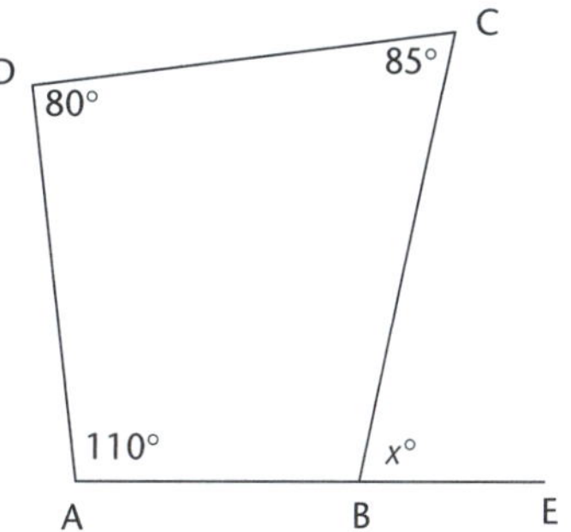

1 $\angle ABC + 85° + 80° + 110° = 360°$

(Angle sum of quadrilateral ABCD.)

$\angle ABC + 275° = 360°$

$\angle ABC = 85°$

$\therefore x + 85 = 180$

($\angle CBE$ is supplementary to $\angle ABC$.)

$\therefore x = 95$

PRACTISE, PRACTISE

Go to p. 258 for quick answers, or to pp. 298–303 for worked solutions.

pp. 167–168

1

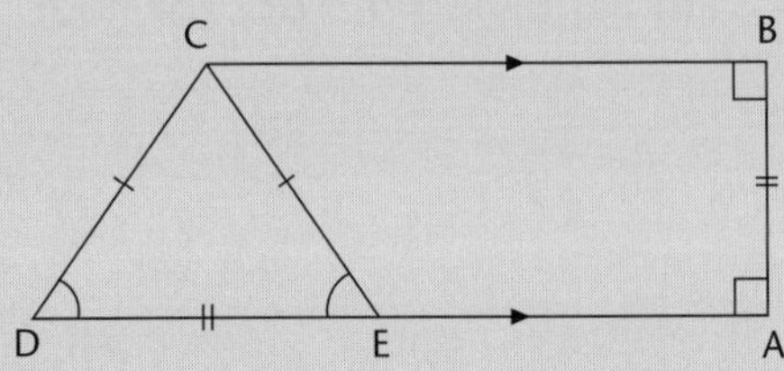

For the given figure, write true (T) or false (F):

a CB // EA **b** CE = DE **c** AB ⊥ BC
d ∠CDE = 90° **e** DC = CE **f** ∠CDE = ∠DEC
g Triangle CDE is isosceles.

pp. 167–171

2 **a** Name the figure ABCDEF.
b How many axes of symmetry does it have?
c What is its order of symmetry?
d Is the figure regular? Why?

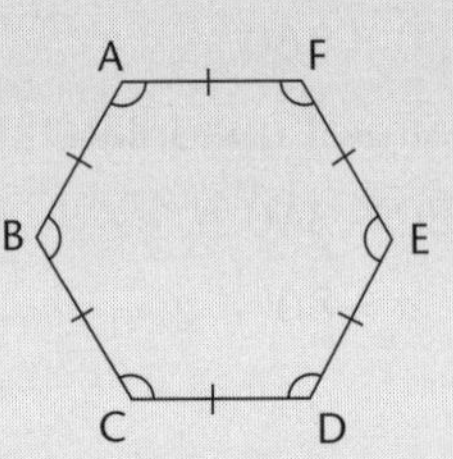

pp. 168–169

3 Name the following polygons:

a 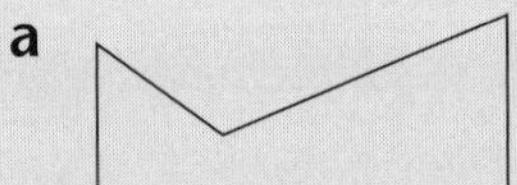 **b** 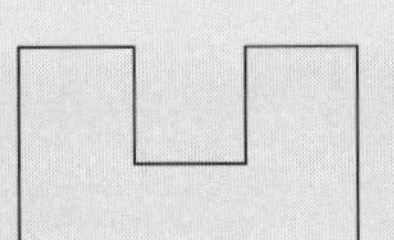 **c** **d**

pp. 168–169

4 Which of the following are regular polygons?

a 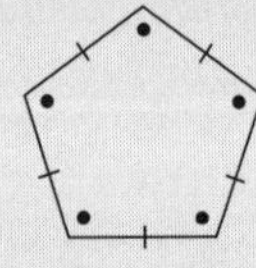 **b** 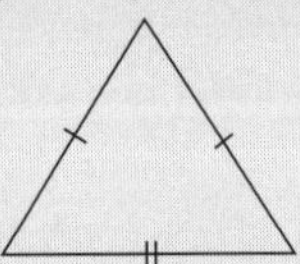 **c** 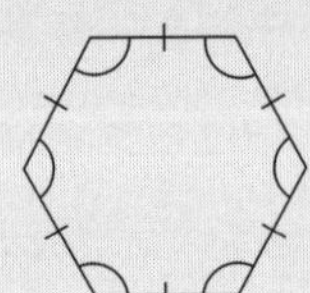

pp. 169–170

5 Name a line which is:
a Parallel to QR
b Perpendicular to SW
c Skew to PT.

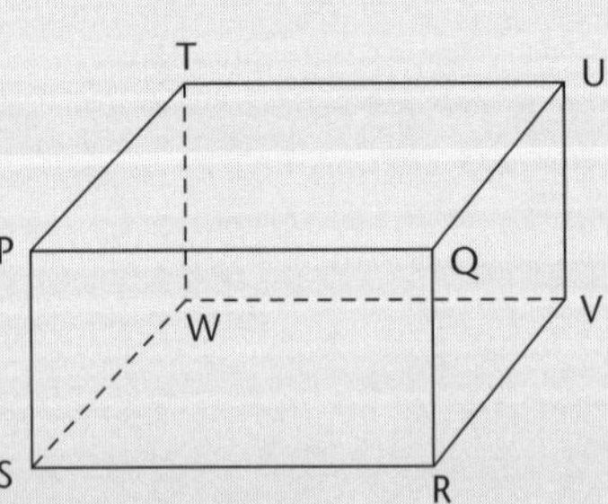

pp. 170–171

6 Copy each of the following shapes and where possible draw all axes of symmetry. State the order of symmetry for each shape:

a 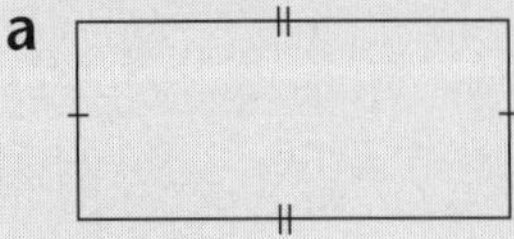 **b** 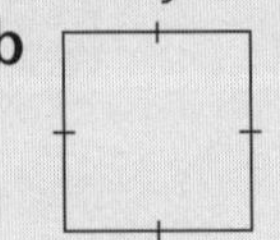 **c** 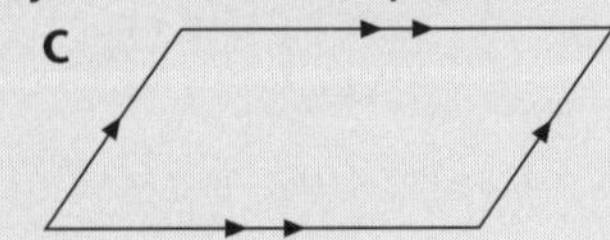 **d** 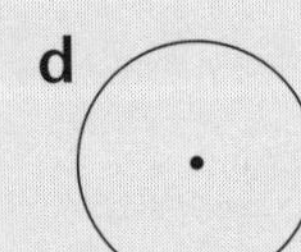

7 Name the following solids:

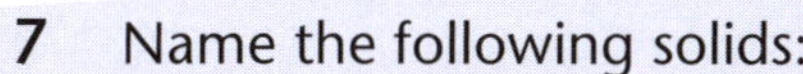

pp. 172–174

a

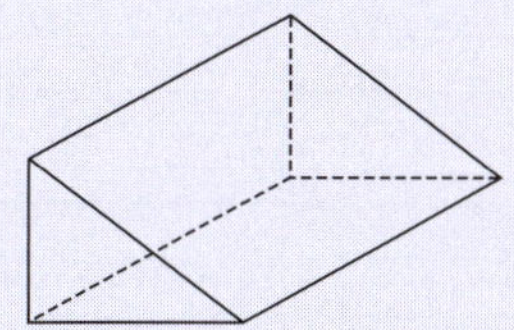

b

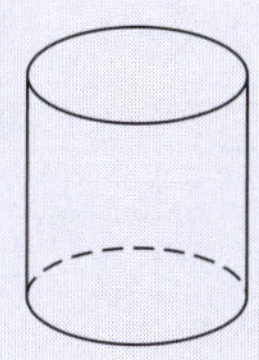

c

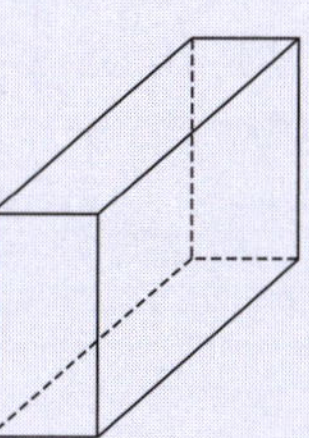

d

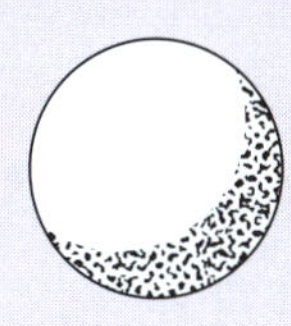

e

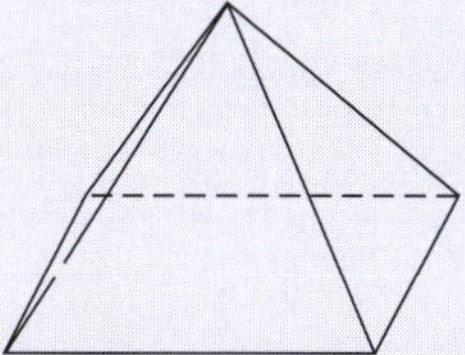

f

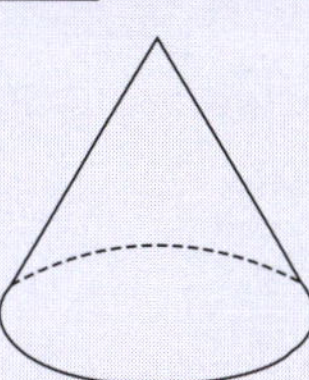

8 Which of the solids in Question 6 are:

pp. 172–174

a Polyhedra?
b Prisms?

9 Name the following quadrilaterals:

pp. 171–172

a

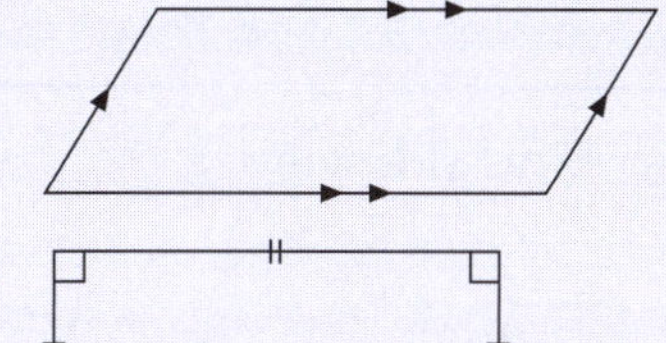

b

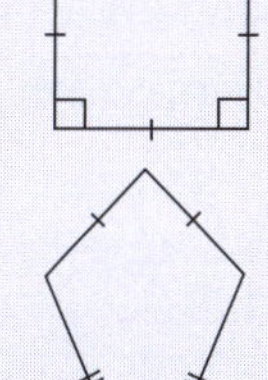

c

d

e

f

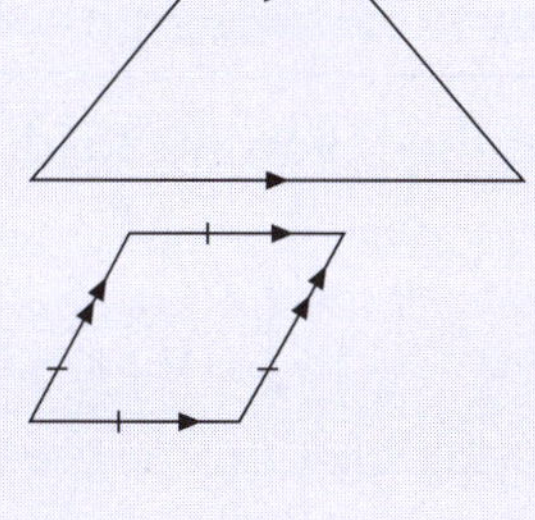

10 Draw this diagram and mark on it the following information:

pp. 169–170

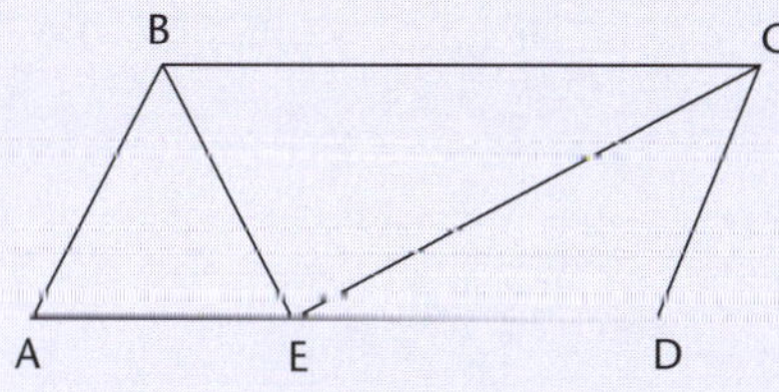

a AB = EB
b BC // AD
c ∠BAE = ∠DCB
d BE ⊥ CE
e AE = CD

11 This figure represents a shed:

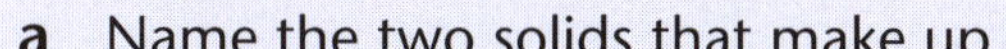

pp. 172–174

a Name the two solids that make up this solid.
b Is this solid a polyhedron? Why?
c Is this solid a prism?
d How many vertices, edges and faces does it have?

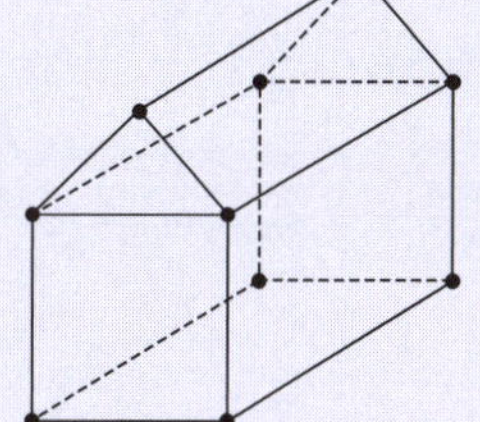

12 ABCD is a rectangle:

p. 172–174

a Name this solid.
b Is it a prism?
c Is it a polyhedron?
d How many edges does it have?

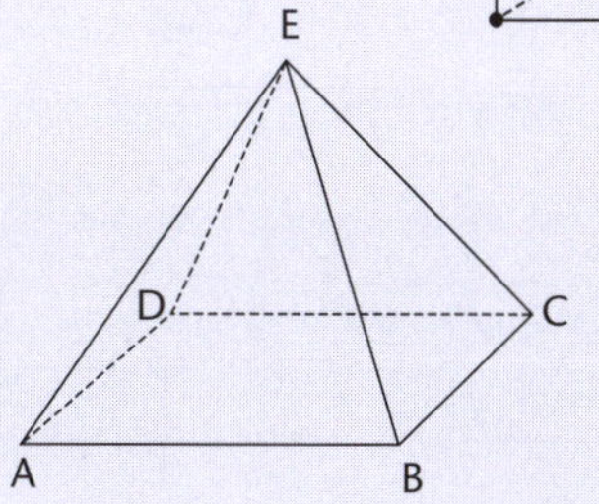

13 **a** Build the following solids using centicubes and then draw them on isometric dot paper: p. 174

b Draw the following solids on isometric dot paper after the shaded cube is removed:

14 **a** For the following solids draw the front view, top view and side view: pp. 174–175

b Draw the front, side and top views of the following solids:

15 **a** Draw the nets of the following solids: p. 175

i Cube **ii** Square pyramid

b Name the solids formed by the following nets:

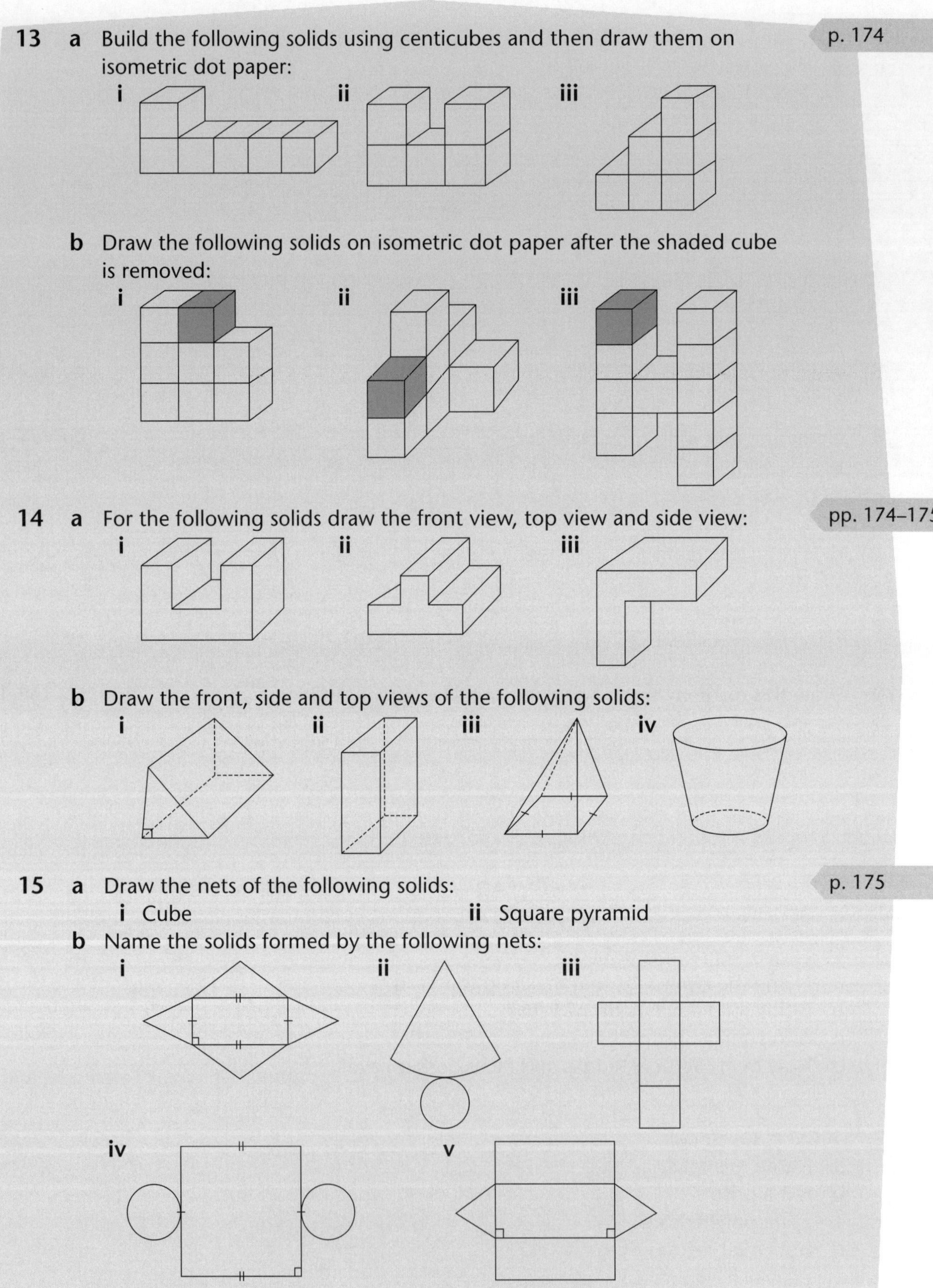

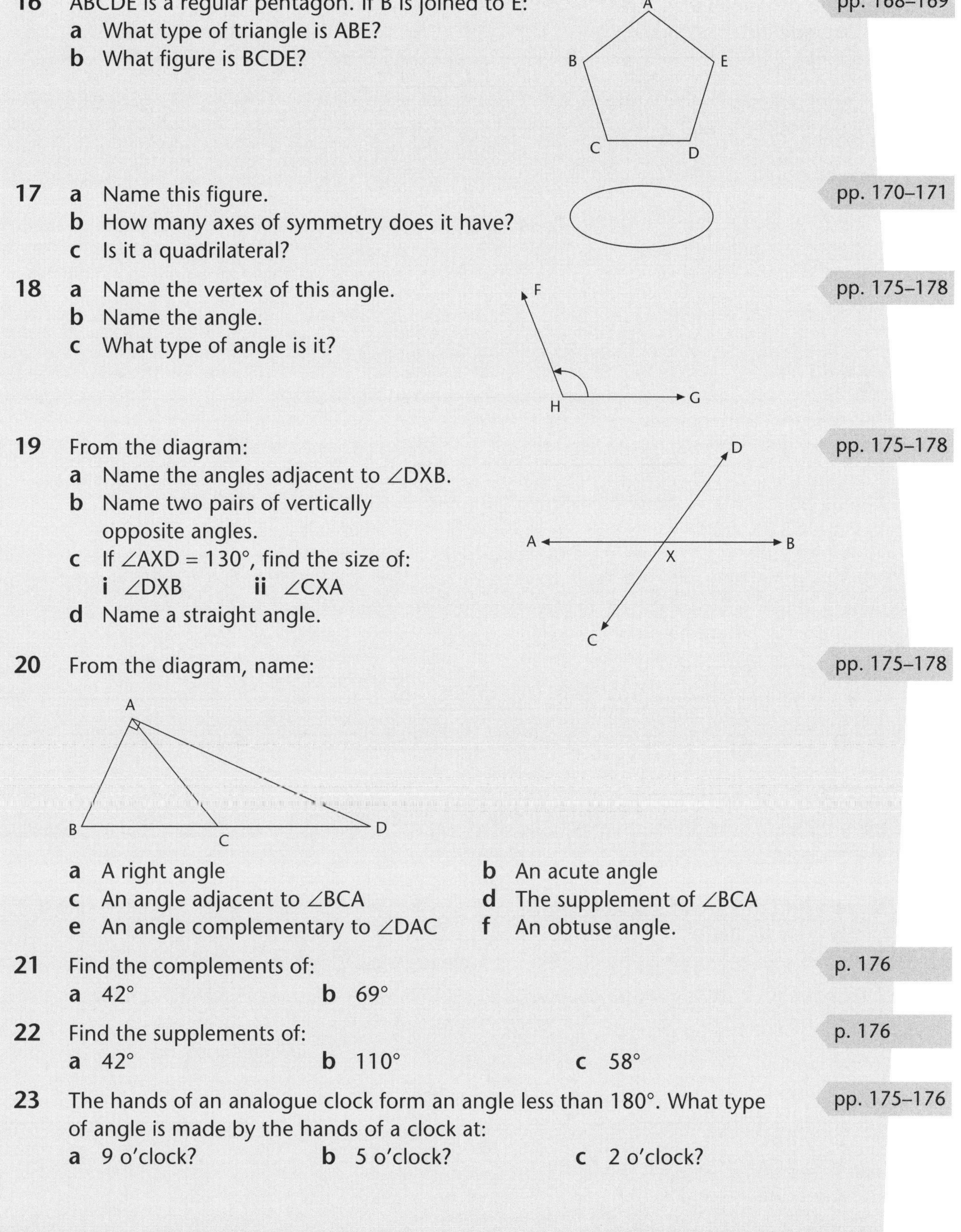

16 ABCDE is a regular pentagon. If B is joined to E: pp. 168–169

a What type of triangle is ABE?

b What figure is BCDE?

17 **a** Name this figure. pp. 170–171

b How many axes of symmetry does it have?

c Is it a quadrilateral?

18 **a** Name the vertex of this angle. pp. 175–178

b Name the angle.

c What type of angle is it?

19 From the diagram: pp. 175–178

a Name the angles adjacent to ∠DXB.

b Name two pairs of vertically opposite angles.

c If ∠AXD = 130°, find the size of:

i ∠DXB **ii** ∠CXA

d Name a straight angle.

20 From the diagram, name: pp. 175–178

a A right angle

b An acute angle

c An angle adjacent to ∠BCA

d The supplement of ∠BCA

e An angle complementary to ∠DAC

f An obtuse angle.

21 Find the complements of: p. 176

a 42° **b** 69°

22 Find the supplements of: p. 176

a 42° **b** 110° **c** 58°

23 The hands of an analogue clock form an angle less than 180°. What type of angle is made by the hands of a clock at: pp. 175–176

a 9 o'clock? **b** 5 o'clock? **c** 2 o'clock?

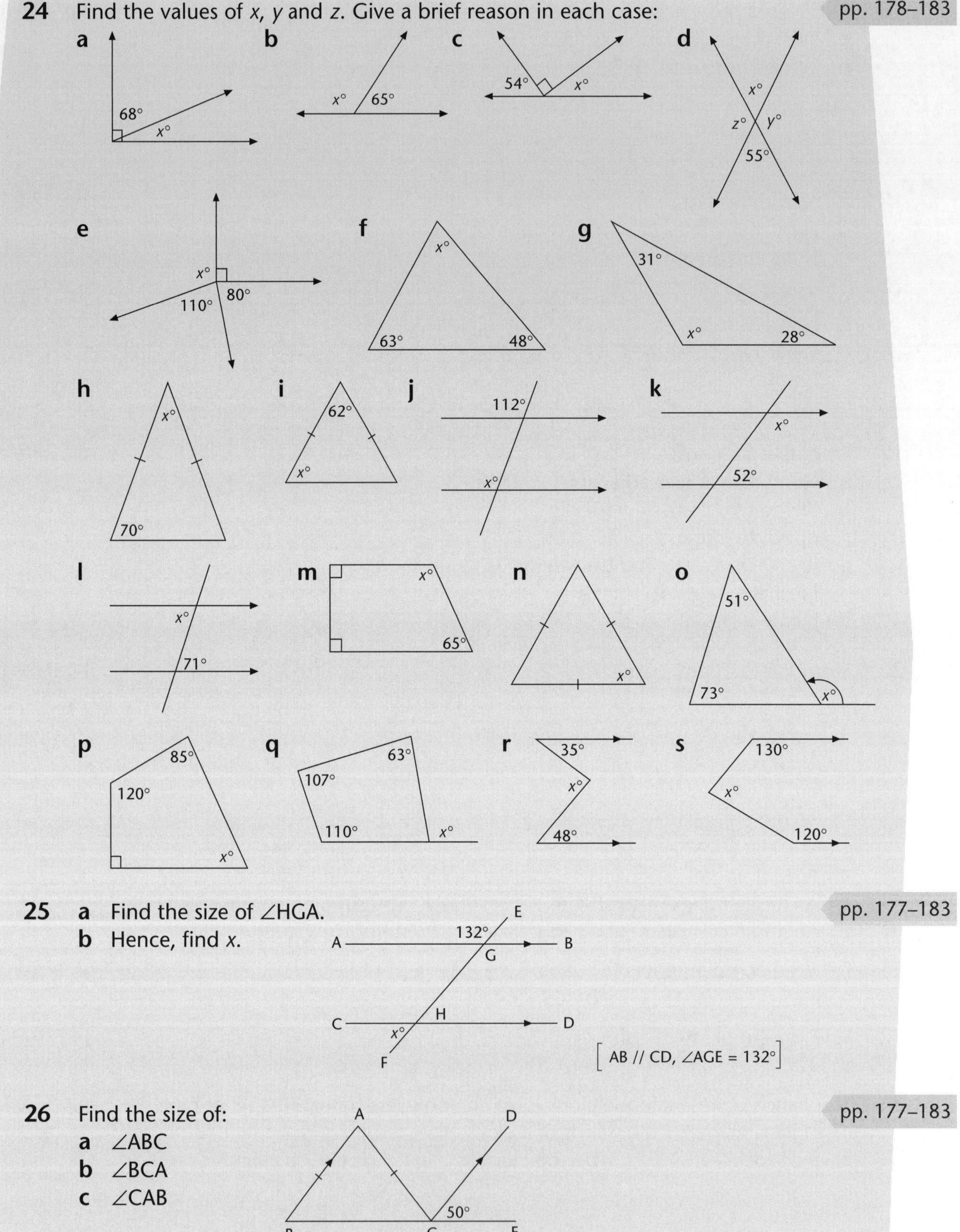

24 Find the values of x, y and z. Give a brief reason in each case:

pp. 178–183

25 **a** Find the size of ∠HGA.
b Hence, find x.

[AB // CD, ∠AGE = 132°]

pp. 177–183

26 Find the size of:
a ∠ABC
b ∠BCA
c ∠CAB

AB // DC, AB = AC and ∠DCE = 50°

pp. 177–183

27 Find x and y. pp. 177–183

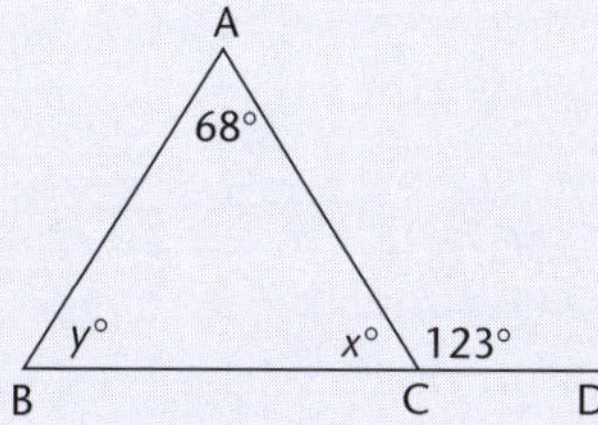

28 Find the size of x. pp. 177–183

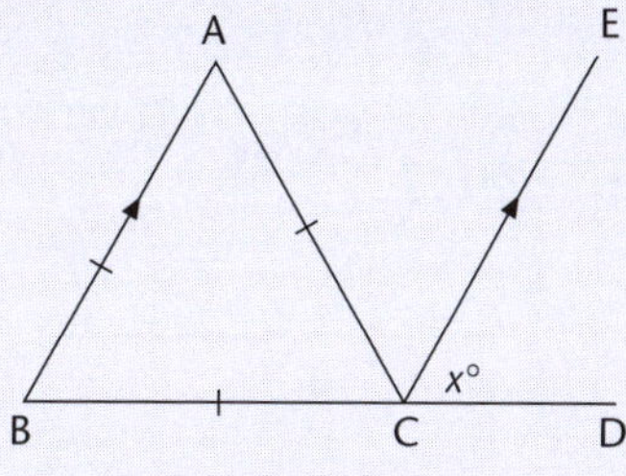

29 AB // CD.
Find x. pp. 178–180

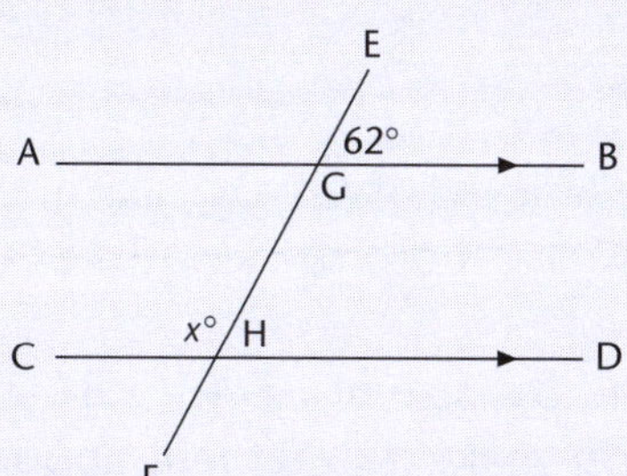

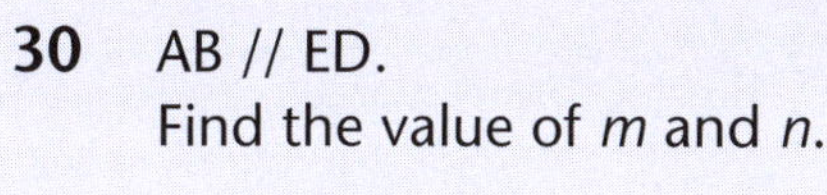

30 AB // ED.
Find the value of m and n. pp. 177–183

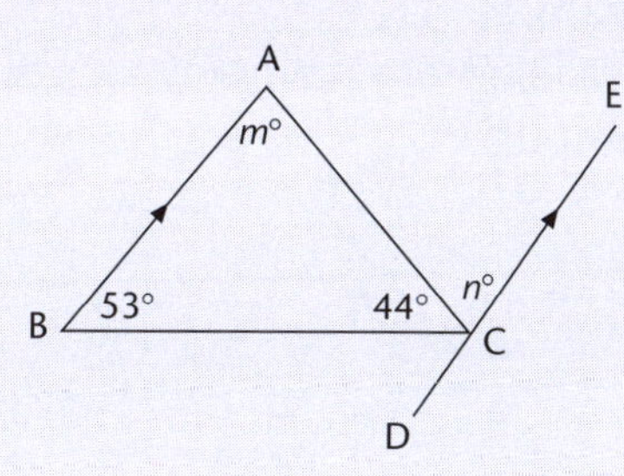

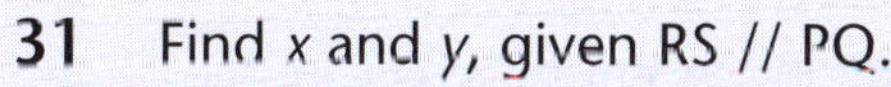

31 Find x and y, given RS // PQ. pp. 177–183

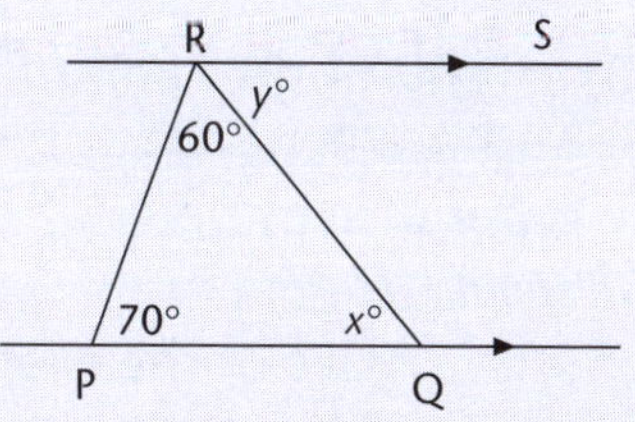

32 Find x. pp. 177–183

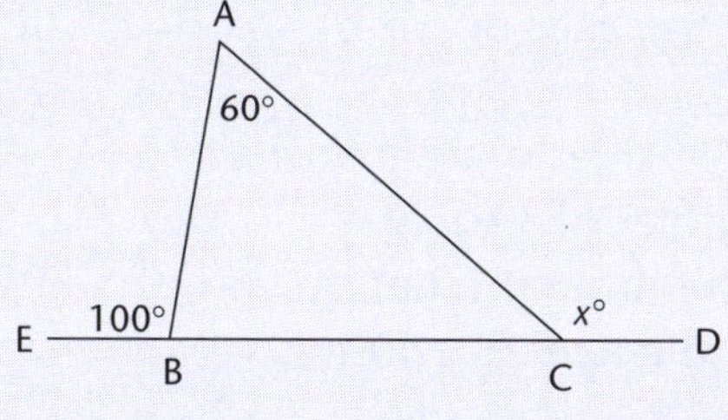

33 From the diagram:

pp. 177–183

a Name the supplement of ∠ADB.
b Name the complement of ∠DBC.
c If ∠ABD = 72°, find the size of ∠DBC.
d If ∠CDB = 126°, find the size of ∠ADB.

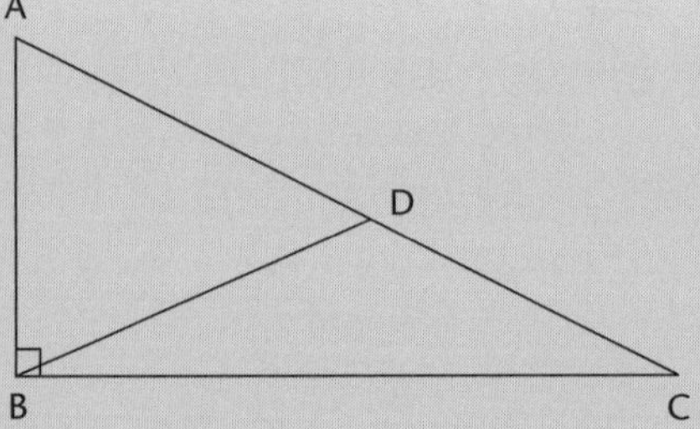

34 **a** **i** Draw a trapezium and describe it in your own words.

pp. 177–183

ii Does it have any axis of symmetry?

b ABCD is a parallelogram:

i Using your ruler, find the lengths of AB, DC, AD and BC. In your own words make a statement about the lengths of the sides of a parallelogram.
ii Using your protractor, find the size of ∠DAB, ∠BCD, ∠ABC and ∠CDA. In your own words, make a statement about the angles of a parallelogram.
iii Does a parallelogram have any axis of symmetry?

c XYZW is a rhombus:

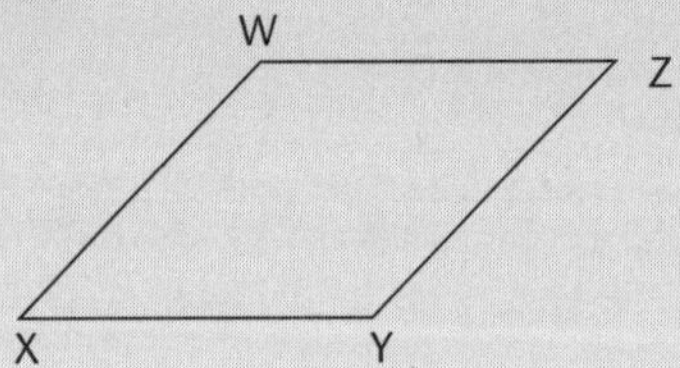

i Measure the length of the sides of rhombus XYZW. In your own words, make a statement about the lengths of sides of a rhombus.
ii What is the order of symmetry of a rhombus? (How many axes of symmetry does it have?)
iii Are the opposite angles of the rhombus equal? (Use a protractor.)
iv Is a rhombus a parallelogram?

d **i** Draw a square and label all the axes of symmetry. How many are there?
ii What do you know about the sides of a square?
iii What is the size of all the angles of a square?
iv Measure the diagonals of the square. What do you conclude?

e **i** Draw a rectangle and label the axes of symmetry. How many are there?
ii What do you know about the sides of a rectangle?
iii What is the size of all the angles of a rectangle?
iv Measure the diagonals of the rectangle. What do you conclude?

f **i** Draw a kite and label any axes of symmetry. How many are there?
ii Measure the lengths of all the sides of the kite. In your own words make a statement about the lengths of the sides of a kite.

g Which of the following quadrilaterals would be classed as trapeziums?

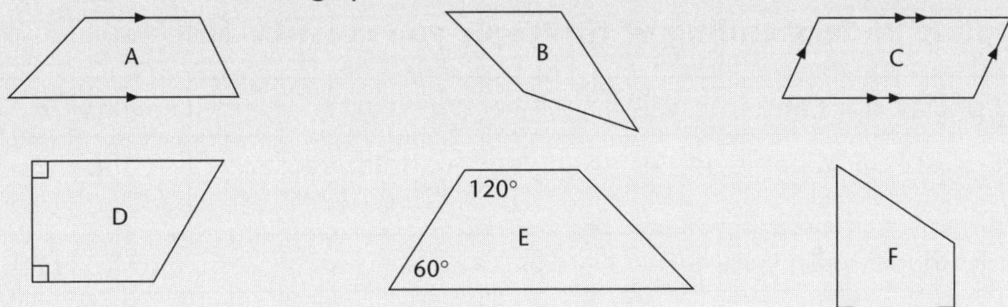

Note: A trapezium is a quadrilateral with one pair of opposite sides parallel.

h Which of the following quadrilaterals would be classed as parallelograms?
Note: A parallelogram is a quadrilateral with both pairs of opposite sides parallel.

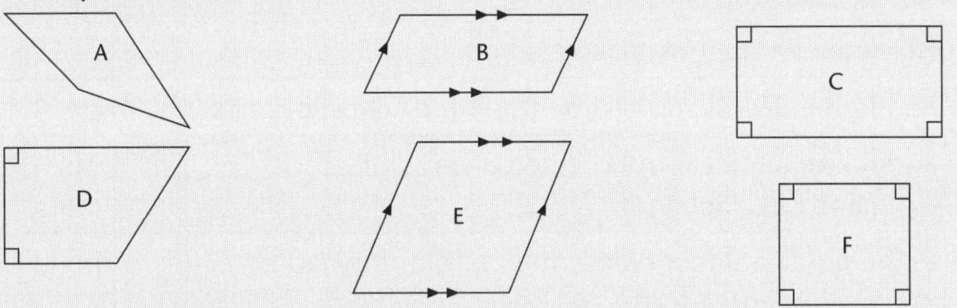

i Answer the following questions as true (T) or false (F):
i A trapezium is a parallelogram.
ii A rhombus is a parallelogram.
iii A square is a parallelogram.
iv A rectangle is a parallelogram.
v A parallelogram is a rhombus.
vi All quadrilaterals have two diagonals.
vii A square is a rhombus.
viii A square is a rectangle.

j Take a rectangular piece of paper ABCD. Label two points X and Y such that BX = DY, as shown in the diagram. Cut along AX and CY to form the quadrilateral AXCY:

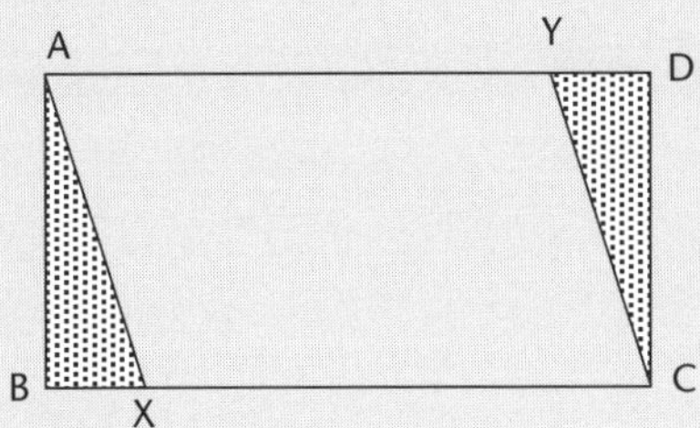

i Name the quadrilateral AXCY.
ii Show by folding that the quadrilateral AXCY has no axis of symmetry.
iii Show that this figure has rotational symmetry of order 2.

Go to p. 258 for quick answers, or to pp. 298–303 for worked solutions.

YOUR CHECKLIST

For a complete understanding of this topic you must be able to:

✓	Label points, lines and intervals		p. 167
✓	Label, classify and name angles, triangles, quadrilaterals and polygons		pp. 167–172
✓	Identify shapes with symmetry		p. 170
✓	Name parts of a circle		p. 172
✓	Classify, describe and name solids according to their geometrical properties		pp. 172–174
✓	Sketch solids on isometric paper		pp. 174–175
✓	Sketch different views of solids		pp. 174–175
✓	Distinguish between the nets of solids		p. 175
✓	Identify and name angles formed by the intersection of lines		pp. 175–178
✓	Identify and name angles formed when a transversal intersects a pair of parallel lines		pp. 178–180
✓	Recognise the equal and supplementary angles formed by a pair of parallel lines and a transversal		pp. 178–180
✓	Use angle properties to identify a pair of parallel lines		pp. 178–180
✓	Apply angle properties of triangles and quadrilaterals.		pp. 180–183

Now you are ready to do the tests!

(50 marks)

1 Name the following plane shapes:

a 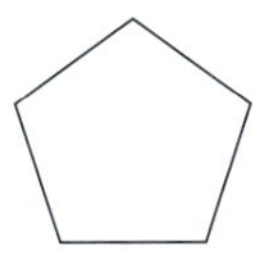 b 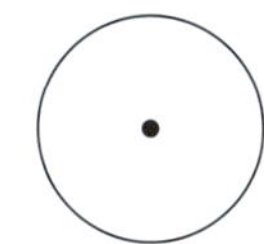 c 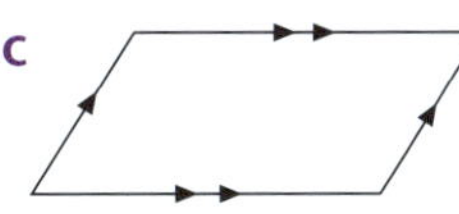 d 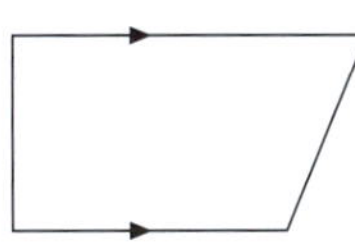

(4 marks)

2 Name the following solids:

a 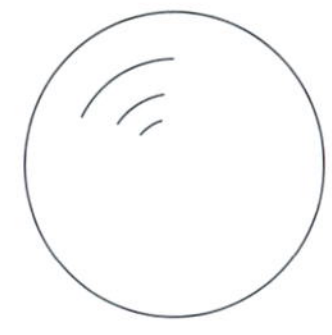 b 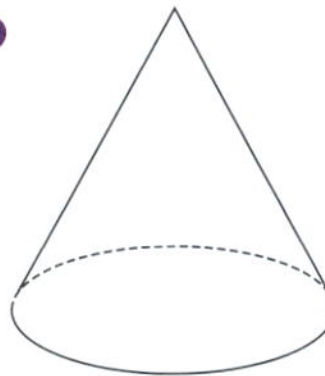 c 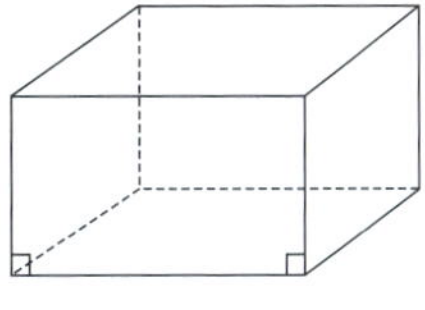 d 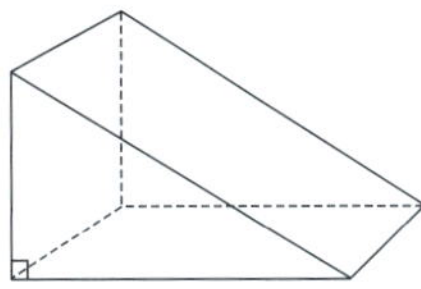

(4 marks)

3 For the following figure, write true (T) or false (F):

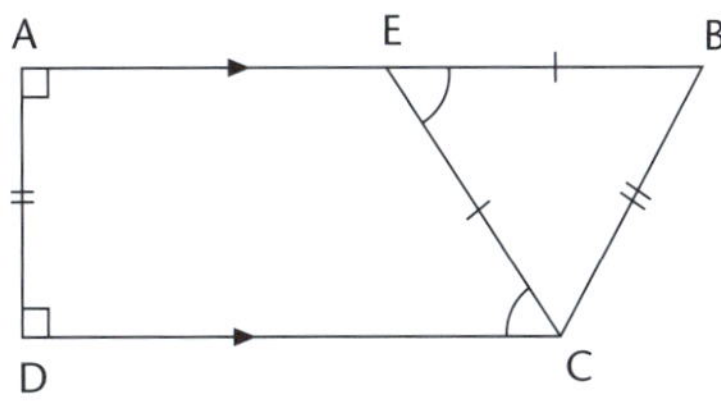

a AB || DC

b BE = BC

c ∠ADC = 90°

d EB = EC

e ∠BEC = ∠DCE

f EB ⊥ EC

(6 marks)

4 Study the solid below and answer the questions:

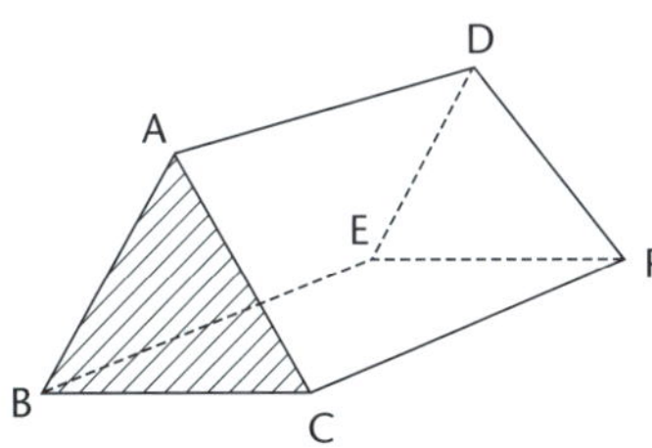
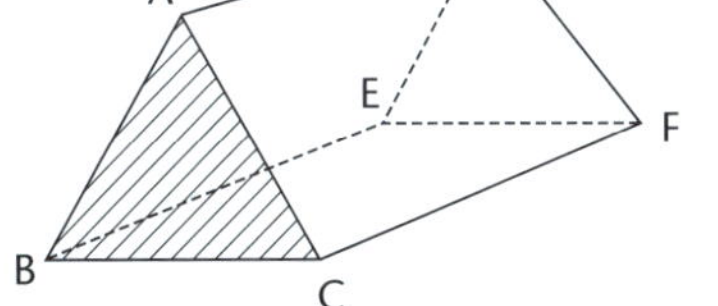

a Name the shape of the shaded cross-section.

b Name the solid.

c Is the solid a polyhedron?

d How many faces does the solid have?

e How many edges does the solid have?

f How many vertices does the solid have?

(6 marks)

5

M

P N

a Name the angle.

b Is the angle acute, obtuse, right or reflex? (2 marks)

6 What special name is given to an eight-sided polygon? (1 mark)

7 How many axes of symmetry does a square have? (1 mark)

8 What is the angle sum of a pair of:

a Supplementary angles? b Complementary angles? (2 marks)

9 Match each figure with one of the descriptions below:

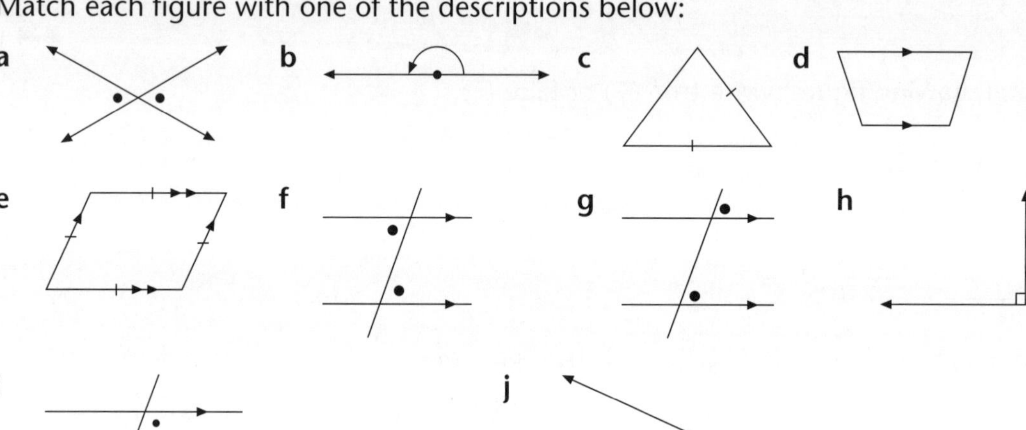

i

j

A	Co-interior angles	B	Alternate angles
C	Corresponding angles	D	Vertically opposite angles
E	Adjacent angles	F	Right angle
G	Straight angle	H	Scalene triangle
I	Isosceles triangle	J	Equilateral triangle
K	Rhombus	L	Trapezium
M	Parallelogram	N	Reflex angle

(10 marks)

10 What is the angle sum of a:

a Triangle? b Quadrilateral? (2 marks)

11 Find the value of x:

a

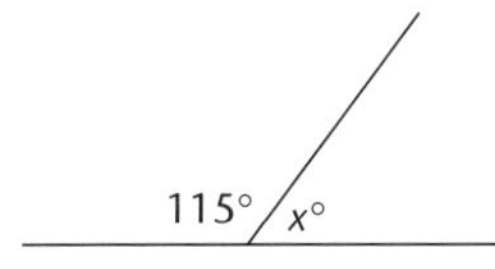

b

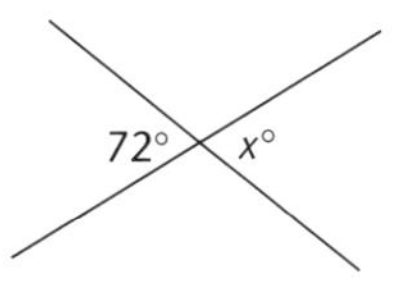

c

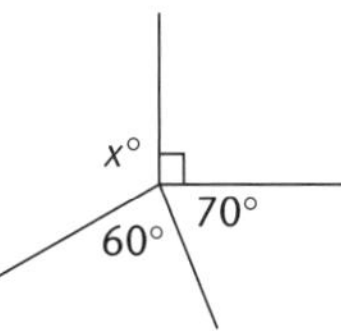

(3 marks)

12 Find the values of the letters:

a

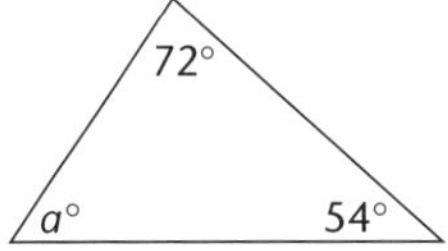

b

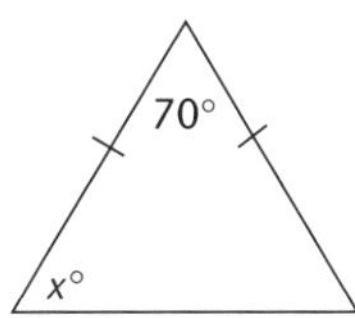

c

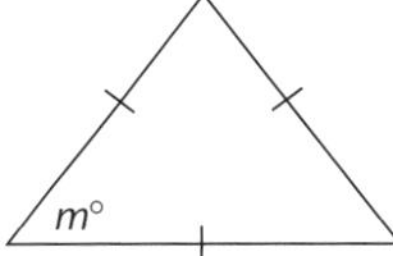

d

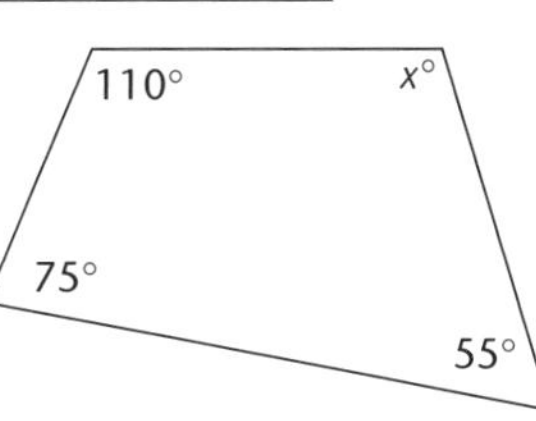

(4 marks)

13 Find the value of x:

a

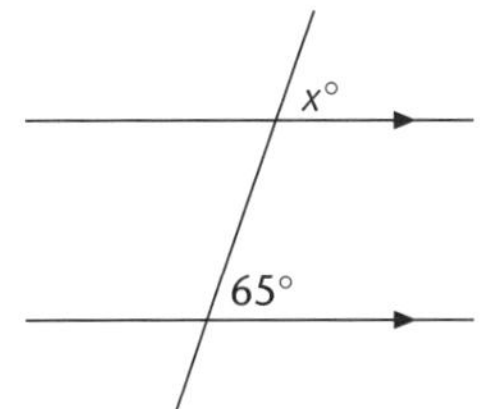

b

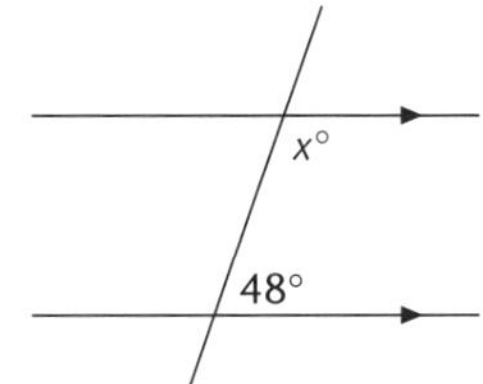

c

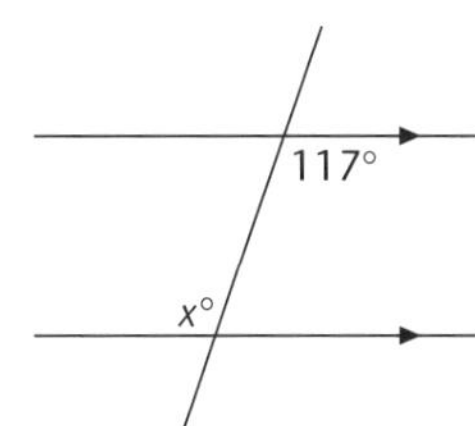

(3 marks)

14 Find the values of x and y:

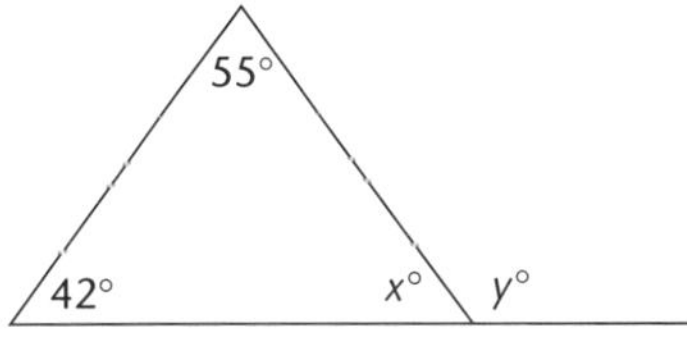

(2 marks)

☞ **Quick answers on page 265**
☞ **Worked solutions on page 332**

Your Feedback $\frac{\square}{50} \times 100\% = \square\%$

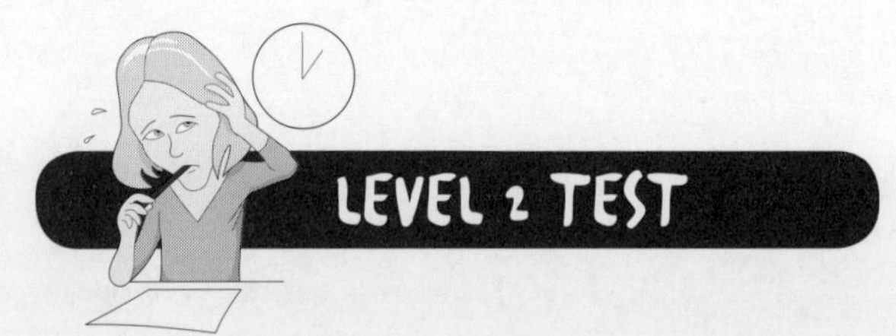

(50 marks)

1 Name the following plane figures:

a 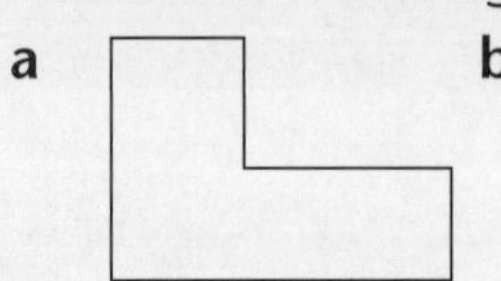 b 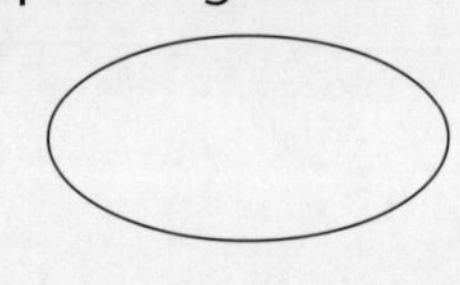 c d 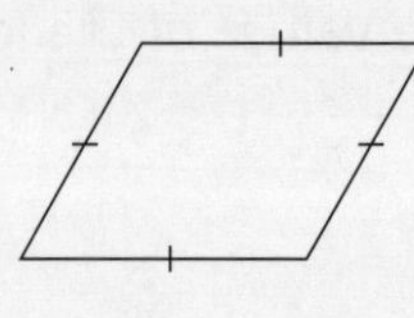

(4 marks)

2 Name the following solids:

a 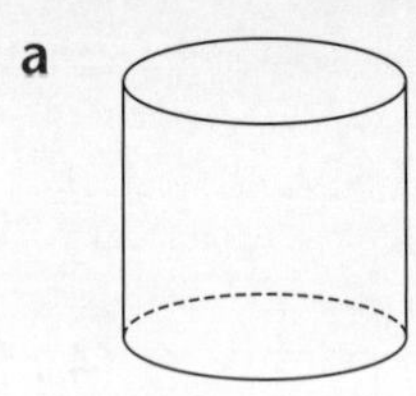 b 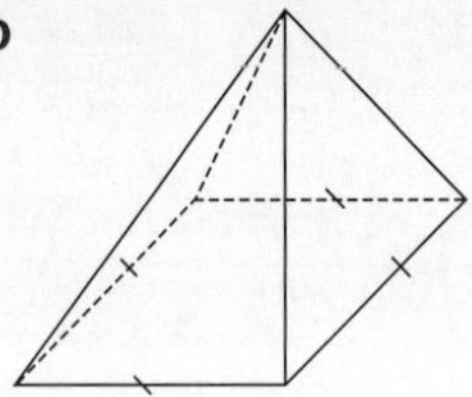 c 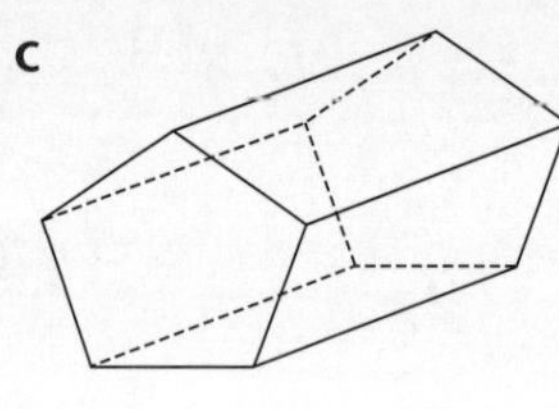 d 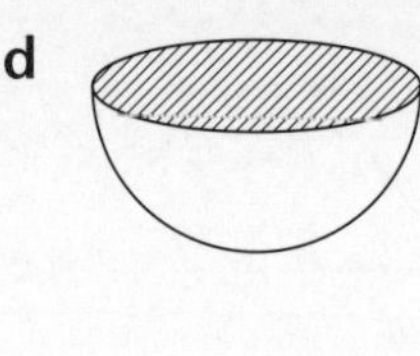

(4 marks)

3 In the figure, ABDE is a rectangle.

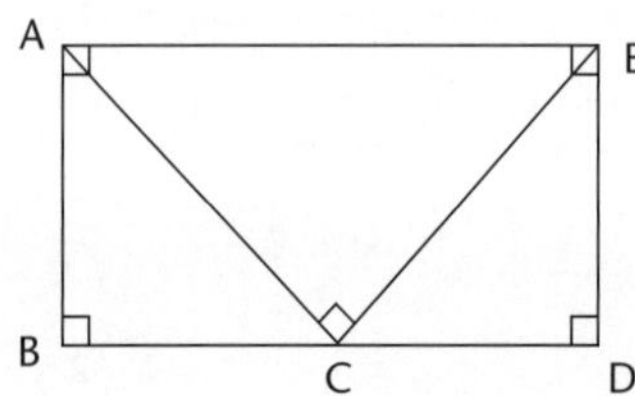

Write true (T) or false (F) for the following:

a $\angle ACE$ is an obtuse angle.

b AE = BD.

c $\angle ECD$ is an acute angle.

d The complement of $\angle EAC$ is $\angle CAB$.

e AB // ED.

f The supplement of $\angle ECD$ is $\angle BCA$.

g $AC \perp EC$.

h $\angle ABD = 90°$.

(8 marks)

4 Draw a regular pentagon. How many axes of symmetry does it have? (1 mark)

5 Draw a cone. Is a cone a polyhedron? (1 mark)

6 Study the solid below and answer the questions:

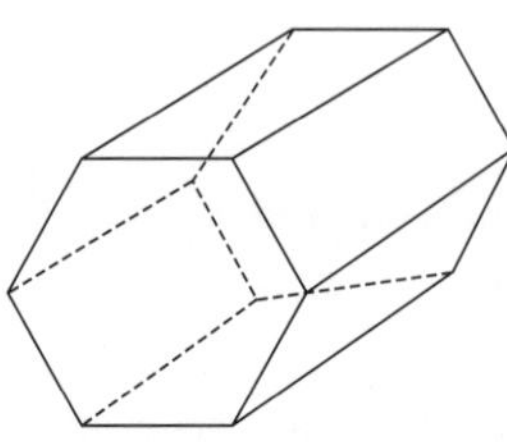

a Is this solid a prism?

b Name the solid.

c How many faces does it have?

d Is this solid a polyhedron?

(4 marks)

7 Shade in the cross-section of the following prism:

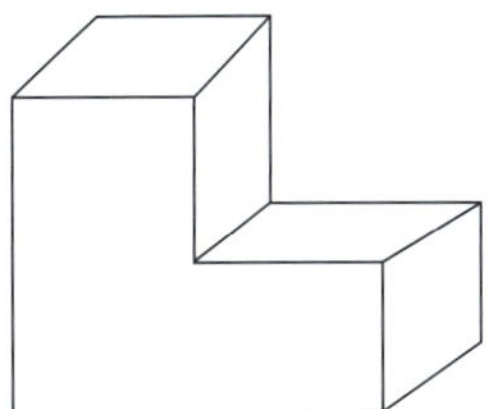

(1 mark)

8 Find the value of the pronumerals:

a

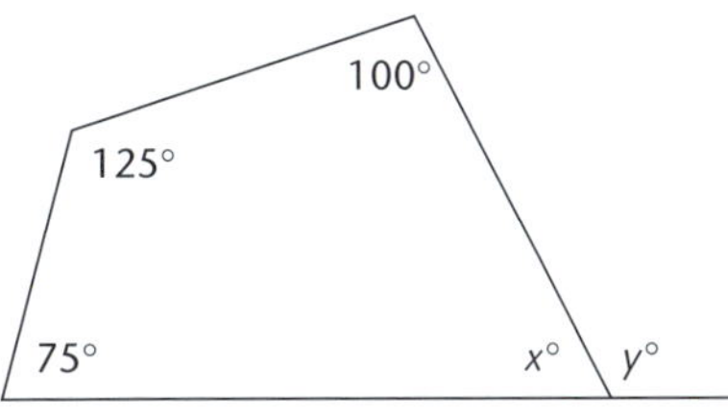

b

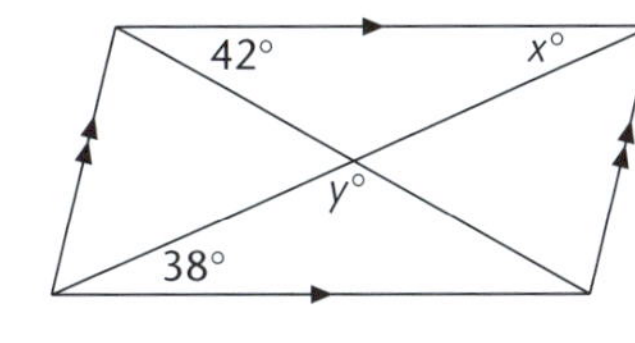

(4 marks)

9 Draw an isosceles triangle and mark on it all important information. (1 mark)

10 What is the:

a Complement of 75°?

b Supplement of 75°? (2 marks)

11 Complete the following statements by using one of the following words—equal, supplementary, complementary:

a Vertically opposite angles are always __________.

b When a pair of parallel lines is cut by a transversal:

i Co-interior angles are __________.

ii Alternate angles are __________.

iii Corresponding angles are __________. (4 marks)

12 Find the values of x:

a

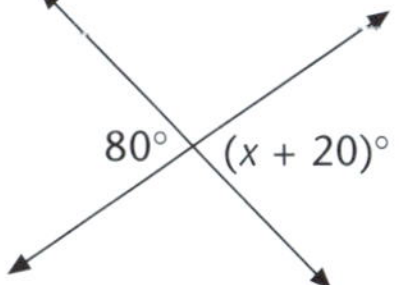

b

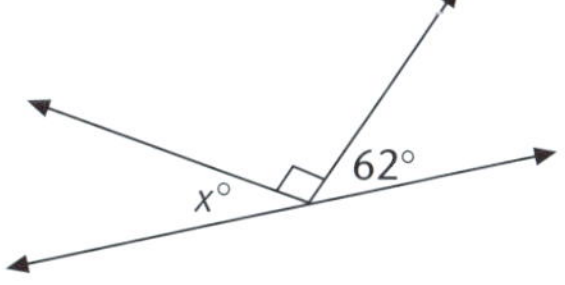

c

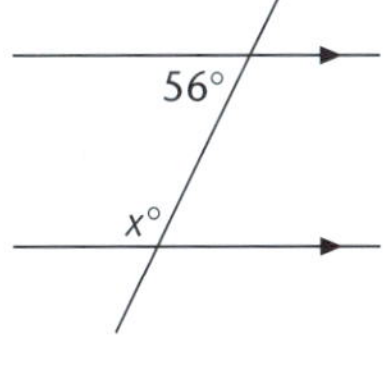

d

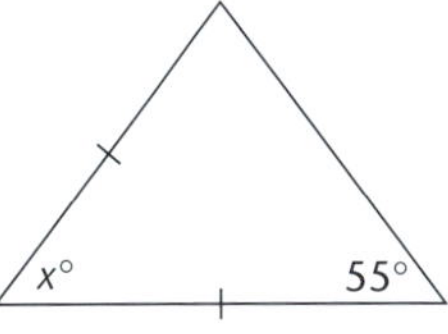

e

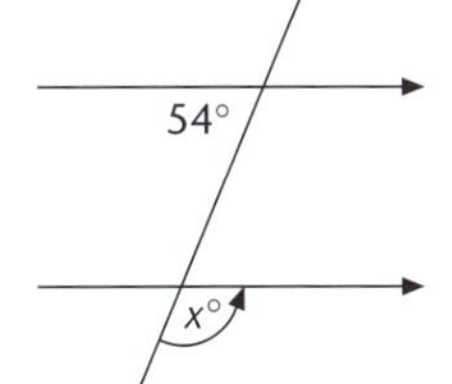

f

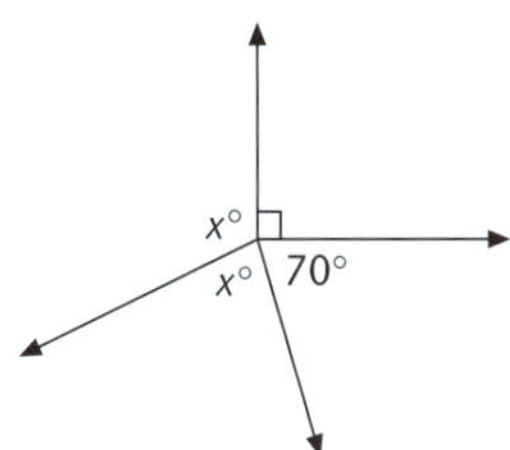

(6 marks)

13

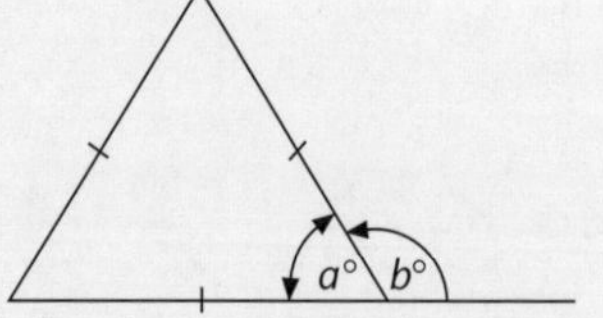

Find the values of a and b. (2 marks)

14

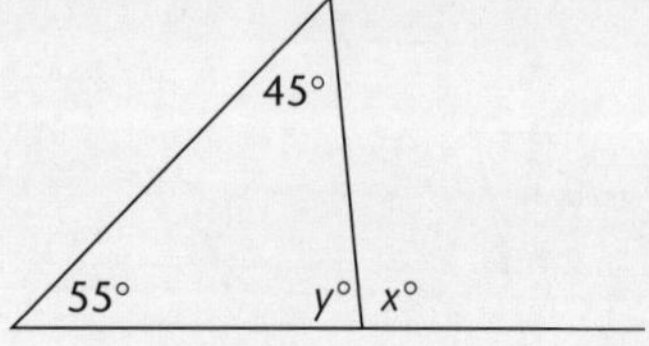

Find the values of x and y. (2 marks)

15

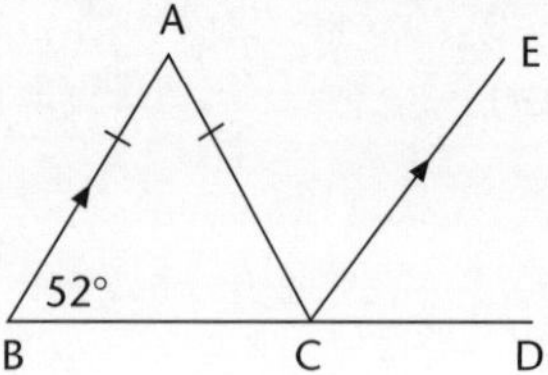

AB // EC and AB = AC

∠ABC = 52°

Complete the following:

a ∠BCA = ______________ Why? ____________________________________

b ∠CAB = ______________ Why? ____________________________________

c ∠ACE = ______________ Why? ____________________________________ (6 marks)

Your Feedback $\frac{\square}{50} \times 100\% = \square\%$

☞ Quick answers on page 265
☞ Worked solutions on page 333

(45 marks)

Questions 1 to 10 are multiple choice. Choose the correct answer: A, B, C or D.

1 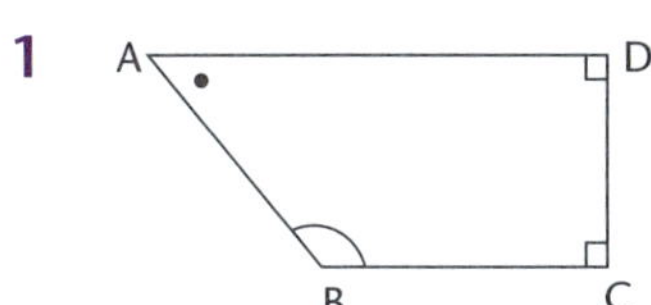

Which of the marked angles (A, B, C or D) is obtuse? (1 mark)

2 One of these shows AB ⊥ CD. Which is it?

A 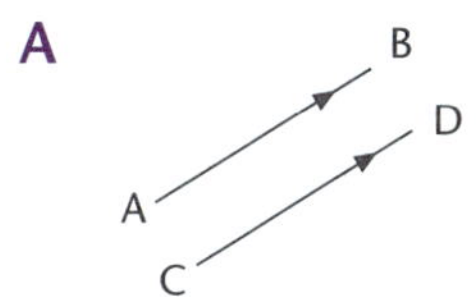

B 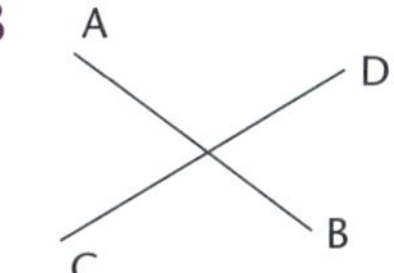

C 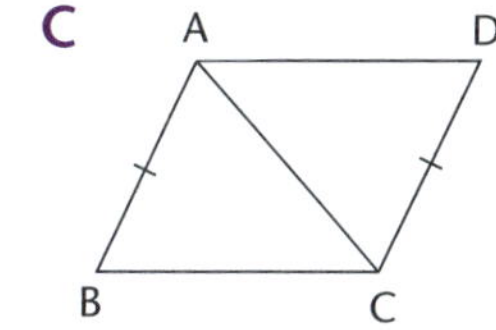

D 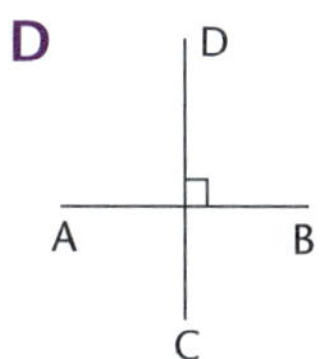

(1 mark)

3 Which of the following is an acute-angled isosceles triangle?

A 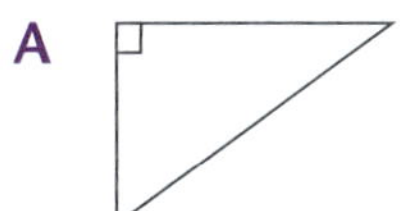

B 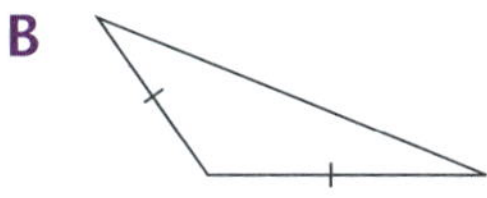

C 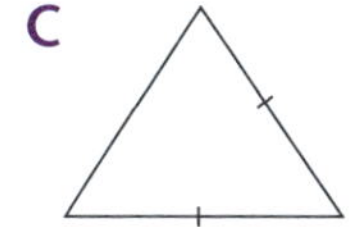

D 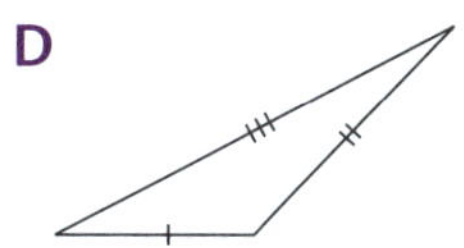

(1 mark)

4 Which road could be parallel to Smith Street?

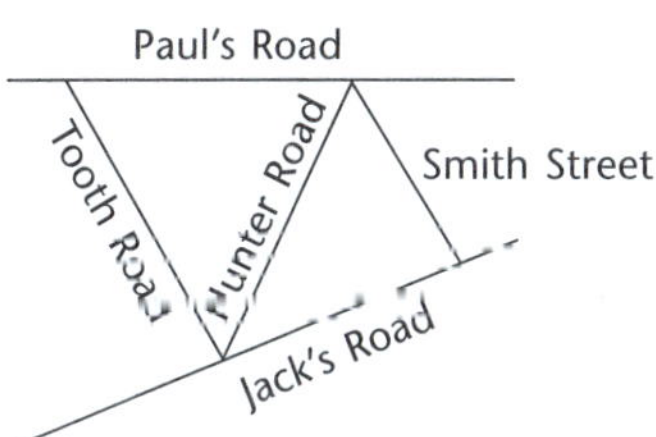

A Jack's Road
B Tooth Road
C Paul's Road
D Hunter Road. (1 mark)

5 Which of the following is a property of a rhombus?

A Angles add up to 270°.
B All angles are equal.
C Diagonals are equal.
D Opposite sides are parallel. (1 mark)

6

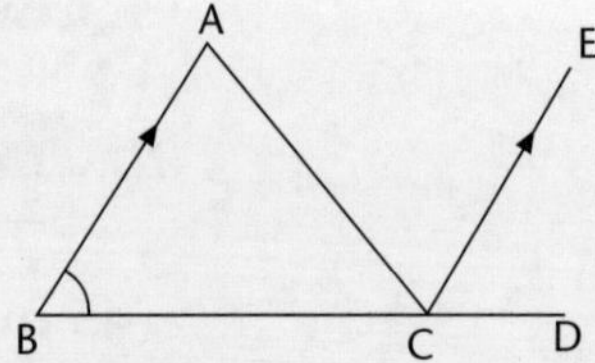

Which angle is corresponding to ∠ABC?

A ∠ACE **B** ∠ECD **C** ∠BCA **D** ∠CBA (1 mark)

7

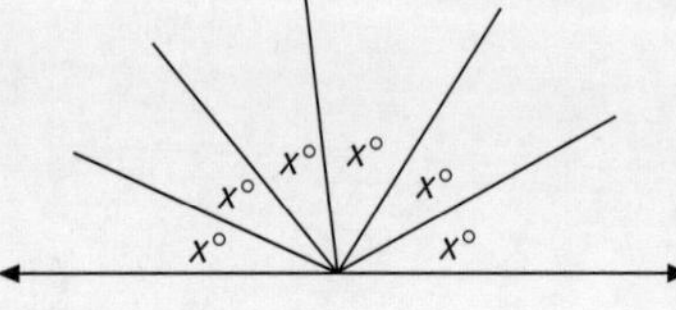

The value of x is:

A 6 **B** 30 **C** 40 **D** 60 (1 mark)

8 In the figure:

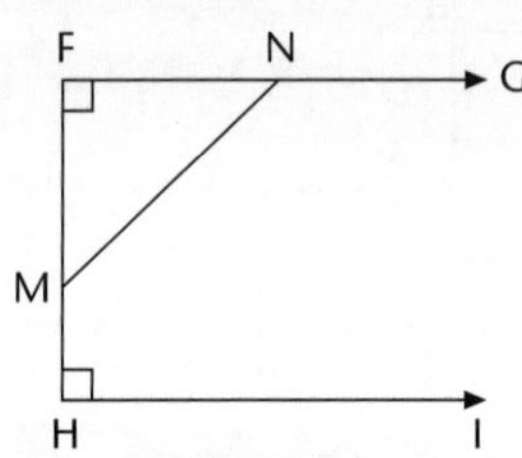

A FN = FM

B The supplement of ∠FNM = ∠GNM

C FG ⊥ HI

D The complement of ∠GFH = ∠FHI. (1 mark)

9 Figure ABCDE is a regular pentagon. ABDE is:

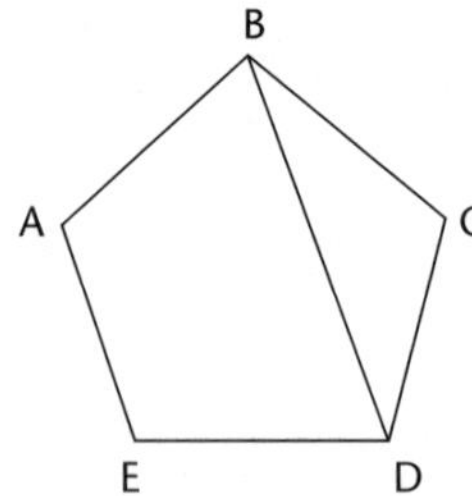

A An isosceles triangle **B** A parallelogram

C A rectangle **D** A trapezium. (1 mark)

10 In the figure, x equals:

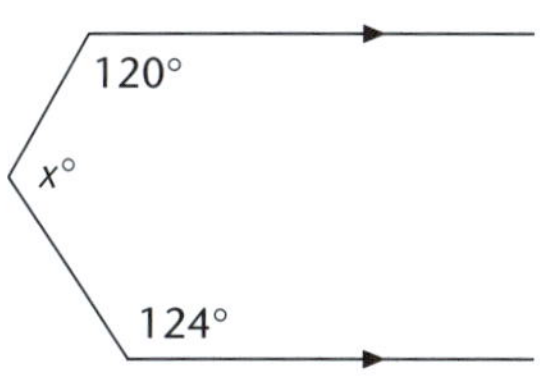

A 116 **B** 120 **C** 124 **D** 144 (1 mark)

11 In the diagram, △ABC is right-angled.

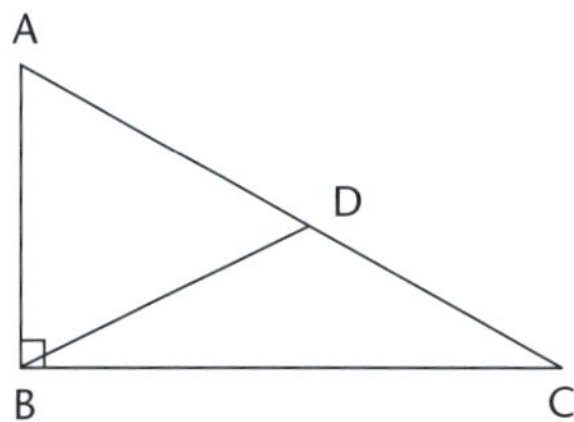

From the diagram, name:

a A right angle

b An obtuse angle

c The supplement of ∠CDB

d The complement of ∠ABD. (4 marks)

12 Copy the following diagram into your book and mark on it the following information:

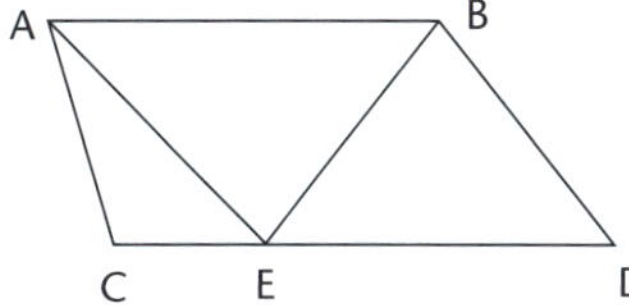

a AB // CD

b EB = BD

c ∠BAE = ∠EDB

d AE ⊥ BE. (4 marks)

13 **a** Draw a rectangular pyramid in your book.

b Is it a prism?

c Is it a polyhedron? (3 marks)

14 Find the values of x and y:

a

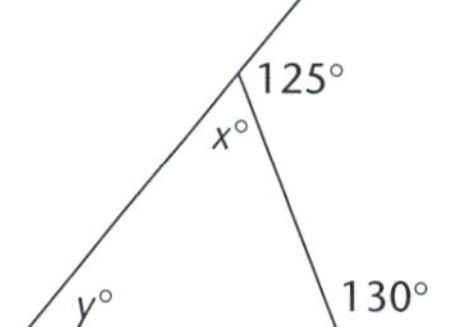

b

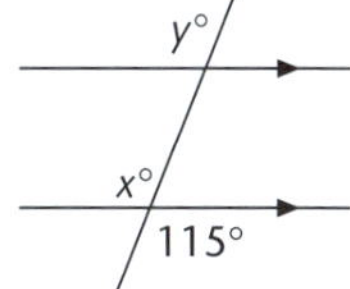

c

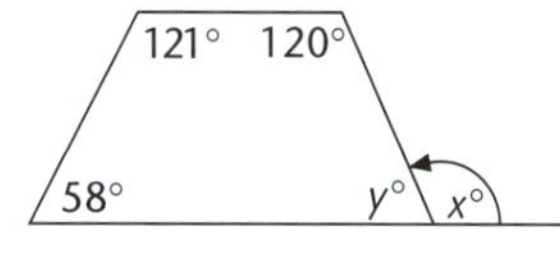

(6 marks)

15 Look at the following diagram and state whether:

a AB // CD

b EF // GH

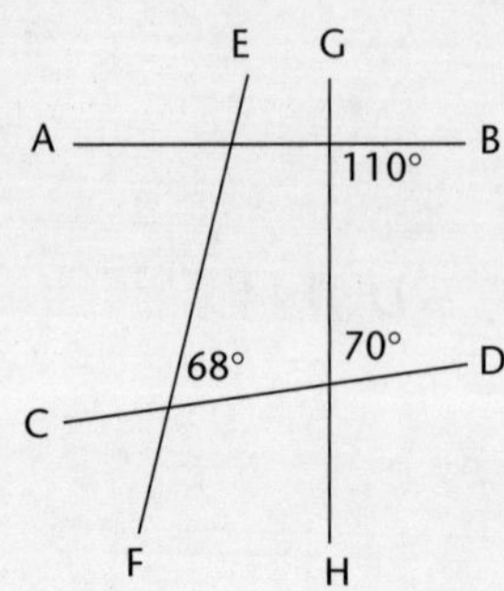

(2 marks)

16 Find the values of a, b, c, x, y and z:

a

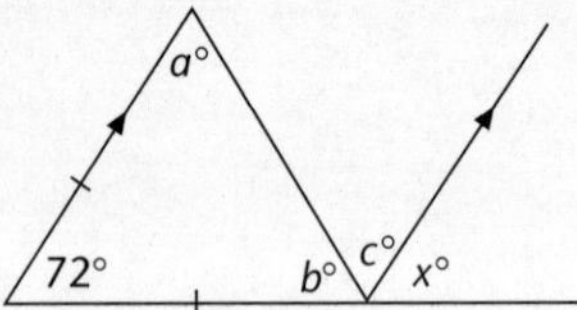

b

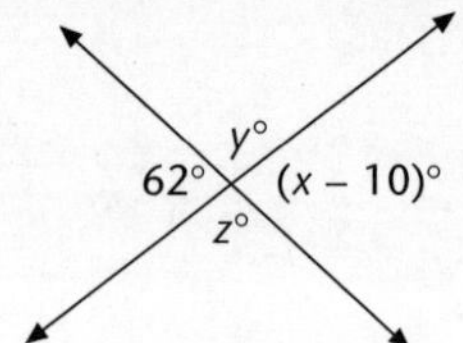

(7 marks)

17 Find the value of x:

a

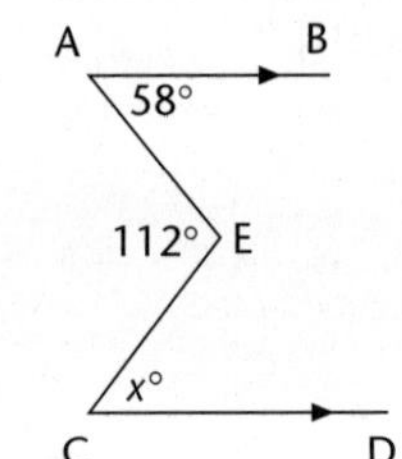

b

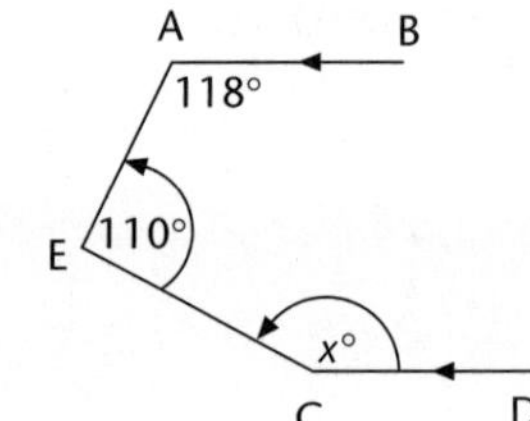

(4 marks)

18 Find the value of x, y and z:

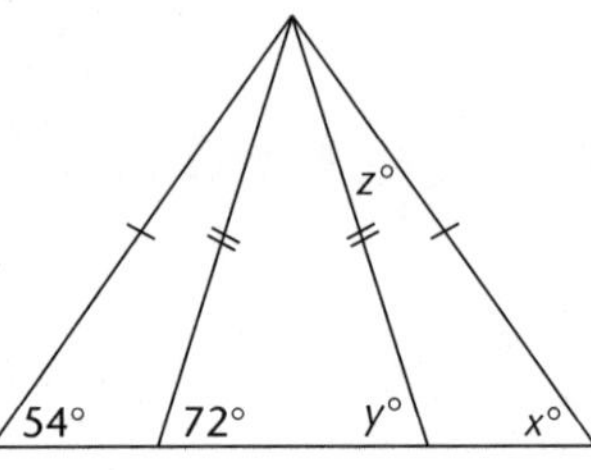

(3 marks)

19 Find the value of x and y:

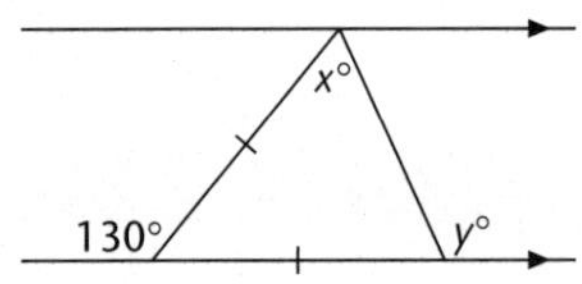

(2 marks)

☞ Quick answers on page 265

☞ Worked solutions on page 334

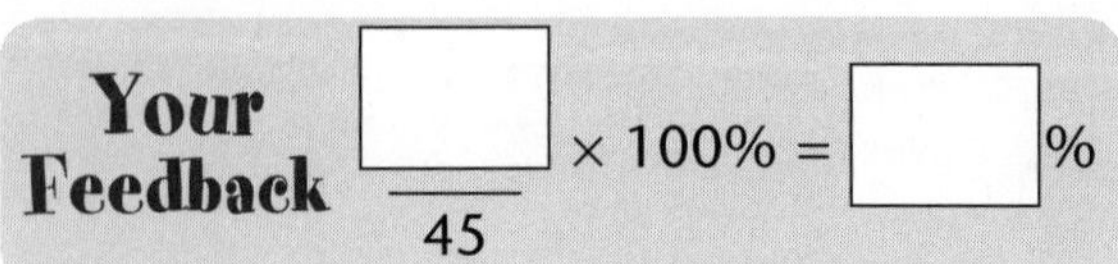

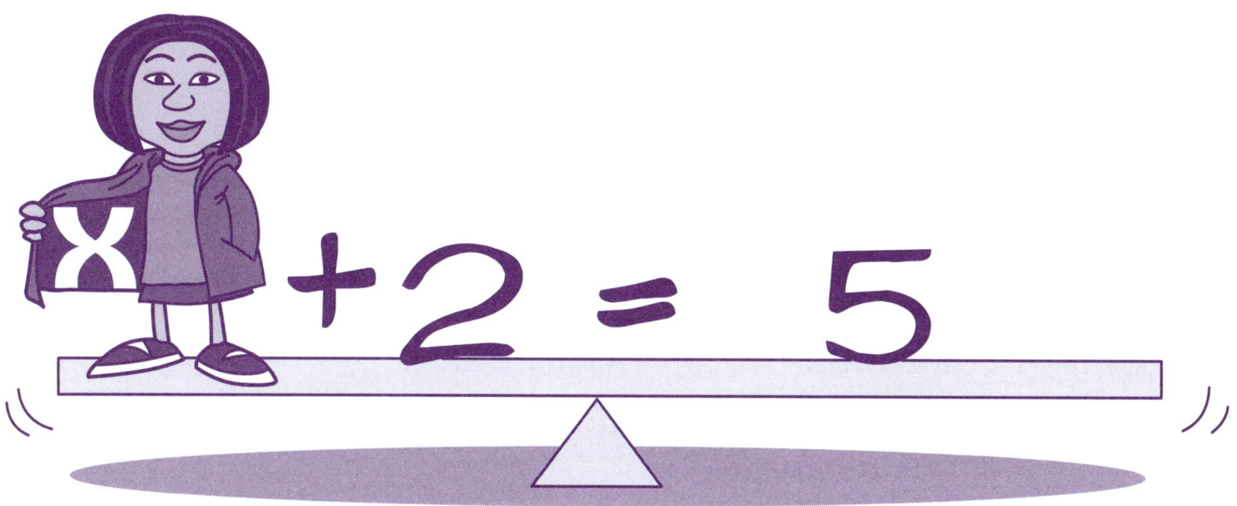

EQUATIONS 10

- Solution of Equations
- Equations in Geometry
- Word Problems Leading to Equations

KEYWORDS

Consecutive	**Rationals**
Equation	**Solution**
Integers	**Substitution**

An equation is an algebraic expression (sentence) with an equals sign. That is, a sentence like $x + 3 = 7$.

Solving an equation means finding the value of the pronumeral that makes the sentence true.

That is, if $x + 3 = 7$

Then $x = 4$

This is the solution of the equation.

Solution of Equations

One-Step Equations

When solving equations we use the following rules.

Rule 1

$x + a = b$ '+' = '−'

$\therefore x = b - a$

The + a on the left-hand side of the equals sign becomes − a when it goes over to the right-hand side of the equal sign.

For Example

1 Solve $x + 4 = 12$.

1 $x + 4 = 12$

$\therefore x = 12 - 4$

$\therefore x = 8$

[The + 4 goes over the equals sign and becomes − 4.]

Rule 2

$x - a = b$

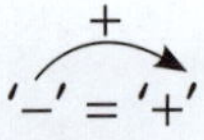

$\therefore x = b + a$

The $-a$ on the left-hand side of the equals sign becomes $+a$ when it goes over to the right-hand side of the equals sign.

For Example

1 Solve $a - 3 = 5$.

1 $a - 3 = 5$

$\therefore a = 5 + 3$

$\therefore a = 8$

[The − 3 goes over the equals sign and becomes + 3.]

Rule 3

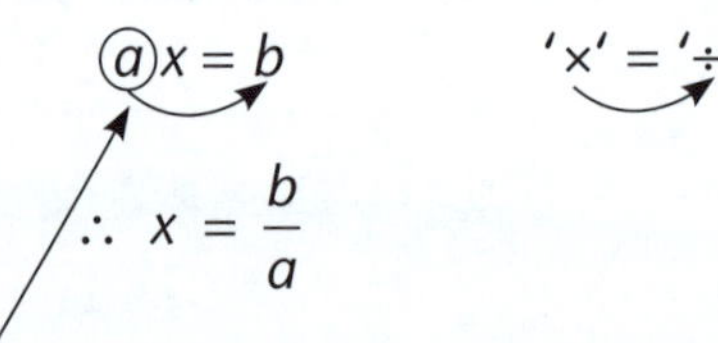

$ax = b$

$\therefore x = \frac{b}{a}$

This means '$a \times x$'.

The a on the left-hand side goes over the equals sign and divides the right-hand side. It does not change sign.

For Example

1 Solve $3x = 12$.

1 $3x = 12$

$\therefore x = \frac{12}{3}$

$\therefore x = 4$

[The 3 goes over the equal sign and divides the 12.]

Rule 4

$$\frac{x}{a} = b$$

$$x = ba$$

'÷ = ×'

This means '$x \div a$'.

The a on the left-hand side goes over the equal sign and multiplies the right-hand side.

For Example

1 Solve $\frac{x}{4} = 12$.

1 $\frac{x}{4} = 12$

$\therefore x = 12 \times 4$

$\therefore x = 48$

[The 4 goes over the equal sign and multiplies the 12.]

Examples Involving Rationals and Integers

The same rules apply.

For Example

Solve the following equations:

1 $x + 7 = 3$

2 $x - 3 = -8$

3 $4x = 12$

4 $x - \frac{1}{2} = \frac{3}{4}$

5 $x + 2\frac{1}{2} = 4\frac{3}{4}$

1 $x + 7 = 3$

$\therefore x = 3 - 7$

$\therefore x = -4$

[+ 7 goes over the equal sign and becomes − 7.]

2 $x - 3 = -8$

$\therefore x = -8 + 3$

$\therefore x = -5$

[− 3 goes over the equal sign and becomes + 3.]

3 $4x = 12$

$\therefore x = \frac{12}{4}$

$\therefore x = 3$

[4 goes over the equal sign and divides 12. It does not become − 4.]

4 $x - \frac{1}{2} = \frac{3}{4}$

$x = \frac{3}{4} + \frac{1}{2}$

$x = 1\frac{1}{4}$

[$-\frac{1}{2}$ goes over the equal sign and becomes $+\frac{1}{2}$.]

5 $x + 2\frac{1}{2} = 4\frac{3}{4}$

$x = 4\frac{3}{4} - 2\frac{1}{2}$

$x = 2\frac{1}{4}$

[$+2\frac{1}{2}$ goes over the equal sign and becomes $-2\frac{1}{2}$.]

Summary

- When a term goes over the equal sign of an equation, its sign is changed to the opposite sign.
- Each side of an equation may be multiplied or divided by the same number.

Two-Step Equations

This type of equation involves the steps of multiplying or dividing and adding or subtracting.

For Example

Solve the following equations:

1 $4x - 2 = 18$

2 $3x + 2 = 23$

3 $\frac{x - 2}{3} = 12$

1 $4x - 2 = 18$ [Taking – 2 over to the other side of the equal sign gives + 2.]

$\therefore 4x = 18 + 2$

$\therefore 4x = 20$

$\therefore x = \frac{20}{4}$ [Dividing by 4.]

$\therefore x = 5$

2 $3x + 2 = 23$ [Taking + 2 over the equal sign gives – 2.]

$\therefore 3x = 23 - 2$

$\therefore 3x = 21$

$\therefore x = \frac{21}{3}$ [Dividing by 3.]

$\therefore x = 7$

3 $\frac{x-2}{3} = 12$

$\therefore x - 2 = 12 \times 3$ [Multiplying by 3.]

$\therefore x - 2 = 36$ [Taking – 2 over the equal sign gives + 2.]

$\therefore x = 36 + 2$

$\therefore x = 38$

Equations with Pronumerals on Both Sides of the Equals Sign

When solving an equation with pronumerals on both sides the pronumerals are moved to the left-hand side of the equal sign, and the numbers to the right-hand side.

For Example

Solve the following equations:

1 $7x - 3 = 5x + 11$

2 $4b - 3 = 3b + 6$

3 $5x + 4 = x + 34$

4 $4y = 21 - 3y$

1 $7x - 3 = 5x + 11$ [Taking – 3 to the RHS gives + 3 and $5x$ (= + $5x$) to the LHS gives – $5x$.]

$\therefore 7x - 5x = 11 + 3$ [Collecting like terms.]

$\therefore 2x = 14$

$\therefore x = \frac{14}{2}$ [Dividing by 2.]

$\therefore x = 7$

2 $4b - 3 = 3b + 6$

$\therefore 4b - 3b = 6 + 3$ [Collecting like terms.]

$\therefore b = 9$

3 $5x + 4 = x + 34$ [Taking + 4 to the RHS gives – 4, and taking x (= + $1x$) to the LHS gives – $1x$.]

$\therefore 5x - x = 34 - 4$ [Collecting like terms.]

$\therefore 4x = 30$

$\therefore x = \frac{30}{4}$ [Dividing by 4.]

$\therefore x = 7\frac{1}{2}$

4 $4y = 21 - 3y$

$\therefore 4y + 3y = 21$

$\therefore 7y = 21$

$\therefore y = \frac{21}{7}$

$\therefore y = 3$

[When solving equations, keep the equal signs underneath each other.]

Equations in Geometry

Equations can be used to solve problems in geometry.

Some Important Geometrical Facts

You need to know these facts so you can form equations:

- Complementary angles add up to 90°.
- Supplementary angles add up to 180°.

- Angles on a straight line add up to 180°.
- Vertically opposite angles are equal.
- Angles at a point always add up to 360°.
- The angle sum of a triangle is 180°.
- The angle sum of a quadrilateral is 360°.
- The base angles of an isosceles triangle are equal.
- Any angle in an equilateral triangle is equal to 60°.
- The exterior angle of a triangle is equal to the sum of the two opposite interior angles.
- For a pair of parallel lines cut by a transversal:
  - Alternate angles are equal
  - Corresponding angles are equal
  - Co-interior angles are supplementary.

For Example

For the following diagrams, use a geometrical fact to form an equation. Solve the equation in each case to find x:

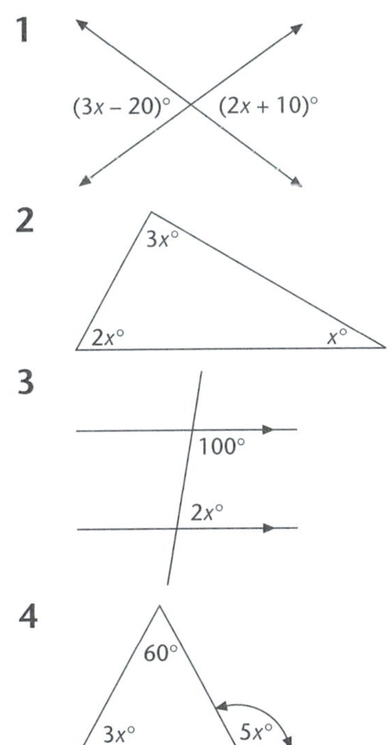

1
$3x - 20 = 2x + 10$
$\therefore 3x - 2x = 10 + 20$
$\therefore x = 30$

[Vertically opposite angles.]

[We do not use degrees in the equation as x is a number—units are not needed in either the question or the answer.]

2
$3x + 2x + x = 180$
$\therefore 6x = 180$
$\therefore x = \frac{180}{6}$
$x = 30$

[Angle sum of a triangle = 180.]

[Collecting terms = $6x$.]

3
$2x + 100 = 180$
$\therefore 2x = 180 - 100$
$\therefore 2x = 80$
$\therefore x = \frac{80}{2}$
$\therefore x = 40$

[Co-interior angles are supplementary when the lines are parallel.]

4
$5x = 3x + 60$
$\therefore 5x - 3x = 60$
$\therefore 2x = 60$
$\therefore x = \frac{60}{2}$
$\therefore x = 30$

[Exterior angle of a triangle is equal to the sum of the opposite interior angles.]

Word Problems Leading to Equations

Problems are often most easily solved by translating the facts given in words into an equation involving numbers and letters.

Procedure for Solving Word Problems

When solving word problems with one unknown, use the following procedure:

a Let the unknown quantity be represented by a pronumeral (say, x).

b Form an equation to represent the facts given in the problem.

c Solve the equation to find the value of the pronumeral.

d Translate the solution calculated into words to answer the question.

For Example

1 Five more than a number is equal to 17. Find the number.

1 Let the number be x. [step **a** of procedure]

Then $x + 5 = 17$ [step **b** of procedure]

$\therefore x = 17 - 5$

$\therefore x = 12$ [step **c** of procedure: solve equation]

$\therefore$ The number is 12. [step **d** of procedure]

For Example

1 If 7 is subtracted from 2 times a certain number, the result is 23. Find the number.

1 Let the number be x. [step **a** of procedure]

$\therefore 2x - 7 = 23$ [step **b** of procedure]

$2x = 23 + 7$

$2x = 30$

$x = 15$ [step **c** of procedure]

$\therefore$ The number is 15. [step **d** of procedure]

For Example

1 The sum of two consecutive integers is 33. (An integer is a whole number.) Find the two integers.

1 Let the consecutive integers be x and $x + 1$.

[step **a** of procedure]

[Note: Consecutive integers follow each other (e.g. 4, 5 ...). So if x is an integer, the next consecutive integer is $x + 1$.]

$\therefore x + x + 1 = 33$ [step **b** of procedure]

$2x + 1 = 33$

$2x = 33 - 1$

$2x = 32$

$x = \frac{32}{2}$

$x = 16$ [step **c** of procedure]

Therefore, the consecutive integers are 16 and 17.

[step **d** of procedure]

Go to p. 259 for quick answers, or to pp. 303–306 for worked solutions.

1 Solve the following equations, showing all necessary working: pp. 204–205

a $x + 4 = 16$ **b** $x - 2 = 5$ **c** $x + \frac{1}{4} = \frac{1}{2}$

d $5x = 25$ **e** $x - 3 = -18$ **f** $x + 3 = 2$

g $\frac{x}{5} = 1$ **h** $\frac{x}{4} = 4$ **i** $\frac{x}{3} = 2$

j $2x = 5$ **k** $5x = 14$ **l** $3x = 7$

m $\frac{x}{2} = 3\frac{1}{2}$

2 Solve the following equations, showing all necessary working: pp. 205–206

a $2x - 1 = 7$ **b** $5x + 2 = 17$

c $3x - 1 = 7$ **d** $2x - 1 = 9$

e $5x - 18 = 7$ **f** $4x + 3 = 35$

g $3 + 2x = 7$ **h** $12 + x = 16$

i $-5 + 3x = 16$ **j** $-8 + 2x = -4$

3 Solve the following equations: pp. 205–206

a $5x - 2 = 3x + 8$ **b** $4x = 2x + 12$

c $6a - 10 = 4a + 6$ **d** $5b - 5 = 2b + 10$

e $5x - 10 = 3x + 12$ **f** $7x + 5 = 3x + 11$

g $12x = 10x + 8$ **h** $2x - 1 = 5 - x$

i $4 - 3x = 5 - 4x$ **j** $2y = 4 - 2y$

4 Solve the following equations: pp. 205–206

a $\frac{x-2}{3} = 5$ **b** $\frac{x+2}{4} = 2$

c $\frac{2x-4}{3} = 6$ **d** $5x = 3x + 7$

e $\frac{x}{4} = 2\frac{1}{2}$ **f** $3x - 14 = 1$

g $4 + a = 17$ **h** $14d - 7 = 4d + 3$

i $4x - 12 = 2x$ **j** $5x - 8 = 3x$

5 Write equations, giving reasons, and solve them to find the value of each pronumeral: pp. 206–207

a

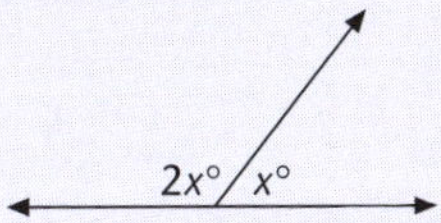

b

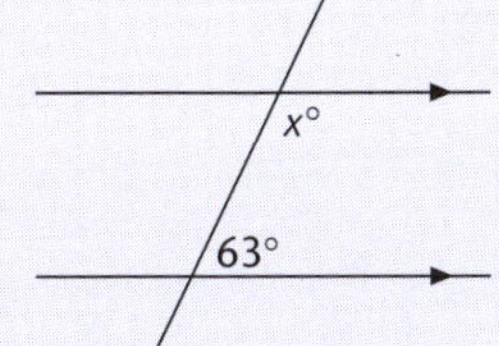

c

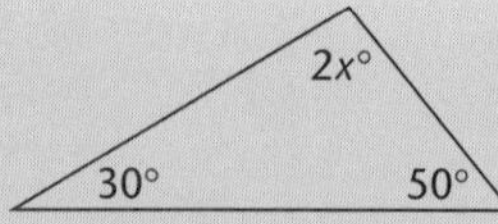

d

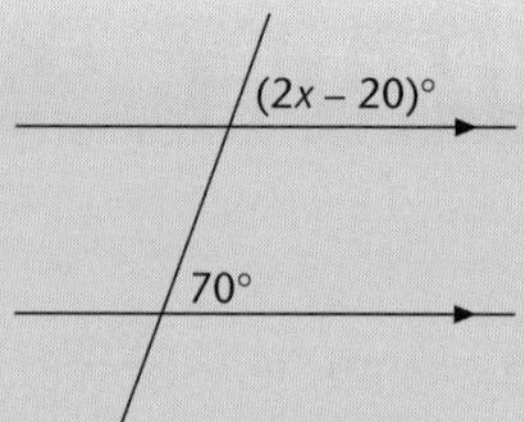

e

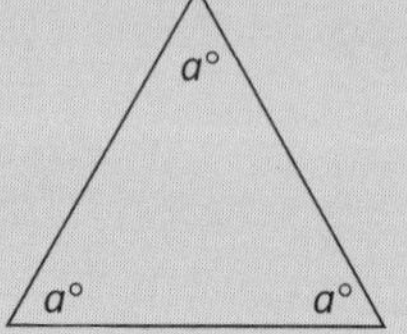

f

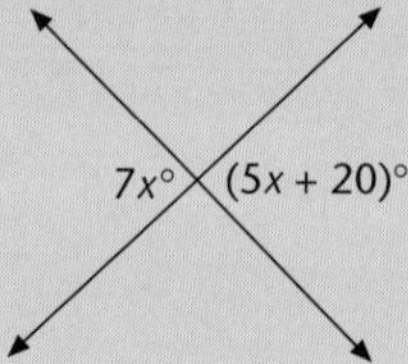

g

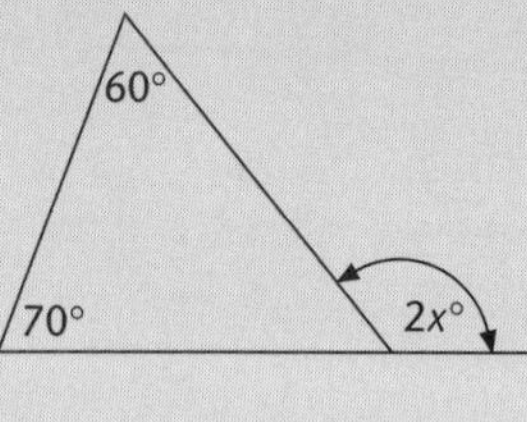

h

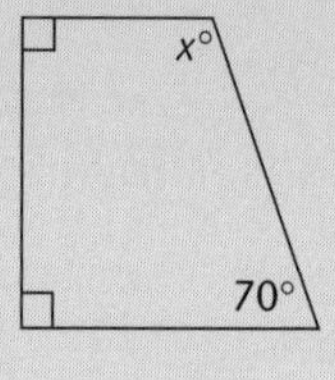

i

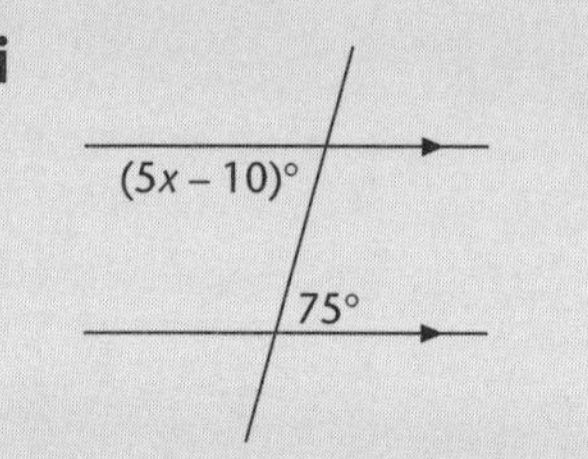

j

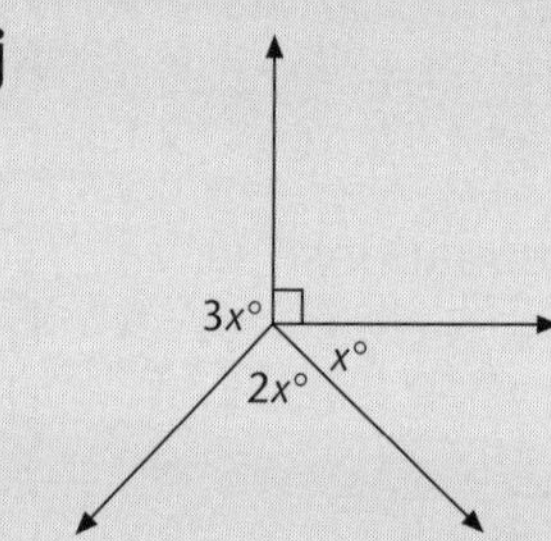

6 For each of the following statements, form an equation and then solve it. pp. 207–208

- **a** Five more than a number is equal to 25. Find the number.
- **b** If 10 is subtracted from a certain number the result is 12. Find the number.
- **c** Three times a number results in 19. Find the number.
- **d** Eight more than a number is equal to 2. Find the number.
- **e** Three times a certain number added to 8 comes to 44. Find the number.
- **f** The sum of two consecutive whole numbers is 43. Find the numbers.
- **g** The sum of three consecutive whole numbers is 39. Find the numbers.
- **h** If you subtract 15 from twice a certain number, the answer is 33. Find the number.
- **i** If you divide a certain number by 5 the result is 8. Find the number.
- **j** If 3 is subtracted from twice a number, the result is the same as when 4 is added to the number. Find the number.

Go to p. 259 for quick answers, or to pp. 303–306 for worked solutions.

YOUR CHECKLIST

For a complete understanding of this topic you must be able to:

✓	Solve equations with basic operations		pp. 204–206
✓	Convert geometry problems into equations and solve them		pp. 206–207
✓	Convert problems given in words into equations and solve them.		pp. 207–208

Now you are ready to do the tests!

(30 marks)

1 State whether the value given in brackets is a solution of the given equation:

a $a + 4 = 10$ [$a = 14$]

b $x - 5 = 3$ [$x = 8$]

c $5x = 4.5$ [$x = 9$]

d $\frac{m}{4} = 3$ [$m = 12$]

e $3x - 1 = 20$ [$x = 7$]

f $\frac{p + 2}{3} = 9$ [$p = 4$] (6 marks)

2 a If $p + 3 = 12$, then $p =$ ____.

b If $a - \frac{1}{2} = 3$, then $a =$ ____.

c If $\frac{m}{10} = 1.23$, then $m =$ ____.

d If $7p = 21$, then $p =$ ____.

e If $2y - 1 = 11$, then $y =$ ____. (5 marks)

3 Solve the following equations:

a $a + 8 = 23$

b $p - 7 = 15$

c $\frac{a - 1}{2} = 9$

d $\frac{x}{5} = 6$

e $y + 7 = 2$ (5 marks)

4 Solve the following equations:

a $2x + 1 = 7$

b $2x - 3 = 7$

c $\frac{a - 2}{3} = 6$

d $2x = x + 3$

e $8 + 3a = 14$ (10 marks)

5 a

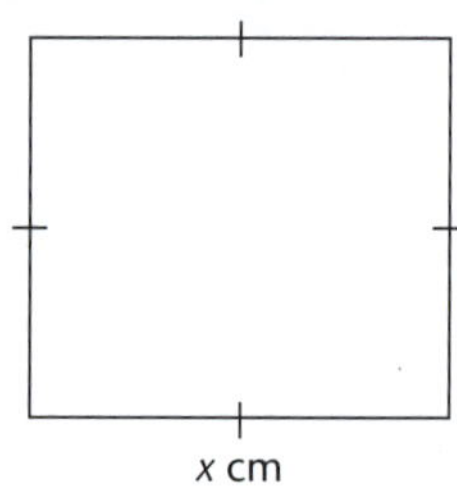

A square has sides with length x cm and perimeter 20 cm:

i Find the value of x.

ii What is the area of the square?

b Michael is x years old. His brother Peter is 4 years younger:

i Find an expression for Peter's age in terms of x.

ii If the sum of their ages is 10 years, how old is Michael? (4 marks)

Your Feedback $\frac{\square}{30} \times 100\% = \square\%$

☞ **Quick answers on page 265**
☞ **Worked solutions on page 334**

LEVEL 2 TEST

(30 marks)

1 State whether $x = 5$ is a solution of the equation $\frac{3x}{2} = 10$. (1 mark)

2 Solve the following equations:

a $x + 2.3 = 7.4$

b $x - 2\frac{1}{2} = \frac{1}{4}$

c $x + 8 = 2$

d $\frac{x + 1}{2} = 4$

e $5x = 6$

f $x - 1 = -9$ (6 marks)

3 Solve the following equations:

a $5x - 3 = 27$

b $3x + 5 = -7$

c $\frac{2x - 2}{3} = 4$

d $5x = 3x + 12$

e $4x - 3 = 2x + 18$

f $\frac{2x - 1}{3} = 5$ (12 marks)

4 For the following diagrams use a geometrical fact to form an equation. Solve the equation and find the value of x in each case:

a 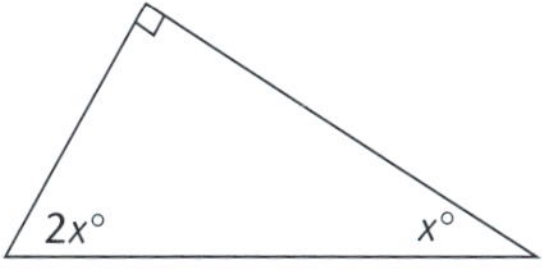

b 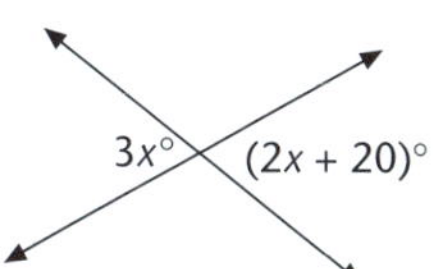

(4 marks)

5 Given $v = u + at$:

a Find v when $u = 5$, $a = 2$ and $t = 3$.

b Find the value of a given $v = 20$, $u = 5$ and $t = 3$. (2 marks)

6 If 3 is subtracted from twice a number the result is 21. Let the number be x, form an equation and find the number. (2 marks)

7 The perimeter of the rectangle is 16 cm:

a Form an equation and find the value of x.

b Hence, find the area of the rectangle. (3 marks)

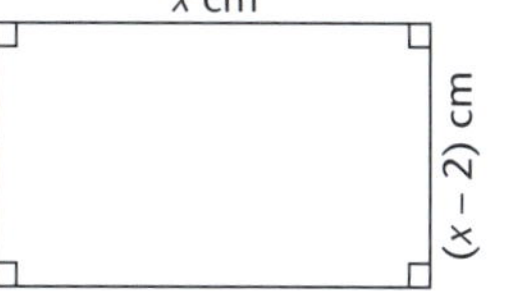

Your Feedback $\frac{\square}{30} \times 100\% = \square\%$

☞ Quick answers on page 265
☞ Worked solutions on page 335

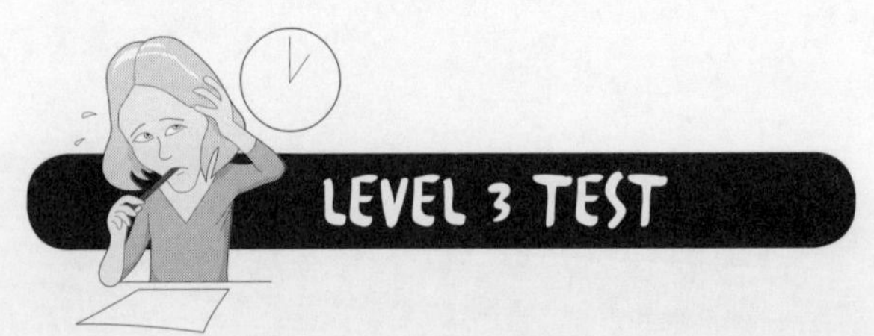

(30 marks)

1 The solution of the equation $2x - 1 = 7$ is:

A 4 B −4 C 3 D −3 (1 mark)

2 Which of the following satisfies the equation $\frac{2x}{3} - 1 = 7$?

A $4\frac{1}{2}$ B 4 C 12 D 8 (1 mark)

3 Given $x = \frac{y}{z}$ and $x = 4$, $y = 32$, the value of z is:

A 8 B 128 C 28 D 36 (1 mark)

4 To solve the equation $3x + 2 = 6 - x$, Lucy wrote the following solution:

$$3x + 2 = 6 - x$$
$$3x - x = 6 - 2 \quad (1)$$
$$2x = 4 \quad (2)$$
$$x = 2 \quad (3)$$

Which of the following is true?

A There is no mistake in the solution.
B There is a mistake only in line (1).
C There are mistakes in lines (2) and (3).
D There are mistakes in lines (1) and (3). (1 mark)

5 The solution of the equation $4x = -2x + 6$ is:

A 1 B 3 C −3 D −1 (1 mark)

6 Solve the following equations:

a $a - 3 = -15$ b $x + 17 = 5$ c $\frac{x}{0.4} = 1.2$

d $7x = 23$ e $\frac{3x}{2} - 1 = 12$ (5 marks)

7 Solve the following equations:

a $5x + 17 = 22$ b $\frac{2x - 3}{5} = 3$

c $12x - 7 = 4x + 49$ d $2x - 6 = -4x$ (8 marks)

8 Given $K = \frac{2B - 1}{A}$ find the value of B if $K = 12.5$ and $A = 4$. (2 marks)

9 If three times a number is subtracted from 17 the result is 2. Form an equation and solve it to find the number. (2 marks)

10 The sum of two consecutive odd integers is 68. Form an equation and solve it to find the numbers. (3 marks)

11 In an isosceles triangle, the base angles have a size of $2x°$ each. The third angle is 52°. Find the size of the base angles. (2 marks)

12 The perimeter of the figure is 18 cm. Find its area:

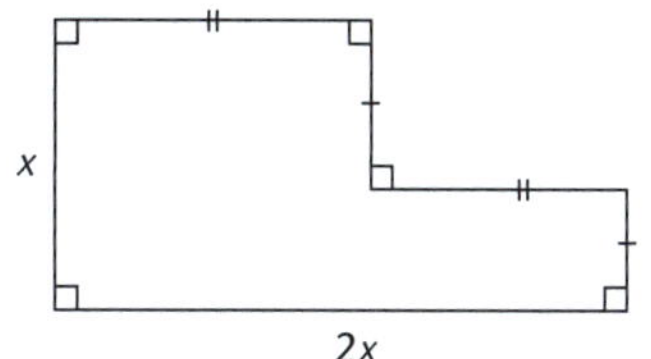

(3 marks)

☞ Quick answers on page 265
☞ Worked solutions on page 336

Your Feedback $\frac{\square}{30} \times 100\% = \square\%$

11

STATISTICS AND PROBABILITY

- Some Explanations
- Census versus Sample
- Biased and Random Samples
- Data Representation
- Data Analysis
- Probability

KEYWORDS

Analyse	**Mean**
Bias	**Median**
Categorical data	**Mode**
Census	**Organisation**
Clustered	**Outlier**
Collection	**Polygon**
Complementary events	**Population sample**
Continuous data	**Probability event outcome**
Data	**Quantitative data**
Discrete data	**Random**
Dot plot	**Range**
Frequency	**Scatter diagram**
Grouped data	**Score**
Histogram	**Spread**
Information	**Stem-and-leaf**
Location	**Tally**

Some Explanations

Statistics involves the collection and organisation of information (**data**) so that:

a Large amounts of information can be easily analysed, and

b Predictions can be made, based on analysis of the data collected.

Tables and graphs allow information to be presented in a clear, concise form. The information can also be readily analysed from the table or graph.

Census versus Sample

A **census** involves surveying every member of the target population, while a **sample** is a limited survey used to get a representative or typical view. The Australian Government conducts a regular census of all Australian households, asking questions about such things as number of residents, ages of residents, occupations etc. The collected census information is analysed and used for future planning by government departments. Things such as new subdivisions, schools, infrastructure and hospitals can be planned in advance, based on information in the census.

Biased and Random Samples

If a sample is **random**, then there is no **bias**. A biased sample is one that does not represent the whole population. For example, if a sample is conducted at a shopping centre on a Thursday morning to find the percentage of Australians who eat Weet Bix for breakfast, it would be a biased sample.

In a random sample participants are chosen 'at random'. For example, to randomly select from 1000 students, a teacher might allocate a three-digit number to each student. She can then use her calculator to generate random numbers to find 20 students to interview.

Data Representation

Data can be **quantitative** or **categorical**. Examples of quantitative data are shoe sizes (**discrete**) or height (**continuous**), while an example of categorical data is hair colour: brown, black, blonde, red. Collected data can be represented in tables or graphs.

Frequency Distribution Table

A table is used to summarise listed data and allows easy analysis of the location and spread of the data.

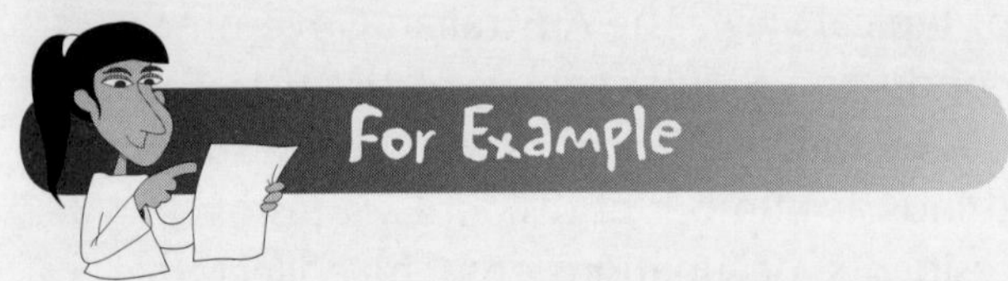

Consider the shoe sizes of 50 11-year-old boys:

6	$4\frac{1}{2}$	7	4	5
$5\frac{1}{2}$	6	(8)	5	7
5	$5\frac{1}{2}$	7	$4\frac{1}{2}$	5
6	$5\frac{1}{2}$	4	6	6
7	$4\frac{1}{2}$	5	5	6
5	5	$5\frac{1}{2}$	$4\frac{1}{2}$	5
$7\frac{1}{2}$	($3\frac{1}{2}$)	5	$4\frac{1}{2}$	$6\frac{1}{2}$
$4\frac{1}{2}$	$5\frac{1}{2}$	6	$7\frac{1}{2}$	6
6	5	6	5	7
6	6	7	$5\frac{1}{2}$	6

1 Draw up a frequency distribution table for this information.

2 How many boys wear a size $5\frac{1}{2}$ shoe?

3 Which size shoe is most commonly worn?

4 If this is a typical example of 11-year-old boys, what fraction of 11-year-old boys in Australia wear a size 6 shoe?

5 If the population of 11-year-olds in Hambelton is 200, how many would you predict wear a size 6 shoe?

1

Score (x)	Tally	Frequency (f)	
$3\frac{1}{2}$	|	1	
4	||	2	
$4\frac{1}{2}$	~~||||~~ |	6	
5	~~||||~~ ~~||||~~ ||	12	
$5\frac{1}{2}$	~~||||~~ |	6	←(b)
6	~~||||~~ ~~||||~~ |||	13	←(c)
$6\frac{1}{2}$	|	1	
7	~~||||~~ |	6	
$7\frac{1}{2}$	||	2	
8	|	1	
	Σf	50	

[Σ means the sum or running total]

The completed table can then be used to answer the questions:

2 Six boys wear size $5\frac{1}{2}$ shoes.

3 The most common size is 6 as 13 boys wear size 6.

4 Fraction of 11-year-old boys with size 6 shoe:

$$= \frac{13}{50} \left(\frac{\text{Number wearing size 6}}{\text{Total number of boys in survey}} \right)$$

5 Number wearing size 6 $\doteqdot \frac{13}{\cancel{50}_1} \times \frac{\cancel{200}^4}{1}$

$\doteqdot 52$

[$\doteqdot$ or » means 'approximately equal to']

You would expect about 52 11-year-olds in Hambelton to wear size 6 shoes.

Frequency Distribution Table for Grouped Data

When data is continuous it is sensible to use class intervals.

For Example

The mass of forty students was measured and the results listed below:

53	51	62	64	63
71	75	79	56	60
53	48	61	64	63
67	59	63	68	44
53	55	59	52	64
67	68	72	75	79
61	65	48	41	47
48	55	57	56	61

1 Arrange the data in a frequency distribution table using class intervals of 41–45, 46–50 etc. Also, find the class centre for each interval.

2 What was the most common class interval of masses (the **modal** class)?

1

Class intervals	Class centre (x)	Tally	Frequency (f)
41–45	43	||	2
46–50	48	||||	4
51–55	53	𝍸 ||	7
56–60	58	𝍸 |	6
61–65	63	𝍸 𝍸 |	11
66–70	68	||||	4
71–75	73	||||	4
76–80	78	||	2
			40

2 Most common class interval of masses was 61–65.

[Note: The scores are **clustered** in the low 60s.]

Stem-and-Leaf Plot

The stem is the *first* digit or digits of a number, whereas the leaf is the *last* digit. The leaf is always a single digit.

For Example

1 The results of a mathematics test were recorded:

19	48	36	40	31	22	18	27
18	20	18	36	49	60	45	13
9	17	22	31	39	26	28	30
44	8	23	19	46	33		

Draw a stem-and-leaf plot for this data.

1

Stem	Leaf
0	98
1	9888379
2	2702683
3	6161903
4	809546
5	
6	0

This is sometimes referred to as the initial plot. However, the plot can be refined by ordering each of the leaves:

Stem	Leaf
0	89
1	3788899
2	0223678
3	0113669
4	045689
5	
6	0

An ordered stem-and-leaf plot can be used to analyse the data. A back-to-back stem-and-leaf plot is used to compare two sets of data.

[Note: The score of 60 is called an **outlier**.]

Frequency Histogram and Polygon

When the data are continuous there are two types of frequency graphs that can be used: a **histogram** and a **polygon**.

Here the table has been used to graph a histogram and frequency polygon:

x	f
3	1
4	3
5	3
6	4
7	5
8	4
9	4
10	1

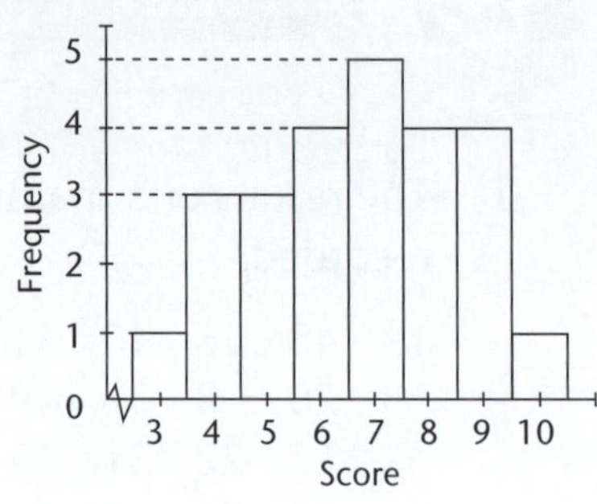

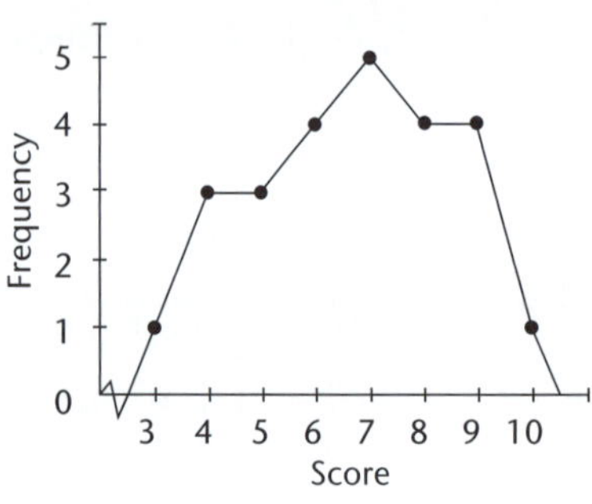

Columns on the histogram are around the scores. Their height is their frequency.

Straight lines on the frequency polygon join dots at the height of the frequency value. The line meets the horizontal axis where the next score would be if there was one. The polygon can also be made by joining the mid-points of the tops of the columns in the histogram.

Dot Plots

A dot plot is an alternative to drawing a histogram and can either be vertical or horizontal.

1 The type of cars passing along Honus Avenue over a half-hour period was noted and the results recorded:

Car manufacturer	Frequency
Ford	8
Holden	11
Toyota	7
Mitsubishi	4
Other	6

Use a dot plot to represent this information.

1 Vehicle survey

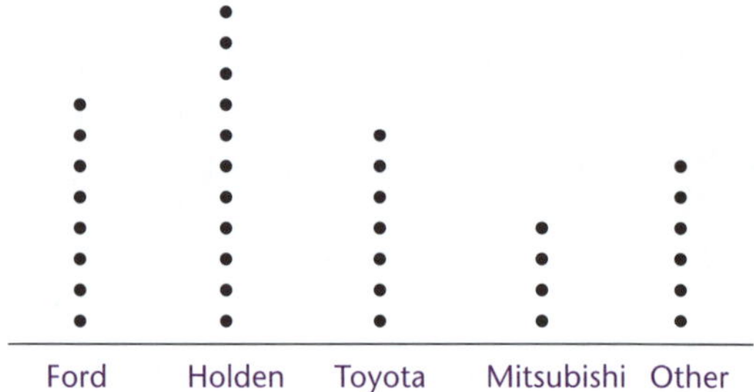

Scatter Diagram

A **scatter diagram** is formed by using points to represent a pair of results.

The results of an English test and a Science test for ten students are detailed:

Student:	A	B	C	D	E	F	G	H	I	J
English:	13	17	14	11	18	17	15	10	20	16
Science:	15	16	13	15	17	18	16	11	14	19

1 Draw a scatter diagram for the data.

2 Which students scored the highest and lowest scores in each test?

1 **Test results**

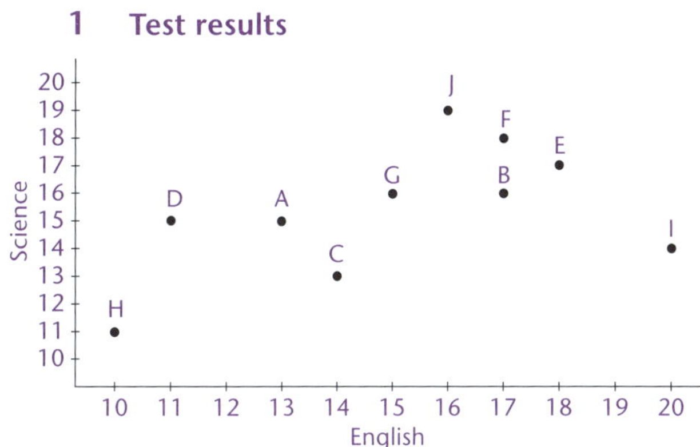

2 English highest: student I; lowest: student H.

Science highest: student J; lowest: student H.

Data Analysis

Measures of Spread

The **range** indicates the size of the distribution of scores; that is, the range = highest score – lowest score.

1 Find the range of the scores 4, 17, 6, –5, 4, 2, 12.

1 Range = 17 – (–5)

= 17 + 5 = 22

Measures of Location

- **Mean** is the *average score*:

  i.e. $\text{mean} = \dfrac{\text{sum of scores}}{\text{number of scores}}$

  [Notation: $\bar{x}$ = mean]

- **Mode** is most common score—the score with the highest frequency.
- **Median** is the middle score when the scores are arranged in order.

Consider these sets of scores for Alison's first season of cricket. Over the first seven innings she scored 17, 14, 10, 10, 9, 21 and 10. Over the next eight innings she scored 0, 19, 10, 26, 19, 35, 0 and 29.

1 For the first seven innings, calculate the range, mode, mean and median.

2 For the next eight innings, calculate the range, mode, mean and median.

3 Taking all scores together, find the season's range, mode, mean and median.

1 17, 14, 10, 10, (9), (21), 10

Range = 21 – 9

= 12

Mode = 10

$$\text{Mean} = \frac{17 + 14 + 10 + 10 + 9 + 21 + 10}{7} = \frac{91}{7} = 13$$

For the median, rewrite the scores in ascending order:

9 10 10 10 14 17 21

↑

3 scores below ⟵ middle score ⟶ 3 scores above

Median = 10

[A suggested method to locate the middle is to cross off the first and last numbers, then the remaining first and last numbers etc. until only one or two middle numbers are left.]

2 0, 19, 10, 26, 19, 35, 0, 29

Range = 35 – 0 = 35

Mode = 0 and 19

$$\text{Mean} = \frac{0 + 19 + 10 + 26 + 19 + 35 + 0 + 29}{8}$$
$$= \frac{138}{8}$$
$$= 17.25$$

For the median, consider:
0, 0, 10, 19, 19, 26, 29, 35.

Median is middle of 4th and 5th scores; that is, 19, 19.
Median = 19

3 17, 14, 10, 10, 9, 21, 10, 0, 19, 10, 26, 19, 35, 0, 29

Range = 35 – 0 = 35

Mode = 10

$$\text{Mean} = \frac{17 + 14 + 10 + 10 + 9 + 21 + 10 + 0 + 19 + 10 + 26 + 19 + 35 + 0 + 29}{15}$$
$$= \frac{229}{15}$$
$$= 15.27 \text{ (to 2 decimal places)}$$

For median: 0, 0, 9, 10, 10, 10, 10, 14, 17, 19, 19, 21, 26, 29, 35

Median is the middle score.

Median = 14

Using a Calculator to Find the Mean

A calculator can be placed in 'statistics' mode to find a number of statistical measures. Refer to your calculator manual for instructions.

Measures of Location and Spread from a Table or Graph

Mean, mode, median and range can be found when data is arranged in a table or graph.

For Example

Score	Frequency
12	5
13	7
14	8
15	7
16	3

1 Find the mean, mode, median and range.

1 To find the mean, another column is added to the table. In this column, we multiply each score by its frequency, i.e. the $x \times f$ column.

Score (x)	Frequency (f)	fx
12	5	60
13	7	91
14	8	112
15	7	105
16	3	48
Total	**30**	**416**

- **Mean:** $(\bar{x}) = \dfrac{\text{Sum of scores}}{\text{Number of scores}} = \dfrac{\text{Sum of } fx}{\text{Sum of } f} = \dfrac{\sum fx}{\sum f}$

  i.e. $\bar{x} = \frac{416}{30}$
  $= 13.87$ (to 2 decimal places)
- **Mode:** 14 (has highest frequency of 8)
- **Median:** As there are 30 scores, then the median is the average of the two middle scores:

  i.e. the average of 15th and 16th scores

  i.e. the average of 14 and 14

  ∴ Median = 14
- **Range:** 16 – 12 = 4

For Example

1 The divided stem-and-leaf plot shows the results for the boys and girls in a maths test:

Boys	Stem	Girls
42	0	9
98430	1	35689
85441	2	555899
62	3	0025

Find the mean, mode, median and range for boys and girls.

1 ■ **Mean**—Boys: $(2 + 4 + 10 + 13 + \dots) \div 14$

[or use calculator in statistics mode]

$= 19.29$ (to 2 decimal places)

—Girls: $(9 + 13 + 15 + \dots) \div 16$

$= 23.625$

■ **Mode**—Boys: 24; Girls: 25

■ **Median**

Boys	Stem	Girls
42	0	9
98430	1	35689
85441	2	555899
62	3	0025

Boys: Middle (average) of 19 and 21

i.e. $\dfrac{19 + 21}{2} = 20$

Girls: Middle of 25 and 25 $\therefore$ 25

■ **Range**—Boys: $36 - 2 = 34$

Girls: $35 - 9 = 26$

For Example

1 A class is surveyed to find the number of mobile phones per household:

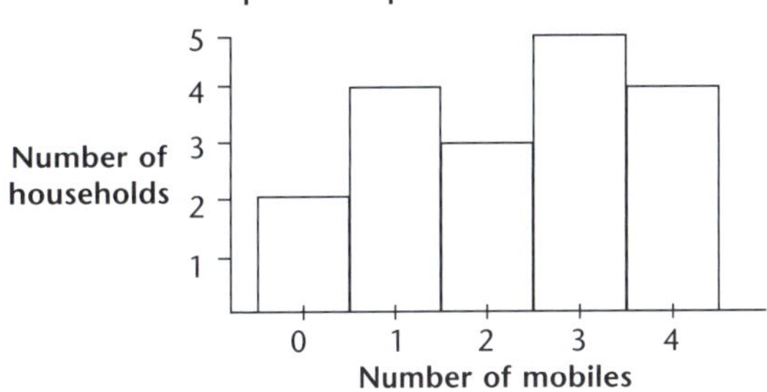

Find the mean, mode and median mobiles.

1 **Mean**: To help find the mean students can either complete a frequency table for the histogram or note the $x \times f$ data on the existing histogram:

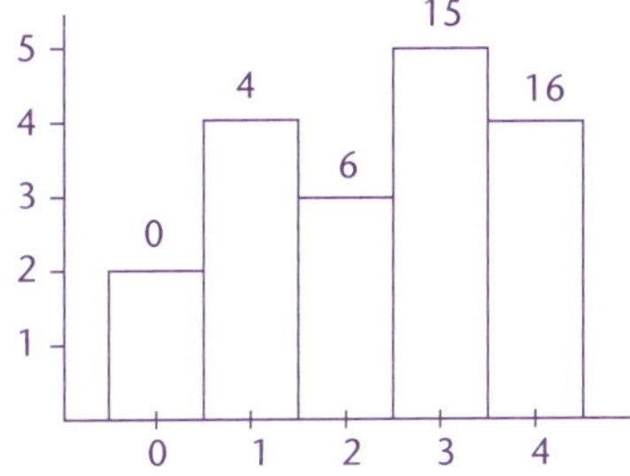

Scores (x)	f	fx
0	2	0
1	4	4
2	3	6
3	5	15
4	4	16
	18	41

$$\bar{x} = \frac{\sum fx}{\sum f}$$

$$= \frac{0 + 4 + 6 + 15 + 16}{2 + 4 + 3 + 5 + 4}$$

$$= \frac{41}{18}$$

$$= 2.2\dot{7}$$

Mode: 3 (highest frequency—tallest column)

Median: As there are 18 scores:

∴ median is middle of 9th and 10th score

i.e. 0, 0, 1, 1, 1, 1, 2, 2, (2, 3)...

∴ median = 2.5

Probability

A coin is tossed once. There is one chance in two that it will land with a head facing up. We say that the probability of throwing a head is one out of two, or:

$P(\text{Head}) = \frac{1}{2}$

Similarly, if a die (singular of dice) is thrown once, there is one chance in 6 of throwing a two:

i.e. $P(2) = \frac{1}{6}$

The probability of an event occurring is defined as:

$$\text{Probability (Event)} = \frac{\text{Number of favourable outcomes}}{\text{Number of possible outcomes}}$$

or $$P(\text{E}) = \frac{n(\text{E})}{\text{Total number possible}}$$

Note: A certainty has probability of one. An impossibility has probability of zero. Therefore probability is expressed as a fraction between and including 0 and 1.

For Example

A die is rolled once. Find the probability of rolling:

1 3

2 5

3 An odd number

4 An even number

5 A number larger than 4.

1 $P(3) = \frac{1}{6}$

[one 3 can occur out of 6 possible numbers]

2 $P(5) = \frac{1}{6}$

3 $P(\text{Odd}) = \frac{3}{6}$ [odd numbers = 3]

$= \frac{1}{2}$

4 $P(\text{Even}) = \frac{3}{6}$

$= \frac{1}{2}$

5 $P(>4) = \frac{2}{6}$

[numbers greater than 4 are 5 and 6, i.e. 2 numbers out of 6]

$= \frac{1}{3}$

Total Probability

If there are two possible things that can happen, then the sum of their probabilities is 1. When a coin is tossed, for example, the two possible outcomes are head or tail:

$P(\text{Head}) = \frac{1}{2}$ $\quad P(\text{Tail}) = \frac{1}{2}$

Now, $P(\text{H}) + P(\text{T}) = \frac{1}{2} + \frac{1}{2} = 1$

This is called total probability. It is true for any number of possible outcomes. If you add up the probabilities of all possible outcomes they will total one.

For Example

1 When a die is thrown there are 6 possible outcomes: a one, two, three, four, five or six.

1 Now, $P(1) = \frac{1}{6}$, $P(2) = \frac{1}{6}$, $P(3) = \frac{1}{6}$,

$P(4) = \frac{1}{6}$, $P(5) = \frac{1}{6}$, $P(6) = \frac{1}{6}$

Total $P = \frac{1}{6} + \frac{1}{6} + \frac{1}{6} + \frac{1}{6} + \frac{1}{6} + \frac{1}{6} = 1$

Complementary Events

Also, because the total of the probabilities of all possible outcomes is 1, then if there are two possible outcomes and we know the probability of one outcome, then we also know the probability of the other.

For Example

If the probability of rain on a Monday is $\frac{2}{5}$, then the probability of no rain on a Monday:

$= 1 - \frac{2}{5}$

$= \frac{3}{5}$

These are called **complementary events**.

For Example

1 If the probability of an archer hitting a tree is 0.8, calculate the probability of her not hitting the tree.

1 $P(\text{Not hitting}) = 1 - 0.8$

$= 0.2$

For Example

A die is thrown. Calculate the probability that the score on the uppermost face is:

1 Not 6

2 Greater than 2.

1 $P(6) = \frac{1}{6}$

$\therefore P\,(\text{Not } 6) = 1 - \frac{1}{6}$

$= \frac{5}{6}$

2 Prob. (1 or 2) $= \frac{2}{6}$

$= \frac{1}{3}$

$P(>2) = 1 - P(1 \text{ or } 2)$

$= 1 - \frac{1}{3}$

$= \frac{2}{3}$

Go to p. 260 for quick answers, or to pp. 306–311 for worked solutions.

1 Johann Bark surveyed the numbers of the letter t in each line of his book. His results were: pp. 217–218

5	3	6	3	4	2	5	3	4	7
2	2	4	4	5	5	4	3	2	5

From this information, draw up a frequency distribution table and answer the following questions.

a How many lines contained:
- **i** 5 ts?
- **ii** 5 or fewer ts?
- **iii** More than 5 ts?

b What fraction of lines contained:
- **i** 5 ts?
- **ii** 5 or fewer ts?
- **iii** More than 5 ts?

c Find the range and mode.

d Draw a histogram to illustrate this information.

2 The height of 30 students was measured and the results listed: pp. 218–219

142	148	156	158	167	169
171	163	158	149	148	155
158	150	153	163	164	169
148	149	157	158	164	165
173	161	165	149	148	155

Complete the frequency distribution table for the data:

Class	Class centre	Tally	Frequency
141–145			
146–150			
...			

3 A survey of students' pulse rates are detailed: p. 219

59	63	82	76	88	83	76	72	65
58	71	69	74	79	81	69	67	68
73	75	81	80	71	69	59	63	71
78	81	76						

Complete a stem-and-leaf plot for the data.

4 Goldilocks surveyed 25 bears in the forest to find out the number of lumps each had in their porridge that morning. This is the set of numbers that she wrote down: p. 220

7 4 0 2 3 6 4 5 3 2 1 3 7
6 6 4 5 2 4 5 5 4 6 7 6

Draw up a frequency distribution table for this information and answer the following questions.

a How many bears reported:
  i 5 lumps?
  ii 5 or fewer lumps?
  iii Less than 5 lumps?
  iv More than 5 lumps?

b What fraction of the bears survey reported:
  i 5 lumps?
  ii 5 or fewer lumps?
  iii Less than 5 lumps?
  iv More than 5 lumps?

c Draw a polygon to illustrate this information.

5 The table shows the number of soccer goals scored by five friends during a season: p. 220

Name	Mitchell	Brendan	Mathew	Jason	Eddie
Goals	6	3	5	4	2

Draw a dot plot to represent the information.

6 Eight students completed a history and a geography quiz and the results are detailed below: pp. 220–221

Student	**A**	**B**	**C**	**D**	**E**	**F**	**G**	**H**
History	5	8	4	7	6	3	9	5
Geography	2	9	3	7	6	4	8	3

Draw a scatter diagram showing the information.

7 Find the range of the scores 8, 6, 4, 7, 6, 2, –1. p. 221

8 Calculate the mean, and find the mode, median and range of these sets of numbers. (Calculate the mean to 1 decimal place where necessary.) pp. 221–222

a 7, 4, 6, 6, 7, 6, 3
b 5, 10, 8, 7, 11, 6, 5, 9
c 7, 7, 4, 4, 4, 1, 1, 1, 7
d 0, 0, 2, 2, 1, 7, 3, 7, 7, 4, 9, 5
e 7, 11, 9, 4, 11, 7, 4, 11, 9, 3, 7, 4, 5, 7, 6

9 George surveyed children buying joggers from his shoe store and discovered this information: pp. 222–224

Shoe size (x)	Number of children (f)	$x \times f$
3	8	
$3\frac{1}{2}$	8	
4	6	
$4\frac{1}{2}$	12	
5	10	
$5\frac{1}{2}$	8	
6	6	
$6\frac{1}{2}$	4	
7	2	
Sum		

By completing this frequency table, answer the following questions:

a How many children were surveyed?

b How many children wore shoes of size:
 i 5 or less? **ii** Greater than 5?

c What fraction (in simplest form) of the children wore shoes of size:
 i 5? **ii** 5 or less?
 iii Greater than 5?

d Find the range and mode.

e Calculate the mean shoe size.

f Find the median.

g Draw a histogram to illustrate this information.

h George wishes to cater for the children of the region. Which size shoe would you advise George to be sure to keep in stock? Which measure (mean, median or mode) is relevant to this question?

10 Professor Kowalski drew a frequency polygon to represent his students' grades out of 10: pp. 222–224

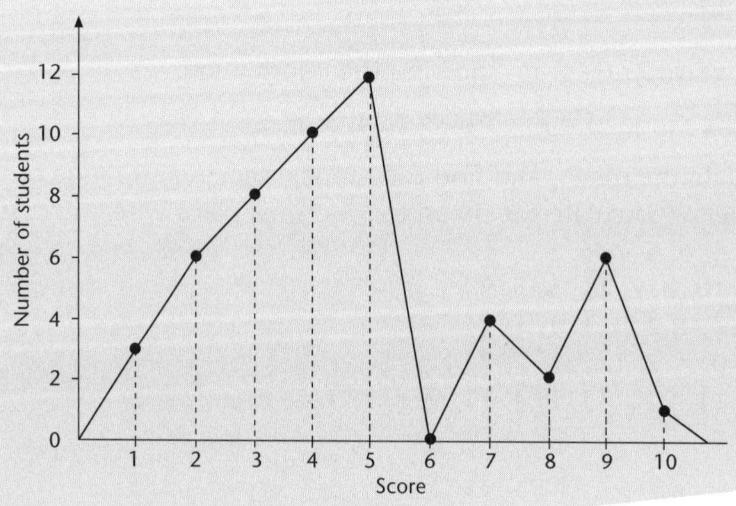

a Use the diagram above to complete the frequency distribution table: pp. 222–224

Score (x)	Frequency (f)	$x \times f$
1		
2		
3		
4		
5	12	60
6		
7		
8		
9		
10		

b Write down the range and mode.

c How many students scored:

  i 5 marks

  ii 5 or fewer marks?

  iii Fewer than 5 marks?

  iv More than 5 marks?

d How many students were in Professor Kowalski's class?

e What fraction of the students scored:

  i 5 marks

  ii 5 or fewer marks?

  iii Fewer than 5 marks?

  iv More than 5 marks?

f Calculate the mean.

g Find the median score.

11 Gai has been playing soccer for six years and has kept a record of her goal scoring. Over these years Gai scored 8, 6, 5, 9, 5 and 3 goals. pp. 221–223

a i Calculate Gai's mean number of goals.

  ii Find the range and the mode.

  iii Find the median number of goals.

b In her seventh year of soccer Gai struck brilliant form and scored 13 goals. How does this affect the mean and median?

c After her eighth season Gai's mean number of goals had slumped to 6.5. How many goals did Gai score in her eighth year?

12 Zoran has to average 70 in his yearly examinations. After getting four results back, his average was 65. How many must Zoran average in his final two papers to attain his overall 70 average? pp. 221–223

13 A survey of families in Nicholas Avenue was conducted to find the number of children in each family. The results were graphed as a histogram: pp. 222–224

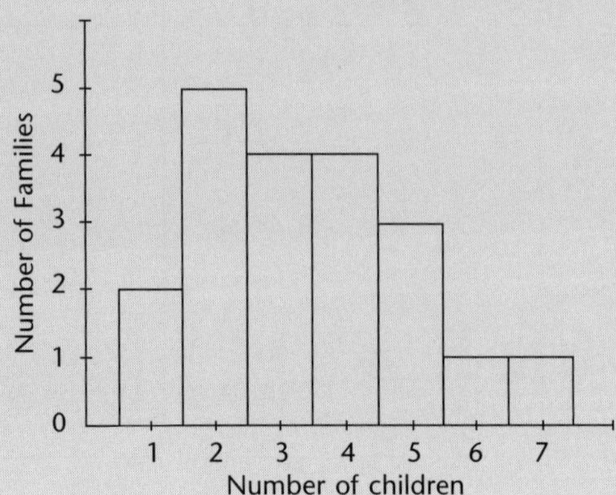

a **i** Complete the frequency distribution table using information from the histogram:

Number of children (Score) x	**Number of families** (Frequency) f	$x \times f$
1		
2		
3		
4		
5		
6		
7		

ii How many families were surveyed?
iii How many children live in Nicholas Avenue?

b Find the range and the mode.

c How many families had:
i 4 children?
ii Less than 4 children?
iii More than 4 children?

d What fraction of the families had:
i 4 children?
ii Less than 4 children?
iii More than 4 children?

e Calculate the mean number of children.

f Find the median number of children.

14 The mass in kg, of students in 7 MP were measured and the results detailed. pp. 221–223

Use the stem-and-leaf plot to answer the following:

Boys	Stem	Girls
	4	9
8743	5	02346
93320	6	03488
9653210	7	1
0	8	

a Find the mean of the boys.
b Find the mean of the entire class.
c Find the mode, median and range of the boys.
d Find the median of the girls.
e Comment on the range of both groups.
f If a new girl is enrolled in the class and her mass is 54 kg, explain the influence on the median of the girl's group.

15 Take the set of numbers 8, 3, 7, 4, 3: pp. 221–223
a Add six to each number in the set. How does this affect the mean?
b Multiply each number by six. How does this alter the mean?

16 A five-sector wheel was spun 100 times and the number of times recorded that each sector stopped at the pointer. pp. 220–223

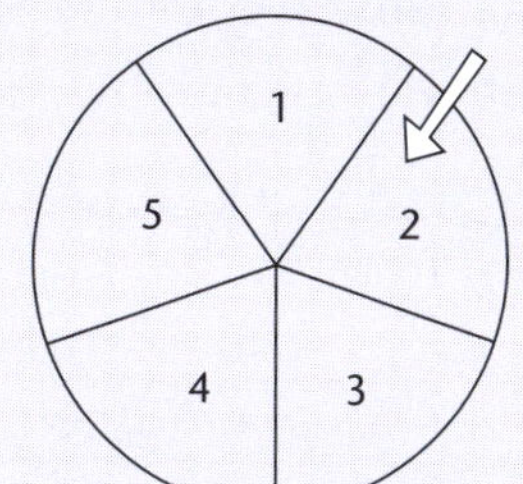

The results were then placed in a table:

Number (x)	Frequency (f)	fx
1	19	
2	21	
3	20	
4	19	
5	21	
	100	

a Complete the $(x \times f)$ column in the table.
b Calculate the times each number occurs as a fraction and then as a decimal.

For example:	Number	Fraction	Decimal
	1	$\frac{19}{100}$	0.19

What do you notice about the decimals? Is there a reason for this?
c Calculate the mean for these scores.
d Find the mode and median.

17 Greg's average over his first five innings was 15.8, but over his next 10 innings he averaged 19.1 runs. Calculate Greg's total runs over the full 15 innings. Find Greg's average score over the 15 innings. If Greg scored 34 runs in his final innings, calculate his season average over 16 innings. pp. 221–222

18 A pack of 52 playing cards with four suits—diamonds and hearts (red), clubs and spades (black)—numbered from 2 to 10, and with King, Queen, Jack and Ace in each suit, is shuffled. One card is drawn at random from the pack. Find the probability that the card is: pp. 224–225

a an Ace
b a King
c a heart
d a spade
e a black card
f a picture card (Jack, Queen or King)
g not a picture card
h a number less than 7
i the six of hearts
j a red card or a 6
k a red six

19 The probability of rain on the June long weekend is $\frac{7}{10}$. What is the probability that it will not rain on the June long weekend? pp. 224–225

20 A jar contains 5 red, 6 black and 4 white jelly babies. If one jelly baby is chosen at random from the jar, find the probability that it is: pp. 224–225

a red
b not red
c black
d red or white
e green
f white
g not white

21 It has been calculated that the probability of a male birth is 0.48: pp. 224–225

a Find the probability of a female birth.
b At a large Perth hospital, 1400 babies were born in 2012. How many male babies would you expect were born that year?

22 Two dice are thrown (faces numbered 1 to 6). The sum of the two uppermost faces is noted. Find the probability that the sum is: pp. 224–225

a 8
b 5
c odd
d greater than 9
e 9 or greater
f not 9
g odd or less than 5
h odd and less than 5

23 A pair of four-sided dice (numbered 1–4) are thrown. The numbers, face up, are added. Find the probability that the sum is: pp. 224–225

a 5
b 7
c 1
d greater than 5
e at least 5

24 In a supermarket promotion 10 gold tokens numbered 1 to 10 are hidden. Rees found one token. Find the probability that Rees found the token with number: pp. 224–225

a 9
b even
c less than 6
d 6 or more

Go to p. 260 for quick answers, or to pp. 306–311 for worked solutions.

YOUR CHECKLIST

For a complete understanding of this topic you must be able to:

✓	Create a frequency distribution table		pp. 218–219
✓	Draw a stem-and-leaf plot		p. 219
✓	Draw a frequency histogram		p. 220
✓	Draw a frequency polygon		p. 220
✓	Draw a dot plot		p. 220
✓	Draw a scatter diagram		pp. 220–221
✓	Find the range as a measure of spread		p. 221
✓	Find the mean, median, mode as measures of location		pp. 221–224
✓	Find the probability of simple events		pp. 224–225
✓	Find the probability of complementary events.		p. 225

Now you are ready to do the tests!

(25 marks)

1

12	17	16	13	15	17	18	19	18	15
14	11	19	16	13	14	16	15	17	16
13	18	17	16	17	18	19	13	14	19
15	16	14	11	12	18	17	19	13	18

For the data above:

a Complete a frequency distribution table using these column headings: score (x), tally and frequency (f).

b Draw a frequency histogram and polygon. (6 marks)

2 **a** Complete the table.

b Find the mean.

c Find the mode, median and range.

d How many scores are less than 24?

e How many scores are even?

Score (x)	Frequency (f)	fx
22	5	
23	7	
24	10	
25	8	
26	5	
Total:		

(8 marks)

3 For the scores 12, 18, 14, 3, 5, 19, 9, 11, 14, 20, find the:

a mean **b** median **c** mode **d** range (4 marks)

4 Use the stem-and-leaf plot to find the:

a outlier **b** mean

c median **d** mode

e range

Stem	Leaf
0	69
1	244789
2	3489
3	
4	5

(5 marks)

5 A die is rolled. What is the probability of rolling:

a a 5? **b** a number less than 3? (2 marks)

☞ Quick answers on page 266
☞ Worked solutions on page 338

Your Feedback $\frac{\square}{25} \times 100\% = \square\%$

(35 marks)

1

68	73	76	61	62	58	54	75	71	68
63	61	72	71	68	59	54	73	70	54
69	64	65	73	69	62	65	75	60	55

a Complete the frequency distribution table for the data above:

Class	Class centre (x)	Tally	Frequency (f)	fx
51–55 56–60 61–65 66–70 71–75 76–80				
		Sum		

b Find the mean.

c Find the modal class.

d Draw a frequency histogram and polygon. (8 marks)

2 For the scores 16, 3, 5, 4, 7, 8, 6, 5, 2, 11, 13, find the:

a mean **b** median **c** mode **d** range (4 marks)

3 The histogram shows the number of people living in each house in Donald Street:

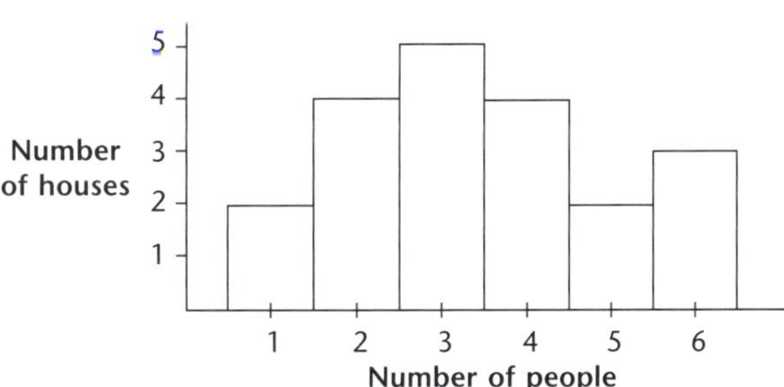

Use the data in the histogram to:

a Complete a frequency table.

b Find the number of people living in the street.

c Find the number of houses in the street.

d Find the mean number of people living in each house. (6 marks)

4 The mean of five scores is 8:

a Find the sum of the five scores.

b When a sixth score is included the mean increases to 10. What is the new score? (3 marks)

5

Stem	Leaf
2	2069
3	84257
4	3428681
5	19

a Draw an ordered stem-and-leaf plot using above the data.

b Find the:

i mean　　ii mode

iii median　　iv range (7 marks)

6 The numbers 1 to 10 are written on ten identical balls and placed in a bag. If one ball is drawn at random from the bag, find the probability that the number is:

a a 6　　b even

c less than 8　　d composite

e a factor of 12　　f a multiple of 3

g not a 7 (7 marks)

☞ **Quick answers on page 266**

☞ **Worked solutions on page 339**

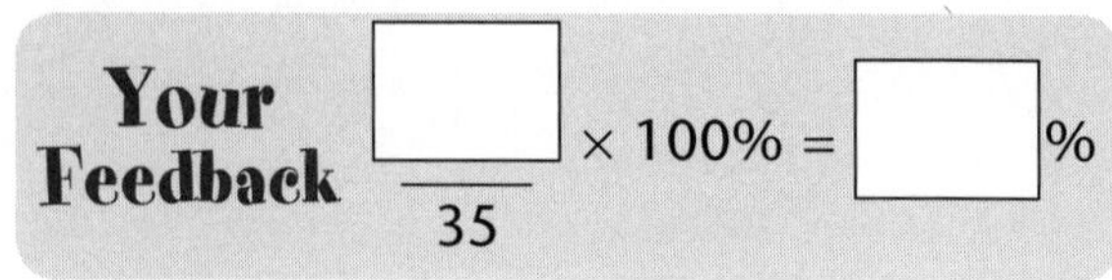

(35 marks)

1 a Complete the frequency distribution table:

Score (x)	Frequency (f)	$f \times x$
3	4	
5		20
8		48
	4	40
12	5	
Sum	23	180

b Find the mean, mode, median and range.

c Draw a frequency polygon. (9 marks)

2

Stem	Leaf
3	24789
4	00246
5	258
6	36
7	□

The stem-and-leaf plot shows the scores in a PDHPE test marked out of 80.
The range is the same as the median:

a Find the median.

b Find the digit represented by □.

c Arrange the data in a grouped frequency table:

Class	Class centre (x)	Frequency (f)	fx
31–35			
36–40			
41–45			
46–50			
51–60			
56–60			
61–65			
66–70			
71–75			

d Compare the mean calculated from the stem-and-leaf plot with the mean from the grouped frequency table. (7 marks)

3

Score	Frequency
x	4
y	6

The mode is 4 and the range is 2. Find a possible mean of the distribution. (2 marks)

4 After five netball games the average number of goals scored by Laura's team is 22. After another game the average dropped by 3. How many goals were scored in the sixth game? (2 marks)

5

Males	Stem	Females
	12	089
874	13	34478
98200	14	25
764	15	

Using measures of location and spread, compare the male and female results from the stem-and-leaf plot above. (4 marks)

6 A survey was conducted at a supermarket checkout to record the number of items in trolleys and baskets. The results were recorded in the histogram:

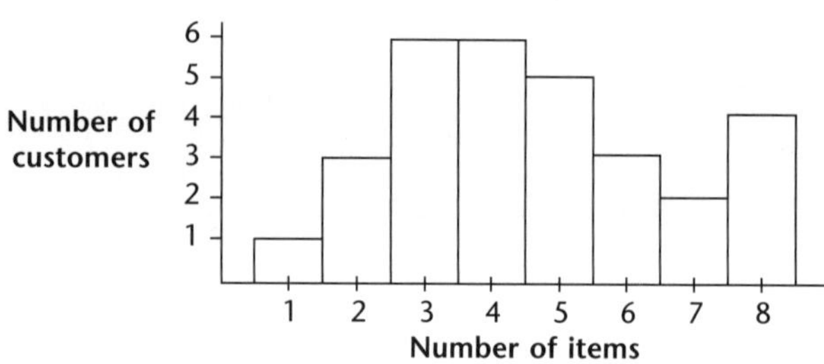

a Find the median, mode and range.

b Find the mean. (5 marks)

7 A normal deck of playing cards is shuffled and a card chosen at random. Find the probability that the card is:

a red   b not red   c not a 7

d a Jack or a Queen   e a black 8   f not a red Jack (6 marks)

☞ Quick answers on page 266
☞ Worked solutions on page 340

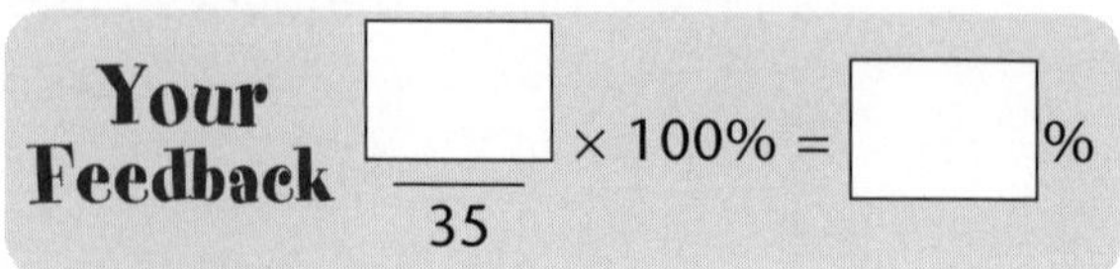

12 SAMPLE EXAMINATIONS

- Sample Examination 1
  - Part A
  - Part B
  - Part C
- Sample Examination 2
  - Part A
  - Part B
  - Part C

SAMPLE EXAMINATION 1

Time Allowed: One Hour Ten Minutes (70 Minutes)

Part A

(50 marks)

Answers only are necessary.

1 Evaluate $17 \times 25 \times 4$. (1 mark)

2 Write in ascending order −3, −4, −2. (1 mark)

3 Simplify $6a - a$. (1 mark)

4 Write as an improper fraction $3\frac{2}{3}$. (1 mark)

5 Write down the multiple of 7 just bigger than 20. (1 mark)

6 Name the shape:

(1 mark)

7 What is 12% of $300? (1 mark)

8 8 is a factor of 32. True or false? (1 mark)

9 $0.6 + 4 = ?$ (1 mark)

10 Find the sum of 17 and 13. (1 mark)

11 Simplify $0.6^2 - 0.5^2$. (1 mark)

12 Find the median of 4, 2, 6, 3, 5, 5. (1 mark)

13 Simplify $4 \times 7y$. (1 mark)

14 $\frac{2}{3} + \frac{1}{3} = \square, \square = ?$ (1 mark)

15 Which is bigger, 0.65 or 0.6? (1 mark)

16 If $7a = 35$, then $a = ?$ (1 mark)

17 Find the product of $3a$ and $9a$. (1 mark)

18 Write in figures: Six hundred and ten thousand and forty eight. (1 mark)

19 Find the length of a rectangle with a width of 2.6 cm and a perimeter of 13.5 cm. (1 mark)

20 Simplify $18a \div 6a$. (1 mark)

21 Write down the complement of 37°. (1 mark)

22 Evaluate 6.5×100. (1 mark)

23 $\frac{3}{4} > \frac{2}{3}$. True or false? (1 mark)

24 Find $\frac{1}{4}$ of $2.80. (1 mark)

25 If $t + 11 = 17$, then $t = ?$ (1 mark)

26 If Ken is 4 years older than Margaret, but Margaret is 12 years younger than Lee, calculate how many years older Lee is than Ken. (1 mark)

27 Evaluate 7^2. (1 mark)

28 Jack wrote down the scores 4, 6, −3, 4, 8, −7. Find the range. (1 mark)

29 A bag contains 4 red balls, 3 green balls and a blue ball. If a ball is chosen at random, what is the probability that it is not green? (1 mark)

30 Evaluate $1 - \frac{3}{4}$. (1 mark)

In Questions 31 to 34, write down the value of the letter.

31

(1 mark)

32

(1 mark)

33

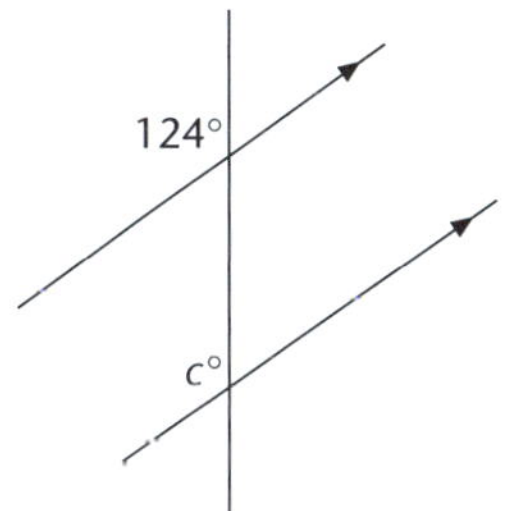

(1 mark)

34

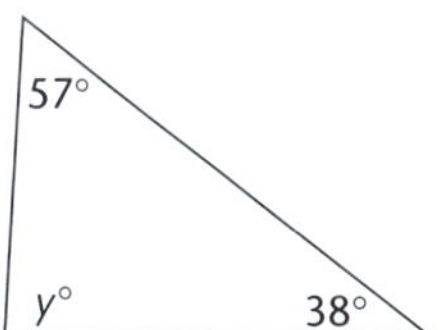

(1 mark)

35 If $\frac{a}{7} = 14$, then $a = ?$ (1 mark)

36 $0.7 + 0.4 = t$, then $t = ?$ (1 mark)

37 Evaluate 0.9×0.4. (1 mark)

38 Find the quotient of 144 and 9. (1 mark)

39 Greg's shoe is 37 cm long. How many millimetres is this? (1 mark)

40 The volume of a rectangular prism is 200 cm^3. If the length is 25 cm and the height is 2 cm, what is the width of the prism? (1 mark)

41 Evaluate $-6 - (-4)$. (1 mark)

42 Evaluate $6a$ when $a = 0.3$. (1 mark)

43 Simplify \$1.50:\$2.50. (1 mark)

44 Round off 620.846 to one decimal place. (1 mark)

45 Evaluate $64 \div 0.8$. (1 mark)

46 Evaluate $-6 - 7$. (1 mark)

47 Evaluate $3 + 7 \times 6$. (1 mark)

48 How many thirds in three? (1 mark)

49 Write $\frac{7}{100}$ as a decimal. (1 mark)

50 Write down the reciprocal of $\frac{2}{5}$ as a fraction in its simplest form. (1 mark)

Part B

(20 marks)

Show all necessary working. (Each question is worth 2 marks)

	WORKING AND ANSWERS
1 A length of cloth 73 cm long is cut from a 5 metre roll. What length of cloth remains on the roll? Answer in metres.	Length remaining = m
2 Write 48 as the product of its prime factors.	48 / \ 48 =
3 Find the values of x and y: $x°$ $y°$ $70°$	

4 Macka wishes to purchase 4 HB pencils. They are advertised at \$4.32 for 6 pencils. How much does Macka pay for his 4 pencils?

Cost = \$

5 Kai averages 17 runs per game for the 12 matches of the season. She averaged 21 over the two finals games. How many runs did Kai score over the full 14 games?

No. of runs =

6 Complete the table and find the mean.

Score (x)	Freq. (f)	$f \times x$
4	7	
5	6	
6	2	

Mean =

7 Fence posts are to be placed along a boundary at $2\frac{1}{2}$ metre intervals:

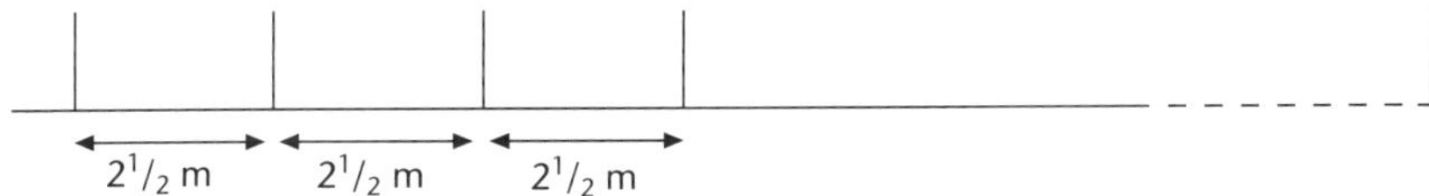

If the boundary is 85 metres long, how many posts are required? (Remember to count the first post.)

No. of posts =

8 Evaluate $3\frac{1}{3} - 2\frac{4}{5}$.

Answer =

9 If $a = 1\frac{1}{3}$, evaluate $9a^2$.	Answer =
10 There are 18 players in a knock-out tennis competition (i.e. when you lose you are out of the competition). How many games must be played to find the single winner?	No. of games =

Part C

(30 marks)

Show all necessary working.

Question One **(15 marks)**

a Solve the equation:

$6x - 17 = 55$ (2 marks)

b Marie's age is 4 times Berend's age plus 3. If Marie is 35 years old, set up an algebraic equation and solve it to find Berend's age. (3 marks)

c The three angles of a triangle are $(3x + 5)°$, $(x + 27)°$ and $28°$:

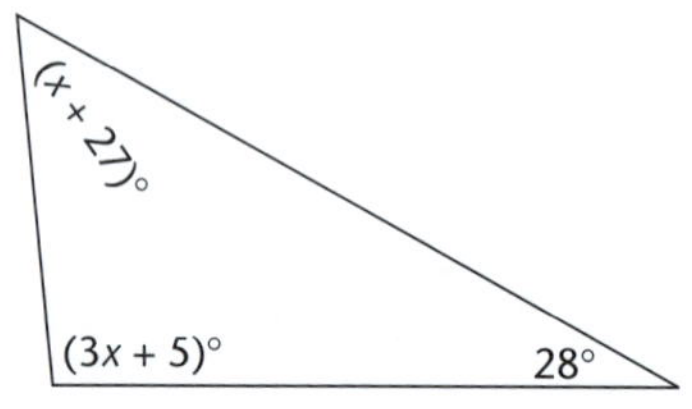

i Show that $4x + 60 = 180$. (2 marks)

ii Hence, or otherwise, find the size of each angle of the triangle. (2 marks)

d Evaluate $2\frac{1}{4} \times 1\frac{3}{4} \times 1\frac{1}{3}$, leaving your answer in simplest form. (3 marks)

e Sackcloth costs $3.75 per metre. If Noah needed 3 lengths of $1\frac{1}{2}$ metres, 2 lengths of 1.8 metres and a single length of 1.9 metres, how much will the sackcloth cost Noah? (3 marks)

Question Two **(15 marks)**

a Calculate the areas of the following shapes:

i

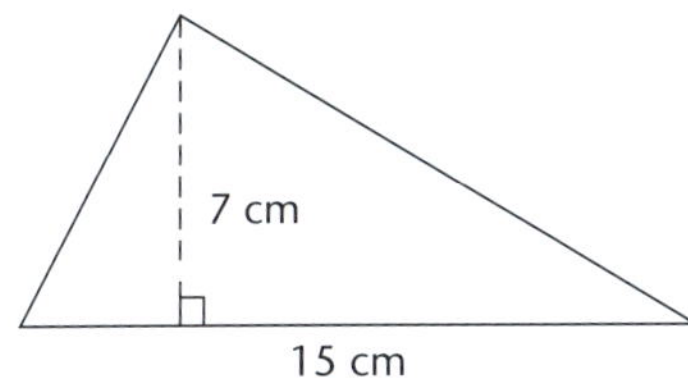

ii

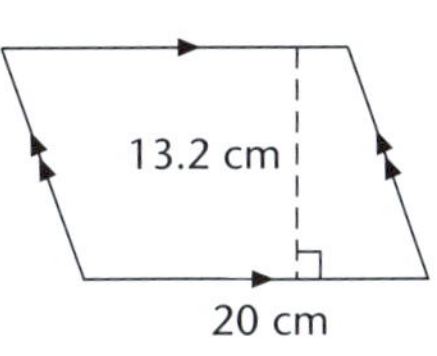

Area = ________ cm^2 Area = ________ cm^2 (4 marks)

b

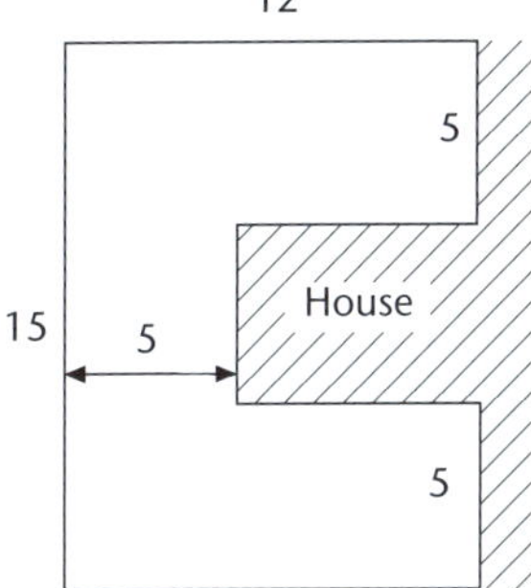

A concrete patio is to be attached to the rear of a house, as in the above diagram (all measurements are in metres):

i Calculate the perimeter of the patio. (2 marks)

ii Calculate the cost of fencing the edges of the patio not adjoining the house if this fencing costs $27 per metre. (3 marks)

c This diagram represents a side view of a backyard swimming pool:

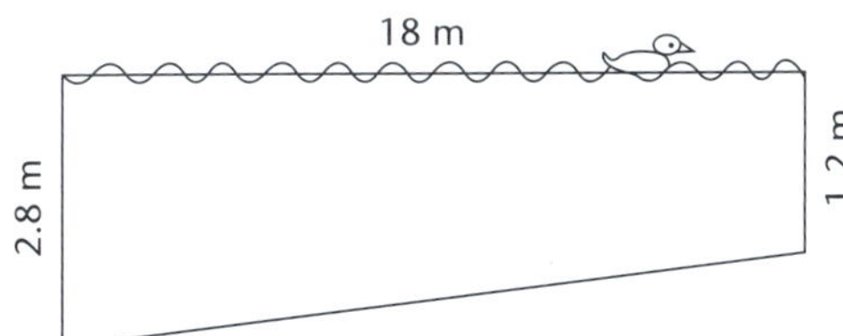

i Calculate this cross-sectional area. (2 marks)

ii If the pool is 8 m wide, calculate the volume of water in the pool in m^3. (Assume that the pool is filled to the top rim.) (1 mark)

iii If 7200 L is drained from the pool, by how far does the water level drop? (Answer in cm.) (3 marks)

☞ Quick answers on page 267
☞ Worked solutions on page 342

Your Total Feedback (Paper 1)

☐ + ☐ + ☐ = ☐/100 = ☐%

Part A Part B Part C

SAMPLE EXAMINATION 2

Time Allowed: One Hour Ten Minutes (70 Minutes)

Part A

(50 marks)

1 Evaluate $5 + 2 \times 3$. (1 mark)

2 Evaluate $7 \times 5 \times 0 \times 3$. (1 mark)

3 $7a - a = ?$ (1 mark)

4 $\sqrt{144} + 2^3 = ?$ (1 mark)

5 If $a = 5$, $2a^2 = ?$ (1 mark)

6 Write 0.07 as a simple fraction. (1 mark)

7 Write $\frac{7}{8}$ as a decimal. (1 mark)

8 Simplify $\frac{3}{4}:1\frac{1}{2}$. (1 mark)

9 $4x - 7y + 3x + 4y = ?$ (1 mark)

10 $2x \times 3x = ?$ (1 mark)

11 If $y + 7 = 15$, $y = ?$ (1 mark)

12 3, 5, 7, x, 10, 11
If the median and range are the same, what is the value of x? (1 mark)

13 Find the remainder when 47 is divided by 9. (1 mark)

14 If the temperature at town A is 5 degrees and falls by 14 degrees, find the new temperature. (1 mark)

15 What is the product of $2a$ and $4b$? (1 mark)

16 $16 \div 4 \times 2 = ?$ (1 mark)

17 Simplify $\frac{25}{45}$. (1 mark)

18 Evaluate $\frac{3}{7} + \frac{4}{7}$ in simplest form. (1 mark)

19 Find the complement of 67°. (1 mark)

20 Increase a by 12. (1 mark)

21 $(-8) - (-4) = ?$ (1 mark)

22 Find $4\frac{1}{2}\%$ of \$200. (1 mark)

23 Find x:

36° x°

(1 mark)

24 $15ab \div 5a = ?$ (1 mark)

25 A ten-sided die is rolled. What is the probability of rolling a number bigger than 5? (1 mark)

26 Express $\frac{4}{25}$ as a decimal. (1 mark)

27 Simplify $5x^2 + 7x - 3x^2 - x$. (1 mark)

28 Evaluate $-2 + 1 - 6$. (1 mark)

29 If $x + 2 = 6$, find the value of $5x$. (1 mark)

30 How many centimetres are there in 2.4 metres? (1 mark)

31 Find $\frac{3}{5}$ of \$60. (1 mark)

32 Find x:

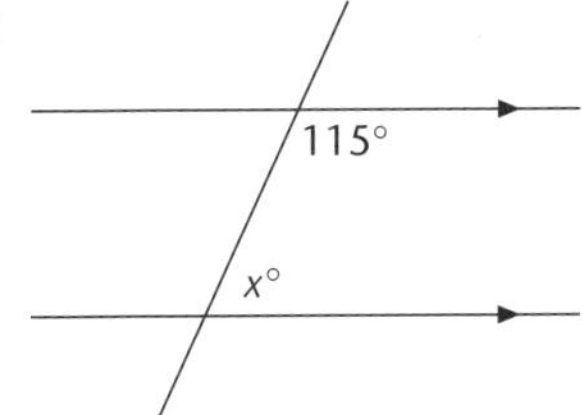

(1 mark)

33 In a maths quiz, Chang got 8 questions incorrect out of a possible 40 questions. What was his mark expressed as a percentage? (1 mark)

34

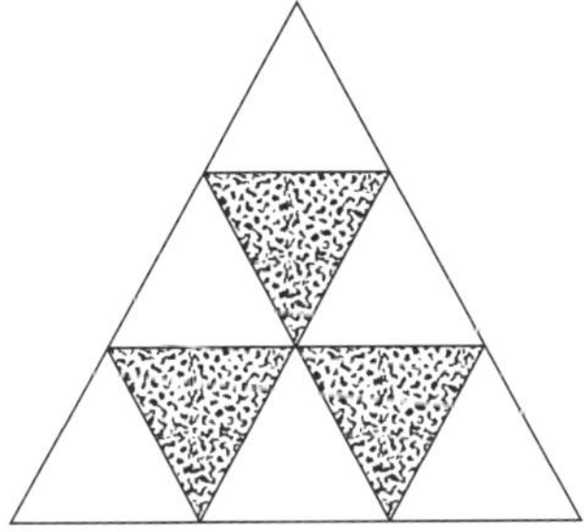

What fraction of the diagram is shaded? (1 mark)

35 Which solid is a prism?

A 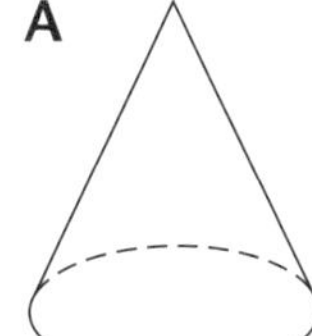 **B** 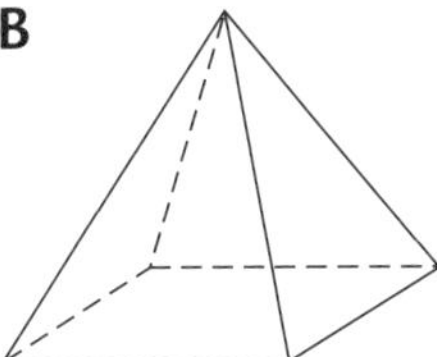 **C** 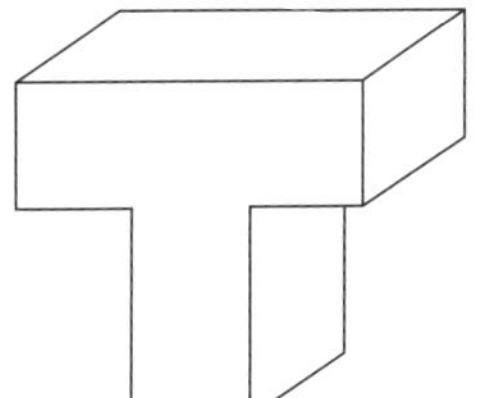 **D** 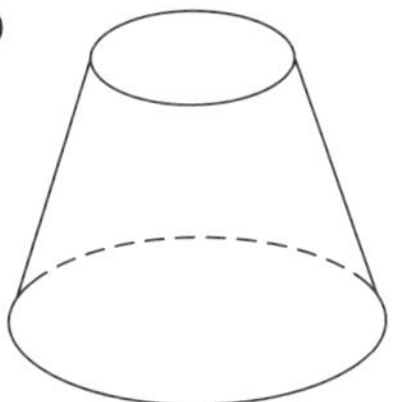 (1 mark)

36 The marked angle is:

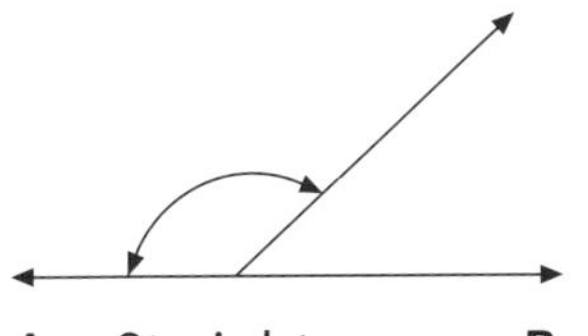

A Straight **B** Acute **C** Reflex **D** Obtuse (1 mark)

37 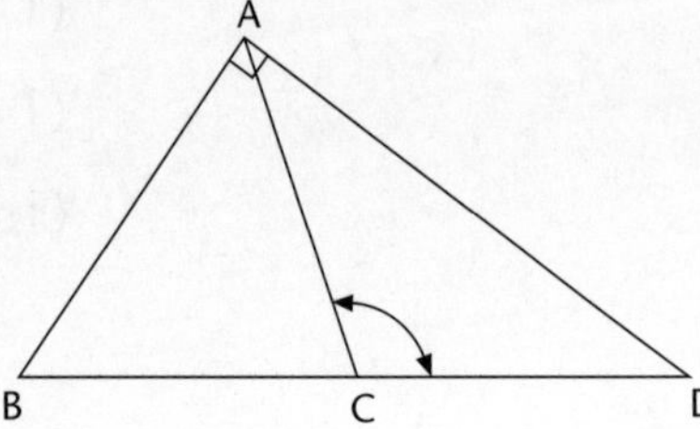

The perimeter of this rectangle is 26 cm. If the width is 5 cm, what is its length? (1 mark)

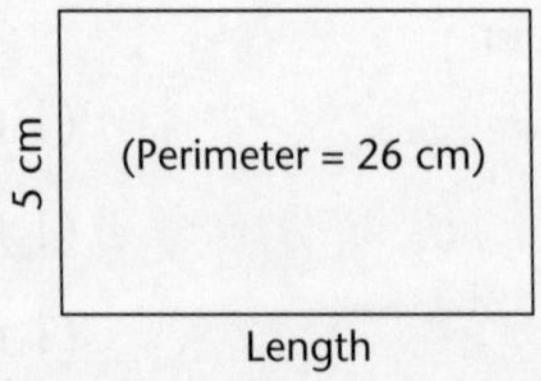

38 Which of these numbers is the smallest: $\frac{1}{4}$, 0.2, 0.22? (1 mark)

39

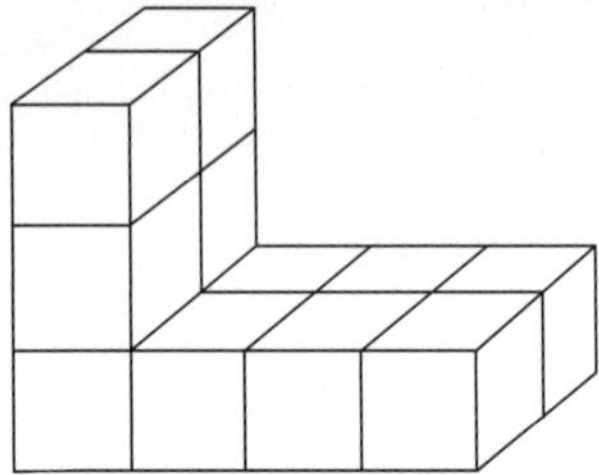

In the diagram, which angle is the supplement of ∠ACD? (1 mark)

40

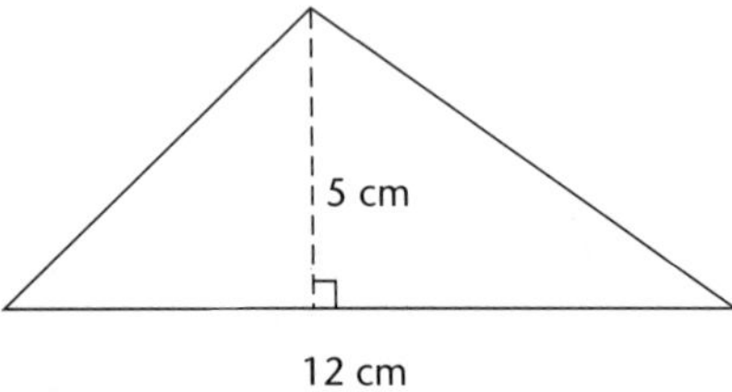

How many cubes are used to make this prism? (1 mark)

41 A paddock is in the shape of a square and has an area of 1 hectare. Find the perimeter of the paddock in metres. (1 mark)

42 Find the area of the triangle: (1 mark)

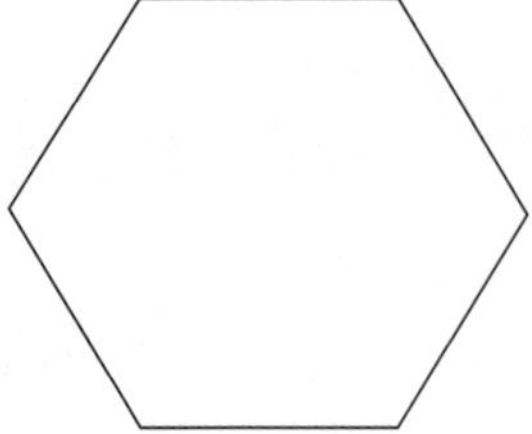

43 Eleni used $\frac{3}{5}$ of a tube of toothpaste in 36 days. At this rate, how long would a full tube last? (1 mark)

44 How many axes of symmetry are there on a regular hexagon?

(1 mark)

45 The average of 5, 8 and x is 8. Find x. (1 mark)

46 Write in symbols:

'The sum of a and the product of 2 and b.' (1 mark)

47 Find the mean of three consecutive whole numbers if the smallest is x. (1 mark)

48 Find the median of the scores in this stem-and-leaf plot:

Stem	Leaf
3	67
4	04457
5	2399

(1 mark)

49 Write the rule linking y and x in the table:

x	0	1	2	3
y	3	5	7	9

(1 mark)

50 Three men can build a wall in 8 hours. How long will it take 4 men to build the same wall? (1 mark)

Part B (32 marks)

1

Menu	
Chips	\$2.50
Hamburger	\$3.50
Soft drink	\$1.50
Coffee	\$1.20

Calculate the total cost of the following order:

2 hamburgers
1 chips
1 soft drink
2 coffees

Total ________

(2 marks)

2 If 12 litres of oil cost \$25.80, find the cost of 7 litres of the same oil. (2 marks)

3 At a shop Eleni bought items for \$1.70, \$8.35, 85 cents and \$2.35. Find the change from a \$20 note. (2 marks)

4 Find the area of this shape:

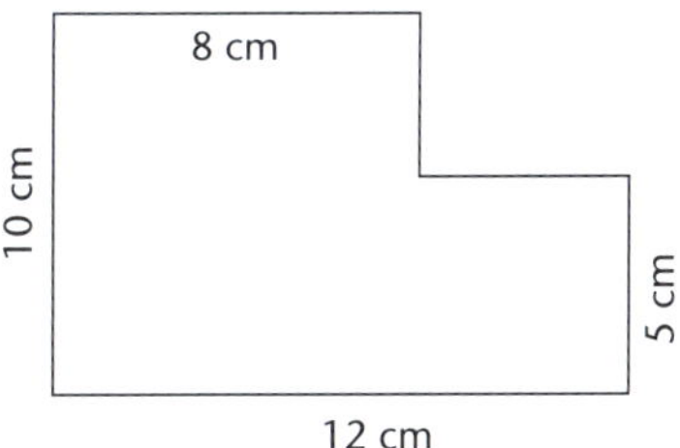

(2 marks)

5 Here is a pattern of circles:

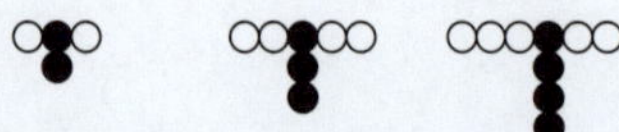

Diagram 1 Diagram 2 Diagram 3

If the pattern is continued, complete the table and find the number of white circles in diagram 17:

Diagram no.	1	2	3	4
No. of whites				
No. of blacks				

Number of white circles in diagram 17: ______________ (2 marks)

6 How long does it take to empty a 1.75 litre bottle if it leaks 5 mL every minute? Give your answer in hours and minutes. (2 marks)

7 For the numbers 27 and 36, find the:

a Set of common factors

b HCF. (2 marks)

8

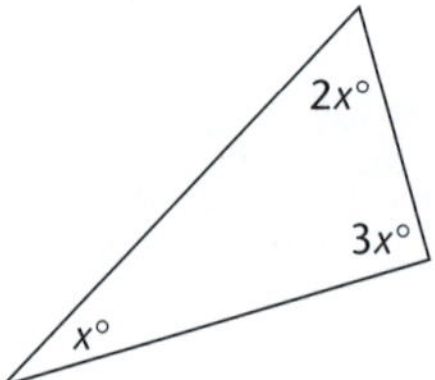

Find the value of x, and hence the size of the largest angle. (2 marks)

9 Solve the equation $2x - 1 = -7$. (2 marks)

10

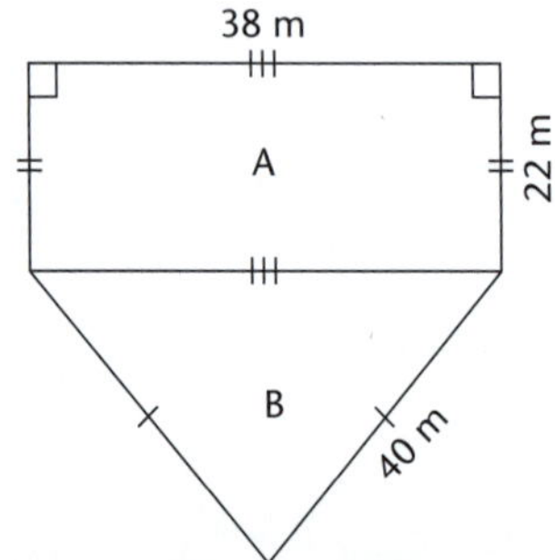

A paddock is divided into two pens, A and B, as shown in the diagram. Find the total cost of fencing off the pens at a cost of $14.25 per metre. (2 marks)

11 Solve the equations:

a $a + 8 = 2$

b $\frac{x + 2}{3} = 5$ (2 marks)

12

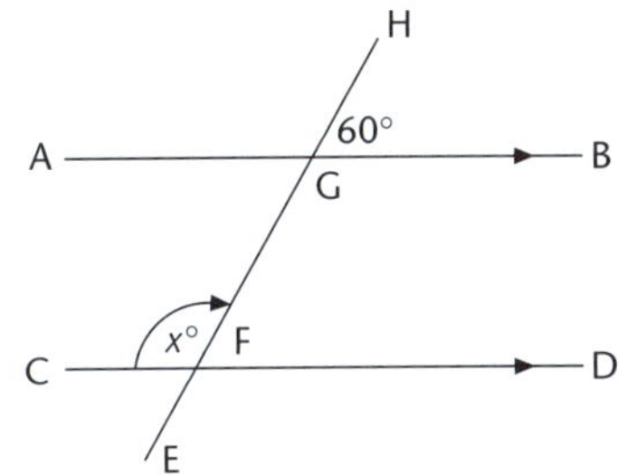

AB // CD and ∠HGB = 60°. Find the size of x, giving reasons for your answer. (2 marks)

13 Use the scores to draw a frequency histogram and polygon:

14	16	17	13	15	18
15	17	16	15	14	17
16	15	17	18	15	14

(2 marks)

14 **a** Write the following numbers in ascending order: 4, –4, –1, $2\frac{1}{2}$, 8, 6.3.

b Which is the third-smallest number? (2 marks)

15 A cyclist left Wallsend at 7.30 a.m. and travelled at an average speed of 20 km/h until he reached Lochinvar 50 kilometres away. At what time did he reach Lochinvar? (2 marks)

16

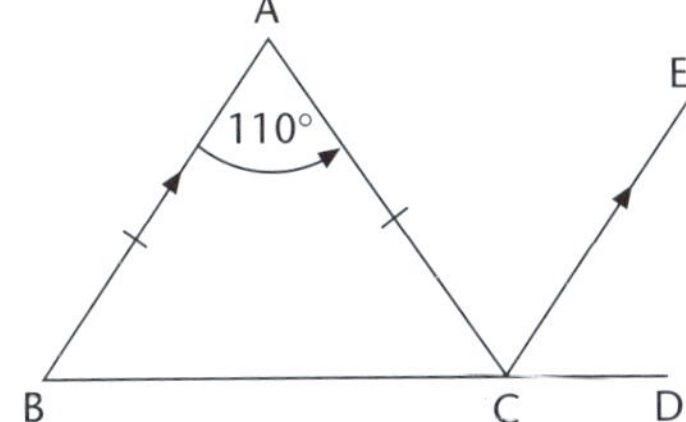

AB // CE and ∠CAB = 110°

a ∠ABC = ?
Why?

b ∠ECD = ?
Why? (2 marks)

Part C (18 marks)

1 Jack has \$42. He gives $\frac{1}{6}$ to Jill and $\frac{1}{3}$ to Paul. Find:

a Paul's share

b Jill's share

c What fraction of Jack's money was left. (3 marks)

2 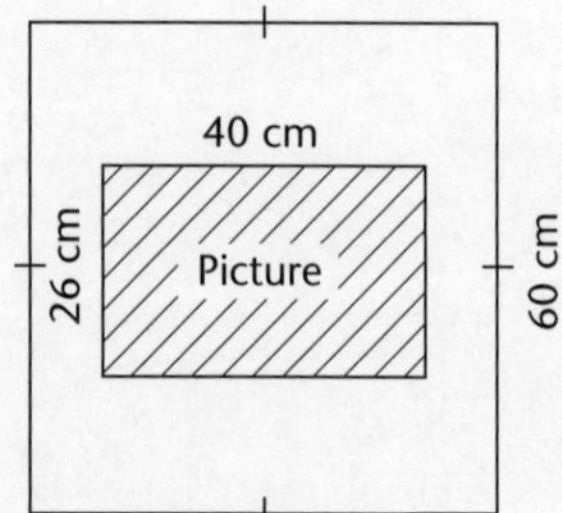

A cardboard square 60 cm long has a picture 40 cm long and 26 cm wide mounted on it:

a Find the area of the cardboard showing around the picture.

b If the picture is mounted in the centre of the cardboard, find the width of the cardboard border on each side of picture. (3 marks)

3 On her social networking page, Miriam has 48 friends. Sixteen friends are workmates, and one-quarter of the remainder she knows from sport:

a How many friends played sport with her?

b What is the ratio of workmates to total friends?

c If 75% of her friends are female, what number are male? (3 marks)

4 If $a = -2$, $b = 3$ and $c = -5$, evaluate:

a $a^2 - b$ b $cb - a$ c $\frac{4a - b + 1}{c}$ (3 marks)

5 Show that $\angle DBF = 90°$:

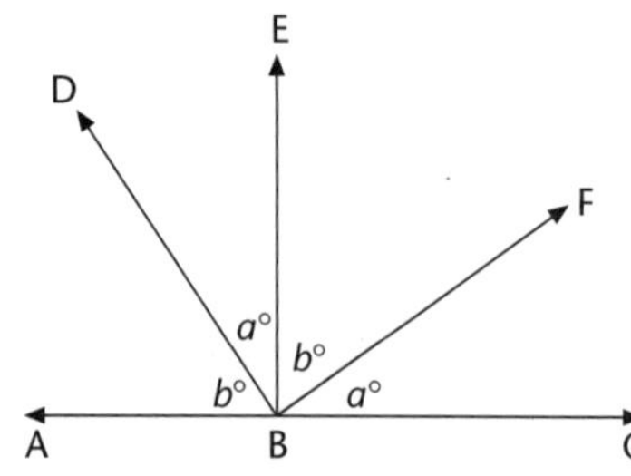

(3 marks)

6 There are eight children in the Jackson family. The average age of the four boys is 12. Two of the girls are eight-year-old twins. One of the girls is 18 years old:

a What is the total age of the boys?

b The average age of the girls is 10. How old are the girls?

c What is the average age of the eight children? (3 marks)

☞ Quick answers on page 267
☞ Worked solutions on page 345

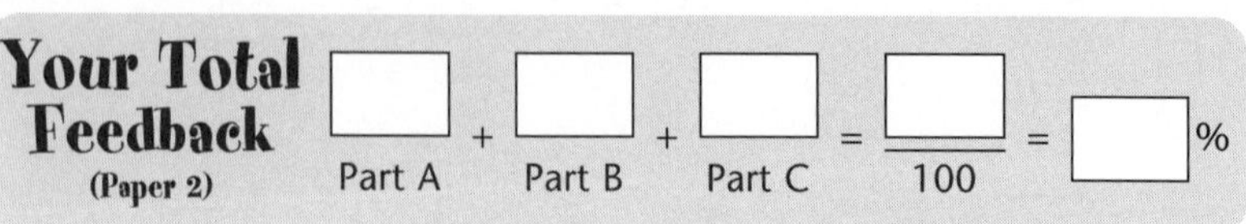

13 ANSWERS AND WORKED SOLUTIONS

- Quick Answers to Practise, Practise
- Quick Answers to Tests
- Quick Answers to Sample Examinations
- Worked Solutions to Practise, Practise
- Worked Solutions to Tests
- Worked Solutions to Sample Examinations

QUICK ANSWERS to Chapter 1 **Practise, Practise**

Chapter 1—Number pp. 21–27

1 a $6 \times 1000 + 1 \times 100 + 4 \times 10 + 7 \times 1$ or $6 \times 10^3 + 1 \times 10^2 + 4 \times 10 + 7 \times 1$ b $7 \times 10000 + 4 \times 1000 + 2 \times 100 + 9 \times 10 + 1 \times 1$ or $7 \times 10^4 + 4 \times 10^3 + 2 \times 10^2 + 9 \times 10 + 1 \times 1$ c $70707 = 7 \times 10000 + 7 \times 100 + 7 \times 1 = 7 \times 10^4 + 7 \times 10^2 + 7 \times 1$ d $317429 = 3 \times 100000 + 1 \times 10000 + 7 \times 1000 + 4 \times 100 + 2 \times 10 + 9 \times 1 = 3 \times 10^5 + 1 \times 10^4 + 7 \times 10^3 + 4 \times 10^2 + 2 \times 10 + 9 \times 1$ e $40030 = 4 \times 10000 + 3 \times 10 = 4 \times 10^4 + 3 \times 10$ **2** a 4732 b 6009 **3** a 5895 b 90870 **4** a 297 b 2970 c 27000 d 2997 e 29700 **5** a $74 > 47$ b $235 < 325$ c $1021 < 1201$ d $3102 > 3012$ e $56174 > 56147$ **6** a T b T c T d F e F f F g T h T i F j T **7** a 36, 47, 63, 74 b 1123, 1132, 2113, 3112 c 5147, 5174, 7145, 7154, 7415, 7514 **8** a 93, 86, 68, 39 b 8844, 8448, 4884, 4848 c 3291, 3219, 3192, 3129, 2391, 2319, 2193, 2139 **9** a 50 b 2 c 60 d 40 e 12 f 60 g 32 **10** a 0 b 28 c 26 d 26 e 18 f 18 g 40 h 11 i 0 j 10 **11** a $(4 + 3) \times 7 = 49$ b $(12 - 5) \times 3 = 21$ c $7 \times (3 + 2) \times 2 = 70$ d $3 \times (2 + 3) + 4 = 19$ e $5 + (4 + 3) \times 2 = 19$ f $25 \div (3 + 2) + 7 = 12$ **12** a 29 b 60 c 11 d 11 e 7 f 1 g 26 h 90 i 26 j 128 **13** \$81 **14** \$45 **15** 200 **16** a 286 b 209 c 1944 d $51\frac{6}{9}$, $51\frac{2}{3}$ e 1679 f 6015 g 152152 h $94\frac{7}{45}$ i $726\frac{8}{69}$ **17** 49 **18** 300 **19** 20 **20** \$1.32, \$13.20 **21** 3 **22** 17 **23** a 61 b 85 c 83 d 17 e 19 f 34 **24** a 101 b 131 c 102 d 19 e 17 f 9 **25** a 190 b 2700 c 1600 d 3100 e 1300 f 1800 **26** a 342 b 513 c 414 d 506 e 858 f 913 **27** a 1160 b 1240 c 1440 d 21 e 12 f 4 **28** a 27 b 118 c 157 **29** a 2, 4, 6, 8, 10, 12, 14, 16, 18 b 19, 21, 23, 25 **30** a even b even c odd d even e odd f even **31** a 1, 3, 6, 10, 15, 21, 28, 36, 45, 55 b 1, 4, 9, 16, 25, 36, 49, 64, 81, 100 **32** a Palindromic b Not Palindromic c Palindromic d Palindromic e Palindromic **33** a 484 Palindromic b 121 Palindromic c 828 Palindromic d 2662 Palindromic e 727 Palindromic **34** 1, 1, 2, 3, 5, 8, 13, 21, 34, 55, 89, 144, 233, 377, 610, 987, 1597, 2584, 4181, 6765 **35** a 5, 7, 9, 11, *13, 15, 17 …* b 6, 8, 10, 12, *14, 16, 18 …* c 2, 4, 8, 16, *32, 64, 128 …* d 2, 5, 8, 11, *14, 17, 20 …* e 2, 3, 5, 8, 12, *17, 23, 30 …* f 2, 6, 18, *54, 162, 486 …* g 2, 4, 7, 11, *16, 22, 29 …* h 2, 4, 12, 48, *240, 1440, 10080 …* i 2, 1, 4, 3, 6, 5, *8, 7, 10 …* j 4, 16, 64, *256, 1024, 4096 …* k 1, 9, 25, *49, 81, 121 …* l 2, 3, 5, 6, 8, 9, *11, 12, 14 …* m 17, 18, 16, 19, 15, 20, *14, 21, 13 …* n 1, 2, 3, 6, 11, 20, *37, 68, 125 …* o 2, 11, 56, 281, *1406, 7031, 35156 …* p 5, 6, 7, 4, 9, 2, *11, 0, 13 …* **36** See worked solutions **37** See worked solutions **38** See worked solutions **39** a 7, 14, 21, 28, 35, 42, 49, 56, 63, 70 b 14, 21, 28, 35 c 2 d 21, 28, 35, 42 e 42, 49, 56, 63 **40** a 4 b 24, 32 c 8, 16, 24, 32, 40, 48, 56, 64, 72 d 40, 80 **41** a 14, 28, 42 b 42, 56 c 3 **42** a 35, 70 b 4 c 70 **43** a 30 b 15 c 28 d 36 e 8 f 12 g 18 h 36 i 12 j 12 k 30 l 60 m 24 n 60 o 28 p 30 q 60 r 12 s 30 t 60 **44** 12 **45** 84 **46** 61 **47** 61 **48** 26 **49** 86 **50** a ①, ②, 3, ④, 6, ⑧, 12, ⑯, 24, 48 b ①, ②, ④, ⑧, ⑯ c 1, 2, 4, 8, 16 **51** a ①, ②, ③, ⑥, ⑨, ⑱, 27, 54 b ①, ②, ③, 4, ⑥, ⑨, 12, ⑱, 36 c 1, 2, 3, 6, 9, 18 **52** a ①, ⑤, 7, 35 b ①, 3, ⑤, 9, 15, 45 c 1, 5 **53** a 2, 3 b 2 c 2, 3 d 5, 7 e 3, 5 f 2, 3 g 2, 3, 7 h 2, 3, 11 i 3, 13 j 3, 17 **54** a 5 b 4 c 4 d 3 e 2 f 1 g 8 h 1 i 4 j 7 k 8 l 5 m 3 n 3 o 17 p 3 q 5 r 15 s 8 t 16 **55** Primes less than 100 are 2, 3, 5, 7, 11, 13, 17, 19, 23, 29, 31, 37, 41, 43, 47, 53, 59, 61, 67, 71, 73, 79, 83, 89, 97 **56** a 23, 29 b 4, 6, 8, 9, 10, 12, 14, 15, 16, 18, 20, 21, 22, 24, 25, 26, 27, 28 c 2 d 2 e 97 f 101 g 45, 49, 51, 55, 57 h 25, 26, 27, 28, 30, 32, 33, 34, 35, 36, 38, 39, 40, 42, 44, 45 i 3, 13, 23, 31, 37, 43, 53, 73, 83 **57** 3, 5 5, 7 11, 13 17, 19 29, 31 41, 43 59, 61 71, 73 **58** 2, 3 **59** $48 = 5 + 43 = 7 + 41 = 11 + 37 = 17 + 31 = 19 + 29$ **60** a $24 = 5 + 19 = 7 + 17 = 11 + 13$ b $36 = 5 + 31 = 7 + 29 = 13 + 23 = 17 + 19$ c $54 = 7 + 47 = 11 + 43 = 13 + 41 = 17 + 37 = 23 + 31$ d $40 = 3 + 37 = 11 + 29 = 17 + 23$ **61** a $24 = 2 \times 2 \times 2 \times 3 = 2^3 \times 3$ b $36 = 2 \times 2 \times 3 \times 3 = 2^2 \times 3^2$ c $42 = 2 \times 3 \times 7$ d $48 = 2 \times 2 \times 2 \times 2 \times 3 = 2^4 \times 3$

QUICK ANSWERS to Chapters 1, 2 and 3 **Practise, Practise**

e $72 = 2 \times 2 \times 2 \times 3 \times 3 = 2^3 \times 3^2$ f $56 = 2 \times 2 \times 2 \times 7 = 2^3 \times 7$ g $84 = 2 \times 2 \times 3 \times 7 = 2^2 \times 3 \times 7$ h $63 = 3 \times 3 \times 7 = 3^2 \times 7$ i $60 = 2 \times 2 \times 3 \times 5 = 2^2 \times 3 \times 5$ j $78 = 2 \times 3 \times 13$ k $150 = 2 \times 3 \times 5 \times 5 = 2 \times 3 \times 5^2$ l $200 = 2 \times 2 \times 2 \times 5 \times 5 = 2^3 \times 5^2$ m $64 = 2 \times 2 \times 2 \times 2 \times 2 \times 2 = 2^6$ n $128 = 2 \times 64 = 2 \times 2^6 = 2^7$ o $81 = 3 \times 3 \times 3 \times 3 = 3^4$ p $162 = 2 \times 81 = 2 \times 3^4$ q $225 = 3 \times 3 \times 5 \times 5 = 3^2 \times 5^2$ r $105 = 3 \times 5 \times 7$ s $300 = 2 \times 2 \times 3 \times 5 \times 5 = 2^2 \times 3 \times 5^2$ t $256 = 2 \times 128 = 2 \times 2^7 = 2^8$ **62** a HCF = 24 LCM = 5040 b HCF = 36 LCM = 3960 c HCF = 40 LCM = 5280 d HCF = 17 LCM = 19 040 **63** a 144 b 196 c 441 d 900 e 125 f 343 g 16 h 7 i 13 j 9 k 6 l 8 **64** a $\sqrt{576} = 24$ b $\sqrt{441} = 21$ c $\sqrt{729} = 27$ d $\sqrt{1296} = 36$ e $\sqrt[3]{1728} = 12$ f $\sqrt[3]{5832} = 18$ g $\sqrt[3]{13824} = 24$ **65** a 324: divisible by 2, 3, also 9 b 645: divisible by 5, 3 c 4173: divisible by 3 d 7491: divisible by 3, 11 e 5967: divisible by 3, also 9 **66** The number is 744, i.e. $a = 4$ **67** Number of centicubes is 46 513 (i.e. 46 512 + 1) **68** 36 marbles **69** a 8 b 100 c 9 d 27 e 64 f 16 g 1 h 36 i 49 j 8 **70** a 36 T b i T ii F iii T iv T v F

Chapter 2—Fractions pp. 46–48

1 a $\frac{1}{3}$ b $\frac{3}{10}$ **2** a $\frac{8}{12}$ b $\frac{12}{15}$ **3** a $\frac{2}{3}$ b $\frac{5}{7}$ c $\frac{4}{9}$ **4** a $\frac{1}{3}, \frac{2}{5}, \frac{1}{2}$ b $\frac{1}{4}, \frac{3}{8}, \frac{2}{5}$ **5** 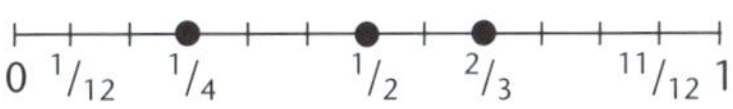

6 a $\frac{5}{6}, \frac{2}{3}, \frac{1}{2}$ b $1\frac{7}{8}, 1\frac{3}{4}, 1\frac{1}{3}$ **7** a $\frac{7}{10} > \frac{2}{3}$ b $\frac{3}{4} < \frac{5}{6}$ **8** a $\frac{9}{2}$ b $\frac{17}{5}$ c $\frac{57}{10}$ **9** a $4\frac{1}{5}$ b $4\frac{1}{3}$ c $4\frac{7}{10}$ **10** a $\frac{3}{4}$ b $\frac{2}{3}$ c $2\frac{1}{2}$ **11** a 1 b $\frac{2}{5}$ c $1\frac{1}{5}$ d $\frac{1}{2}$ e 4 f $7\frac{2}{5}$ **12** a $1\frac{1}{2}$ b $\frac{3}{10}$ c $\frac{1}{10}$ d $\frac{11}{12}$ e $5\frac{7}{10}$ f $2\frac{1}{20}$ g $\frac{5}{6}$ h $1\frac{8}{15}$ i $2\frac{1}{5}$ j $1\frac{13}{30}$ **13** a $\frac{4}{9}$ b $\frac{15}{32}$ c $\frac{27}{50}$ d $\frac{9}{40}$ e $3\frac{1}{8}$ f $26\frac{2}{3}$ **14** a $\frac{3}{5}$ b 8 c 16 d $\frac{16}{25}$ e 1 f $12\frac{1}{4}$ g 4 h 10 i 80 j 40 k $\frac{1}{8}$ **15** a $1\frac{1}{15}$ b $\frac{49}{50}$ c 4 d $2\frac{1}{4}$ e $53\frac{1}{3}$ f 32 g 2 h 2 **16** a 1 b $4\frac{4}{5}$ c 1 d $\frac{1}{3}$ e 9 f $\frac{9}{10}$ g $2\frac{3}{10}$ h $2\frac{1}{2}$ i $\frac{3}{4}$ **17** a \$4.80 b \$1.80 c 1400 mL d 80 seconds e $\frac{9}{10}$ metre or 90 cm f $2\frac{1}{4}$ km **18** a $\frac{1}{5}$ b $\frac{1}{10}$ c $\frac{1}{7}$ d $\frac{3}{5}$ **19** a Next term is 1 b Next term is $\frac{1}{8}$ **20** $\frac{5}{12}$ **21** $10\frac{3}{4}$ mins needed to wash and wipe up **22** a $\frac{1}{5}$ b 96 breaks **23** $\frac{5}{12}$ **24** 280 plum trees **25** 6 oranges **26** a $\frac{5}{48}$ b 17 hours **27** 9 brown-haired girls

Chapter 3—Decimals pp. 63–65

1 a 3 tenths b 3 hundredths c 3 thousandths **2** a $3 \times 1 + 2 \times \frac{1}{10} + 1 \times \frac{1}{100}$ b $7 \times 10 + 6 \times 1 + 4 \times \frac{1}{100} + 7 \times \frac{1}{1000}$ **3** a 3 b 4 **4** a $4 + 2 \times \frac{1}{10} + 5 \times \frac{1}{100}$ b $8 + \frac{1}{100} + \frac{7}{1000}$ c $\frac{1}{10} + \frac{2}{1000} + \frac{2}{10000}$ **5** a 4.17 b 6.037 c 0.73 d 74.113 **6** a $\frac{3}{10}$ b $\frac{1}{2}$ c $\frac{4}{5}$ d $\frac{9}{20}$ e $\frac{21}{50}$ f $\frac{1}{8}$ g $\frac{14}{25}$ h $6\frac{2}{5}$ i $1\frac{41}{100}$ j $4\frac{1}{4}$ k $10\frac{7}{20}$ l $1\frac{1}{500}$ **7** a 0.21 b 0.9 c 2.1 d 0.03 e 0.003 f 0.6 g 0.35 h 0.28 i 0.14 j 0.085 **8** a $0.\dot{3}$ b 0.875 c $0.\dot{2}\dot{7}$ d $0.\dot{8}$ e $0.\dot{1}4\dot{8}$ **9** a 0.07, 0.14, 0.3 b 1.2, 1.217, 1.27 c 0.23, $\frac{37}{100}$, 0.8 d 4.27, 4.3, $4\frac{1}{2}$ **10** a $7.2 < 7.3$ b $0.48 > 0.46$ c $2\frac{1}{2} < 2.54$ d $0.07 < 0.17$ **11** a 4.71 b 20.86 c 11.06 d 12.80 **12** a 4.248 b 200.415 **13** a 11.1 b 0.4 **14** a \$4.76 b \$21.44 **15** a 3.95 b 26.525 c 1.6 d 1.667 e 3.7 f 3.82 g 9.19 h 1.4 **16** a 2.4, 2.8 b 2.2, 1.7 c 4.19, 4.23 d 4.7, 3.8 **17** a 47.6 b 3147.6 c 2140 d 0.76 **18** a 1.261 b 17.64 c 0.0947 d 0.000 21 **19** a 6.8 b 4.41 c 1.12 d 0.12 e 1.206 f 0.25 g 0.0144 h 860 **20** a 0.7 b 7 c 40 d 61 e 200 f 0.2 **21** a 3 b 2.7 c 20 d 0.6 **22** a 7.21 b 1.73 c 1.52 d 21 **23** a 8.05 b 1.62 **24** \$79.95 **25** 72.1 **26** \$0.85 **27** 4.2 **28** 2.5 **29** 47 690.1 **30** a 10 h b 20 h c 40 h **31** a 12.3 km/L b \$13.75 **32** a \$US 144.40 b 15 882 yen c \$NZ 246

QUICK ANSWERS to Chapters 4 and 5 **Practise, Practise**

Chapter 4—Percentages pp. 78–80

1 a 40% b 45% c 52% d 17.5% e 41% f $33\frac{1}{3}\%$ g $83\frac{1}{3}\%$ h 325% **2** a 28.6% b 45.5% c 226.7% **3** a 30% b 3% c 0.3% d 19% e 57.6% f 112% g 470% h 7.5% **4** a $\frac{3}{5}$ b $\frac{3}{100}$ c $\frac{19}{20}$ d $1\frac{27}{100}$ e 3 f $\frac{11}{200}$ g $\frac{49}{400}$ h $\frac{19}{500}$ **5** a 0.16 b 0.07 c 0.6 d 0.076 e 0.125 f 1.76 g 1.046 h 0.0025 i 0.0775 j $0.\dot{3}$ **6** a 162 b $22.40 c $25.20 d 420 m e 360 mL f 30 min g 15 g h $435 i $35 200 j 635 **7** a $22 b 826 c 58.5 L d 3240 g **8** a 208 b $61.75 c 2625 d $34.80 **9** $91 **10** a 48% b 10% c 0.8% d 12.5% e $8\frac{1}{3}\%$ f 5% **11** a 10% b 0.175% c 200% d 6000% **12** a 3500 b 500 **13** 3500 **14** a $71.25 b $28.50 **15** a $2.75 b $19.25 **16** 12.5% **17** $54.60 **18** $2549.85 **19** 40% **20** a 36% b 8% **21** a 20% b 25% **22** $626.08 **23** 80% **24** 900 **25** a $170 b $34 **26** $12.5% **27** a 68.89% b 1820

Chapter 5—Integers pp. 94–98

1 a I am overdrawn $65. b −4°C c −2°C **2** a −5 < −4 b −3 > −4 c 0 < 5 d 0 > −5 e 55 > −55 f −99 > −100 **3** a F b F c T d F e F f F **4** a −9, −4, −1, 2 b −5, −4, 0, 2, 6 c −14, −7, 0, 5, 8, 14 d −7, −5, −3, −1, 1, 3 e −12, −10, −6, −4, 0, 1, 5 f −151, −141, −131, −115, −114, −113, −112 **5** a 0, −2, −5, −6 b 3, 2, 0, −2, −3 c 5, 0, −2, −4, −5 d 6, 2, −4, −5, −8, −10 e −5, −6, −7, −8, −9, −10, −11, −12 f −112, −114, −115, −123, −124, −126 **6** a −3 b 2 c −12 d 2 e −13 f 0 g 0 h −10 i −8 j −6 k −42 l 1 m 2 n −1 o 11 **7** a −3 b −3 c −1 d −22 e 0 f 0 g −18 h 0 i −12 j 0 **8** a 14 b −5 c −7 d 6 e 4 f 10 g −14 h 6 i −5 j 9 k −6 l −7 m 0 n −3 **9** a −6 b −10 c 4 d −18 e 0 **10** a −3 b −20 c 0 d −18 e 0 **11** a −10 b −2 c −8 d −14 e −8 f 10 g 9 h 2 i −5 j 4 k −12 l −22 m −14 n −30 o 40 **12** a −13 b −10 c −7 d −20 e 4 f −7 g −2 h −4 i −2 j 30 k −20 l 8 **13** a 7 b 7 c 2 **14** a −16 b 20 **15** a −45 b 56 c −32 d −21 e 40 f −24 g −27 h −36 i 36 j −7 k 9 l −45 m 49 n 64 o −120 **16** a 105 b 48 c −48 d 30 e −8 f −27 g 45 h 75 i 36 j 120 **17** a −48 b −48 c 63 d 200 **18** a −9 b −7 c −7 d −7 e 6 f −4 g −7 h 8 i 9 j −4 **19** a −5 b 5 c −7 d 6 **20** a −5 b −24 c −11 d 24 e −5 f −11 g 144 h −24 i −10 j −14 k −30 l 0 m −45 n −3 o 4 **21** a −11 b 2 c 25 d 2 e −5 f −8 g 12 h 100 i 0 j −3 k 18 l −12 m −6 n −6 **22** a $-6a$ b $-5t$ c $-14y$ d $-6y$ e $6w$ f $8t$ g $-8y$ h $-3n^2$ i $-m^2$ j z k a l $-5a$ m $-6y$ n 0 o 0 p $-14x$ q $-3a$ r $-8y^2$ s $-a$ t $-2b$ **23** a $7 + 3a$ b $3y + 3b$ c $2a + 4b$ d $-13y + z$ e $-6t + 8s$ f $5a - 2c$ g $3a - 6b$ h $-6t + 5u$ i $-2y + 4$ j $-2a + 5b$ **24** a $-7a$ b $-12a$ c $-24y$ d $36b$ e $16t^2$ f $-42a^2$ g -4 h $-3a$ i $-4y$ j 6 k $-35ab$ l $72ac$ **25** a $18a$ b 0 c $-14a^2$ d $21y$ e $15yt$ **26** a 1 b −9 c 9 d −20 e 2 f 20 g −7 h 8 i −1 j 83 k −11 l −7 m 16 n −48 o 10 p −9 q −4 r −4 s −36 t −36 u −9 v −9 w −22 x −30 y −1 **27** A (2, 2) B (2, 4) C (−2, 2) D (5, 3) E (−3, 3) F (−4, 1) G (0, 4) H (2, 0) I (5, 0) J (0, 2) K (−2, −2) L (−4, −1) M (−3, −4) N (2, −3) P (3, −2) Q (4, −4) R (−4, 0) S (0, −3) T (4, 0) U (0, −1) V (5, −3) W (−4, −3) **28** a b See worked solutions **29** Temperature at 5 p.m. = −2°C **30** (−1, 2) **31** a $\frac{-1}{2}$ b $\frac{-1}{5}$ c $\frac{-3}{4}$ d $\frac{-2}{5}$ e −0.2 f −2.8 g −0.2 h $\frac{-5a}{12}$

QUICK ANSWERS to Chapters 6 and 7 **Practise, Practise**

Chapter 6—Ratios pp. 112–114

1 a 5:3 b 3:7 c 1:2:3 d 2:7 e 1:1000 f 3:5 **2** a $x = 3$ b $x = 3$ c $x = 2$ d $x = 49$ e $x = 39$
3 a 7:3 b 10:7 c 2:1 d 13:17 e 3:2 f 4:5 g 2:5 h 9:10 i 4:11 **4** a 15:4 b 3:2 c 1:6
d 1:10:1000 e 7:2 f 2:1 g 1:1 000 000 h 3:20 **5** a $2b$:1 b 1:5 c 1:2 d 1:p^2 **6** a \$200
b i 7:9 ii 2:7 **7** a 25 b i 31:25 ii 56:25 **8** 5:12 **9** a 3:5 b 1:1 **10** a i 3 cm ii 8 cm
b i 3:8 ii 2:3 iii 8:7 **11** 1:2 **12** a 1:3 b 1:9 c 1:27 **13** 5:1 **14** a 60° b i 3:5 ii 1:1
iii 1:4 **15** 16 **16** a 18 cm b 60 cm **17** \$840 **18** 10 cm **19** a 80:4:1 b 960 students, 12 assistants
20 \$120 000 **21** \$55, \$20 **22** \$120 **23** a F b B c G d J **24** a \$189 000 b \$234 000 c 21:26
25 40 minutes **26** a 3:7 b 12 000

Chapter 7—Algebra pp. 127–130

1 a

Number of triangles	1	2	3	4	5	6	7	8
Number of matchsticks	3	6	9	12	15	18	21	24

b The number of matchsticks = 3 times the number of triangles c $T = 3 \times t$ or $T = 3t$
d i 30 ii 150 iii 300 e 95

2 a

Number of squares	1	2	3	4	5	6	7	8
Number of matchsticks	4	7	10	13	16	19	22	25

b The number of matchsticks = 3 times the number of squares plus one c $x = 3 \times y + 1$ or $x = 3y + 1$
d i 46 ii 157 iii 1201 e i 20 ii 104 iii 300

3 a $y = 2x$ b $y = x + 1$ c $y = x - 2$ d $y = x^2$ e $y = 2x + 1$ f $y = 3x - 1$

4 a $y = x + 3$

x	1	2	3	4
y	4	5	6	7

b $T = t - 1$

t	1	2	3	4
T	0	1	2	3

c $n = 6 - m$

m	1	2	3	4
n	5	4	3	2

d $Q = 2 \times P$

P	1	2	3	4
Q	2	4	6	8

e $y = 3 \times x - 1$

x	2	3	4	5
y	5	8	11	14

f $w = 12 - 3 \times v$

v	1	2	3	4
w	9	6	3	0

5 a ab b $3y$ c $2t$ d $4ts$ e $7y^2$ **6** a $x \times y$ b $3 \times y \times b$ c $4 \times b \times b$ d $2 \times m \times n \times q$ e $2 \times m \times m \times n \times q \times q$ f $9 \times b \times b$ **7** a 8 b 16 c 6 d 7 e 25 f 3 g 40 h 9 i 8 j 8 k 5 l 12 m 36 n 1
8 a $\frac{1}{4}$ b $1\frac{1}{12}$ c 3 d 2 e $\frac{9}{16}$ f $2\frac{1}{2}$ g 1 h $\frac{1}{12}$ **9** 32 **10** 26 **11** 18 **12** 36 **13** 44 **14** 75
15 $4\frac{4}{5}$ **16** 4 **17** a m^2 b $5n$ c $4a^2$ d pq e $2y$ f $2x$ g $5ab$ h $6p$ i $12ba$ j xyz k $6a^2$
l $10y^2x$ m $8ca$ n $4a^3$ o $4a^2b$ p $25p^2$ q 4 r $6b$ **18** a a^{11} b a^{10} c b^{15} d a^4b^4 e y^7 f $6xy^3$
g $2a^{10}$ h a^5 i $36a^2$ j $9a^8$ k x^9y^6 l $\frac{2b^3}{a^2}$ m $\frac{3x}{2y}$ **19** a 64 b 100 c 32 d 1 e 216 f 32
g 64 **20** a $15t$ b $9a$ c $8x$ d $8x$ e $8a - 2$ f $5a + 9$ g $4ab$ h $3x^2 + 4x$ i $2a^2 + 6a$ j $3y + 4$
k $2x + 6$ l $3p$ m $4x + 5y$ n $3p + 10$ o $5x + 7$ **21** a $4a$ b $4a$ c a^4 d $2b$ e a^3b^2 f $7x + 3y$
g $4x$ h $8x^3$ i $\frac{3x}{2y^2}$ j $\frac{3y}{2}$ k $2a + 13$ l $8a^2b$ m $24a^3$ n $7b$ o $16x^6$ p $4x^2 - x$ q $5x^2 + 2x$ r $\frac{2}{3}$
s $6a + b$

QUICK ANSWERS to Chapters 7, 8 and 9 **Practise, Practise**

22 a 0 b $2t$ c $-8x$ d $-2a$ e $3x-2y$ f $2x^2-3x$ g $7x-2$ h $2-2p$ i $5x-3$ j $7x-11$ k $8y-3x$ l $-2p-5$ m $-3x-2$ n $2x-10y$ o x^2-2x p $-a$ q $-3x$ r $2x-10y$ s $-4-t$

Chapter 8—Measurement pp. 152–156

1 a 2 km b 4210 m c 370 mm d 35 mm e 42 cm **2** a B b C c B d B **3** a 24 mm b 31 mm c 35 mm **4** a 11.5 cm to 12.5 cm b 15.5 m to 16.5 m **5** a Perimeter is 84 cm. b Perimeter is 51 cm. c Perimeter is 56 cm. **6** 20 mm **7** a 82 cm b 74 cm **8** a 18 m b 4 rolls c rolls cost $87.60 **9** a 6 units^2 b $12\frac{1}{2}$ units^2 **10** a 105 cm^2 b 144 mm^2 c 8000 cm^2 d 11.56 m^2 e 66.5 cm^2 f 36 cm^2 g 44 cm^2 h 86 cm^2 i 460 cm^2 j 120 cm^2 **11** a 6.4 ha b 48 m^2 **12** a 224 cm^2 b 350 m^2 **13** 9 cm **14** 1.12 ha **15** a 6 m b 1 m c 9 m^2 d 112 m^2 e $180 f $33.60 **16** a 110 cm, 80 cm b 380 cm c $45.60 d 8800 cm^2 e $61.60 **17** a 600 m^2 b 0.25 m^2 c 2400 **18** a 32 000 m^2 b 1.6 ha c 0.37 m^2 d 58 400 cm^2 **19** a 756 cm^3 b 64 cm^3 c 2240 cm^3 d 150 m^3 e 736 cm^3 f 1680 cm^3 g 240 m^3 **20** a 7000 mm^3 b 45 cm^3 **21** 30 m^3 **22** 9.6 m^3 **23** a 1.25 L b 14 830 L c 3.7 kL **24** a 1000 cm^3 b 1 L

Chapter 9—Geometry pp. 184–191

1 a T b F c T d F e T f T g T **2** a Hexagon b 6 c 6 d Yes, all sides are equal in length and all angles are equal. **3** a Pentagon b Quadrilateral c Octagon d Pentagon **4** a and c **5** a PS, TW, UV b PS, RS, TW, VW c QR, UV, SR, WV

6 a

order = 2

b

order = 4

c

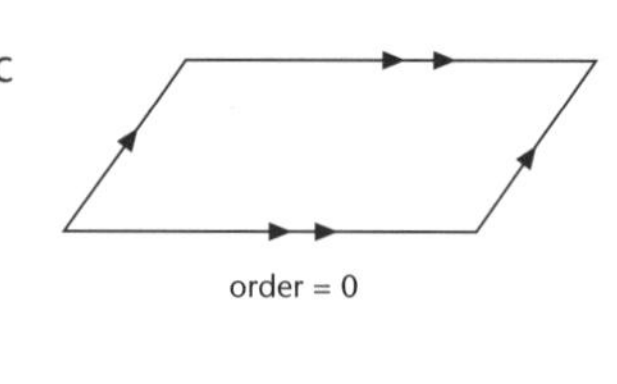

d

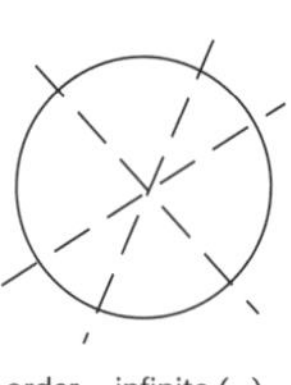

7 a Triangular prism b Cylinder c Rectangular prism d Sphere e Rectangular pyramid f Cone **8** a a, c and e are polyhedra b a and c are prisms **9** a Parallelogram b Square c Trapezium d Rectangle e Kite f Rhombus **10**

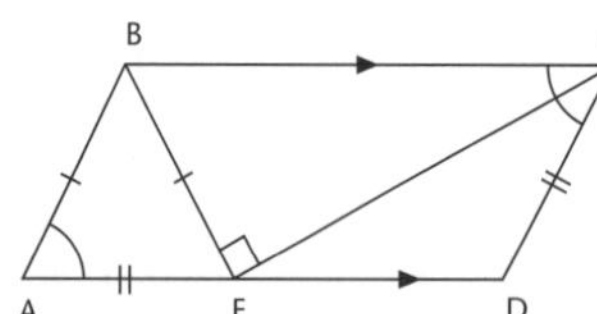

QUICK ANSWERS to Chapters 9 and 10 **Practise, Practise**

11 a Square prism and triangular prism b Yes, because all its surfaces are flat. c Yes, it is a prism. d It has 10 vertices, 15 edges and 7 faces. **12** a Rectangular pyramid b No, it is not a prism. c Yes, because all its surfaces are flat. d 8 edges **13** See worked solutions **14** See worked solutions **15** See worked solutions **16** a △ABE is isosceles. b BCDE is a trapezium. **17** a Ellipse b 2 c No **18** a H b ∠FHG or ∠GHF c Obtuse **19** a ∠AXD and ∠BXC b ∠DXB, ∠CXA and ∠AXD, ∠BXC c i 50° ii 50° d ∠AXB or ∠CXD are straight angles. **20**

A B C D

a ∠DAB b One of ∠ABC, ∠BCA, ∠CAB, ∠DAC and ∠CDA c ∠ACD d ∠ACD e ∠CAB f ∠ACD **21** a 48° b 21° **22** a 138° b 70° c 122° **23** a Right b Obtuse c Acute **24** a $x = 22$ (complementary angles) b $x = 115$ (angles in a straight line) c $x = 36$ (angles in a straight line) d $x = 55$ (vertically opposite); $y = 125$ (supplementary to $x = 55$); $z = 125$ (vertically opposite to $y = 125$) e $x = 80$ (angles at a point) f $x = 69$ (angle sum of a triangle) g $x = 121$ (angle sum of a triangle) h $x = 40$ (angle sum of triangle and isosceles triangle) i $x = 59$ isosceles triangle j $x = 112$ (corresponding angles in parallel lines) k $x = 128$ (co-interior angles in parallel lines) l $x = 71$ (alternate angles in parallel lines) m $x = 115$ (angle sum of a quadrilateral) n $x = 60$ (angle in equilateral triangle) o $x = 124$ (exterior angle of triangle) p $x = 65$ (angle sum of quadrilateral) q $x = 100$ r $x = 83$ s $x = 110$ **25** a 48 b 48 **26** a 50 b 50 c 80 **27** $x = 57, y = 55$ **28** $x = 60$ **29** 118 **30** $m = n = 83$ **31** $x = 50, y = 50$ **32** $x = 140$ **33** a ∠CDB b ∠ABD c 18° d 54° **34** a i A trapezium is a quadrilateral with one pair of opposite sides parallel. ii Not necessarily. b i Opposite sides are equal. ii Opposite angles are equal. iii Not unless it's a rhombus. c i All sides are equal. ii 2 iii Yes iv Yes, a special type of parallelogram. d i 4 ii Equal iii 90° iv They are equal. e i 2 ii Opposite sides are equal. iii 90 iv Diagonals are equal. f i 1 ii 2 pairs of adjacent sides equal g A, C, D, E, F h B, C, E, F i i F ii T iii T iv T v F vi T vii T viii T j i Parallelogram

Chapter 10—Equations pp. 209–210

1 a $x = 12$ b $x = 7$ c $x = \frac{1}{4}$ d $x = 5$ e $x = -15$ f $x = -1$ g $x = 5$ h $x = 16$ i $x = 6$ j $x = 2\frac{1}{2}$ k $x = 2\frac{4}{5}$ l $x = 2\frac{1}{3}$ m $x = 7$ **2** a $x = 4$ b $x = 3$ c $x = 2\frac{2}{3}$ d $x = 5$ e $x = 5$ f $x = 8$ g $x = 2$ h $x = 4$ i $x = 7$ j $x = 2$ **3** a $x = 5$ b $x = 6$ c $a = 8$ d $b = 5$ e $x = 11$ f $x = 1.5$ g $x = 4$ h $x = 2$ i $x = 1$ j $y = 1$ **4** a $x = 17$ b $x = 6$ c $x = 11$ d $x = 3\frac{1}{2}$ e $x = 10$ f $x = 5$ g $a = 13$ h $d = 1$ i $x = 6$ j $x = 4$ **5** a $x = 60$ b $x = 117$ c $x = 50$ d $x = 45$ e $a = 60$ f $x = 10$ g $x = 65$ h $x = 110$ i $x = 17$ j $x = 45$ **6** a 20 b 22 c $6\frac{1}{3}$ d −6 e 12 f 21 and 22 g 12, 13 and 14 h 24 i 40 j 7

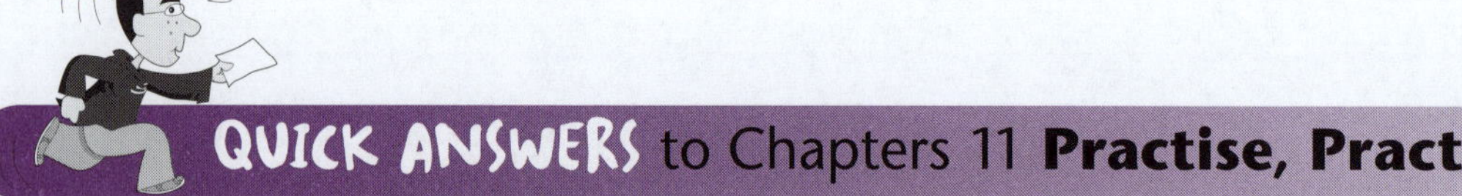

Chapter 11—Statistics and Probability pp. 226–232

1 a i 5 ii 18 iii 2 b i $\frac{1}{4}$ ii $\frac{9}{10}$ iii $\frac{1}{10}$ c Range 5; mode 4 and 5 d See solutions **2** See solutions **3** See solutions **4** a i 4 ii 17 iii 13 iv 8 b i $\frac{4}{25}$ ii $\frac{17}{25}$ iii $\frac{13}{25}$ iv $\frac{8}{25}$ c See solutions **5** See solutions **6** See solutions **7** 9 **8** a 5.6 6 6 4 b 7.6 5 7.5 6 c 4 1, 4 and 7 4 6 d 3.9 7 3.5 9 e 7 7 7 8 **9** a 64 b i 44 ii 20 c i $\frac{5}{32}$ ii $\frac{11}{16}$ iii $\frac{5}{16}$ d 4 $4\frac{1}{2}$ e 4.7 f $4\frac{1}{2}$ g See worked solutions h $4\frac{1}{2}$ **10** a See solutions b 9 5 c i 12 ii 39 iii 27 iv 13 d 52 e i $\frac{3}{13}$ ii $\frac{3}{4}$ iii $\frac{27}{52}$ iv $\frac{1}{4}$ f 4.75 g 4 **11** a i 6 ii 6 5 iii 5.5 b Mean increased by 1, median increased by $\frac{1}{2}$ c 3 goals **12** 80 **13** a i See solutions ii 20 iii 68 b 6 2 c i 4 ii 11 iii 5 d i $\frac{1}{5}$ ii $\frac{11}{20}$ iii $\frac{1}{4}$ e 3.4 f 3 **14** a 66.76 b 63.55 c 63, 69, 27 d 58 e Boys' range larger than girls' range f Median drop to 56 **15** a Increase by 6 b Multiplied by 6 **16** a b See solutions c 3.02 d 2 and 5, 3 **17** 270, 18, 19 **18** a $\frac{1}{13}$ b $\frac{1}{13}$ c $\frac{1}{4}$ d $\frac{1}{4}$ e $\frac{1}{2}$ f $\frac{3}{13}$ g $\frac{10}{13}$ h $\frac{5}{13}$ i $\frac{1}{52}$ j $\frac{7}{13}$ k $\frac{1}{26}$ **19** $\frac{3}{10}$ **20** a $\frac{1}{3}$ b $\frac{2}{3}$ c $\frac{2}{5}$ d $\frac{3}{5}$ e 0 f $\frac{4}{15}$ g $\frac{11}{15}$ **21** a 0.52 b 672 **22** a $\frac{5}{36}$ b $\frac{1}{9}$ c $\frac{1}{2}$ d $\frac{1}{6}$ e $\frac{5}{18}$ f $\frac{8}{9}$ g $\frac{11}{18}$ h $\frac{1}{18}$ **23** a $\frac{1}{4}$ b $\frac{1}{8}$ c 0 d $\frac{3}{8}$ e $\frac{5}{8}$ **24** a $\frac{1}{10}$ b $\frac{1}{2}$ c $\frac{1}{2}$ d $\frac{1}{2}$

QUICK ANSWERS to Chapters 1, 2 and 3 **Tests**

Chapter 1—Number

Level 1 Test pp. 29–30

1 37 **2** 1700 **3** 17 **4** 72 **5** 20 **6** 4 **7** 4 **8** 26 **9** 44 **10** 140 **11** 7000 **12** 3207, 3702, 7203, 7302 **13** a $(6 + 4) \times (7 - 5)$ b $6 + [4 \times (7 - 5)]$ c $(6 + 4) \times 7 - 5$ **14** 41 **15** a 2240 b $14\frac{11}{27}$ **16** 1055 **17** a 14, 16, 18, 20, 22 b 4, 8, 12, 16, 20, 24, 28, 32, 36 c 1, 2, 3, 4, 6, 12 d 16, 25, 36, 49 **18** a 25 b 192 **19** 4 **20** 24 **21** 2 and 3 **22** a 6^7 b T **23** 9 **24** a 3×2^4 b 2×3^3 c 6, 432 **25** a T b F

Level 2 Test p. 31

1 a 30 b 19 000 **2** 6 **3** 6 **4** 10 **5** 8 **6** 40 000 **7** 62 169, 49 612, 46 129, 42 169, 21 496 **8** a 48 b 65 c 140 **9** a 3572 b $24\frac{24}{36}$ or $24\frac{5}{6}$ c 4700 d 753 e 1290 f 243 g 297 **10** $8.00 **11** a 11, 13, 15, 17, 19, 21, 23 b 15, 20, 25, 30, 35, 40, 45, 50, 55 c 1, 2, 3, 4, 6, 9, 12, 18, 36 d 1, 3, 6, 10, 15, 21, 28 **12** 6 **13** 36 **14** a $2 \times 2 \times 2 \times 3 \times 3$ b $2 \times 2 \times 3 \times 7$ c HCF = 12, LCM = 504 **15** $3 \times 3 \times 3 \times 3 \times 3 \times 3 \times 3 \times 3$ **16** $3^5 \times 4^3$ **17** 81 **18** a 7 b 4 **19** 9, 3, No

Level 3 Test pp. 32–33

1 T **2** 27 **3** 1028 **4** a 16 b 44 c 39 d 510 **5** $(4 + 8) \times (3 + 7)$ **6** 97 **7** 100 **8** 58 **9** 112 **10** 101 **11** 3 **12** 216 **13** a $2 \times 2 \times 2 \times 2 \times 2 \times 3$ b $2 \times 2 \times 3 \times 3 \times 3$ c HCF = 12, LCM = 864 **14** $3 \times 3 \times 3 \times 3 \times 3 \times 5 \times 5 \times 5$ **15** $7^7 \times 8^5$ **16** a 11 b 256 c 13 **17** a Yes b Yes c No d 24 **18** 21

Chapter 2—Fractions

Level 1 Test pp. 50–51

1 $\frac{2}{3}$ **2** $1\frac{2}{3}$ **3** $\frac{9}{4}$ **4** a $\frac{1}{4} < \frac{1}{3}$ b $\frac{2}{5} = \frac{4}{10}$ **5** $1\frac{1}{4}$ **6** a $\frac{1}{5}$ b $1\frac{3}{20}$ c $1\frac{3}{4}$ d $9\frac{11}{12}$ **7** a $\frac{9}{20}$ b $\frac{1}{3}$ c 2 d $\frac{49}{50}$ e $\frac{5}{6}$ **8** $\frac{4}{5}$ **9** a $11 b $33 **10** $\frac{1}{5}$ **11** $\frac{1}{2}$ **12** 56 hours **13** 90 out of 100 **14** $80

Level 2 Test p. 52

1 $\frac{3}{5}$ **2** $4\frac{1}{4}$ **3** $\frac{29}{5}$ **4** $\frac{1}{2}, \frac{2}{3}, \frac{3}{4}$ **5** $\frac{3}{5}$ **6** a $1\frac{1}{6}$ b $1\frac{2}{3}$ c $4\frac{2}{15}$ d $4\frac{4}{5}$ **7** a $\frac{3}{14}$ b $1\frac{1}{2}$ c $\frac{14}{15}$ d 25 **8** $1\frac{1}{4}$ **9** $16 **10** $\frac{11}{20}$ **11** 15 **12** 60 out of 100 **13** $\frac{4}{15}$ **14** $60

Level 3 Test pp. 53–54

1 $4\frac{3}{4}$ **2** $\frac{60}{7}$ 3 $1\frac{3}{4}, \frac{8}{5}, 1\frac{7}{12}$ **4** $1\frac{11}{24}$ **5** $\frac{2}{3}$ **6** $1\frac{3}{7}$ **7** a $3\frac{1}{3}$ b $2\frac{19}{30}$ **8** $1\frac{3}{5}$ **9** $60 **10** 2 **11** $7.50 **12** 31 lengths, $1\frac{3}{5}$ cm left **13** a $\frac{9}{10}$ b $0 < \triangle < \frac{9}{10}$ **14** 10 **15** 30

Chapter 3—Decimals

Level 1 Test pp. 67–68

1 4.76 > 4.577 **2** 9 hundredths **3** 0.031 **4** $\frac{1}{25}$ **5** 47.6 **6** 2.475 **7** 0.283 **8** 0.35 **9** a 4.033 b 5.6 c 0.2 **10** 4.2 **11** a 17.05 b 60.47 c 1.221 **12** a 2.6 b 0.1843 c 1.2 **13** $1.68 **14** 21.05 m **15** 0.94 **16** 1.6 **17** 3.28 m **18** $8.82 **19** 40

QUICK ANSWERS to Chapters 3, 4 and 5 **Tests**

Level 2 Test p. 69

1 4.17, 4.7, 4.701, 4.71 **2** 0.875 **3** $\frac{12}{25}$ **4** 8.04 **5** 3.18 **6** a 11.175 b 2.91 **7** 2.96 **8** a 1.806 b 3148.6 **9** a 9 b 0.06476 c 51.02 **10** 5 **11** 3.27 **12** \$26.20 **13** 2.2 **14** 0.25 L **15** 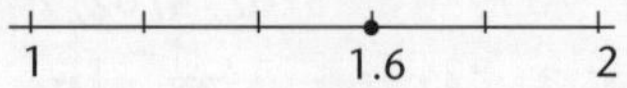

16 a 2.75 b 0.94 **17** \$8.50/week

Level 3 Test p. 70

1 3.074, 3.407, 3.47, 3.471 **2** $0.6\dot{3}$ **3** $8\frac{11}{200}$ **4** 0.63 **5** a 20.94 b 3.969 **6** 4.06 **7** a 1.44 b 7476 c 0.0018 **8** a 41 b 26.1 c 0.00047 **9** 2.25 units **10** 0.4 **11** 34 **12** 2.4 **13** 9.4 **14** \$16.50 **15** 9.716 **16** 20 **17** \$54.37 **18** 12.6°

Chapter 4—Percentages

Level 1 Test p. 82

1 a 60% b 150% c 45% d 80% e 6% f 35.5% **2** a $\frac{27}{100}$ b $\frac{2}{25}$ c $\frac{2}{5}$ **3** a 0.04 b 0.075 c 2.15 **4** a 25% b 80% c 48% **5** a \$3 b \$48 c \$2.40 **6** \$36 **7** 25% **8** \$120 **9** \$150 **10** \$540

Level 2 Test p. 83

1 a $\frac{2}{5}$, 41%, 0.42 b 37%, $\frac{3}{8}$, 0.4 **2** a \$11.60 b \$57.60 c \$595 d 210 mm **3** a 10% b 0.25% c 5% d 1.75% **4** a \$27.50 b \$702 **5** a \$598 b \$3620 **6** a \$6.75 b \$496 **7** a \$20 b \$140 **8** $\frac{3}{5}$ **9** 480 **10** \$720

Level 3 Test p. 84

1 a $66.\dot{6}\%$ b 87.5% **2** a 0.085 b 0.034 c 0.1275 **3** a \$8.75 b 40c **4** a 118.8 **5** $\frac{2}{5}$ **6** a 25% b 12.5% **7** \$480 **8** \$984 **9** a \$2.34 b 10% **10** \$1500 **11** \$159.26 **12** \$246.40

Chapter 5—Integers

Level 1 Test p. 100

1 a $-5 < 0$ b $-3 > -4$ **2** a 4 b −12 c −2 d −18 e 5 **3** 3 **4** a −2 b −5 **5** a −6 b −4 c 15 d −3 **6** a $-2t$ b $-4a$ c $-5m$ d $-12a$ e −5 **7** a −2 b 12 c 9 d −12 e −9 **8** −11°C **9** a (−3, 1) b (1, −3) **10** a $\frac{1}{4}$ b −2.3

Level 2 Test p. 101

1 −8, −1, 0, **2** a 3 b −8 c −14 d −6 e 1 f −12 **3** a 22 b −12 **4** −15 **5** a 24 b −5 c −10 d 2 e 12 f 36 **6** a $-6t$ b $-6t$ c $6b$ d 0 e 0 f $-2a + 16$ **7** a $-1\frac{1}{4}$ b −8 c −11 d −4 **8** a −11 b 10 c 5 d 28 e 49 f −1 **9** a $\frac{-2}{15}$ b −0.8 c −3.7 **10** −8°C **11** a −5 b −2 c −6

QUICK ANSWERS to Chapters 5, 6 and 7 **Tests**

Level 3 Test pp. 102–103

1 5, –1, –7, –10 **2** a P(–3, –2), Q(4, –3) b $(\frac{1}{2}, 2\frac{1}{2})$ **3** a –40 b –15 c –5 d 0 **4** –16 **5** 13 **6** a –13 b –1 c 0 **7** a –32 b –3 **8** a 14 b 4 **9** –7 **10** a $-8m$ b $-6y$ c $-21a$ d $16y^2$ e $-2a$ f –10 **11** a 5 b –17 c 17 d –1 e 16 **12** a $1\frac{1}{3}$ b $-1\frac{3}{8}$ c $1\frac{9}{10}$ **13** 13 **14** a $-x+2$ b $8a-3$ c $11+a$ **15** a 5 b –2

Chapter 6—Ratios

Level 1 Test p. 116

1 a 1:4 b 9:7 c 10:1 **2** a 1:1 b 1:4 c 1:1 **3** a 8:7 b 7:15 **4** a 2:1 b 3:2 **5** 90 mL **6** 290 **7** 18 m **8** a 1:2 b 2:3 **9** 12 boys **10** 990 students

Level 2 Test p. 117

1 a 4:5 b 5:4:2 c 1:10 000 **2** a 4 b 10 **3** 3:2 **4** 8 **5** 420 g, 120 g **6** 0.6 m **7** 200 ha **8** 25 black jelly beans **9** a 1:3 b 1:3 c 1:9 **10** 35 kg

Level 3 Test p. 118

1 a 3:4 b 1:4 c 8:9 **2** a $x=3$ b $x=14$ **3** 2:1 **4** 3:20 **5** \$72, \$108, \$180 **6** 16 cm × 24 cm **7** 4.2 kg **8** \$2.90 **9** a 2:3 b 4:9 **10** 52 **11** \$6250 **12** a 15:10:4 b 180

Chapter 7—Algebra

Level 1 Test pp. 132–133

1 a 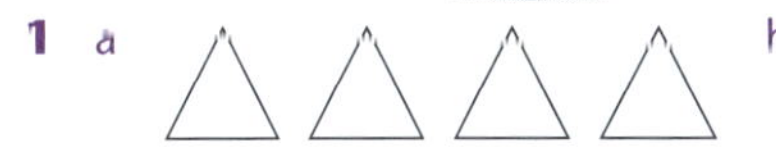

b

Number of triangles	1	2	3	4	5	6
Number of matchsticks	3	6	9	12	15	18

c Number of matches used = 3 × Number of triangles formed d 45 e 10

2 a 6 b 11 c 40 d 21

3 a $y=x+2$

x	1	2	3	4
y	3	4	5	6

b $T=3\times t$

t	1	2	3	4
T	3	6	9	12

4 a $y=x+2$ b $y=x-1$ **5** a $5a$ b $2b$ c $12x$ d c^2 **6** a $4\times k$ b $a\times b$ c $5\times x\times y$ d $y\times y$ **7** a $3y$ b $2b$ **8** a 8 b 3 c 10 d 5 **9** a 8 b 2 c 24 d 16 **10** 6 **11** $5a^2$ **12** a $10b$ b $4y$ c $7a+3b$ d $5t+4n$ **13** a 2^3 b a^4 c $3a^3$ **14** 16 **15** 5

QUICK ANSWERS to Chapters 7 and 8 **Tests**

Level 2 Test pp. 134–135

1 a

b

Number of triangles	1	2	3	4	5	6	7	8
Number of matchsticks	3	5	7	9	11	13	15	17

c The number of matchsticks = 2 × number of triangles + 1 d $M = 2T + 1$ e i 41 ii 128

2 a 6, 3 b 25, 36 c 13, 21 d 18, 9

3 a

x	1	2	3	4
y	1	3	5	7

b

x	2	3	4	5
y	11	9	7	5

4 a $y = 3x + 2$ b $y = 2x - 3$ **5** a $10a$ b $6abc$ c $4a^2$ d 12 e 5 f $10a^2b$ g $9b^2$ h $4a$ i $12a + 4b$ j $7a^2 + a$ **6** a $6a^3$ b a^7 c a^6 d a^{10} **7** a 17 b 18 c 2 d 3 **8** 2 **9** 3 **10** a $3x$ b $6xy + 6x$ c $21x^5y^5$ d $8a^{12}$ e $6ab$ f $6b^2a$

Level 3 Test pp. 136–137

1 a 28 b

Number of squares	1	3	5	7	9	11
Number of matchsticks	4	10	16	22	28	34

c The number of matchsticks = 3 × number of squares + 1 d $m = 3T + 1$ e 172 f 200 **2** a 10, 12, 14, add 2 to get the next term b 15, 11, 7, subtract 4 to get the next term c 13, 21, 34, add the previous two terms

3 a

x	1	2	3	4
y	1	7	17	31

b

x	1	5	21	50
y	1	13	61	148

4 a $y = x^2 - 3$ b $y = 5 - x^2$ **5** a $2xy + 4x$ b $15t^3m^5$ c $2x$ d $5a^3$ e $\frac{2x}{3}$ f $8a^3$ g $9a^8b^4$ h $4ab$ **6** a 152 b 19 c 8 **7** a $\frac{5t^2}{2}$ b $\frac{1}{2}$ **8** a 6 b 36 **9** a 12 b 115 **10** a $125a^3 - 1$ b $\frac{2a}{b}$

Chapter 8—Measurement

Level 1 Test pp. 158–160

1 9.6 cm **2** 23 km **3** a 56 m b 22 m c 46 m **4** 434.5 cm – 435.5 cm **5** 4 cm **6** $128 **7** 68 cm **8** 99 m **9** a 84 cm^2 b 400 mm^2 c 84 cm^2 **10** a 1440 cm^3 b 280 cm^3 **11** a 144 cm^2 b 116 cm^2 **12** 116 cm^2 **13** 4.8 cm **14** 960 cm^3 **15** 512 mL

Level 2 Test pp. 161–162

1 5 cm **2** 24 m **3** a 62 m b 100 mm **4** 67.5 cm – 68.5 cm **5** 8 cm **6** 94 cm **7** 24 cm **8** 31 ha **9** a 528 cm^2 b 225 cm^2 c 244 cm^2 **10** Rectangle 6 cm × 2 cm or 4 cm × 3 cm etc. **11** a 288 m^2 b 244 m^2 **12** 92 cm^2 **13** 33.6 kg **14** 11.76 m^2 **15** a 360 m^3 b 1152 cm^3

Level 3 Test pp. 163–165

1 4 km **2** 10 m **3** 15 m by 5 m **4** 920 cm **5** 42 cm **6** 64 cm **7** 25 cm **8** e.g. triangle with base 4 cm, height 4 cm **9** 6 814 000 mm **10** 2.4 m^2 **11** 0.96 ha **12** 120 cm^2 **13** 162 cm^2 **14** $60 **15** 9 cm^2 **16** a 3000 cm^3 b 380 cm^3

QUICK ANSWERS to Chapters 9 and 10 **Tests**

Chapter 9—Geometry

Level 1 Test pp. 193–195

1 a Pentagon b Circle c Parallelogram d Trapezium **2** a Sphere b Cone c Rectangular prism d Triangular prism **3** a T b F c T d T e T f F **4** a Triangle b Triangular prism c Yes d 5 e 9 f 6 **5** a ∠MPN or ∠NPM b Acute **6** Octagon **7** 4 **8** a 180° b 90° **9** a D b G c I d L e K f B g C h F i A j E **10** a 180° b 360° **11** a 65 b 72 c 140 **12** a 54 b 55 c 60 d 120 **13** a 65 b 132 c 117 **14** $x = 83$ $y = 97$

Level 2 Test pp. 196–198

1 a Hexagon b Ellipse c Isosceles triangle d Rhombus **2** a Cylinder b Square pyramid c Pentagonal prism d Hemisphere **3** a F b T c T d T e T f F g T h T **4** 5 **5** No

6 a Yes b Hexagonal prism c 8 d Yes **7** **8** a $x = 60$, $y = 120$ b $x = 38$, $y = 100$

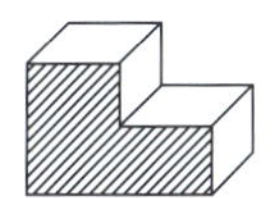

9 **10** a 15° b 105° **11** a Equal b i Supplementary ii Equal iii Equal

12 a 60 b 28 c 124 d 70 e 126 f 100 **13** $a = 60$, $b = 120$ **14** $y = 80$, $x = 100$ **15** a ∠BCA = 52° = ∠ABC (base angles of isosceles △ABC) b ∠CAB = 76° (angle sum of △ABC) c ∠ACE = 76° (alternate to ∠CAB and AB // EC).

Level 3 Test pp. 199–202

1 B **2** D **3** C **4** B **5** D **6** B **7** B **8** B **9** D **10** A **11** a ∠ABC or ∠CBA b ∠CDB or ∠BDC c ∠BDA or ∠ADB d ∠DBC or ∠CBD **12** A B C E D **13** a b No c Yes

14 a $x = 55$, $y = 75$ b $x = 115$, $y = 115$ c $x = 119$, $y = 61$ **15** a Yes b No **16** a $a = 54$, $b = 54$, $c = 54$, $x = 72$ b $x = 72$, $y = 118$, $z = 118$ **17** a $x = 54$ b $x = 132$ **18** $x = 54$, $y = 72$, $z = 18$ **19** $x = 65$, $y = 115$

Chapter 10—Equations

Level 1 Test p. 212

1 a No b Yes c No d Yes e Yes f No **2** a 9 b $3\frac{1}{2}$ c 12.3 d 3 e 6 **3** a 15 b 22 c 19 d 30 e −5 **4** a 3 b 5 c 20 d 3 e 2 **5** a i 5 ii 25 cm^2 b i $x - 4$ ii 7

Level 2 Test p. 213

1 no **2** a 5.1 b $2\frac{3}{4}$ c −6 d 7 e $1\frac{1}{5}$ f −8 **3** a 6 b −4 c 7 d 6 e $10\frac{1}{2}$ f 8 **4** a 30 b 20 **5** a 11 b 5 **6** 12 **7** a 5 b 15 cm^2

Level 3 Test pp. 214–215

1 A **2** C **3** A **4** B true **5** A **6** a −12 b −12 c 0.48 d $3\frac{2}{7}$ e $8\frac{2}{3}$ **7** a 1 b 9 c 7 d 1 **8** 25.5 **9** 5 **10** 33, 35 **11** 64° **12** 13.5 cm^2

QUICK ANSWERS to Chapter 11 **Tests**

Chapter 11—Statistics and Probability

Level 1 Test p. 234

1 See solutions **2** a See solutions b 24.03 c 24, 24, 4 d 12 e 20 **3** a 12.5 b 13 c 14 d 17 **4** a 45 b 19.85 c 18 d 14 e 39 **5** a $\frac{1}{6}$ b $\frac{1}{3}$

Level 2 Test pp. 235–236

1 a See solutions b $65.\dot{3}$ c 61–65 and 71–75 d See solutions **2** a $7.\dot{2}\dot{7}$ b 6 c 5 d 14 **3** a See solutions b 69 c 20 d 3.45 **4** a 40 b 20 **5** a See solutions b i 38.61 ii 48 iii 39.5 iv 39 **6** a $\frac{1}{10}$ b $\frac{1}{2}$ c $\frac{7}{10}$ d $\frac{1}{2}$ e $\frac{1}{2}$ f $\frac{3}{10}$ g $\frac{9}{10}$

Level 3 Test pp. 237–238

1 a See solutions b 7.83, 8, 8, 9 c See solutions **2** a 43 b 5 c See solutions d See solutions **3** 3.2 or 4.8 **4** 4 goals **5** See solutions **6** a 4, 3 and 4, 7 b 4.6 **7** a $\frac{1}{2}$ b $\frac{1}{2}$ c $\frac{12}{13}$ d $\frac{2}{13}$ e $\frac{1}{26}$ f $\frac{25}{26}$

QUICK ANSWERS to Sample Examinations

Sample Examination 1

Part A pp. 240–242

1 1700 **2** −4, −3, −2 **3** $5a$ **4** $\frac{11}{3}$ **5** 21 **6** Trapezium **7** $36 **8** T **9** 4.6 **10** 30 **11** 0.11 **12** 4.5 **13** $28y$ **14** 1 **15** 0.65 **16** 5 **17** $27a^2$ **18** 610 048 **19** 4.15 cm **20** 3 **21** 53° **22** 650 **23** True **24** $0.70 **25** 6 **26** 8 **27** 49 **28** 15 **29** $\frac{5}{8}$ **30** $\frac{1}{4}$ **31** 68 **32** 49 **33** 124 **34** 85 **35** 98 **36** 1.1 **37** 0.36 **38** 16 **39** 370 mm **40** 4 cm **41** −2 **42** 1.8 **43** 3:5 **44** 620.8 **45** 80 **46** −13 **47** 45 **48** 9 **49** 0.07 **50** $2\frac{1}{2}$

Part B pp. 242–244

1 Length remaining is 4.27 m **2** $48 = 2 \times 2 \times 2 \times 2 \times 3$ or $2^4 \times 3$ **3** $x = 70, y = 70$ **4** $2.88 **5** 246 **6** $4\frac{2}{3}$ **7** 35 **8** $\frac{8}{15}$ **9** 16 **10** 17

Part C pp. 244–245

1 a $x = 12$ b 8 years old c i $4x + 60 = 180$ ii 95°, 57° and 28° d $5\frac{1}{4}$ e $37.50 **2** a i $52\frac{1}{2}$ cm^2 ii 264 cm^2 b i 68 m ii $1053 c i 36 m^2 ii 288 m^3 iii 5 cm

Sample Examination 2

Part A pp. 246–249

1 11 **2** 0 **3** $6a$ **4** 20 **5** 50 **6** $\frac{7}{100}$ **7** 0.875 **8** 1:2 **9** $7x - 3y$ **10** $6x^2$ **11** 8 **12** $x = 9$ **13** 2 **14** −9° **15** $8ab$ **16** 8 **17** $\frac{5}{9}$ **18** 1 **19** 23° **20** $a + 12$ **21** −4 **22** $9 **23** 54 **24** $3b$ **25** $\frac{1}{2}$ **26** 0.16 **27** $2x^2 + 6x$ **28** −7 **29** 20 **30** 240 cm **31** $36 **32** 65 **33** 80% **34** $\frac{1}{3}$ **35** C **36** D: obtuse **37** 8 cm **38** 0.2 **39** ∠ACB **40** 12 **41** 400 m **42** 30 cm^2 **43** 60 days **44** 6 axes **45** 11 **46** $a + 2b$ **47** $x + 1$ **48** 45 **49** $y = 2x + 3$ **50** 8 hours

Part B pp. 249–251

1 $13.40 **2** $15.05 **3** $6.75 **4** 100 cm^2 **5** 34 whites **6** 5 hours 50 mins **7** a 1, 3, 9 b 9 **8** 90° **9** $x = -3$ **10** $2850 **11** a $a = -6$ b $x = 13$ **12** $x = 120$ **13** See worked solutions **14** a −4, −1, $2\frac{1}{2}$, 4, 6.3, 8 b $2\frac{1}{2}$ **15** 10.00 a.m. **16** a = 35° (base angles of isosceles △) b ∠ECD = 35° (corresponding to ∠ABC, AB // CE)

Part C pp. 251–252

1 a $14 b $7 c $\frac{1}{2}$ **2** a 2560 cm^2 b 10 cm on sides and 17 cm above and below **3** a 8 b 1:3 c 12 **4** a 1 b −13 c 2 **5** See worked solutions **6** a 48 b Girls' ages are 6, 8, 8, 18. c 11 years

WORKED SOLUTIONS to Chapter 1 **Practise, Practise**

Chapter 1—Number

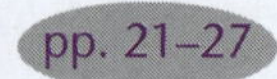

1 **a** $6147 = 6 \times 1000 + 1 \times 100 + 4 \times 10 + 7 \times 1$
$= 6 \times 10^3 + 1 \times 10^2 + 4 \times 10 + 7 \times 1$

b $74\,291 = 7 \times 10\,000 + 4 \times 1000 + 2 \times 100 + 9 \times 10 + 1 \times 1$
$= 7 \times 10^4 + 4 \times 10^3 + 2 \times 10^2 + 9 \times 10 + 1 \times 1$

c $70\,707 = 7 \times 10\,000 + 7 \times 100 + 7 \times 1$
$= 7 \times 10^4 + 7 \times 10^2 + 7 \times 1$

d $317\,429 = 3 \times 100\,000 + 1 \times 10\,000 + 7 \times 1000 + 4 \times 100 + 2 \times 10 + 9 \times 1$
$= 3 \times 10^5 + 1 \times 10^4 + 7 \times 10^3 + 4 \times 10^2 + 2 \times 10 + 9 \times 1$

e $40\,030 = 4 \times 10\,000 + 3 \times 10$
$= 4 \times 10^4 + 3 \times 10$

2 **a** $4 \times 10^3 + 7 \times 10^2 + 3 \times 10 + 2 \times 1$
$= 4 \times 1000 + 7 \times 100 + 3 \times 10 + 2 \times 1$
$= 4000 + 700 + 30 + 2 = 4732$

b $6 \times 10^3 + 9 \times 1$
$= 6 \times 1000 + 9 \times 1$
$= 6000 + 9 = 6009$

3 **a** $5 \times 1000 + 8 \times 100 + 9 \times 10 + 5 \times 1$
$= 5000 + 800 + 90 + 5 = 5895$

b $9 \times 10^4 + 8 \times 10^2 + 7 \times 10$
$= 9 \times 10\,000 + 8 \times 100 + 7 \times 10$
$= 90\,000 + 800 + 70 = 90\,870$

4 **a** $300 - 3 = 297$ **b** $3000 - 30 = 2970$
c $30\,000 - 3000 = 27\,000$
d $3000 - 3 = 2997$
e $30\,000 - 300 = 29\,700$

5 **a** $74 > 47$ **b** $235 < 325$
c $1021 < 1201$ **d** $3102 > 3012$
e $56\,174 > 56\,147$

6 **a** T $(426 < 462)$ **b** T $(72 = 72)$
c $46 \neq 36$ T **d** $7 < 6$ F $(7 > 6)$
e $22 > 22$ F $(22 = 22)$
f $13 \nless 14$ F $(13 < 14)$
g $24 \ngtr 24$ T $(24 = 24)$
h $8 \leq 8$ T $(8 = 8)$
i $3 < 3$ F $(3 = 3)$ **j** $48 = 48$ T

7 **a** 36, 47, 63, 74
b 1123, 1132, 2113, 3112
c 5147, 5174, 7145, 7154, 7415, 7514

8 **a** 93, 86, 68, 39
b 8844, 8448, 4884, 4848
c 3291, 3219, 3192, 3129, 2391, 2319, 2193, 2139

9 **a** $10 \times 5 = 50$ **b** $8 \div 4 = 2$
c $3 \times 5 \times 4 = 60$
d $15 + 5 \times 5$
$= 15 + 25$
$= 40$
e $[25 \div 5] + 7$
$= 5 + 7$
$= 12$
f $100 - \{80 - [60 - 20]\}$
$= 100 - \{80 - 40\}$
$= 100 - 40$
$= 60$
g $8 \times [21 - (17)]$
$= 8 \times 4$
$= 32$

10 **a** $27 - 8 - 9 - 10 = 19 - 9 - 10 = 10 - 10 = 0$
b $8 + 14 - 8 + 14 = 8 - 8 + 14 + 14 = 28$
c $27 - 14 + 26 - 13 = 13 + 26 - 13 = 39 - 13 = 26$
d $6 + 4 \times 5 = 6 + 20 = 26$
e $27 - 36 \div 4 = 27 - 9 = 18$
f $5 + 3 \times 2 + 7 = 5 + 6 + 7 = 18$
g $6 \times 4 + 4 \times 4 = 24 + 16 = 40$
h $36 \div 9 + 49 \div 7 = 4 + 7 = 11$
i $21 - 6 - 15 = 0$
j $14 - 2 \times 4 + 4 = 14 - 8 + 4 = 10$

11 **a** $(4 + 3) \times 7 = 49$ **b** $(12 - 5) \times 3 = 21$
c $7 \times (3 + 2) \times 2 = 70$
d $3 \times (2 + 3) + 4 = 19$
e $5 + (4 + 3) \times 2 = 19$
f $25 \div (3 + 2) + 7 = 12$

12 **a** $15 + 14 = 29$ **b** $15 \times 4 = 60$
c $15 - 4 = 11$ **d** $15 - 4 = 11$
e $42 \div 6 = 7$
f $73 \div 9 = 8$ remainder 1
$(8 \times 9 = 72)$; remainder is 1
g $\text{Average} = \frac{19 + 21 + 36 + 40 + 14}{5}$
$= \frac{130}{5} = 26$
h $\text{Total} = 11 + 28 + 14 + 9 + 23 + 5$
$= 90$

WORKED SOLUTIONS to Chapter 1 **Practise, Practise**

i Sum = 19 + 13 = 32
Difference = 19 − 13 = 6
Answer = 32 − 6 = 26

j Sum = 7 + 9 = 16
Difference = 17 − 9 = 8
∴ Product = 16 × 8 = 128

13 8 T-shirts cost \$72
1 T-shirt costs \$72 ÷ 8 = \$9
∴ 9 T-shirts cost \$9 × 9 = \$81

14 7 hours work earns \$78.75
1 hour work earns \$78.75 ÷ 7 = \$11.25
∴ 4 hours work earns \$11.25 × 4 = \$45.00

15 6 cartons contain 150 packages
1 carton contains 150 ÷ 6 = 25 packages
∴ 8 cartons contain 25 × 8 = 200 packages

16
a
```
 72 +
 27
 98
 89
286
```

b
```
407 −
198
209
```

c
```
  72 ×
  27
 504
1440
1944
```

d
```
   5 1 6/9
9)46¹5
```
Answer:
$51\frac{6}{9} = 51\frac{2}{3}$

e
```
  75 +
 871
   4
 637
  92
1679
```

f
```
7004 −
 989
6015
```

g
```
   532 ×
   286
  3192
 42560
106400
152152
```

h
```
     94
45)4237
   405
    187
    180
      7
```
Answer: $94\frac{7}{45}$

i
```
     726
69)50102
   483
    180
    138
     422
     414
       8
```
Answer: $726\frac{8}{69}$

17 Number of chess pieces = 17 + 11 + 9 + 12 = 49

18 Number of cards = 15 × 20 = 300

19 $\text{Average} = \dfrac{17 + 23 + 42 + 8 + 10}{5} = \dfrac{100}{5} = 20$

20 Each packet costs \$10.56 ÷ 8 = \$1.32
10 packets cost \$1.32 ×10 = \$13.20

21 Number of rows = 87 ÷ 14 = 6 remainder 3
There are 6 complete rows but there are only 3 trees in the seventh row.

22 Number of hours = 765 ÷ 45
= 17
Mitchell needs 17 hours.
```
    17
45)765
   45
   315
   315
     0
```

23
a 23 + 35 = 23 + 30 + 5 = 56 + 5 = 61
b 48 + 37 = 48 + 30 + 7 = 78 + 7 = 85
c 27 + 56 = 27 + 50 + 6 = 77 + 6 = 83
d 33 − 16
16 + 10 + 7 = 33
This means the answer is 17.
e 48 − 29
29 + 10 + 9 = 48
This means the answer is 19.
f 82 − 48
48 + 30 + 4 = 82
This means the answer is 34.

24
a 33 + 68 = 30 + 60 + 3 + 8 = 90 + 11 = 101
b 76 + 55 = 70 + 50 + 6 + 5 = 120 + 11 = 131
c 83 + 19 = 80 + 10 + 3 + 9 = 90 + 12 = 102
d 46 − 27 = 46 − 20 − 7 = 26 − 7 = 19
e 53 − 36 = 53 − 30 − 6 = 23 − 6 = 17
f 48 − 39 = 48 − 30 − 9 = 18 − 9 = 9

25
a 5 × 19 × 2 = 5 × 2 × 19 = 10 × 19 = 190
b 25 × 27 × 4 = 25 × 4 × 27 = 100 × 27 = 2700
c 5 × 16 × 20 = 5 × 20 × 16 = 100 × 16 = 1600
d 10 × 31 × 10 = 10 × 10 × 31 = 100 × 31 = 3100
e 20 × 13 × 5 = 20 × 5 × 13 = 100 × 13 = 1300
f 18 × 25 × 4 = 18 × 100 = 1800

WORKED SOLUTIONS to Chapter 1 **Practise, Practise**

26 a $38 \times 9 = 38 \times 10 - 38 = 380 - 38 = 342$
b $57 \times 9 = 57 \times 10 - 57 = 570 - 57 = 513$
c $46 \times 9 = 46 \times 10 - 46 = 460 - 46 = 414$
d $46 \times 11 = 46 \times 10 + 46 = 460 + 46 = 506$
e $78 \times 11 = 78 \times 10 + 78 = 780 + 78 = 858$
f $83 \times 11 = 83 \times 10 + 83 = 830 + 83 = 913$

27 a $58 \times 20 = 58 \times 2 \times 10 = 116 \times 10 = 1160$
b $62 \times 20 = 62 \times 2 \times 10 = 124 \times 10 = 1240$
c $48 \times 30 = 48 \times 3 \times 10 = 144 \times 10 = 1440$
d $630 \div 30 = 630 \div 10 \div 3 = 63 \div 3 = 21$
e $480 \div 40 = 480 \div 10 \div 4 = 48 \div 4 = 12$
f $360 \div 90 = 360 \div 10 \div 9 = 36 \div 9 = 4$

28 a $4 + 17 + 6 = 4 + 6 + 17 = 10 + 17 = 27$
b $43 + 18 + 57 = 43 + 57 + 18 = 100 + 18 = 118$
c $36 + 57 + 64 = 36 + 64 + 57 = 100 + 57 = 157$

29 a 2, 4, 6, 8, 10, 12, 14, 16, 18
b 19, 21, 23, 25
(Note: Between 17 and 27 means 17 and 27 are not included.)

30 a 2374 (even) b 7234 (even)
c 3427 (odd) d 4372 (even)
e 4723 (odd) f 40 690 (even)

31 a 1, 3, 6, 10, 15, 21, 28, 36, 45, 55
b 1, 4, 9, 16, 25, 36, 49, 64, 81, 100

32 a 727 backwards 727, Palindromic
b 4848 backwards 8484, not Palindromic
c 4884 Palindromic
d 1 001 001 Palindromic
e 43 234 Palindromic

33 a 58 + 85 = 143; 143 + 341 = 484 Palindromic
b 74 + 47 = 121 Palindromic
c 117 + 711 = 828 Palindromic
d 724 + 427 = 1151; 1151 + 1511 = 2662 Palindromic
e 512 + 215 = 727 Palindromic

34 Fibonacci sequence: 1, 1, 2, 3, 5, 8, 13, 21, 34, 55, 89, 144, 233, 377, 610, 987, 1597, 2584, 4181, 6765

35 a 5, 7, 9, 11, *13, 15, 17* … (Add 2)
b 6, 8, 10, 12, *14, 16, 18* … (Add 2)
c 2, 4, 8, 16, *32, 64, 128* … (× 2)
d 2, 5, 8, 11, *14, 17, 20* … (Add 3)
e 2, 3, 5, 8, 12, *17, 23, 30* … (Difference increases by 1.)
f 2, 6, 18, *54, 162, 486* … (× 3)
g 2, 4, 7, 11, *16, 22, 29* … (Difference increases by 1.)
h 2, 4, 12, 48, *240, 1440, 10 080* … (× 2, × 3, × 4, × 5 etc.)
i 2, 1, 4, 3, 6, 5, *8, 7, 10* … (Even numbers, odd numbers.)
j 4, 16, 64, *256, 1024, 4096* … (× 4)
k 1, 9, 25, *49, 81, 121* … (Odd numbers, squared.)
l 2, 3, 5, 6, 8, 9, *11, 12, 14* … (Consecutive numbers, miss one, consecutive numbers etc.)
m 17, 18, 16, 19, 15, 20, *14, 21, 13* … (Odd positions 17, 16, 15 … Even positions 18, 19, 20 …)
n 1, 2, 3, 6, 11, 20, *37, 68, 125* … ($T_1 + T_2 + T_3 = T_4$, $T_2 + T_3 + T_4 = T_5$ etc.) [T_1 means term one]
o 2, 11, 56, 281, *1406, 7031, 35 156* … (× 5 + 1)
p 5, 6, 7, 4, 9, 2, *11, 0, 13* … (Odd terms 5, 7, 9 … Even terms 6, 4, 2 …)

36

$\frac{1}{1} = 1$	$\frac{144}{89} = 1.617\,978$
$\frac{2}{1} = 2$	$\frac{233}{144} = 1.618\,056$
$\frac{3}{2} = 1.500\,000$	$\frac{377}{233} = 1.618\,026$
$\frac{5}{3} = 1.666\,667$	$\frac{610}{377} = 1.618\,037$
$\frac{8}{5} = 1.600\,000$	$\frac{987}{610} = 1.618\,033$
$\frac{13}{8} = 1.625\,000$	$\frac{1597}{987} = 1.618\,034$
$\frac{21}{13} = 1.615\,385$	$\frac{2584}{1597} = 1.618\,034$
$\frac{34}{21} = 1.619\,048$	$\frac{4181}{2584} = 1.618\,034$

$\frac{55}{34} = 1.617647$ $\frac{6765}{4181} = 1.618034$

$\frac{89}{55} = 1.618182$

Yes, the single value is 1.618 034. (This is actually known as the Golden Ratio. The symbol for this is ϕ [phi]. The Golden Ratio is related to the Golden Rectangle, both of which occur frequently in the history of mathematics. They have applications in architecture, art, music and many other areas.)

37 1, 1, 4, 9, 25, 64, 169, 441, 1156, 3025, 7921, 20 736 …

Add consecutive terms 2, 5, 13, 34, 89, 233, 610, 1597, 4181, 10 946 … The sequence generated is every second term of the Fibonacci sequence beginning at 2.

38 **a** Sum 10 = 143 **b** Sum 8 = 54
c Sum 6 = 20 **d** Sum 12 = 376
e Sum 10 = (term 12) − 1
Sum 8 = (term 10) − 1
Sum 6 = (term 8) − 1
Sum 12 = (term 14) − 1
f Sum 18 = (term 20) − 1 = 6764
Sum 20 = (term 22) − 1 = 17 711 − 1 = 17 710

39 **a** 7, 14, 21, 28, 35, 42, 49, 56, 63, 70
b 14, 21, 28, 35 **c** 2
d 21, 28, 35, 42
(Wording 'between' implies that end points 14 and 49 are **not** included.)
e 42, 49, 56, 63
(Wording 'from … to' implies that end points **are** included.)

40 Multiples of 8? 8, 16, 24, 32, 40, 48, 56, 64, 72, 80, 88, 96 …
a 4 (i.e. 24, 32, 40, 48) **b** 24, 32
c 8, 16, 24, 32, 40, 48, 56, 64, 72
d 40, 80

41 Common multiples of 2 and 7: 14, 28, 42, 56, 70 …
a 14, 28, 42 **b** 42, 56
c 3 (i.e. 14, 28, 42)

42 Common multiples of 5 and 7: 35, 70, 105, 140, 175 …
a 35, 70 **b** 4 (i.e. 35, 70, 105, 140)
c 70

43

a 30	**b** 15	**c** 28	**d** 36
e 8	**f** 12	**g** 18	**h** 36
i 12	**j** 12	**k** 30	**l** 60
m 24	**n** 60	**o** 28	**p** 30
q 60	**r** 12	**s** 30	**t** 60

44 The question requires you to find the least number that both 4 and 6 divide into evenly; that is, the LCM of 4 and 6.
LCM of 4, 6 = 12 There were 12 coconuts.

45 LCM of 4, 6, 7 required.
LCM of 4, 6 and 7 = 84
There were 84 coconuts.

46 The question requires you to find the smallest number that both 15 and 12 divide into evenly. The answer will be 1 more than this as there is a remainder of 1 when the jelly beans are divided up.
LCM of 15, 12 = 60
Add 1 gives 61. The number of jelly beans was 61.

47 LCM of 15, 12, 10 required. LCM of 15, 12, 10 = 60. Add 1 gives 61. Number of jelly beans is still 61.

48 Consider the multiples of 5 and 6.
Multiples of 5: 5, 10, 15, 20, (25), 30, 35, 40, 45, 50, 55 …
Multiples of 6: 6, 12, 18, (24), 30, 36, 42, 48, 54, 60
Now, we need a multiple of 5 that is **1** more than a multiple of 6. Why? Because, bananas were divided evenly among 5, but there was 1 over when shared among 6.
These are 25 and 24.
Now working backwards, 25 were divided among 6 pirates—4 bananas each + 1 for orangutang. The 25 had been divided among 5 pirates but there also was 1 left over.
$\therefore$ Number of bananas is (25 + 1) = 26
There were 26 bananas to start with.

WORKED SOLUTIONS to Chapter 1 **Practise, Practise**

49 Once again we need multiples of 5, 6, 7. We need a multiple of 6 that is 1 less than a multiple of 5, but this multiple of 6 must also be a multiple of 7 (common multiple of 6, 7). Why?

Because there are none left over after dividing the bananas 7 ways.

Common multiples of 6 and 7 are: 42, 84, 126, 168 ...

Now 85 is a multiple of 5, and 84 is 1 less than 85. Check: 84 bananas could be divided among 7 evenly. 84 could be divided evenly among 6 and, by giving one to the orangutang, they had 85. Also, 85 could be divided evenly among 5. However, there was 1 left over here as well.

∴ Number of bananas was (85 + 1) = 86.

There were 86 bananas.

50 **a** (1), (2), 3, (4), 6, (8), 12, (16), 24, 48
b (1), (2), (4), (8), (16)
c 1, 2, 4, 8, 16

51 **a** (1), (2), (3), (6), (9), (18), 27, 54
b (1), (2), (3), 4, (6), (9), 12, (18), 36
c 1, 2, 3, 6, 9, 18

52 **a** (1), (5), 7, 35
b (1), 3, (5), 9, 15, 45
c 1, 5

53 **a** 2, 3 **b** 2 **c** 2, 3
d 5, 7 **e** 3, 5 **f** 2, 3
g 2, 3, 7 **h** 2, 3, 11 **i** 3, 13
j 3, 17

54 **a** 5 **b** 4 **c** 4 **d** 3
e 2 **f** 1 **g** 8 **h** 1
i 4 **j** 7 **k** 8 **l** 5
m 3 **n** 3 **o** 17 **p** 3
q 5 **r** 15 **s** 8 **t** 16

55 Sieve of Eratosthenes

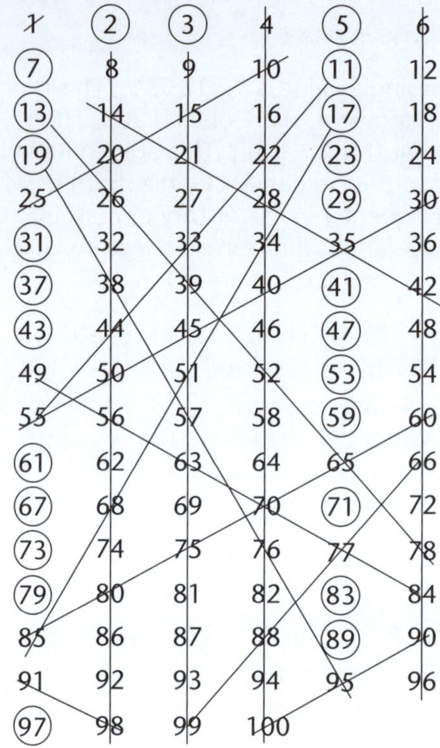

In order: Circle 2. Cross out multiples of 2 (straight down)
Circle 3. Cross out multiples of 3 (straight down)
Circle 5. Cross out multiples of 5 (diagonal ↙)
Circle 7. Cross out multiples of 7 (diagonal ↘)
Circle 11. Cross out multiples of 11 (diagonal ↙)
Circle 13. Cross out multiples of 13 (diagonal ↘)
Circle 17. Cross out multiples of 17 (diagonal ↙)

At this stage any numbers not crossed out are primes. All cross-out lines begin at a prime or a multiple of the prime. Lines for 2 start at 2 and 4 and 6. Lines for 5 start at 5, 30, 60, 90, and so on.

Primes less than 100 are 2, 3, 5, 7, 11, 13, 17, 19, 23, 29, 31, 37, 41, 43, 47, 53, 59, 61, 67, 71, 73, 79, 83, 89, 97.

WORKED SOLUTIONS to Chapter 1 **Practise, Practise**

56 a 23, 29

b 4, 6, 8, 9, 10, 12, 14, 15, 16, 18, 20, 21, 22, 24, 25, 26, 27, 28

c 2 d 2 e 97 f 101

g 45, 49, 51, 55, 57

h 25, 26, 27, 28, 30, 32, 33, 34, 35, 36, 38, 39, 40, 42, 44, 45

i 3, 13, 23, 31, 37, 43, 53, 73, 83

57 3, 5 (Separated by 4)

5, 7 (Separated by 6)

11, 13 (Separated by 12)

17, 19 (Separated by 18)

29, 31 (Separated by 30)

41, 43 (Separated by 42)

59, 61 (Separated by 60)

71, 73 (Separated by 72)

58 2, 3

59 $48 = 5 + 43$
$= 7 + 41$
$= 11 + 37$
$= 17 + 31$
$= 19 + 29$

60 a $24 = 5 + 19$
$= 7 + 17$
$= 11 + 13$

b $36 = 5 + 31$
$= 7 + 29$
$= 13 + 23$
$= 17 + 19$

c $54 = 7 + 47$
$= 11 + 43$
$= 13 + 41$
$= 17 + 37$
$= 23 + 31$

d $40 = 3 + 37$
$= 11 + 29$
$= 17 + 23$

61 a

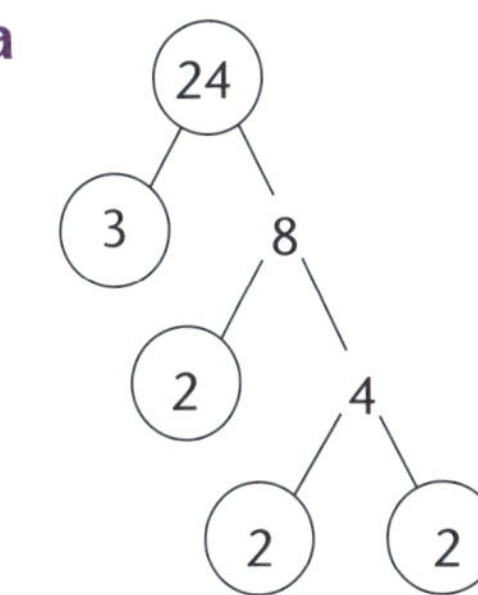

$24 = 2 \times 2 \times 2 \times 3$
$= 2^3 \times 3$

b

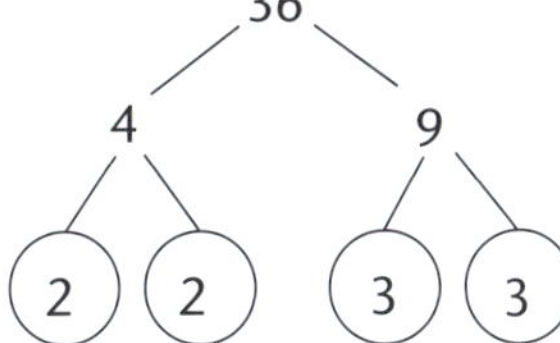

$36 = 2 \times 2 \times 3 \times 3$
$= 2^2 \times 3^2$

c

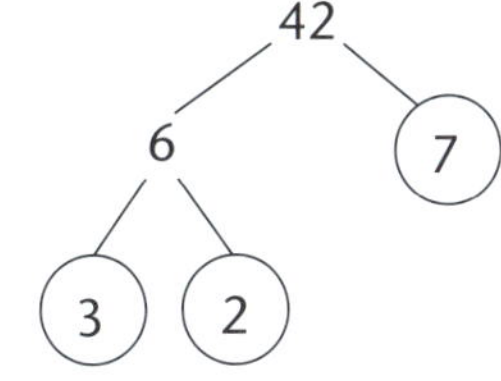

$42 = 2 \times 3 \times 7$

d

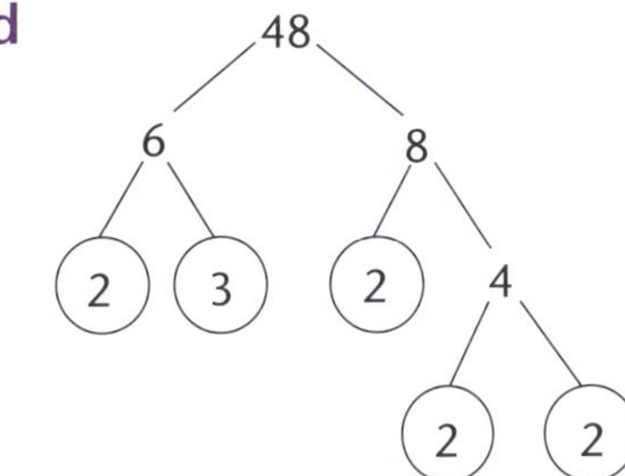

$48 = 2 \times 2 \times 2 \times 2 \times 3$
$= 2^4 \times 3$

e

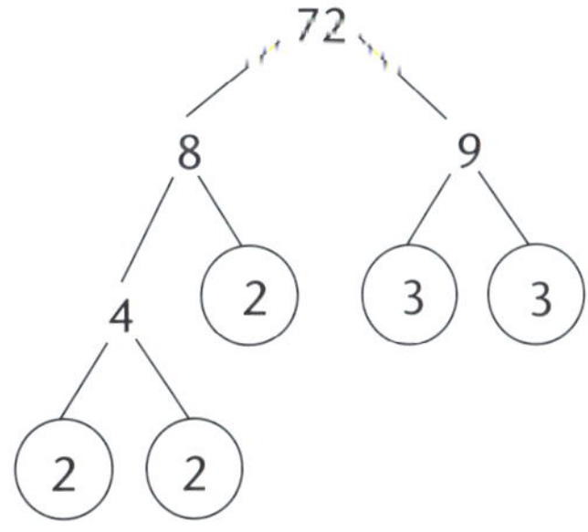

$72 = 2 \times 2 \times 2 \times 3 \times 3$
$= 2^3 \times 3^2$

f

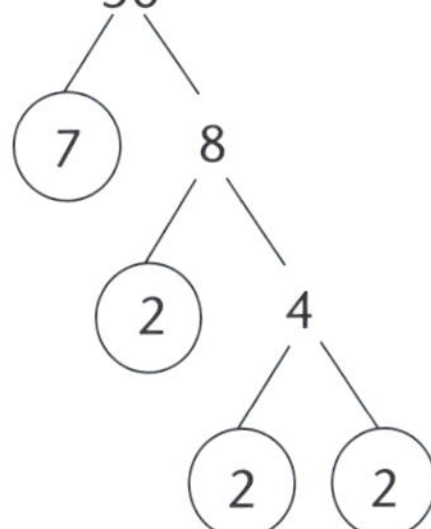

$56 = 2 \times 2 \times 2 \times 7$
$= 2^3 \times 7$

WORKED SOLUTIONS to Chapter 1 **Practise, Practise**

g

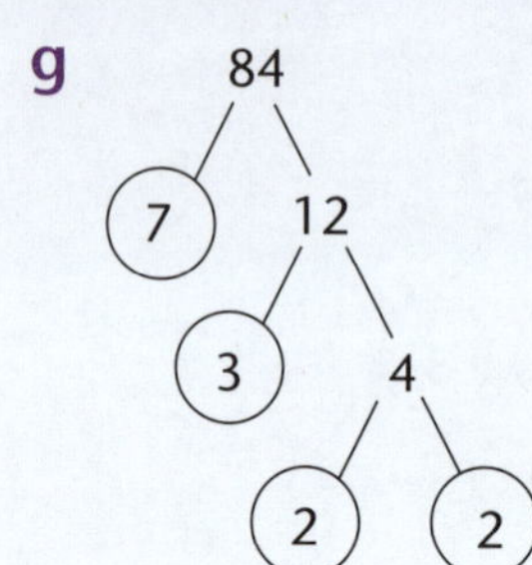

$84 = 2 \times 2 \times 3 \times 7$
$= 2^2 \times 3 \times 7$

h

$63 = 3 \times 3 \times 7$
$= 3^2 \times 7$

i

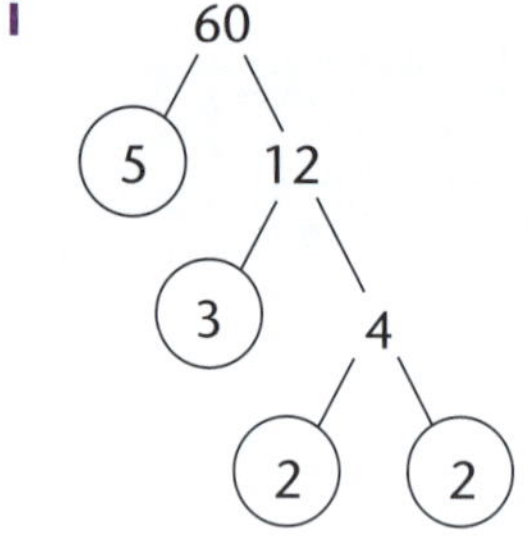

$60 = 2 \times 2 \times 3 \times 5$
$= 2^2 \times 3 \times 5$

j

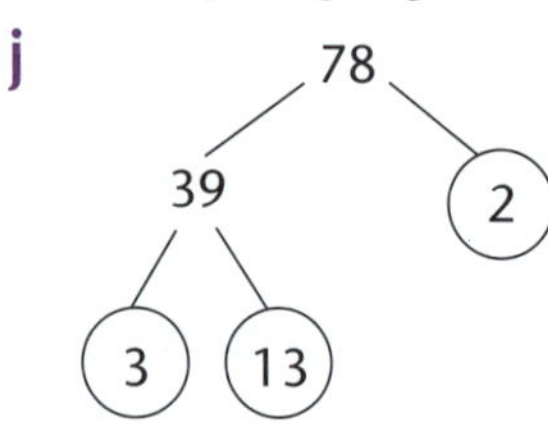

$78 = 2 \times 3 \times 13$

k

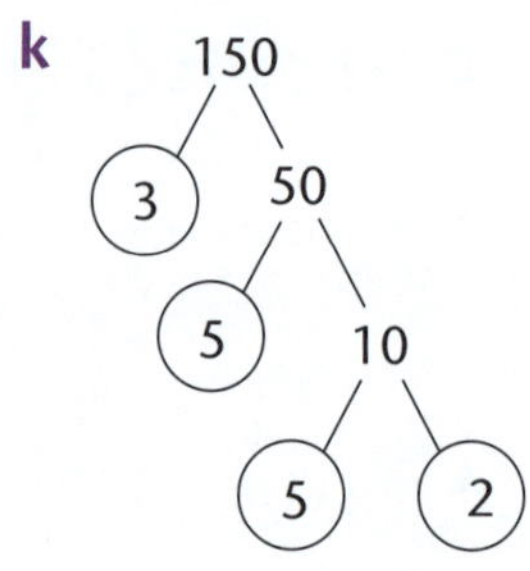

$150 = 2 \times 3 \times 5 \times 5$
$= 2 \times 3 \times 5^2$

l

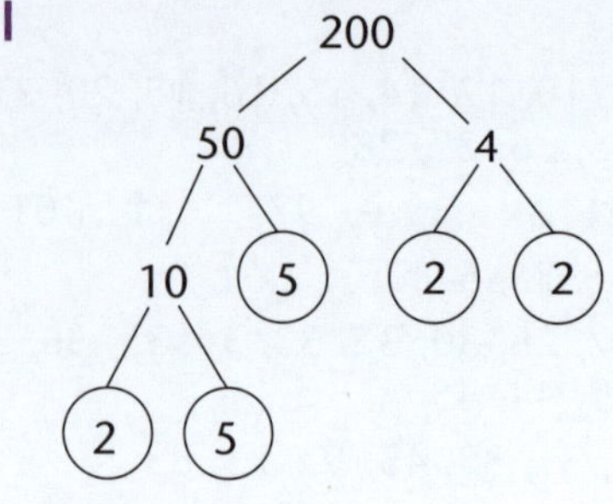

$200 = 2 \times 2 \times 2 \times 5 \times 5$
$= 2^3 \times 5^2$

m

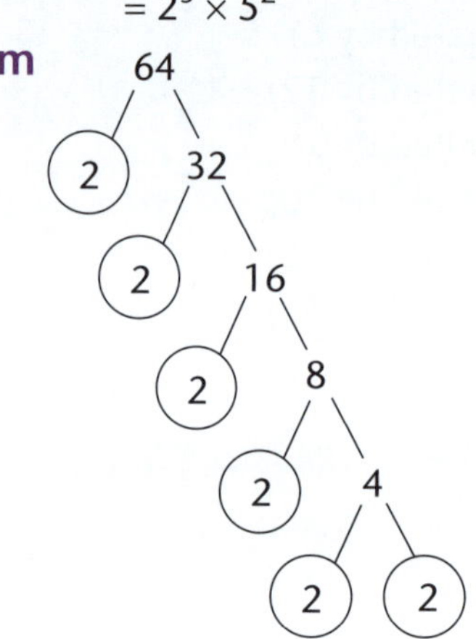

$64 = 2 \times 2 \times 2 \times 2 \times 2 \times 2$
$= 2^6$

n $128 = 2 \times 64$
$= 2 \times 2^6$ (see **m**) $= 2^7$

o

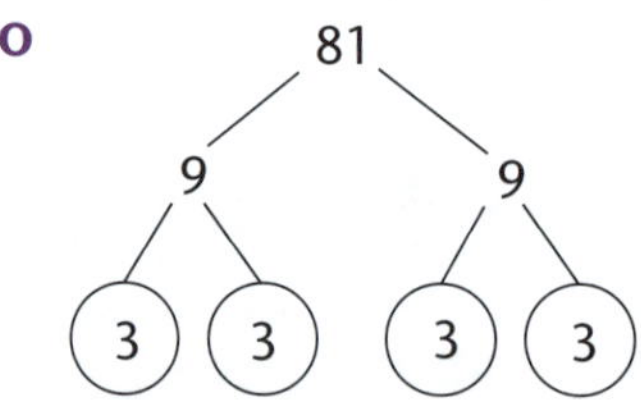

$81 = 3 \times 3 \times 3 \times 3$
$= 3^4$

p $162 = 2 \times 81$
$= 2 \times 3^4$ (see **o**)

q

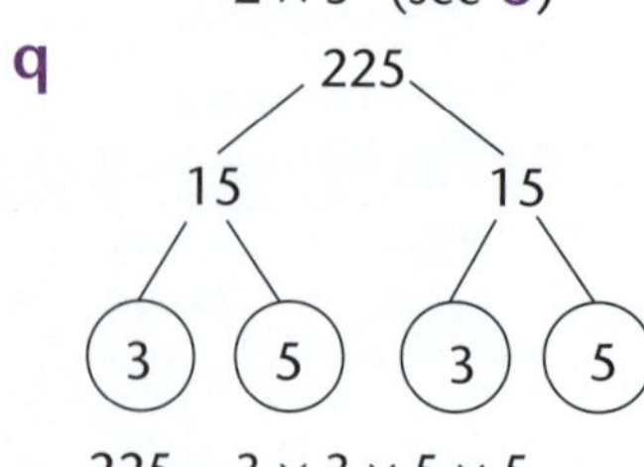

$225 = 3 \times 3 \times 5 \times 5$
$= 3^2 \times 5^2$

r

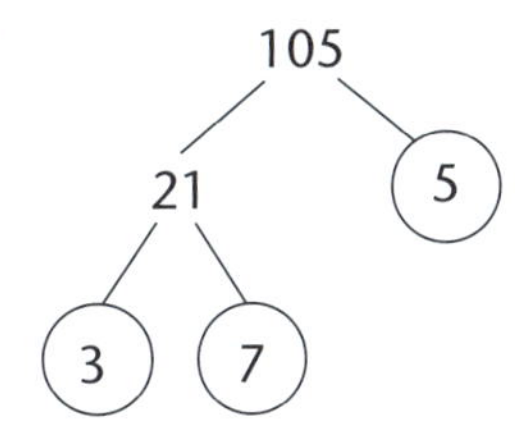

$105 = 3 \times 5 \times 7$

s

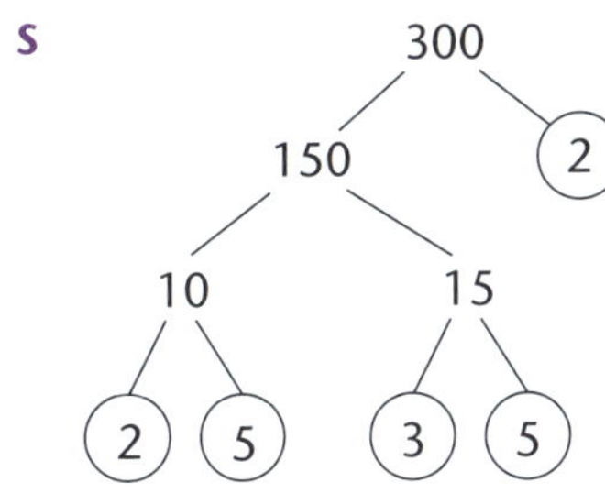

$300 = 2 \times 2 \times 3 \times 5 \times 5$
$= 2^2 \times 3 \times 5^2$

t $256 = 2 \times 128$
$= 2 \times 2^7$ (see **n**) $= 2^8$

62 **a**

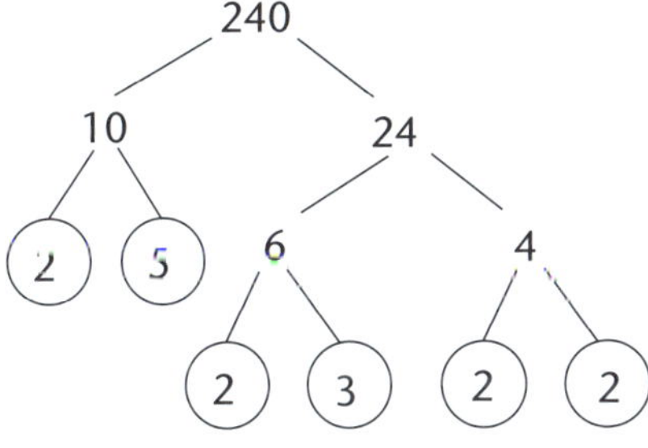

$240 = \underline{2 \times 2 \times 2} \times 2 \times \underline{3} \times 5$

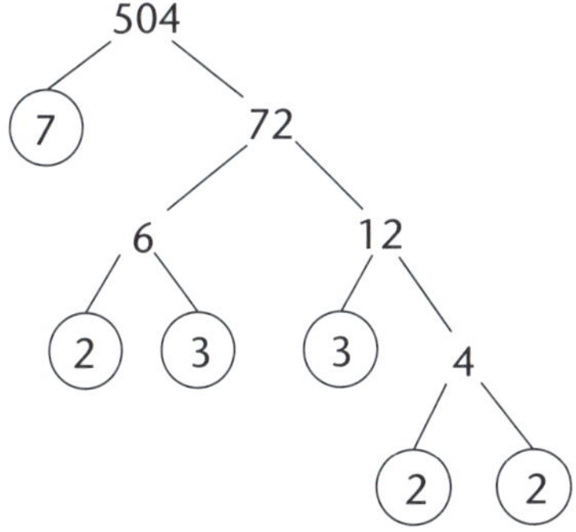

$504 = \underline{2 \times 2 \times 2} \times \underline{3} \times 3 \times 7$
$\text{HCF} = 2 \times 2 \times 2 \times 3 = 24$
(Factors common to both numbers.)

$\text{LCM} = \underbrace{2 \times 2 \times 2 \times 3 \times 3 \times 7}_{504} \times \underbrace{2 \times 5}_{10}$

Factors from 240 not already included.

$= 5040$

b

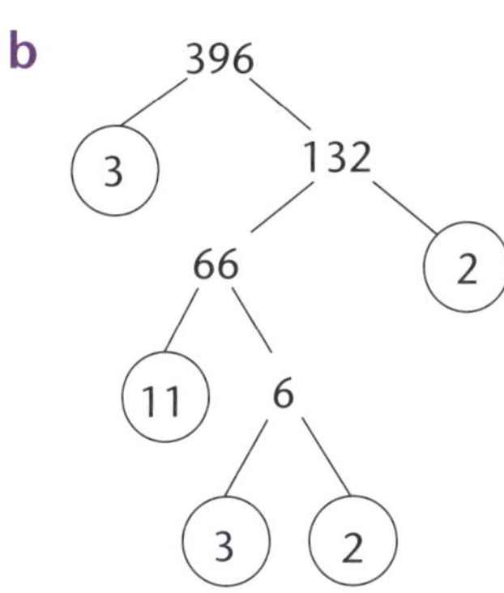

$396 = \underline{2 \times 2} \times \underline{3 \times 3} \times 11$

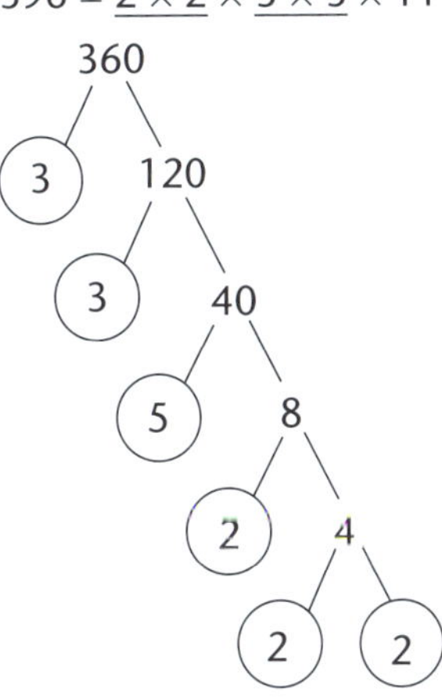

$360 = \underline{2 \times 2} \times 2 \times \underline{3 \times 3} \times 5$
$\text{HCF} = 2 \times 2 \times 3 \times 3 = 36$
$\text{LCM} = (2 \times 2 \times 3 \times 3 \times 11) \times 2 \times 5$
$= 3960$

c

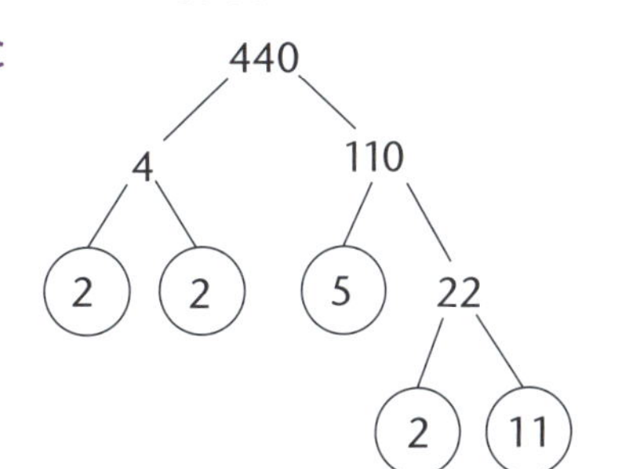

$440 = \underline{2 \times 2 \times 2} \times \underline{5} \times 11$

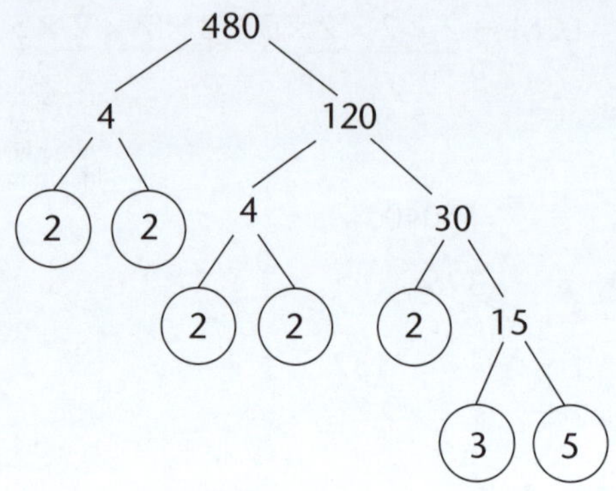

$480 = \underline{2 \times 2 \times 2} \times 2 \times 2 \times 3 \times \underline{5}$

HCF $= 2 \times 2 \times 2 \times 5 = 40$

LCM $= (2 \times 2 \times 2 \times 2 \times 2 \times 3 \times 5) \times (11)$
$= 5280$

d

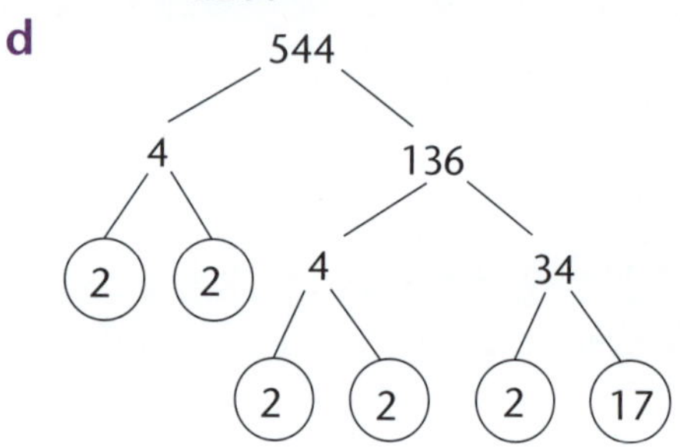

$544 = 2 \times 2 \times 2 \times 2 \times 2 \times \underline{17}$

595 → 5, 119; 119 → 7, 17

$595 = 5 \times 7 \times \underline{17}$

HCF $= 17$

LCM $= (5 \times 7 \times 17) \times (2 \times 2 \times 2 \times 2 \times 2)$
$= 19\,040$

63

a 144	**b** 196	**c** 441	**d** 900
e 125	**f** 343	**g** 16	**h** 7
i 13	**j** 9	**k** 6	**l** 8

64 The following products are found using factor trees.

a $576 = 2 \times 2 \times 2 \times 2 \times 2 \times 2 \times 3 \times 3$
$= (2 \times 2 \times 2 \times 3) \times (2 \times 2 \times 2 \times 3)$
$\therefore \sqrt{576} = 2 \times 2 \times 2 \times 3$
$= 24$

b $441 = 3 \times 3 \times 7 \times 7$
$= (3 \times 7) \times (3 \times 7)$
$\therefore \sqrt{441} = 3 \times 7$
$= 21$

c $\underline{729} = 3 \times 3 \times 3 \times 3 \times 3 \times 3$
$\therefore \sqrt{729} = 3 \times 3 \times 3$
$= 27$

d $1296 = 2 \times 2 \times 2 \times 2 \times 3 \times 3 \times 3 \times 3$
$= (2 \times 2 \times 3 \times 3) \times (2 \times 2 \times 3 \times 3)$
$\therefore \sqrt{1296} = 2 \times 2 \times 3 \times 3$
$= 36$

e

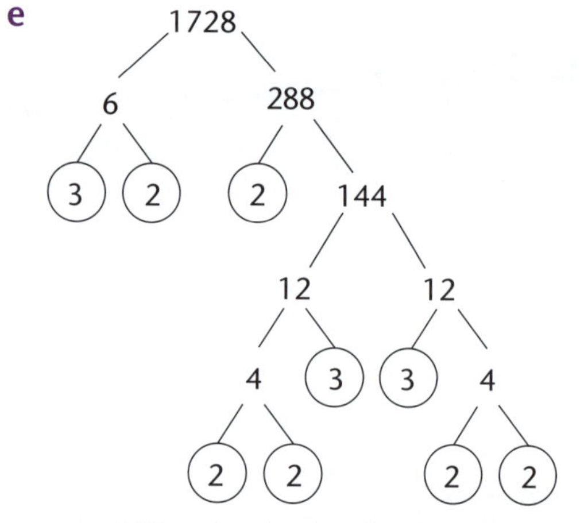

$1728 = 2 \times 2 \times 2 \times 2 \times 2 \times 2 \times 3 \times 3 \times 3$
$= (2 \times 2 \times 3) \times (2 \times 2 \times 3)$
$\times (2 \times 2 \times 3)$
$\therefore \sqrt[3]{1728} = 2 \times 2 \times 3$
$= 12$

f $5832 = 2 \times 2 \times 2 \times 3 \times 3 \times 3 \times 3 \times 3 \times 3$
$= (2 \times 3 \times 3) \times (2 \times 3 \times 3)$
$\times (2 \times 3 \times 3)$
$\therefore \sqrt[3]{5832} = 2 \times 3 \times 3$
$= 18$

g

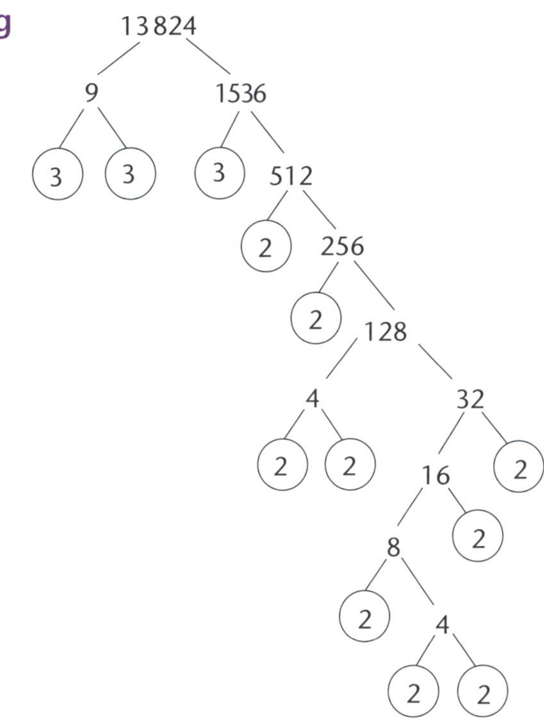

$$13824 = 2 \times 2 \times 2 \times 2 \times 2 \times 2 \times 2 \times 2 \times 2 \times 3 \times 3 \times 3$$
$$= (2 \times 2 \times 2 \times 3) \times (2 \times 2 \times 2 \times 3) \times (2 \times 2 \times 2 \times 3)$$
$$\therefore \sqrt[3]{13824} = 2 \times 2 \times 2 \times 3 = 24$$

65 **a** 324: even, divisible by 2
$3 + 2 + 4 = 9$, divisible by 3 also by 9.

b 645: obviously divisible by 5
$6 + 4 + 5 = 15$, divisible by 3.

c 4173: $4 + 1 + 7 + 3 = 15$, divisible by 3.

d 7491
$7 + 4 + 9 + 1 = 21$, divisible by 3
$\left.\begin{array}{l} 7 + 9 = 16 \\ 4 + 1 = 5 \end{array}\right\} 16 - 5 = 11$
$\therefore$ Divisible by 11.

e 5967: $5 + 9 + 6 + 7 = 27$, divisible by 3 also 9.

66 $74a$ divisible by 6. For this to happen, a must be even or 0 (i.e. divisible by 2), and $(7 + 4 + a)$ must be a multiple of 3.

Consider $a = 0$, $7 + 4 = 11$ (No)
$a = 2$, $7 + 4 + 2 = 13$ (No)
$a = 4$, $7 + 4 + 4 = 15$ (Yes)
$a = 6$, $7 + 4 + 6 = 17$ (No)
$a = 8$, $7 + 4 + 8 = 19$ (No)

The number is 744; that is, $a = 4$.

67 Once again the question is about LCM. We need the lowest number that 16, 17, 18 and 19 will divide into evenly. We then add 1 to the LCM as there is a remainder of 1.

Consider each number:
19 is prime
17 is prime
$16 = 2 \times 8$
$18 = 2 \times 9$

The LCM must contain products of 19, 17, 8, 9 and 2

i.e. $\text{LCM} = 19 \times 17 \times 8 \times 9 \times 2$

$\left.\begin{array}{l} 2 \times 8 \text{ is contained} \\ 2 \times 9 \text{ is contained} \end{array}\right\}$ in LCM

i.e. 17, 19, 16, 18 divide evenly into LCM.

LCM = 46 512

Number of centicubes is 46 513 (i.e. 46 512 + 1).

68 The question requires the LCM of 9 and 12.
LCM = 36
There were 36 marbles.

69

a 8	**b** 100	**c** 9	**d** 27
e 64	**f** 16	**g** 1	**h** 36
i 49	**j** 8		

70 **a** $(2 \times 3)^2 = 6^2 = 36$; $2^2 \times 3^2 = 4 \times 9 = 36$; True

b **i** $(3 \times 4)^2 = 12^2 = 144$; $3^2 \times 4^2 = 9 \times 16 = 144$; T

ii $(2 + 3)^2 = 5^2 = 25$; $2^2 + 3^2 = 4 + 9 = 13$; F

iii $\sqrt{9 \times 4} = \sqrt{36} = 6$; $\sqrt{9} \times \sqrt{4} = 3 \times 2 = 6$; T

iv $2^3 = 2 \times 2 \times 2 = 8$; T

v $\sqrt{9 + 4} = \sqrt{13}$; $\sqrt{9} + \sqrt{4} = 3 + 2 = 5$, F

Chapter 2—Fractions

pp. 46–48

1 **a** $\frac{1}{3}$ **b** $\frac{3}{10}$

2 **a** $\frac{2}{3} = \frac{\square}{12}$ (×4) $\therefore \frac{2}{3} = \frac{8}{12}$ (×4)

b $\frac{4}{5} = \frac{12}{\square}$ (×3) $\therefore \frac{4}{5} = \frac{12}{15}$ (×3)

WORKED SOLUTIONS to Chapter 2 **Practise, Practise**

3 a $\frac{6}{9} = \frac{2}{3}$ b $\frac{25}{35} = \frac{5}{7}$ c $\frac{16}{36} = \frac{4}{9}$

4 a $\frac{1}{3} = \frac{10}{30}, \frac{1}{2} = \frac{15}{30}, \frac{2}{5} = \frac{12}{30}$ order is $\frac{1}{3}, \frac{2}{5}, \frac{1}{2}$

b $\frac{3}{8} = \frac{15}{40}, \frac{2}{5} = \frac{16}{40}, \frac{1}{4} = \frac{10}{40}$ order is $\frac{1}{4}, \frac{3}{8}, \frac{2}{5}$

5 $\frac{1}{4} = \frac{3}{12}, \frac{1}{2} = \frac{6}{12}, \frac{2}{3} = \frac{8}{12}$

0 1/12 1/4 1/2 2/3 11/12 1

6 a $\frac{2}{3} = \frac{4}{6}, \frac{1}{2} = \frac{3}{6}$ order is $\frac{5}{6}, \frac{4}{6}, \frac{3}{6}$ or $\frac{5}{6}, \frac{2}{3}, \frac{1}{2}$

b $1\frac{1}{3} = 1\frac{8}{24}, 1\frac{7}{8} = 1\frac{21}{24}, 1\frac{3}{4} = 1\frac{18}{24}$

order is $1\frac{21}{24}, 1\frac{18}{24}, 1\frac{8}{24}$ or $1\frac{7}{8}, 1\frac{3}{4}, 1\frac{1}{3}$

7 a $\frac{7}{10} = \frac{21}{30}, \frac{2}{3} = \frac{20}{30}$

$\therefore \frac{21}{30} > \frac{20}{30}$ i.e. $\frac{7}{10} > \frac{2}{3}$

b $\frac{3}{4} = \frac{9}{12}, \frac{5}{6} = \frac{10}{12}$

$\therefore \frac{9}{12} < \frac{10}{12}$ or $\frac{3}{4} < \frac{5}{6}$

8 a $4\frac{1}{2} = \frac{9}{2}$ b $3\frac{2}{5} = \frac{17}{5}$ c $5\frac{7}{10} = \frac{57}{10}$

9 a $\frac{21}{5} = 4\frac{1}{5}$ b $\frac{13}{3} = 4\frac{1}{3}$ c $\frac{47}{10} = 4\frac{7}{10}$

10 a $\frac{3}{4}$ b $1\frac{1}{2} = \frac{3}{2}$, reciprocal $= \frac{2}{3}$

c $\frac{5}{2} = 2\frac{1}{2}$

11 a $\frac{3}{3} = 1$ b $\frac{4}{10} = \frac{2}{5}$

c $\frac{6}{5} = 1\frac{1}{5}$ d $\frac{5}{4} - \frac{3}{4} = \frac{2}{4} = \frac{1}{2}$

e $3 + (\frac{2}{3} + \frac{1}{3}) = 3 + 1 = 4$

f $6 + \frac{7}{5} = 6 + 1\frac{2}{5} = 7\frac{2}{5}$

12 a $\frac{3}{4} + \frac{3}{4} = \frac{6}{4} = 1\frac{2}{4} = 1\frac{1}{2}$

b $\frac{4}{5} - \frac{1}{2} = \frac{8}{10} - \frac{5}{10} = \frac{3}{10}$

c $\frac{7}{10} - \frac{3}{5} = \frac{7}{10} - \frac{6}{10} = \frac{1}{10}$

d $\frac{2}{3} + \frac{1}{2} - \frac{1}{4} = \frac{8}{12} + \frac{6}{12} - \frac{3}{12} = \frac{11}{12}$

e $3\frac{1}{2} + 2\frac{1}{5} = 5 + \frac{5}{10} + \frac{2}{10} = 5\frac{7}{10}$

f $4\frac{1}{4} - 2\frac{1}{5} = 2 + (\frac{1}{4} - \frac{1}{5})$
$= 2 + (\frac{5}{20} - \frac{4}{20})$
$= 2\frac{1}{20}$

g $3\frac{1}{2} - 2\frac{2}{3} = 1 + (\frac{1}{2} - \frac{2}{3})$
$= 1 + \frac{1}{2} - 1 + \frac{1}{3}$
$= \frac{1}{2} + \frac{1}{3}$
$= \frac{3}{6} + \frac{2}{6}$
$= \frac{5}{6}$

h $5\frac{1}{3} - 3\frac{4}{5} = 2 + (\frac{1}{3} - \frac{4}{5})$
$= 2 + \frac{1}{3} - 1 + \frac{1}{5}$
$= 1 + (\frac{5}{15} + \frac{3}{15})$
$= 1\frac{8}{15}$

i $5 - 2\frac{4}{5} = 3 - \frac{4}{5} = 2\frac{1}{5}$

j $7 - 3\frac{2}{3} - 1\frac{9}{10} = 3 - \frac{2}{3} - \frac{9}{10}$
$= 3 - 1 + \frac{1}{3} - 1 + \frac{1}{10}$
$= 1 + \frac{1}{3} + \frac{1}{10}$
$= 1 + \frac{10}{30} + \frac{3}{30}$
$= 1\frac{13}{30}$

13 a $\frac{4}{9}$ b $\frac{15}{32}$ c $\frac{27}{50}$ d $\frac{9}{40}$

e $\frac{5}{5} \times \frac{5}{4} = \frac{25}{8} = 3\frac{1}{8}$ f $\frac{20}{1} \times \frac{4}{3} = \frac{80}{3} = 26\frac{2}{3}$

14 a $\frac{3}{\cancel{4}_1} \times \frac{\cancel{4}_1}{5} = \frac{3}{5}$ b $\frac{2}{\cancel{3}_1} \times \frac{\cancel{12}_4}{1} = 8$

c $\frac{4}{5}$ of $20 = \frac{4}{\cancel{5}_1} \times \frac{\cancel{20}_4}{1}$
$= 16$

d $(\frac{4}{5})^2 = \frac{4}{5} \times \frac{4}{5}$
$= \frac{16}{25}$

e $1\frac{1}{2} \times \frac{2}{3} = \frac{\cancel{3}_1}{\cancel{2}_1} \times \frac{\cancel{2}_1}{\cancel{3}_1}$
$= 1$

f $(3\frac{1}{2})^2 = 3\frac{1}{2} \times 3\frac{1}{2}$
$= \frac{7}{2} \times \frac{7}{2}$
$= \frac{49}{4}$
$= 12\frac{1}{4}$

g $2\frac{1}{2} \times 1\frac{3}{5} = \frac{\cancel{5}_1}{\cancel{2}_1} \times \frac{\cancel{8}_4}{\cancel{5}_1}$
$= 4$

h $6\frac{1}{4} \times 1\frac{3}{5} = \frac{\cancel{25}_5}{\cancel{4}_1} \times \frac{\cancel{8}_2}{\cancel{5}_1}$
$= 10$

i $60 \times 1\frac{1}{3} = \frac{\cancel{60}_{20}}{1} \times \frac{4}{\cancel{3}_1}$
$= 80$

j $20 \times 2\frac{1}{2} \times \frac{4}{5} = \frac{\cancel{20}^{10}}{1} \times \frac{\cancel{5}^{1}}{\cancel{2}_{1}} \times \frac{4}{\cancel{5}_{1}}$
$= 40$

k $\frac{1}{2} \times \frac{1}{2} \times \frac{1}{2} = \frac{1}{8}$

15 **a** $\frac{4}{5} \div \frac{3}{4} = \frac{4}{5} \times \frac{4}{3}$
$= \frac{16}{15} = 1\frac{1}{15}$

b $\frac{7}{10} \div \frac{5}{7} = \frac{7}{10} \times \frac{7}{5}$
$= \frac{49}{50}$

c $\frac{4}{5} \div \frac{1}{5} = \frac{4}{\cancel{5}_{1}} \times \frac{\cancel{5}^{1}}{1}$
$= 4$

d $1\frac{1}{2} \div \frac{2}{3} = \frac{3}{2} \times \frac{3}{2}$
$= \frac{9}{4} = 2\frac{1}{4}$

e $40 \div \frac{3}{4} = \frac{40}{1} \times \frac{4}{3}$
$= \frac{160}{3}$
$= 53\frac{1}{3}$

f $40 \div 1\frac{1}{4} = \frac{40}{1} \div \frac{5}{4}$
$= \frac{\cancel{40}^{8}}{1} \times \frac{4}{\cancel{5}_{1}}$
$= 32$

g $3\frac{1}{2} \div 1\frac{3}{4} = \frac{7}{2} \div \frac{7}{4}$
$= \frac{\cancel{7}^{1}}{\cancel{2}_{1}} \times \frac{\cancel{4}^{2}}{\cancel{7}_{1}}$
$= 2$

h $2\frac{1}{2} \div 1\frac{1}{4} = \frac{5}{2} \div \frac{5}{4}$
$= \frac{\cancel{5}^{1}}{\cancel{2}_{1}} \times \frac{\cancel{4}^{2}}{\cancel{5}_{1}}$
$= 2$

16 **a** $\frac{1}{2} + \frac{\cancel{3}^{1}}{\cancel{4}_{2}} \times \frac{\cancel{2}^{1}}{\cancel{3}_{1}} = \frac{1}{2} + \frac{1}{2} = 1$

b $\frac{4}{5} + \frac{2}{3} \div \frac{1}{6} = \frac{4}{5} + \frac{2}{\cancel{3}_{1}} \times \frac{\cancel{6}^{2}}{1}$
$= \frac{4}{5} + 4$
$= 4\frac{4}{5}$

c $\dfrac{\frac{3}{4} - \frac{1}{2}}{\frac{1}{4}} = (\frac{3}{4} - \frac{1}{2}) \div \frac{1}{4}$
$= (\frac{3}{4} - \frac{2}{4}) \div \frac{1}{4}$
$= \frac{1}{4} \times \frac{4}{1}$
$= 1$

d $\dfrac{\frac{1}{2} - \frac{1}{4}}{\frac{1}{2} + \frac{1}{4}} = (\frac{1}{2} - \frac{1}{4}) \div (\frac{1}{2} + \frac{1}{4})$
$= \frac{2}{4} - \frac{1}{4} \div (\frac{2}{4} + \frac{1}{4})$
$= \frac{1}{4} \div \frac{3}{4}$
$= \frac{1}{\cancel{4}_{1}} \times \frac{\cancel{4}^{1}}{3}$
$= \frac{1}{3}$

e $(\frac{3}{4} + \frac{3}{5}) \div (\frac{3}{4} - \frac{3}{5}) = (\frac{15}{20} + \frac{12}{20}) \div (\frac{15}{20} - \frac{12}{20})$
$= \frac{27}{20} \div \frac{3}{20}$
$= \frac{\cancel{27}^{9}}{\cancel{20}_{1}} \times \frac{\cancel{20}^{1}}{\cancel{3}_{1}}$
$= 9$

f $\frac{3}{5} + (\frac{\cancel{2}^{1}}{5} \times \frac{3}{\cancel{4}_{2}}) = \frac{3}{5} + \frac{3}{10}$
$= \frac{6}{10} + \frac{3}{10}$
$= \frac{9}{10}$

g $\frac{2}{3} + (\frac{\cancel{3}^{1}}{5} \times \frac{4}{\cancel{3}_{1}}) + \frac{5}{6} = \frac{2}{3} + \frac{4}{5} + \frac{5}{6}$
$= \frac{20}{30} + \frac{24}{30} + \frac{25}{30}$
$= \frac{69}{30} = 2\frac{9}{30} = 2\frac{3}{10}$

h $1\frac{1}{3} + 1\frac{1}{4} - \frac{3}{4} + \frac{2}{3} = (1\frac{1}{3} + \frac{2}{3}) + (1\frac{1}{4} - \frac{3}{4})$
$= 2 + (\frac{5}{4} - \frac{3}{4})$
$= 2\frac{2}{4} = 2\frac{1}{2}$

i $6 - 3 - 3 + \frac{1}{2} + \frac{1}{4} = \frac{1}{2} + \frac{1}{4} = \frac{3}{4}$

17 **a** $\frac{4}{5}$ of \$6 $= \frac{4}{\cancel{5}_{1}} \times \frac{\cancel{600}^{120}}{1} = 480$

$\therefore$ answer is \$4.80

b $\frac{3}{4}$ of \$2.40 $= \frac{3}{\cancel{4}_{1}} \times \frac{\cancel{240}^{60}}{1} = 180$

$\therefore$ answer is \$1.80

c $\frac{7}{10} \times 2$ L $= \frac{7}{\cancel{10}_{1}} \times \frac{\cancel{2000}^{200}}{1} = 1400$

$\therefore$ answer is 1400 mL

d $\frac{2}{3}$ of 2 min $= \frac{2}{\cancel{3}_{1}} \times \frac{\cancel{120}^{40}}{1} = 80$

$\therefore$ answer is 80 seconds

e $\frac{3}{5}$ of $1\frac{1}{2} = \frac{3}{5} \times \frac{3}{2} = \frac{9}{10}$

$\therefore$ answer is $\frac{9}{10}$ metre = 90 cm

f $\frac{9}{10}$ of $2\frac{1}{2} = \frac{9}{\cancel{10}_{2}} \times \frac{\cancel{5}^{1}}{2} = \frac{9}{4} = 2\frac{1}{4}$

$\therefore 2\frac{1}{4}$ km

18 **a** $\frac{4}{20} = \frac{1}{5}$ **b** $\frac{20}{200} = \frac{1}{10}$

c $\frac{3}{21} = \frac{1}{7}$ **d** $\frac{21}{35} = \frac{3}{5}$

19 **a** Change to $\frac{1}{4}$

$\therefore \frac{1}{4}, \frac{2}{4}, \frac{3}{4} \ldots$

Next term is $\frac{4}{4} = 1$

b Change to $\frac{1}{8}$

$\therefore \frac{8}{8}, \frac{4}{8}, \frac{2}{8} \ldots$

Next term is $\frac{1}{8}$

20 Change to twelfths: $\frac{1}{2} = \frac{6}{12}, \frac{1}{3} = \frac{4}{12}$

$\therefore \frac{5}{12}$ is in the middle

21 $6\frac{1}{2} + 4\frac{1}{4} = 10 + \frac{1}{2} + \frac{1}{4}$
$= 10 + \frac{2}{4} + \frac{1}{4}$
$= 10\frac{3}{4}$

$\therefore$ $10\frac{3}{4}$ mins is needed to wash and wipe up

22 **a** $\frac{12}{60} = \frac{1}{5}$

b No. of breaks $= \frac{12}{3}$
$\therefore$ 4 breaks per hour
Over 24 hours = 24×4 breaks, i.e. 96
$\therefore$ 96 breaks

23 Total dist. = 350 + 250 = 600 km

$\therefore$ Fraction $= \dfrac{\text{Dist. to travel}}{\text{Total dist.}}$
$= \frac{250}{600}$
$= \frac{5}{12}$

24 Fraction plum $= 1 - (\frac{1}{3} + \frac{1}{5})$
$= 1 - (\frac{5}{15} + \frac{3}{15})$
$= 1 - \frac{8}{15}$
$= \frac{7}{15}$

No. plum $= \frac{7}{15} \times 600$
$= \frac{7}{\cancel{15}_1} \times \frac{\cancel{600}_{40}}{1}$
$= 280$

$\therefore$ 280 plum trees

25 Total pieces needed = $8 \times 3 = 24$
$\therefore$ 24 quarters consumed, i.e. $24 \times \frac{1}{4} = 6$
$\therefore$ 6 oranges needed

26 **a** $\frac{2\frac{1}{2}}{24} = \frac{5}{48}$ [Multiply top and bottom by 2.]

b Hours $= 5 \times 2\frac{1}{2} + 4\frac{1}{2}$
$= \frac{5}{1} \times \frac{5}{2} + 4\frac{1}{2}$
$= \frac{25}{2} + 4\frac{1}{2}$
$= 12\frac{1}{2} + 4\frac{1}{2}$
$= 17$

$\therefore$ 17 hours of TV watched

27 No. of brown-haired girls $= \frac{3}{4}$ of $\frac{1}{2}$ of 24
$= \frac{3}{\cancel{4}_1} \times \frac{1}{\cancel{2}_1} \times \frac{\cancel{24}_{\cancel{6}3}}{1}$
$= 9$

$\therefore$ 9 brown-haired girls

Chapter 3—Decimals

pp. 63–65

1 **a** 0.317 $\therefore$ 3 tenths ($\frac{3}{10}$)

b 28.031 $\therefore$ 3 hundredths ($\frac{3}{300}$)

c 104.763 $\therefore$ 3 thousandths ($\frac{3}{1000}$)

2 **a** $3.21 = 3 \times 1 + 2 \times \frac{1}{10} + 1 \times \frac{1}{100}$

b $76.047 = 7 \times 10 + 6 \times 1 + 4 \times \frac{1}{100}$
$+ 7 \times \frac{1}{1000}$

3 **a** 3.476: 3 decimal places
b 21.5048: 4 decimal places

4 **a** $4.25 = 4 \times 1 + 2 \times \frac{1}{10} + 5 \times \frac{1}{100}$

or $4 + \frac{2}{10} + \frac{5}{100}$

b $8.017 = 8 \times 1 + 0 \times \frac{1}{10} + 1 \times \frac{1}{100} + 7 \times \frac{1}{1000}$

or $8 + \frac{1}{100} + \frac{7}{1000}$

c $0.1022 = \frac{1}{10} + \frac{2}{1000} + \frac{2}{10000}$

5 **a** 4.17 **b** 6.037 **c** 0.73 **d** 74.113

6 **a** $\frac{3}{10}$ **b** $\frac{5}{10} = \frac{1}{2}$ **c** $\frac{8}{10} = \frac{4}{5}$
d $\frac{45}{100} = \frac{9}{20}$ **e** $\frac{42}{100} = \frac{21}{50}$ **f** $\frac{125}{1000} = \frac{1}{8}$
g $\frac{56}{100} = \frac{14}{25}$ **h** $6.4 = 6\frac{4}{10} = 6\frac{2}{5}$

i $1.41 = 1\frac{41}{100}$ **j** $4.25 = 4\frac{25}{100} = 4\frac{1}{4}$
k $10.35 = 10\frac{35}{100} = 10\frac{7}{20}$
l $1.002 = 1\frac{2}{1000} = 1\frac{1}{500}$

7 **a** 0.21 **b** 0.9 **c** 2.1 **d** 0.03
e 0.003 **f** $\frac{3}{5} = \frac{6}{10} = 0.6$
g $\frac{7}{20} = \frac{35}{100} = 0.35$ **h** $\frac{7}{25} = \frac{28}{100} = 0.28$

i $\frac{7}{50} = \frac{14}{100} = 0.14$ j $\frac{17}{200} = \frac{85}{1000} = 0.085$

8 a $\frac{1}{3} = 0.\dot{3}$ $3\overline{)1.00\ldots}$ = 0.33 ...

b $\frac{7}{8} = 0.875$ $8\overline{)7.000}$ = 0.875

c $\frac{3}{11} = 0.\dot{2}\dot{7}$ $11\overline{)3.0000\ldots}$ = 0.2727 ...

d $\frac{8}{9} = 0.\dot{8}$ $9\overline{)8.000\ldots}$ = 0.888 ...

e $\frac{4}{27} = 0.\dot{1}4\dot{8}$ $27\overline{)4.00000\ldots}$ = 0.14814 ...

9 a 0.30, 0.14, 0.07
New order: 0.07, 0.14, 0.30
i.e. 0.07, 0.14, 0.3

b 1.200, 1.270, 1.217
New order: 1.200, 1.217, 1.270
i.e. 1.2, 1.217, 1.27

c 0.80, 0.37, 0.23
New order: 0.23, 0.37, 0.8
$= 0.23, \frac{37}{100}, 0.8$

d 4.50, 4.27, 4.30
New order: 4.27, 4.3, 4.5
$= 4.27, 4.3, 4\frac{1}{2}$

10 a $7.2 < 7.3$
b $0.48 > 0.46$
c 2.5 2.54 $\therefore 2.5 < 2.54$
d $0.07 < 0.17$

11 a 4.7084 = 4.71 (2 dec. places)
(third digit greater than 5)
b 20.8649 = 20.86 (2 dec. places)
(third digit less than 5)
c 11.0594 = 11.06 (2 dec. places)
(third digit greater than 5)
d 12.795 13 = 12.80 (2 dec. places)
(third digit 5)

12 To three decimal places:
a 4.247 61 = 4.248
(to nearest thousandth)
b 200.415 29 = 200.415
(to nearest thousandth)

13 To one decimal place:
a 11.076 = 11.1 (to nearest tenth)
b 0.42 = 0.4 (to nearest tenth)

14 To two decimal places:
a \$4.762 94 = \$4.76 (to nearest cent)
b \$21.439 78 = \$21.44 (to nearest cent)

15
a 3.06 + 0.89 = 3.95

b 3.700 + 1.205 + 21.620 = 26.525

c 4.0 − 2.4 = 1.6

d 4.467 − 2.800 = 1.667

e 1.2 + 3.1 = 4.3 $\therefore$ 4.3 − 0.6 = 3.7

f 1.40 − 0.72 = 0.68 $\therefore$ 0.68 + 3.14 = 3.82

g 1.20 + 2.61 = 3.81 $\therefore$ 13.00 − 3.81 = 9.19

h 4.1 − 2.0 = 2.1 $\therefore$ 2.1 − 0.7 = 1.4

16 a adding 0.4
$\therefore$ 2.4, 2.8
b subtracting 0.5
$\therefore$ 2.2, 1.7
c adding 0.04
$\therefore$ 4.19, 4.23
d subtracting 0.9
$\therefore$ 4.7, 3.8

17 a $4.76 \times 10 = 47.6$
b $31.476 \times 100 = 3147.6$
c $21.40 \times 100 = 2140$
d $0.076 \times 10 = 0.76$

18 a $12.61 \div 10 = 1.261$
b $1764 \div 100 = 17.64$
c $09.47 \div 100 = 0.0947$
d $0.0021 \div 10 = 0.000\,21$

19 a $34 \times 2 = 68$, and one dec. place
$\therefore 3.4 \times 2 = 6.8$
b $147 \times 3 = 441$, and 2 dec. places
$\therefore 1.47 \times 3 = 4.41$
c $28 \times 4 = 112$, and 2 dec. places
$\therefore 2.8 \times 0.4 = 1.12$
d $4 \times 3 = 12$, and 2 dec. places
$\therefore 4 \times 0.03 = 0.12$
e $201 \times 6 = 1206$, and 3 dec. places
$\therefore 2.01 \times 0.6 = 1.206$

f $(0.5)^2 = 0.5 \times 0.5$
$5 \times 5 = 25$, and 2 dec. places
$\therefore (0.5)^2 = 0.25$

g 1.2×0.1
$\therefore 12 \times 1 = 12$, and 2 dec. places
$\therefore 1.2 \times 0.1 = 0.12$
$\therefore (1.2 \times 0.1)^2 = (0.12)^2 = 0.12 \times 0.12$
$\therefore 12 \times 12 = 144$, and 4 dec. places
$\therefore (0.12)^2 = 0.0144$

h $4.3 \times 200 = 4.30 \times 100 \times 2$
$= 430 \times 2$
$= 860$

20 **a** $1.4 \div 2 = 0.7$ $\quad 2\overline{)1.4}$ = 0.7
b $1.4 \div 0.2 = 14 \div 2 = 7$
c $48 \div 1.2 = 480 \div 12 = 40$
d $3.05 \div 0.05 = 305 \div 5 = 61$
e $4.00 \div 0.02 = 400 \div 2 = 200$
f $0.18 \div 0.9 = 1.8 \div 9 = 0.2$ $\quad 9\overline{)1.8}$ = 0.2

21 **a** $1.2 + 3 \times 0.6 = 1.2 + 1.8 = 3$
b $1.5 \div 5 + 2 \times 1.2 = 0.3 + 2.4 = 2.7$
c $\frac{3.4 + 0.6}{0.2} = \frac{4}{0.2} = 4 \div 0.2 = 40 \div 2 = 20$
d $\frac{1.2 \times 0.4}{1.2 - 0.4} = \frac{0.48}{0.8} = 0.48 \div 0.8$
$= 4.8 \div 8 = 0.6$

22 **a** $3.2 + 4.01 = 7.21$
b $3.8 - 2.07 = 1.73$
c $7.6 \times 0.2 = 1.52$
d $8.4 \div 0.4 = 84 \div 4 = 21$

23 **a** Av. $= \frac{12.2 + 4.7 + 3.9 + 11.4}{4}$
$= \frac{32.2}{4}$
$= 8.05$
b $3.7 - 2.08 = 3.70 - 2.08 = 1.62$

24 Rewrite 16.69 cents = \$0.1669
$\therefore$ Cost $= 0.1669 \times 479$
$= 79.9451$
$= 79.95$ (to 2 dec. places)
$\therefore$ Electricity cost is \$79.95.

25 Disregard 8.4 (lowest), 9.4 (highest)
$\therefore$ Total $= 8.9 + 9.2 + 9.2 + 9.0 + 8.7 + 9.0 + 9.1 + 9.0$
$= 72.1$
$\therefore$ Gabrielle's score is 72.1.

26 Cost = \$3.40 ÷ 4 = \$0.85
$\therefore$ Each pineapple costs 85c.

27 Number $= 25.2 \div 6 = 4.2$
$\therefore$ The other number is 4.2.

28 Total of 4 numbers $= 3.7 + 2.9 + 11 + 4.9 = 22.5$
$\therefore$ missing number $= 25 - 22.5 = 2.5$
$\therefore$ The missing number is 2.5.

29 $47\,214.8 + 475.3 = 47\,690.1$

30 **a** Time $= 40 \div 4 = 10$
$\therefore$ 10 hours
b Time $= 130 \div 6.50 = 1300 \div 65 = 20$
$\therefore$ 20 hours
c Time $= 300 \div 7.5 = 600 \div 15 = 40$
$\therefore$ 40 hours

31 **a** Distance $= 98.4 \div 8 = 12.3$
$\therefore$ 12.3 kilometres/litre
b 171.9c = \$1.719
$\therefore$ Cost $= 1.719 \times 8$
$= 13.752$
$= 13.75$ (to 2 dec. places)
$\therefore$ Petrol cost is \$13.75.

32 **a** US dollars $= 72.2 \times 200$
$= 7220 \times 2$
$= 14\,440$
$\therefore$ 14 440 cents
$\therefore$ \$144.40 (US)
b Japanese yen $= 79.41 \times 200$
$= 7941 \times 2$
$= 15\,882$
$\therefore$ 15 882 yen
c NZ dollars $= 1.23 \times 200$
$= 123 \times 2$
$= 246$
$\therefore$ \$246 (New Zealand)

Chapter 4—Percentages

pp. 78–80

1 **a** $\frac{2}{5} \times 100\% = 40\%$
b $\frac{9}{20} \times 100\% = 45\%$

WORKED SOLUTIONS to Chapter 4 **Practise, Practise**

c $\frac{13}{25} \times 100\% = 52\%$

d $\frac{7}{40} \times 100\% = 17.5\%$

e $\frac{41}{100} \times 100\% = 41\%$

f $\frac{1}{3} \times 100\% = 33\frac{1}{3}\%$

g $\frac{5}{6} \times 100\% = 83\frac{1}{3}\%$

h $3\frac{1}{4} \times 100\% = 325\%$

2 a $\frac{2}{7} \times 100\% = 28.\dot{5}7142\dot{8}$
$= 28.6\%$ (to 1 decimal place)

b $\frac{5}{11} \times 100\% = 45.\dot{4}\dot{5}$
$= 45.5\%$ (to 1 decimal place)

c $2\frac{4}{15} \times 100\% = 226.\dot{6}$
$= 226.7\%$ (to 1 decimal place)

3 a $0.3 \times 100\% = 30\%$
b $0.03 \times 100\% = 3\%$
c $0.003 \times 100\% = 0.3\%$
d $0.19 \times 100\% = 19\%$
e $0.576 \times 100\% = 57.6\%$
f $1.12 \times 100\% = 112\%$
g $4.7 \times 100\% = 470\%$
h $0.075 \times 100\% = 7.5\%$

4 a $\frac{60}{100} = \frac{3}{5}$

b $\frac{3}{100} = \frac{3}{100}$

c $\frac{95}{100} = \frac{19}{20}$

d $\frac{127}{100} = 1\frac{27}{100}$

e $\frac{300}{100} = 3$

f $\frac{5\frac{1}{2}}{100} = \frac{\frac{11}{2}}{100}$ [Numerator as improper fraction.]
$= \frac{11}{200}$ [Multiply numerator and denominator by 2.]

g $\frac{12\frac{1}{4}}{100} = \frac{\frac{49}{4}}{100}$ [Numerator as improper fraction.]
$= \frac{49}{400}$ [Multiply numerator and denominator by 4.]

h $\frac{3\frac{4}{5}}{100} = \frac{\frac{19}{5}}{100}$ [Numerator as improper fraction.]
$= \frac{19}{500}$ [Multiply numerator and denominator by 5.]

5 a $16 \div 100 = 0.16$
b $7 \div 100 = 0.07$
c $60 \div 100 = 0.6$
d $7.6 \div 100 = 0.076$
e $12.5 \div 100 = 0.125$
f $176 \div 100 = 1.76$
g $104.6 \div 100 = 1.046$
h $\frac{1}{4} \div 100 = 0.0025$
i $7\frac{3}{4} \div 100 = 0.0775$
j $33\frac{1}{3} \div 100 = 0.\dot{3}$

6 a $0.27 \times 600 = 162$

b $0.16 \times 140 = 22.4$
$\therefore$ \$22.40

c $0.6 \times 42 = 25.2$
$\therefore$ \$25.20

d $0.07 \times 6000 = 420$
$\therefore$ 420 metres [change to metres]

e $0.045 \times 8000 = 360$
$\therefore$ 360 mL [change to mL]

f $0.125 \times 240 = 30$
$\therefore$ 30 minutes

g $0.0025 \times 6000 = 15$
$\therefore$ 15 g [change to grams]

h $1.45 \times 300 = 435$
$\therefore$ \$435

i $0.352 \times 100\,000 = 35\,200$
$\therefore$ \$35 200

j $0.15875 \times 4000 = 635$

7 a Increase = 10% of \$20 = $0.1 \times 20 = 2$
$\therefore$ increase of \$2
$\therefore$ new amount = \$20 + \$2
$\therefore$ The new amount is \$22.

Alternative method used for rest of question:

b 118% of 700 = 1.18×700
$= 826$ [100 + 18]
$\therefore$ The new amount is 826.

c 130% of 45 = 1.3×45
$= 58.5$ [100 + 30]
$\therefore$ The new amount is 58.5 L.

d 108% of 3000 = 1.08×3000
$= 3240$ [in grams]
$\therefore$ The new amount is 3240 g.

WORKED SOLUTIONS to Chapter 4 **Practise, Practise**

8 **a** Decrease = 20% of 260 = 52
∴ decrease of 52
∴ new amount = 260 − 52 = 208
∴ The new amount is 208.

Alternative method used for rest of question:

b 95% of 65 = 0.95 × 65
= 61.75 [100 − 5]
∴ The new amount is $61.75.

c $87\frac{1}{2}$% of 3000 = 0.875 × 300
= 2625 [100 − $12\frac{1}{2}$]
∴ The new amount is 2625.

d 58% of $60 = 0.58 × 60
= 34.8 [100 − 42]
∴ The new amount is $34.80.

9 ↑ by 30% = 130% [100 + 30]
↓ by 30% = 70% [100 − 30]
130% of 100 = 1.3 × 100 = 130
Now, 70% of 130 = 0.7 × 130 = 91
∴ The result is $91.
Note the two percentages do not cancel out each other.

10 **a** $\frac{12}{25}$ × 100% = 48%

b 20c, $2 = 200c
∴ $\frac{20}{200}$ × 100% = 10%

c 4 cm, 500 cm
∴ $\frac{4}{500}$ × 100% = 0.8%

d 250 mL, 2000 mL
∴ $\frac{250}{2000}$ × 100% = 12.5%

e 20 seconds, 240 seconds
∴ $\frac{20}{240}$ × 100% = $8\frac{1}{3}$%

f 45c, 900c
∴ $\frac{45}{900}$ × 100% = 5%

11 **a** $\frac{4}{40}$ × 100% = 10%

b 3.5 g, 2000 g
$\frac{3.5}{2000}$ × 100% = 0.175%

c 400 cm, 200 cm
$\frac{400}{200}$ × 100% = 200%

d 180 seconds, 3 seconds
$\frac{180}{3}$ × 100% = 6000%

12 **a** 20% of amount = 700
100% of amount = 700 × 5
= 3500 [100% = 5 × 20%]
∴ The amount is 3500.

[This was quicker than going down to 1% and then back up to 100%.]

b 9% of amount = 45
1% of amount = 45 ÷ 9 = 5
100% of amount = 5 × 100 = 500
∴ The amount is 500.

13 15% of object = 525
5% of object = 525 ÷ 3 = 175
100% of object = 175 × 20 = 3500
∴ The whole object weighs 3500 g.

14 25% off
∴ discounted price = 75% of original price

a 75% of 95 = 0.75 × 95 = 71.25
∴ The discounted price is $71.25.

b 75% of 38 = 0.75 × 38 = 28.5
∴ The discounted price is $28.50.

15 **a** Discount $12\frac{1}{2}$% of 22 = 0.125 × 22 = 2.75
∴ It gives a saving of $2.75.

b Cost of shovel = $22 − $2.75 = $19.25
∴ The shovel will cost $19.25.

16 Discount = $800 − $700 = $100

$$\therefore \text{Percentage discount} = \frac{\text{Discount}}{\text{Original price}} \times 100\% = \frac{100}{800} \times 100\% = 12.5\%$$

∴ The percentage discount is 12.5%.

17 Discount = 35%
∴ new price is 65% of old price
∴ new price = 65% of 84 = 54.6
∴ The new price is $54.60.

18 Savings = 15% of 16 999
= 0.15 × 16 999
= 2549.85
∴ The savings are $2549.85.

WORKED SOLUTIONS to Chapters 4 and 5 **Practise, Practise**

19 Won 12

∴ lost 8

$$\therefore \text{Percentage} = \frac{\text{Lost}}{\text{Games played}} \times 100\%$$
$$= \frac{8}{20} \times 100\%$$
$$= 40\%$$

∴ Adam's team lost 40% of the games.

20 **a** 18 students have 2 TVs

$$\therefore \text{Percentage} = \frac{18}{50} \times 100\% = 36\%$$

∴ 36% of students have 2 TVs.

b 4 students have more than 4 TVs

$$\therefore \text{Percentage} = \frac{4}{50} \times 100\% = 8\%$$

∴ 8% of students have more than 4 TVs.

21 **a** Kathy must have 8 yellow jelly beans (40 – 32)

$$\therefore \text{Percentage} = \frac{8}{40} \times 100\% = 20\%$$

∴ 20% of the jelly beans are yellow.

b Remaining jelly beans = 40 – 8 = 32

$$\therefore \text{Percentage} = \frac{8}{32} \times 100\% = 25\%$$

∴ 25% of remaining jelly beans are yellow.

22 100 + 4 = 104

∴ new pay = 104% of 602
= 1.04 × 602
= 626.08

∴ Jacylyn's new pay is $626.08.

23 Number that failed to finish = 80 – 16 = 64

$$\therefore \text{Percentage} = \frac{64}{80} \times 100\% = 80\%$$

∴ 80% of the cars failed to finish.

24 22% of students = 198
1% of students = 198 ÷ 22 = 9
100% of students = 9 × 100 = 900
∴ There are 900 students at the school.

25 **a** 40% of total cost = 68
1% of total cost = 68 ÷ 40 = 1.7
100% of total cost = 1.7 × 100 = 170
∴ The complete set costs $170.

b 20% is half of 40%
i.e. 20% is half of $68
∴ Ken contributes $34.

[Or 20% of $170 = 0.2 × 170 = 34.]

26 Increase = 270 – 240 = 30

$$\therefore \text{Percentage} = \frac{30}{240} \times 100\% = 12.5\%$$

∴ The town had a 12.5% increase in population.

27 **a** Decrease = 4500 – 1400 = 3100

$$\therefore \text{Percentage} = \frac{3100}{4500} \times 100\%$$
$$= 68.\dot{8}\%$$
$$= 68.89\% \text{ (to 2 decimal places)}$$

∴ The percentage decrease was 68.89%.

b 30% increase on 1400

∴ No. of sheep = 130% of 1400
= 1.3 × 1400
= 1820

∴ Trevor had 1820 sheep by the end of 1995.

Chapter 5—Integers

pp. 94–98

1 **a** $40 – $105 = –$65
I am overdrawn $65.

b Temperature falls
(3 × 2°) = 6°C
New temperature = 2° – 6°C = –4°C
Temperature at 4 a.m. is –4°C.

c New temperature = –8° + 6 °C = –2°C

2 **a** –5 < –4 **b** –3 > –4 **c** 0 < 5
d 0 > –5 **e** 55 > –55 **f** –99 > –100

3 **a** F **b** F **c** T
d F **e** F **f** F

4 Ascending = Going ↑

a –9, –4, –1, 2
b –5, –4, 0, 2, 6
c –14, –7, 0, 5, 8, 14
d –7, –5, –3, –1, 1, 3
e –12, –10, –6, –4, 0, 1, 5
f –151, –141, –131, –115, –114, –113, –112

WORKED SOLUTIONS to Chapter 5 **Practise, Practise**

5 **a** 0, −2, −5, −6
b 3, 2, 0, −2, −3
c 5, 0, −2, −4, −5
d 6, 2, −4, −5, −8, −10
e −5, −6, −7, −8, −9, −10, −11, −12
f −112, −114, −115, −123, −124, −126

6 **a** −3 **b** 2
c −12 **d** 2

[Don't forget—add by using the number line.]

e −13 **f** 0 **g** 0
h −10 **i** −8 **j** −6
k −42 **l** 1 **m** 2
n −1 **o** 11

7 **a** −8 + 3 + 2 = −5 + 2 = −3
b (−6) + (−4) + 7 = (−10) + 7 = −3
c −9 + 14 + (−6) = 5 + (−6) = −1
d (−9) + (−5) + (−8) = −22
e −14 + 6 + 8 = 0
f 7 + (−9) + 2 = −2 + 2 = 0
g −54 + 17 + 19 = −37 + 19 = −18
h −20 + (−18) + 38 = −38 + 38 = 0
i (−8) + (−6) + (−8) + 14 + (−4)
= −14 + (−8) + 14 + (−4)
= −22 + 14 + (−4)
= −8 + (−4)
= −12
j 16 + (−9) + 17 + (−20) + (−14) + 10
= 7 + 17 + (−20) + (−14) + 10
= 24 + (−20) + (−14) + 10
= 4 + (−14) + 10
= −10 + 10
= 0

8 **a** −5 + 14 = 9 **b** 7 + (−5) = 2
c 4 + (−7) = −3 **d** −6 + 6 = 0
e −8 + 4 = −4 **f** −3 + 10 = 7
g 5 + (−14) = −9 **h** 6 + (−8) = −2
i (−5) + (−3) = −8 **j** −5 + 9 = 4
k −4 = 2 + (−6) **l** −2 = 5 + (−7)
m 0 + (−2) = −2 **n** (−3) + 3 = 0

9 **a** −8 + 2 = −6 **b** −6 + (−4) = −10
c 4 + 7 + (−7) = 4
d −4 + (−6) + (−8) = −18
e 9 + (−4) + 4 + (−9) = 0

10 **a** −7 + (−4) + 3 + 5 = −3
b −8 + (−2) + (−6) + (−4) = −20
c 9 + (−9) + 4 + (−4) + 8 + (−8) = 0
d −112 + 99 + (−4) + 5 + (−6) = −18
e 114 + (−53) + (−114) + 71 + 53 + (−71) = 0

11 **a** −4 − 6 = −4 + (−6) = −10
b 11 − 13 = 11 + (−13) = −2
c 6 − 14 = 6 + (−14) = −8
d −9 − 5 = −9 + (−5) = −14
e 5 − 13 = 5 + (−13) = −8
f 6 − (−4) = 6 + 4 = 10
g 5 − (−4) = 5 + 4 = 9
h −5 − (−7) = −5 + 7 = 2
i −8 − (−3) = −8 + 3 = −5
j 7 − 3 = 7 + (−3) = 4
k 5 − 17 = 5 + (−17) = −12
l −4 − 18 = −4 + (−18) = −22
m 31 − 45 = 31 + (−45) = −14
n −18 − 12 = −18 + (−12) = −30
o 21 − (−19) = 21 + 19 = 40

12 **a** −6 + (−3) + (−4) = −13
b 7 + (−12) + (−5) = −10
c 5 + (−8) + (−4) = −7
d −8 + (−3) + (−9) = −20
e 5 − (−5) − 6 = 5 + 5 + (−6) = 4
f −6 − (−4) − 5 = −6 + 4 + (−5) = −7
g 3 + (−9) + 4 = −2
h −11 − (−4) + 3 = −11 + 4 + 3 = −4
i −6 − 4 + 8 = −6 + (−4) + 8 = −2
j 13 − (−8) + 9 = 13 + 8 + 9 = 30
k 7 − 15 − 9 + (−3) = 7 + (−15) + (−9) + (−3)
= −20
l −4 + (−3) + 11 + 4 = 8

13 **a** Difference = 5 − (−2) or
= 5 + 2
= 7

−2 −1 0 1 2 3 4 5
7

WORKED SOLUTIONS to Chapter 5 **Practise, Practise**

b Difference $= 3 - (-4)$ or
$= 3 + 4$
$= 7$

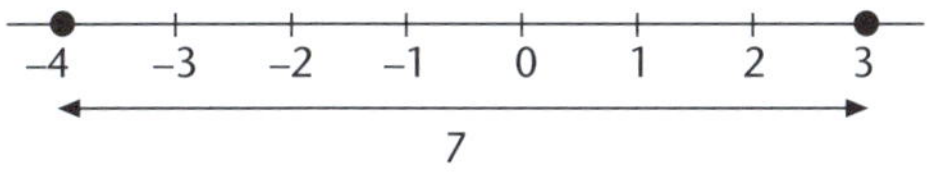

c Difference $= -3 - (-5)$ or
$= -3 + 5$
$= 2$

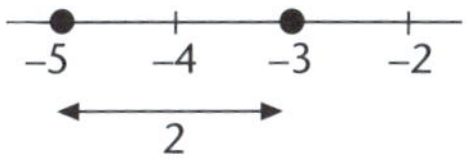

14 **a** $(-7) + (-9) = -16$
b $4 - (-16) = 4 + 16 = 20$

15 **a** -45 **b** 56
c -32 **d** -21
e 40 **f** -24
g -27 **h** -36
i 36 **j** -7
k 9 **l** -45
m 49 (-7×-7) **n** 64
o -120

$- \times - = +$
$- \times + = -$
$+ \times - = -$

16 **a** $(-3) \times (-5) \times 7 = 15 \times 7 = 105$
b $(-2) \times 4 \times (-6) = -8 \times -6 = 48$
c $8 \times (-3) \times 2 = -24 \times 2 = -48$
d $(-5) \times 3 \times (-2) = -15 \times -2 = 30$
e $(-2)^3 = -2 \times -2 \times -2 = 4 \times -2 = -8$
f $(-9) \times (-3) \times (-1) = 27 \times -1 = -27$
g $5 \times (-3)^2 = 5 \times [(-3) \times (-3)] = 5 \times 9 = 45$
h $(-5)^2 \times 3 = -5 \times -5 \times 3 = 25 \times 3 = 75$
i $(-2)^2 \times (-3)^2 = [(-2) \times (-2)] \times [(-3) \times (-3)] = 4 \times 9 = 36$
j $(-3) \times 5 \times (-4) \times 2 = -15 \times -8 = 120$

17 **a** $-8 \times 6 = -48$ **b** $4 \times -12 = -48$
c $-9 \times -7 = 63$ **d** $-10 \times -20 = 200$

18 **a** -9 **b** -7 **c** -7 **d** -7
e 6 **f** -4 **g** -7 **h** 8
i 9 **j** -4

19 **a** $40 \div -8 = -5$ **b** $-30 \div -6 = 5$
c $84 \div -12 = -7$ **d** $-24 \div -4 = 6$

20 **a** -5 **b** -24 **c** -11
d 24 **e** -5 **f** -11
g 144 **h** -24 **i** -10
j -14 **k** -30 **l** 0
m -45 **n** -3 **o** 4

21 **a** $4 - 3 \times 5 = 4 - 15 = -11$
b $-8 + 10 = 2$
c $10 - (-15) = 10 + 15 = 25$
d $20 - (-6) \times (-3) = 20 - 18 = 2$
e $-20 + 15 = -5$
f $20 - 28 = -8$
g $-3 \times -4 = 12$
h $-10 \times -10 = 100$
i $4 + 3 \times (-4) + 8 = 4 + (-12) + 8 = 0$
j $15 \div (-5) = -3$
k $15 - 27 \div (-9) = 15 - (-3) = 15 + 3 = 18$
l $-3 \times 4 = -12$
m $-24 \div 4 = -6$
n $3 - 36 \div 4 = 3 - 9 = -6$

22 **a** $-6a$ $(1 - 7 = -6)$
b $-5t$ $(-9 + 4 = -5)$
c $-14y$ $(-8 - 6 = -14)$
d $-6y$ $(5 - 11 = -6)$
e $6w$ $(3 - (-3) = 6)$
f $8t$ $(-5 + 13 = 8)$
g $-8y$ $(9 - 17 = -8)$
h $-3n^2$ $(-9 + 6 = -3)$
i $-m^2$ $(5 - 6 = -1)$
j z $(-12 + 13 = 1)$
k a $(-1 + 2 = 1)$
l $-5a$ $(-1 - 4 = -5)$
m $-6y$ $(2 - 3 - 5 = -6)$
n 0 $(-8 + 4 + 4 = 0)$
o 0 $(3 - 8 + 5 = 0)$
p $-14x$ $(4 - 11 - 7 = -14)$
q $-3a$ $(-1 - 1 - 1 = -3)$
r $-8y^2$ $(1 - 2 - 7 = -8)$
s $-a$ $(1 + 2 - 3 - 1 = -1)$
t $-2b$ $(7 - 11 + 5 - 3 = -2)$

WORKED SOLUTIONS to Chapter 5 **Practise, Practise**

23 **a** $14 + 3a - 7 = 14 - 7 + 3a = 7 + 3a$
b $7y + 3b - 4y = 7y - 4y + 3b = 3y + 3b$
c $7a + 4b - 5a = 7a - 5a + 4b = 2a + 4b$
d $-9y + z - 4y = -9y - 4y + z = -13y + z$
e $5t + 6s - 11t + 2s = 5t - 11t + 6s + 2s$
$= -6t + 8s$
f $9a - 7c - 4a + 5c = 9a - 4a - 7c + 5c$
$= 5a - 2c$
g $a + a + a - 4b - 2b = 3a - 6b$
h $t + 3u - 7t + 2u = t - 7t + 3u + 2u$
$= -6t + 5u$
i $4y - 8 - 6y + 12 = 4y - 6y - 8 + 12$
$= -2y + 4$
j $a - b - 2a + 4b - a + 2b$
$= a - 2a - a - b + 4b + 2b = -2a + 5b$

> Note how like terms can be moved so that they are next to each other. The sign in front of the term moves with the term.

24

a	$-7a$	**b**	$-12a$
c	$-24y$	**d**	$36b$
e	$-4t \times -4t = 16t^2$	**f**	$-42a^2$
g	-4	**h**	$-3a$
i	$-4y$	**j**	6
k	$-35ab$	**l**	$72ac$

25 **a** $6a + 4 \times 3a = 6a + 12a = 18a$
b $12y - 3 \times 4y = 12y - 12y = 0$
c $2a \times 3a - 4a \times 5a = 6a^2 - 20a^2 = -14a^2$
d $24y - 18y^2 \div 6y = 24y - 3y = 21y$
e $(5y - 8y) \times (3t - 8t) = -3y \times -5t = 15yt$

26

> $x = -4$
> $y = 5$
> $z = -2$

a $-4 + 5 = 1$
b $-4 - 5 = -9$
c $5 - (-4) = 5 + 4 = 9$
d $-4 \times 5 = -20$
e $\frac{-4}{-2} = 2$
f $-5 \times (-4) = 20$
g $\frac{-35}{5} = -7$
h $-4 \times -2 = 8$
i $-4 + 5 + -2 = -1$
j $-4 + 5 - (-2) = 1 + 2 = 3$
k $-4 - 5 + (-2) = -9 + (-2) = -11$
l $-4 - (5 + -2) = -4 - 3 = -7$
m $(-4)^2 = 16$
n $-3x^2 = -3 \times (-4)^2 = -3 \times 16 = -48$
o $6 - (-4) = 6 + 4 = 10$
p $-5 + (-4) = -9$
q $(-4) \times 5 + (-4)^2 = -20 + 16 = -4$
r $-4 \times (5 + -4) = -4 \times 1 = -4$
s $(-4) \times 5 - (-4)^2 = -20 - 16 = -36$
t $-4 \times (5 - [-4]) = -4 \times (5 + 4) = -4 \times 9$
$= -36$
u $(-4)^2 - (5)^2 = 16 - 25 = -9$
v $(-4 - 5) \times (-4 + 5) = (-9) \times 1 = -9$
w $3 \times (-4) - 2 \times 5 = -12 - 10 = -22$
x $3 \times (-4) - 2 \times 5 + 4 \times (-2)$
$= -12 - 10 + (-8)$
$= -12 + (-10) + (-8) = -30$
y $(-4 + 5) \times (-2 - 5) \div (5 + 2)$
$= 1 \times (-7) \div 7 = (-7) \div 7 = -1$

27

A	(2, 2)	I	(5, 0)	Q	(4, −4)
B	(2, 4)	J	(0, 2)	R	(−4, 0)
C	(−2, 2)	K	(−2, −2)	S	(0, −3)
D	(5, 3)	L	(−4, −1)	T	(4, 0)
E	(−3, 3)	M	(−3, −4)	U	(0, −1)
F	(−4, 1)	N	(2, −3)	V	(5, −3)
G	(0, 4)	P	(3, −2)	W	(−4, −3)
H	(2, 0)				

28 **a**

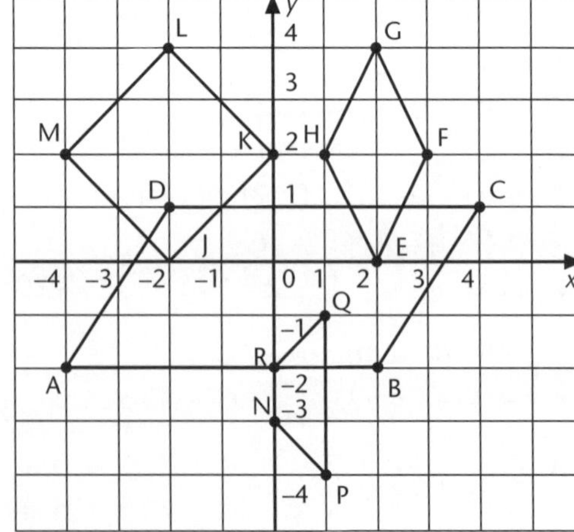

ABCD—Parallelogram
EFGH—Rhombus
MJKL—Square
NPQR—Trapezium

b

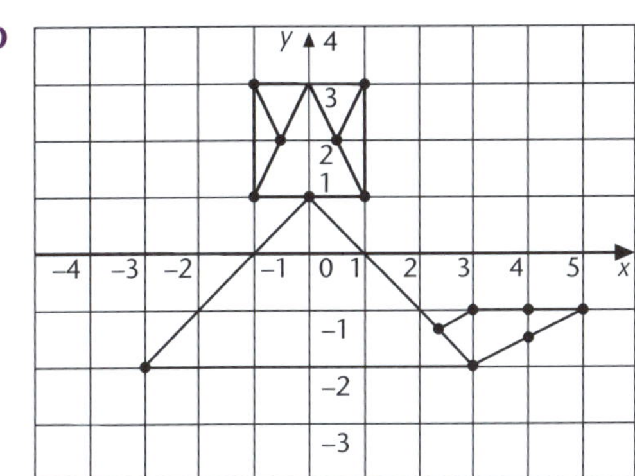

29 Temperature drop = $2° \times 5 = 10°C$
New temperature = $8° - 10° = -2°C$
Temperature at 5 p.m. = $-2°C$

30

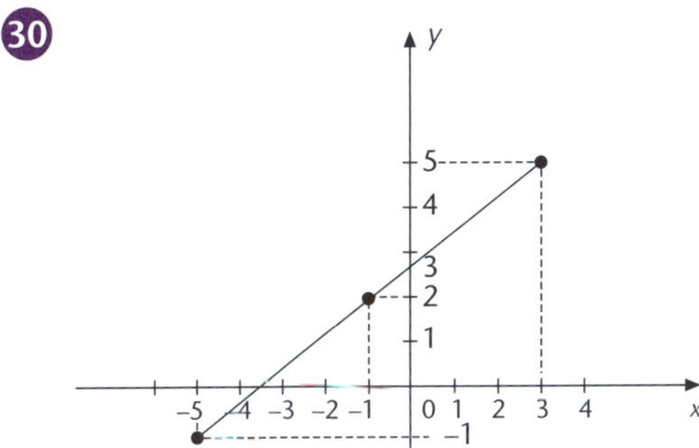

For x: Halfway between -5 and $3 = -1$
For y: Halfway between -1 and $5 = 2$
Mid-point = $(-1, 2)$

31

a $\frac{1}{2} - 1 = \frac{1}{2} - \frac{2}{2} = \frac{-1}{2}$ $\quad [1 - 2 = -1]$

b $\frac{3}{5} - \frac{4}{5} = \frac{-1}{5}$ $\quad [3 - 4 = -1]$

c $1\frac{1}{2} - 2\frac{1}{4} = \frac{3}{2} - \frac{9}{4}$ $\quad \left[1\frac{1}{2} - 3 + \frac{3}{4} = -1\frac{1}{2} + \frac{3}{4} = \frac{-3}{4}\right]$

$= \frac{6}{4} - \frac{9}{4}$ $\quad [6 - 9 = -3]$

$= \frac{-3}{4}$

d $\frac{-3}{5} + \frac{1}{5} = \frac{-2}{5}$ $\quad [-3 + 1 = -2]$

e $0.6 - 0.8 = -0.2$ $\quad [6 - 8 = -2]$

f $4 - 6.8 = -2.8$

g $-0.8 + 0.6 = -0.2$ $\quad [-8 + 6 = -2]$

h $\frac{1}{3}a - \frac{3}{4}a = \frac{4}{12}a - \frac{9}{12}a$

$= \frac{-5}{12}a$

$= \frac{-5a}{12}$

Chapter 6—Ratios

pp. 112–114

1

a 25:15 = 5:3 (divide by 5)

b 30:70 = 3:7 (with the zeros struck out)

c 9:18:27 = 1:2:3 (divide by 9)

d 14:49 = 2:7 (divide by 7)

e 100:100 000 = 1:1000 (with the zeros struck out)

f 75:125 = 3:5 (divide by 25)

2

a 6:2 = 3:1
$\therefore x = 3$

b 15:45 = 1:3
$\therefore x = 3$

c 12:8 = 3:2
$\therefore x = 2$

d 7:5 = 49:35
$\therefore x = 49$

e 13:3 = 39:9
$\therefore x = 39$

3

a 3.5:1.5 = 35:15
= 7:3

b 2:1.4 = 20:14
= 10:7

c 8.4:4.2 = 84:42
= 2:1

d 3.25:4.25 = 325:425
= 13:17

e $\frac{3}{4}:\frac{1}{2} = \frac{3}{4}:\frac{2}{4}$
= 3:2

f $\frac{2}{3}:\frac{5}{6} = \frac{4}{6}:\frac{5}{6}$
= 4:5

g $\frac{3}{5}:1\frac{1}{2} = \frac{3}{5}:\frac{3}{2}$
$= \frac{6}{10}:\frac{15}{10}$
= 6:15
= 2:5

h $2\frac{1}{4}:2\frac{1}{2} = \frac{9}{4}:\frac{5}{2}$
$= \frac{9}{4}:\frac{10}{4}$
= 9:10

i $\frac{4}{5}$:2.2 = 0.8:2.2
= 8:22
= 4:11

4

a \$1.50:40c = 150c:40c
= 150:40
= 15:4

b \$2.40:\$1.60 = 240c:160c
= 240:160
= 3:2

c 8 hours:2 days = 8 hours:48 hours
= 8:48
= 1:6

d 1 mm:1 cm:1 m = 1 mm:10 mm:1000 mm
= 1:10:1000

e 3 weeks:6 days = 21 days:6 days
= 21:6
= 7:2

WORKED SOLUTIONS to Chapter 6 **Practise, Practise**

f $1\frac{1}{2}$ mins:45 secs = 90 secs:45 secs
= 90:45
= 2:1

g 1 mL:1 kL = 1 mL:1 000 000 mL
= 1:1 000 000

h 3 decades:2 centuries
= 3 decades:20 decades
= 3:20

5 **a** $\frac{\cancel{12}^2 \cancel{a}^1 b}{\cancel{6}_1 \cancel{a}_1} : \frac{\cancel{6}^1 \cancel{a}^1}{\cancel{6}_1 \cancel{a}_1} = 2b:1$

b $\frac{\cancel{3}^1 x^1 y^1}{\cancel{3}_1 x_1 y_1} : \frac{\cancel{15}^5 x^1 y^1}{\cancel{3}_1 x_1 y_1} = 1:5$

c $2a^2:(2a)^2 = \frac{\cancel{2a^2}}{\cancel{2a^2}} : \frac{\cancel{4}^2 \cancel{a^2}}{\cancel{2} \cancel{a^2}} = 1:2$

d $\frac{3}{p}:3p = 3:3p^2$ (multiply by p)
$= \frac{\cancel{3}}{\cancel{3}} : \frac{\cancel{3}p^2}{\cancel{3}}$
$= 1:p^2$

6 **a** Profit = \$900 – \$700 = \$200
b **i** CP = \$700, SP = \$900
∴ \$700:\$900 = 7~~00~~:9~~00~~
= 7:9
ii Profit = \$200, CP = \$700
∴ \$200:\$700 = 2~~00~~:7~~00~~
= 2:7

7 **a** Female = 56 – 31 = 25
∴ There are 25 female teachers.
b **i** male:female = 31:25
ii total:female = 56:25

8 10:24 = 5:12

9 **a** 6 black, 10 red
∴ 6:10 = 3:5
b 10 red, 10 non-red
∴ 10:10 = 1:1

10 **a** **i** AB = 3 cm
ii AC = 8 cm
b **i** AB:AC = 3 cm:8 cm = 3:8
ii CD:AB = 2 cm:3 cm = 2:3
iii AC:BD = 8 cm:7 cm = 8:7

11 Bad apples = $\frac{1}{3}$, good apples = $\frac{2}{3}$
∴ bad:good = $\frac{1}{3}:\frac{2}{3}$ = 1:2

12 **a** Sides: 2 cm:6 cm = 2:6 = 1:3
b Areas: (2×2) cm^2:(6×6) cm^2
= 4 cm^2:36 cm^2
= 4:36
= 1:9
c Volumes: $(2 \times 2 \times 2)$ cm^3:$(6 \times 6 \times 6)$ cm^3
= 8 cm^3:216 cm^3
= 8:216
= 1:27

13 Perimeter = 45 cm
∴ length of side = 45 ÷ 5 = 9 cm
∴ ratio = 45 cm:9 cm = 45:9 = 5:1

14 **a** Netball + soccer = 360° – (90 + 150)
= 360° – 240°
= 120°
∴ Netball and soccer are both 60° sectors.
b **i** 90°:150° = 9~~0~~:15~~0~~
= 9:15
= 3:5
ii 60°:60° = 60:60
= 1:1
iii 90°:360° = 9~~0~~:36~~0~~
= 9:36
= 1:4

15 Adults:children = 2:5
∴ 5 parts = 40
1 part = 8
2 parts = 16
∴ There are 16 adults.

16 **a** 3:7 = smaller:longer
∴ 7 parts = 42 cm
1 part = 6 cm
3 parts = 18 cm
∴ The smaller piece of timber is 18 cm.
b Original length = 42 cm + 18 cm = 60 cm
∴ The original length of timber is 60 cm.

WORKED SOLUTIONS to Chapters 6 and 7 **Practise, Practise**

17 James:Nick:Ben = 4:3:2
3 parts = \$630
∴ 1 part = \$210
∴ 4 parts = \$840
∴ James contributed \$840.

18 Old width:new width = 2:5
5 parts = 25 cm
1 part = 5 cm
∴ 2 parts = 10 cm
The original width of the photo was 10 cm.

19 **a** Students:teachers = 20:1 = 80:4
Teachers:assistants = 4:1
∴ Students:teachers:assistants = 80:4:1

b 80:4:1
4 parts = 48
∴ 1 part = 12 ← assistants
80 parts = 960 ← students
∴ 960 students and 12 school assistants.

20 Josh = \$2.00 – \$0.80 = \$1.20
∴ Jo—\$0.80, Josh—\$1.20
∴ Ratio of contributions = Jo:Josh = 80c:\$1.20
= 80c:120c
= 8~~0~~:12~~0~~
= 2:3
∴ Total parts = 2 + 3 = 5
∴ $\frac{3}{5} \times 200\,000 = 120\,000$

21 ∴ Josh will win \$120 000.
Total parts = 11 + 4 = 15
∴ 15 parts = \$75
∴ $\frac{11}{15} \times 75 = 55$
$\frac{4}{15} \times 75 = 20$
∴ \$55 and \$20 are the parts.

22 Total parts = 2 + 4 + 3 = 9
∴ 9 parts = \$270
∴ $\frac{4}{9} \times 270 = 120$
∴ The largest share is \$120.

23 **a** 5:7 ∴ F **b** 1:11 ∴ B
c 1:1 = 6:6 ∴ G **d** 3:1 = 9:3 ∴ J

24 **a** Cost price:profit = 21:5
5 parts = \$135 000
∴ 1 part = \$27 000
∴ 21 parts = \$567 000
∴ Cost price is \$567 000.

b Selling price = cost price + profit
= \$567 000 + \$135 000
= \$702 000
∴ Selling price is \$702 000.

c Cost:selling = \$567 000:\$702 000
= 567 ~~000~~:702 ~~000~~
= 567:702
= 21:26 (divide by 9)

[Note: Easier way is 21:21 + 5 (profit) = 21:26]

25 Total parts = 4 + 11 = 15
Also, $2\frac{1}{2}$ hours = 150 minutes
∴ 15 parts = 150 mins
∴ $\frac{4}{15} \times 150 = 40$
∴ 40 minutes is spent watching advertising.

26 **a** Voted for:voted against = 3:7
b Total parts = 3 + 7 = 10
∴ 10 parts = 40 000
$\frac{3}{10} \times 40\,000 = 12\,000$
∴ 12 000 voted people for Kentish.

Chapter 7—Algebra

pp. 127–130

1 **a**

Number of triangles	1	2	3	4	5	6	7	8
Number of matchsticks	3	6	9	12	15	18	21	24

b The number of matchsticks = 3 times the number of triangles.

c $T = 3 \times t$ or $T = 3t$

d **i** $T = 3 \times 10 = 30$
ii $T = 3 \times 50 = 150$
iii $T = 3 \times 100 = 300$

e $285 = 3t$, $t = 95$
∴ 95 triangles

WORKED SOLUTIONS to Chapter 7 **Practise, Practise**

2 **a**

Number of squares	1	2	3	4	5	6	7	8
Number of matchsticks	4	7	10	13	16	19	22	25

b The number of matchsticks = 3 times the number of squares plus one.

c $x = 3 \times y + 1$ or $x = 3y + 1$

d **i** $x = 3 \times 15 + 1 = 46$
ii $x = 3 \times 52 + 1 = 157$
iii $x = 3 \times 400 + 1 = 1201$

e **i** $61 = 3 \times y + 1, y = 20$
ii $313 = 3 \times y + 1, y = 104$
iii $901 = 3 \times y + 1, y = 300$

3 **a** $y = 2x$ **b** $y = x + 1$
c $y = x - 2$ **d** $y = x^2$
e $y = 2x + 1$ **f** $y = 3x - 1$

4 **a** $y = x + 3$

x	1	2	3	4
y	4	5	6	7

b $T = t - 1$

t	1	2	3	4
T	0	1	2	3

c $n = 6 - m$

m	1	2	3	4
n	5	4	3	2

d $Q = 2 \times P$

P	1	2	3	4
Q	2	4	6	8

e $y = 3 \times x - 1$

x	2	3	4	5
y	5	8	11	14

f $w = 12 - 3 \times v$

v	1	2	3	4
w	9	6	3	0

5 **a** ab **b** $3y$ **c** $2t$
d $4ts$ **e** $7yy = 7y^2$
[Number is usually written before the letter.]

6 **a** $xy = x \times y$ **b** $3yb = 3 \times y \times b$
c $4b^2 = 4 \times b \times b$
d $2mnq = 2 \times m \times n \times q$
e $2m^2nq^2 = 2 \times m \times m \times n \times q \times q$
f $(3b)^2 = 3b \times 3b = 9 \times b \times b$

7 **a** $ab = 4 \times 2 = 8$ **b** $4a = 4 \times 4 = 16$
c $a + b = 4 + 2 = 6$
d $a - b + c = 4 - 2 + 5 = 7$
e $c^2 = 5^2 = 5 \times 5 = 25$
f $2a - c = 2 \times 4 - 5 = 8 - 5 = 3$
g $abc = 4 \times 2 \times 5 = 40$
h $c^2 - a^2 = 5^2 - 4^2 = 25 - 16 = 9$
i $a(c - 3) = 4(5 - 3) = 4 \times 2 = 8$
j $ac - 3a = 4 \times 5 - 3 \times 4 = 20 - 12 = 8$
k $\frac{4bc}{2a} = \frac{4 \times 2 \times 5}{2 \times 4} = \frac{40}{8} = 5$
l $3b^2 = 3 \times 2^2 = 3 \times 4 = 12$
m $(3b)^2 = (3 \times 2)^2 = 6^2 = 36$
n $\frac{4c^2}{25a} = \frac{4 \times 5^2}{25 \times 4} = \frac{4 \times 25}{25 \times 4} = 1$

8 **a** $by = \frac{1}{\cancel{3}_1} \times \frac{\cancel{3}^1}{4} = \frac{1}{4}$

b $b + y = \frac{1}{3} + \frac{3}{4} = \frac{4}{12} + \frac{9}{12} = \frac{13}{12} = 1\frac{1}{12}$

c $4y = \frac{\cancel{4}^1}{1} \times \frac{3}{\cancel{4}_1} = 3$

d $6b = \frac{\cancel{6}^2}{1} \times \frac{1}{\cancel{3}_1} = 2$

e $y^2 = (\frac{3}{4})^2 = \frac{9}{16}$

f $2y + 3b = \frac{\cancel{2}^1}{1} \times \frac{3}{\cancel{4}_2} + \frac{\cancel{3}^1}{1} \times \frac{1}{\cancel{3}_1} = \frac{3}{2} + 1 = 1\frac{1}{2} + 1 = 2\frac{1}{2}$

g $9b^2 = 9 \times (\frac{1}{3})^2 = 9 \times \frac{1}{9} = 1$

h $y - 2b = \frac{3}{4} - 2 \times \frac{1}{3} = \frac{3}{4} - \frac{2}{3} = \frac{9}{12} - \frac{8}{12} = \frac{1}{12}$

9 $A = nb$
$= n \times b$
$= 8 \times 4$
$= 32$

10 $P = 2(n + b)$
$= 2(10 + 3)$
$= 2(13)$
$= 2 \times 13$
$= 26$

11 $T = 4n - 2$
$= 4 \times n - 2$
$= 4 \times 5 - 2$
$= 20 - 2$
$= 18$

12 $N = 4(n + 1)$
$= 4(8 + 1)$
$= 4(9)$
$= 4 \times 9$
$= 36$

13 $V = u + at$
$= u + a \times t$
$= 32 + 2 \times 6$
$= 32 + 12$
$= 44$

14 $E = mc^2$
$= m \times c \times c$
$= 3 \times 5 \times 5$
$= 75$

15 $D = \dfrac{M}{V}$
$= \frac{24}{5}$
$= 4\frac{4}{5}$

16 $Q = \dfrac{10A}{P + 12}$
$= \frac{10 \times 6}{3 + 12} = \frac{60}{15}$
$= 4$

17 **a** $m \times m = m^2$ **b** $n \times 5 = 5n$
c $4 \times a \times a = 4a^2$ **d** $p \times q = pq$
e $y + y = 2y$
$[y + y = 1y + 1y = 2y]$

f $4x \div 2 = \dfrac{4x}{2}$
$= \dfrac{\cancel{4}_2 \times x}{\cancel{2}_1}$
$= \dfrac{2x}{1}$
$= 2x$

g $5a \times b = 5ab$ **h** $p \times 6 = 6p$
i $4b \times 3a = 12ba$ **j** $x \times y \times z = xyz$
k $3a \times 2a = 6a^2$ **l** $2y \times 5xy = 10y^2x$
m $c \times 2a \times 4 = 8ca$ **n** $4a \times a \times a = 4a^3$
o $a \times 4 \times b \times a = 4a^2b$
p $(5p)^2 = 5p \times 5p = 25p^2$

q $16x \div 4x = \dfrac{16x}{4x}$
$= \dfrac{\cancel{16}_4 \times \cancel{x}_1}{\cancel{4}_1 \times \cancel{x}_1}$
$= 4$

r $18ab \div 3a = \dfrac{18ab}{3a}$
$= \dfrac{\cancel{18}_6 \times \cancel{a}_1 \times b}{\cancel{3}_1 \times \cancel{a}_1}$
$= \dfrac{6b}{1}$
$= 6b$

18 **a** $a^3 \times a^8 = a^{11}$ **b** $a^{12} \div a^2 = a^{10}$
c $(b^3)^5 = b^{15}$ **d** $(ab)^4 = a^4b^4$
e $y \times y^4 \times y^2 = y^7$ **f** $3xy \times 2y^2 = 6xy^3$
$[y = y^1]$

g $4a^{15} \div 2a^5 = 2a^{10}$ **h** $\dfrac{a^4 \times a^6}{a^5} = \dfrac{a^{10}}{a^5}$
$= a^5$

i $(6a)^2 = 6a \times 6a$
$= 36a^2$

j $(3a^4)^2 = 3^2 \times (a^4)^2$
$= 9 \times a^8$
$= 9a^8$

k $(x^3y^2)^3 = (x^3)^3 \times (y^2)^3$
$= x^9 \times y^6$
$= x^9y^6$

l $12a^4b^5 \div 6a^6b^2 = \dfrac{12a^4b^5}{6a^6b^2} = \dfrac{2b^3}{a^2}$

WORKED SOLUTIONS to Chapter 7 **Practise, Practise**

m $15x^2y \div 10xy^2 = \dfrac{15x^2y}{10xy^2} = \dfrac{\not{15}_3 \times \not{x}_1 \times x \times \not{y}_1}{\not{10}_2 \times \not{x}_1 \times \not{y}_1 \times y} = \dfrac{3x}{2y}$

19 **a** $4^3 = 4 \times 4 \times 4 = 64$
b $10^2 = 10 \times 10 = 100$
c $2^5 = 2 \times 2 \times 2 \times 2 \times 2 = 32$
d $5^0 = 1$
e $6^3 = 6 \times 6 \times 6 = 216$
f $2^2 \times 2^3 = 2^5 = 2 \times 2 \times 2 \times 2 \times 2 = 32$
g $(2^3)^2 = 2^6 = 2 \times 2 \times 2 \times 2 \times 2 \times 2 = 64$

20 **a** $8t + 10t - 3t = 15t$
b $10a - a = 9a$
c $5x + 2y + 3x - 2y = 8x$
d $4x + 7x - 3x = 8x$
e $5a - 2 + 3a = 8a - 2$
f $4a + 2 + a + 7 = 5a + 9$
g $5ab + 2ba - 3ab = 4ab$ [$ab = ba$]
h $5x^2 + 3x - 2x^2 + x = 3x^2 + 4x$
i $3a^2 + 2a + 4a - a^2 = 2a^2 + 6a$
j $4y + 5 - y - 1 = 3y + 4$
k $5x + 8 - 3x - 2 = 2x + 6$
l $12p - 8p - p = 3p$
m $x + 9y + 3x - 4y = 4x + 5y$
n $4p + 8 - p + 2 = 3p + 10$
o $4x + 10 + x - 3 = 5x + 7$

21 **a** $a + a + a + a = 4a$
b $4 \times a = 4a$
c $a \times a \times a \times a = a^4$
d $\dfrac{4ab}{2a} = \dfrac{\not{4}_2 \times \not{a}_1 \times b_1}{\not{2}_1 \times \not{a}_1} = \dfrac{2b}{1} = 2b$
e $a \times b \times a \times b \times a = a^3b^2$
f $7x + 3y = 7x + 3y$

[$7x$ and $3y$ cannot be added as they are unlike terms.]

g $8x - 3x - x = 4x$

h $(2x)^3 = 2x \times 2x \times 2x = 8x^3$

i $15x^2y \div 10xy^3 = \dfrac{15x^2y}{10xy^3} = \dfrac{\not{15}_3 \not{x}x\not{y}}{\not{10}_2 \not{x}\not{y}yy} = \dfrac{3x}{2y^2}$

j $12xy \div 8x = \dfrac{12xy}{8x} = \dfrac{\not{12}_3 \times \not{x}_1 \times y}{\not{8}_2 \times \not{x}_1} = \dfrac{3y}{2}$

k $4a + 3 - 2a + 10 = 2a + 13$
l $2a \times 4b \times a = 8a^2b$
m $2a \times 3a \times 4a = 24a^3$
n $8b - b = 7b$
o $(4x^3)^2 = 4^2 \times (x^3)^2 = 16 \times x^6 = 16x^6$
p $5x^2 - x - x^2 = 4x^2 - x$
q $2x^2 + 5x - 3x + 3x^2 = 5x^2 + 2x$

[x and x^2 are **not** like terms.]

r $2b \div 3b = \dfrac{2b}{3b} = \dfrac{2 \times \not{b}_1}{3 \times \not{b}_1} = \dfrac{2}{3}$

s $6 \times a + b = 6a + b$

[Order of operations.]

22 **a** $8x + 2x - 4x - 6x = 0$
b $5t - 7t + 4t = 2t$
c $-12x + 4x = -8x$
d $-a - a = -1a - 1a = -2a$
e $2x - 5y + x + 3y = 3x - 2y$
f $4x^2 + 7x - 2x^2 - 10x = 2x^2 - 3x$
g $10x - 4 - 3x + 2 = 7x - 2$
h $2p + 5 - 4p - 3 = -2p + 2$ or $2 - 2p$
i $2x - 7 + 3x + 4 = 5x - 3$
j $2x - 4 + 5x - 7 = 7x - 11$
k $x + 9y - 4x - y = -3x + 8y$ or $8y - 3x$
l $4p - 8 - 6p + 3 = -2p - 5$

m $-x + 5 - 2x - 7 = -3x - 2$
n $x - 9y + x - y = 2x - 10y$
o $2x^2 - x - x^2 - x = x^2 - 2x$
p $a - a + a - 2a = -a$
q $x^2 - 5x - x^2 + 2x = -3x$
r $x - 8y + x - 2y = 2x - 10y$
s $2 - 7t - 6 + 6t = -4 - t$

Chapter 8—Measurement

pp. 152–156

1 **a** $2000 \div 1000 = 2$ km
b $4.21 \times 1000 = 4210$ m
c $37 \times 10 = 370$ mm
d $3\frac{1}{2} \times 10 = 35$ mm
e $420 \div 10 = 42$ cm

2 **a** B **b** C **c** B **d** B

3 **a** 24 mm **b** 31 mm **c** 35 mm

4 **a** 11.5 cm to 12.5 cm
b 15.5 m to 16.5 m

5 **a** $P = 30 + 12 + 30 + 12 = 84$
$\therefore$ Perimeter is 84 cm.
b $P = 11 + 17 + 23 = 51$
$\therefore$ Perimeter is 51 cm.
c Octagon: 8 sides
$\therefore P = 8 \times 7 = 56$
$\therefore$ Perimeter is 56 cm.

6 Rectangle's perimeter = 96 mm
$\therefore$ Length + width = 96 mm ÷ 2 = 48 mm
But length = 28 mm
$\therefore$ 28 + width = 48
$\therefore$ width = 20
$\therefore$ Width is 20 mm.

7 **a**

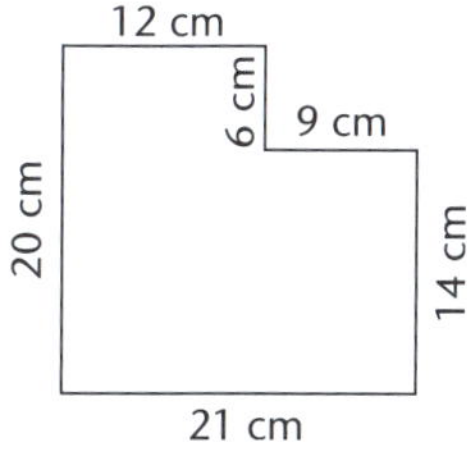

$\therefore P = 20 + 21 + 14 + 9 + 6 + 12 = 82$
$\therefore$ Perimeter is 82 cm.

b

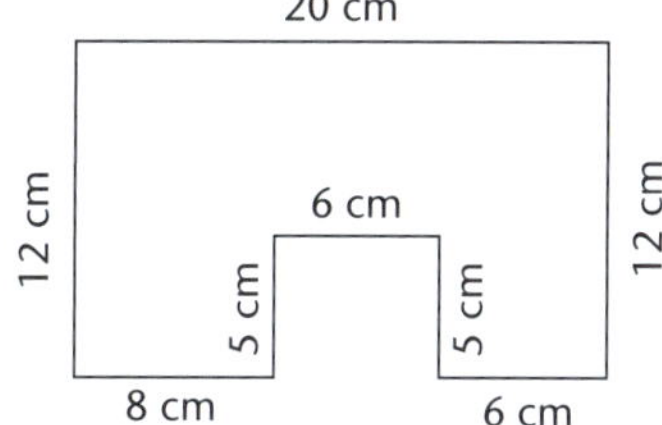

$\therefore P = 12 + 8 + 5 + 6 + 5 + 6 + 12 + 20 = 74$
$\therefore$ Perimeter is 74 cm.

8 **a**

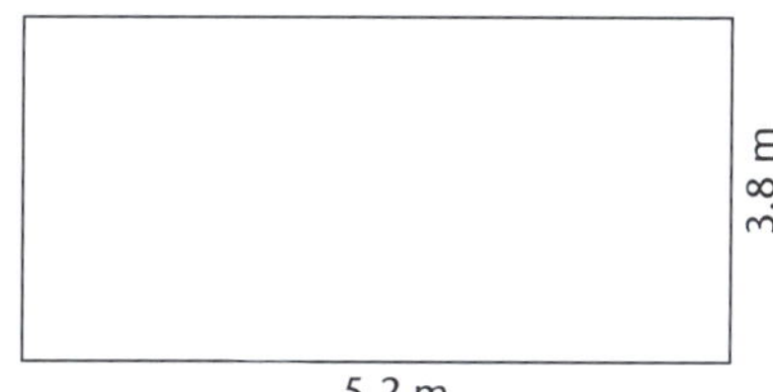

$P = 5.2 + 3.8 + 5.2 + 3.8$
$= 2(5.2 + 3.8) = 2(9)$
$= 18$
$\therefore$ Perimeter is 18 m.
b Rolls $= 18 \div 5 = 3\frac{3}{5}$
$\therefore$ 4 rolls must be purchased.
c Cost = \$21.90 × 4 = \$87.60
$\therefore$ Rolls cost \$87.60.

9 **a** 6 units^2
b 9 whole squares + 7 half squares
$\therefore 9 + 3\frac{1}{2} = 12\frac{1}{2}$
$\therefore$ Approximately $12\frac{1}{2}$ units^2.

10 **a** $A = lb = 15 \times 7 = 105$
$\therefore$ Area is 105 cm^2.
b $A = lb = 12 \times 12 = 144$
$\therefore$ Area is 144 mm^2.
c Firstly 2 m = 200 cm
$\therefore A = lb = 200 \times 40 = 8000$
$\therefore$ Area is 8000 cm^2.
d Firstly 3 m 40 cm = 3.4 m
$\therefore A = 3.4 \times 3.4$ $[A = lb]$
$= 11.56$ m^2
e $A = \frac{1}{2}bh$
$= \frac{1}{2} \times 19 \times 7$
$= 66.5$
$\therefore$ Area is 66.5 cm^2.

f $A = \frac{1}{2}bh$

$= \frac{1}{2} \times 9 \times 8$

$= 36$

$\therefore$ Area is 36 cm^2.

g

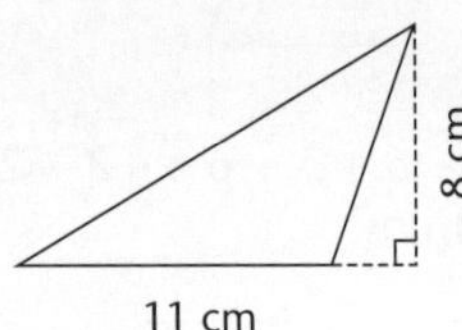

$A = \frac{1}{2}bh$

$= \frac{1}{2} \times 11 \times 8$

$= 44$

$\therefore$ Area is 44 cm^2.

h

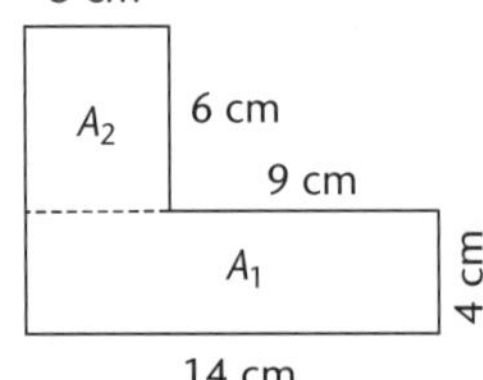

$A = A_1 + A_2$ $\quad [A = lb]$

$= 14 \times 4 + 6 \times 5$

$= 56 + 30$

$= 86$

$\therefore$ Area is 86 cm^2.

i $A = bh$

$= 23 \times 20$

$= 460$

$\therefore$ 460 cm^2

j $A = bh$

$= 10 \times 12$

$= 120$

$\therefore$ 120 cm^2

11 **a** $A = 320 \times 200$ $\quad [A = lb]$

$= 64\,000$

$\therefore$ Area is 64 000 m^2

$\therefore$ Area is (64 000 ÷ 10 000) ha = 6.4 ha.

b $A = \frac{1}{2} \times 1200 \times 800$ $\quad [A = \frac{1}{2}bh]$

$= 480\,000$

$\therefore$ Area is 480 000 cm^2

$\therefore$ Area is (480 000 ÷ 10 000) m^2 = 48 m^2.

12 **a** Area = Area of rectangle – Area of triangle

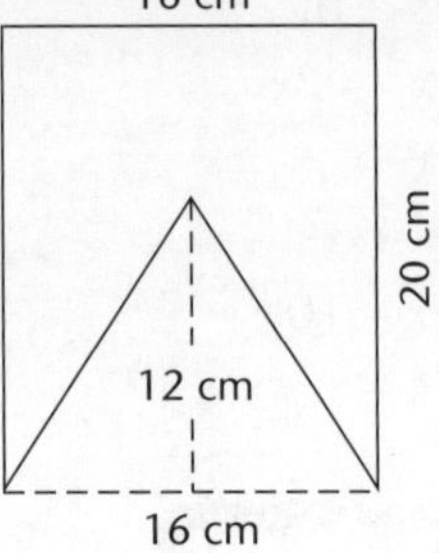

$\therefore A = 16 \times 20 - \frac{1}{2} \times 16 \times 12$

$= 320 - 96$

$= 224$

$\therefore$ Area is 224 cm^2.

b

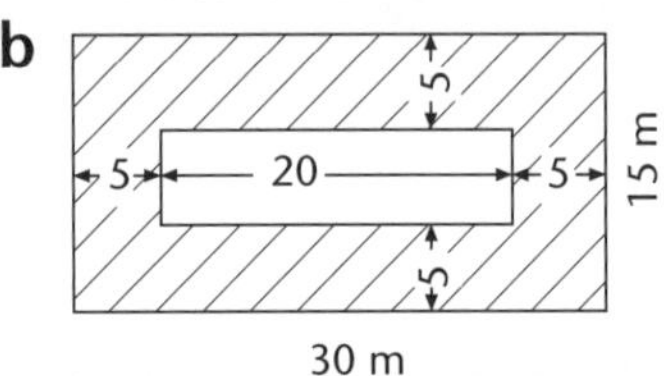

Area = Area of large rectangle – area of small rectangle

$\therefore A = 30 \times 15 - 20 \times 5$

$= 450 - 100$

$= 350$

$\therefore$ Area is 350 m^2.

13 Area = length × width

$\therefore 126 = 14 \times \text{width}$

$\therefore \text{Width} = \frac{126}{14} = 9$

$\therefore$ Width is 9 cm.

14

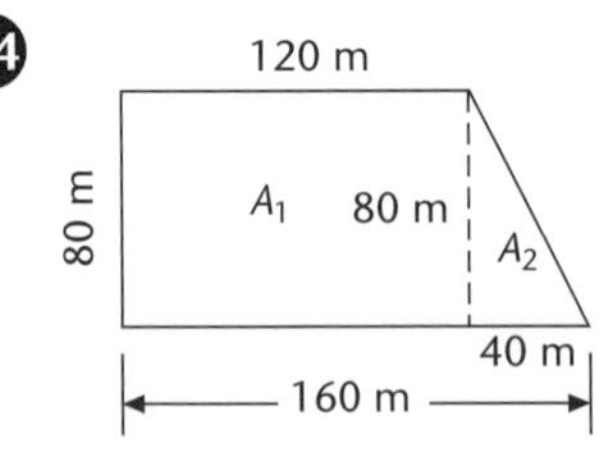

$A_1 = lb$

$A_2 = \frac{1}{2}bh$

80 A_2 40

$A = A_1 + A_2$

$= 120 \times 80 + \frac{1}{2} \times 40 \times 80$

$= 9600 + 1600$

$= 11\,200$

$\therefore$ Area is 11 200 m^2

$\therefore$ Area is (11 200 ÷ 10 000) ha = 1.12 ha.

WORKED SOLUTIONS to Chapter 8 **Practise, Practise**

15 **a** Length = 6 metres

b Width = 1 metre

c $A = 3 \times 3 = 9$ $[A = lb]$
$\therefore$ Area of dining room is 9 m^2.

d Length = 6 + 3 + 5 = 14 m
Width = 4 + 1 + 3 = 8 m
Area = $14 \times 8 = 112$ $[A = lb]$
$\therefore$ Area is 112 m^2.

e Width bathroom $= 14 - (3 + 3 + 2 + 4)$
$= 14 - 12$
$= 2$ m
Area $= lb = 3 \times 2 = 6\ m^2$
$\therefore$ Cost = $30 × 6 = $180
$\therefore$ Cost of tiling is $180.

f Areas $= 3 \times 4 + 5 \times 4 + 4 \times 4$
$= 12 + 20 + 16$
$= 48$
$\therefore$ Total area is 48 m^2
$\therefore$ Cost = 48 × $0.70 = $33.60
$\therefore$ Cost of compound is $33.60.

16

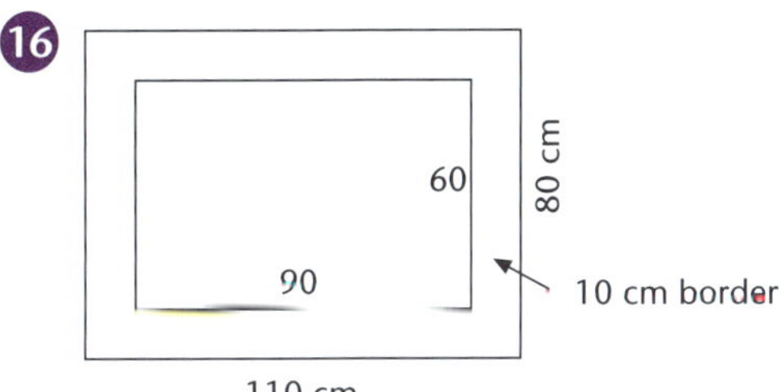

a Length is (90 + 10 + 10) cm and width is (60 + 10 + 10) cm
i.e. length = 110 cm, width = 80 cm

b $P = 2(110 + 80)$
$= 2(190)$
$= 380$
$\therefore$ Perimeter is 380 cm.

c Now, 380 cm = 3.8 m
$\therefore$ Cost = $12 × 3.8 = $45.60
$\therefore$ Cost of frame is $45.60.

d $A = 110 \times 80 = 8800$
$\therefore$ Area is 8800 cm^2.

e Now, 8800 cm^2 = (8800 ÷ 10 000) m^2
= 0.88 m^2
$\therefore$ Cost = 0.88 × $70 = $61.60
$\therefore$ Cost of glass is $61.60.

17 **a** $A = 30 \times 20 = 600$
$\therefore$ Area is 600 m^2.

b 50 cm = 0.5 m
Area of 1 carpet square = $0.5 \times 0.5 = 0.25$
Area of each square is 0.25 m^2.

c No. of squares = 600 ÷ 0.25 = 2400
$\therefore$ 2400 carpet squares.

18 **a** $3.2 \times 10\,000 = 32\,000$
i.e. 32 000 m^2

b 16 000 ÷ 10 000 = 1.6
i.e. 1.6 ha

c 3700 ÷ 10 000 = 0.37
i.e. 0.37 m^2

d $5.84 \times 10\,000 = 58\,400$
i.e. 58 400 cm^2

19 **a** $V = lbh$
$= 12 \times 7 \times 9$
$= 756$
$\therefore$ Volume is 756 cm^3.

b $V = lbh$ (or $V = s^3$)
$= 4 \times 4 \times 4$
$= 64$
$\therefore$ Volume is 64 cm^3.

c $V = Ah$
$= (\frac{1}{2} \times 14 \times 16) \times 20$ $[A = \frac{1}{2}bh]$
$= 2240$
$\therefore$ Volume is 2240 cm^3.

d $V = Ah$
$= 25 \times 6$
$= 150$
$\therefore$ Volume is 150 m^3.

e

5
A_1 4
7
10
A_2 6
12

$A = A_1 + A_2$
$= 5 \times 4 + 12 \times 6$
$= 20 + 72$
$= 92$

$\therefore V = Ah$
$= 92 \times 8$
$= 736$

$\therefore$ Volume is 736 cm^3.

f $V = Ah$
$= 140 \times 12$
$= 1680$
$\therefore$ Volume is 1680 cm^3.

g

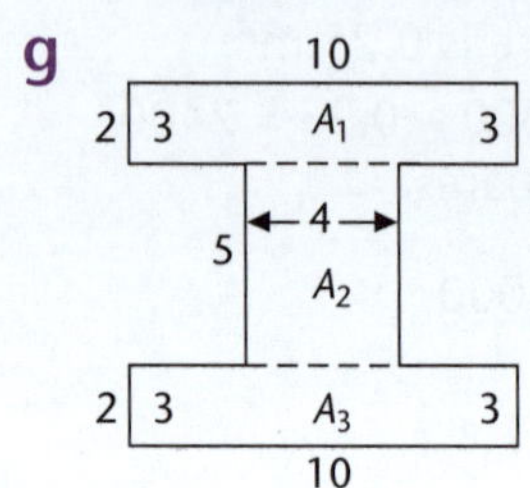

$A = A_1 + A_2 + A_3$
$= 10 \times 2 + 5 \times 4 + 10 \times 2$
$= 20 + 20 + 20$
$= 60$

Now, $V = Ah$
$= 60 \times 4$
$= 240$
$\therefore$ Volume is 240 m^3.

20 **a** $7 \times 1000 = 7000$
$\therefore$ 7 cm^3 = 7000 mm^3

b $45\,000 \div 1000 = 45$
$\therefore$ 45 000 mm^3 = 45 cm^3

21 $V = lbh$
$= 4 \times 3 \times 2.5$
$= 30$
$\therefore$ Volume is 30 m^3.

22 20 cm = 0.2 m
$\therefore V = lbh$
$= 8 \times 6 \times 0.2$
$= 9.6$
$\therefore$ Volume is 9.6 m^3.

23 **a** $1250 \div 1000 = 1.25$
i.e. 1.25 L

b $14.83 \times 1000 = 14\,830$
i.e. 14 830 L

c $3700 \div 1000 = 3.7$
i.e. 3.7 kL

24 **a** $V = lbh$
$= 20 \times 10 \times 5$
$= 1000$
$\therefore$ Volume is 1000 cm^3

b Capacity = $1000 \div 1000 = 1$
$\therefore$ Capacity is 1 litre.

Chapter 9—Geometry

pp. 184–191

1 **a** T **b** F **c** T **d** F
e T **f** T **g** T

2 **a** Hexagon **b** 6 **c** 6
d Yes, all sides are equal in length and all angles are equal.

3 **a** Pentagon **b** Quadrilateral
c Octagon **d** Pentagon

4 a and c

5 **a** PS, TW, UV
b PS, RS, TW, VW
c QR, UV, SR, WV

6 **a**

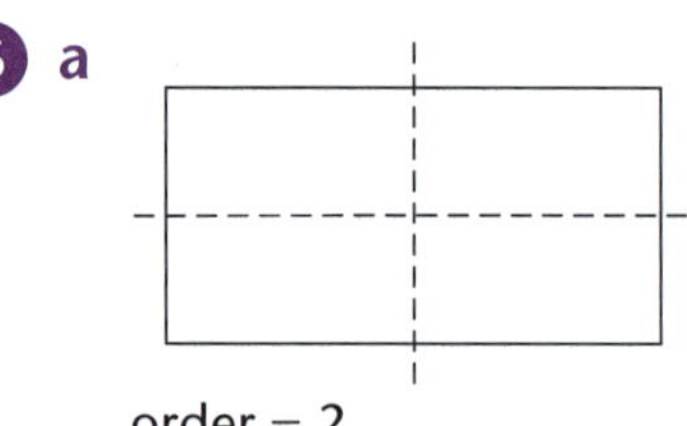

order = 2

b

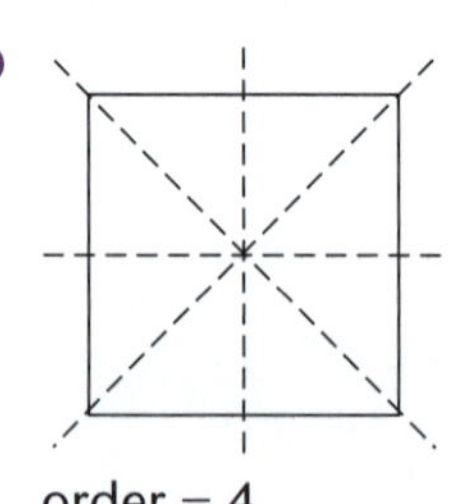

order = 4

c

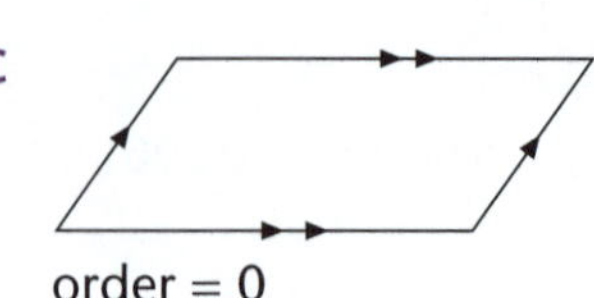

order = 0

d

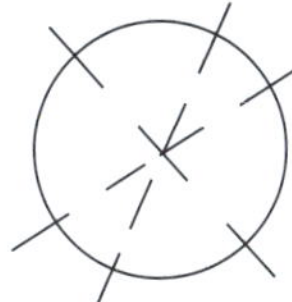

order = infinite (∞)

7 a Triangular prism
b Cylinder
c Rectangular prism
d Sphere
e Rectangular pyramid
f Cone

8 a Shapes (a), (c) and (e) are polyhedra.
b Shapes (a) and (c) are prisms.

9 a Parallelogram
b Square
c Trapezium
d Rectangle
e Kite
f Rhombus

10

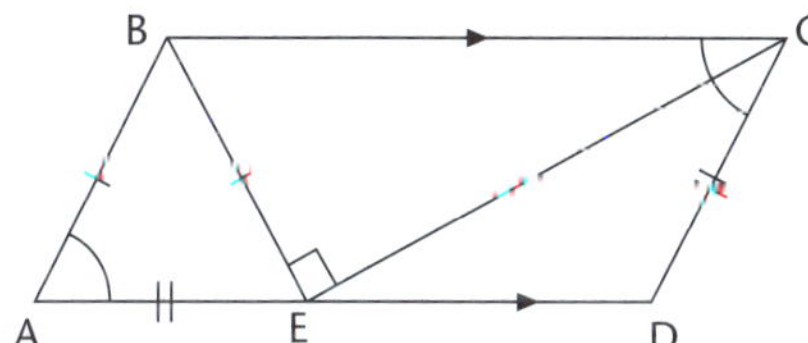

11 a Square prism and triangular prism.
b Yes, because all its surfaces are flat.
c Yes, it is a prism, because it has a constant cross-section.
d It has 10 vertices, 15 edges and 7 faces.

12 a Rectangular pyramid.
b No, it is not a prism, because it does not have a constant cross-section.
c Yes, because all its surfaces are flat.
d There are 8 edges.

13 a i

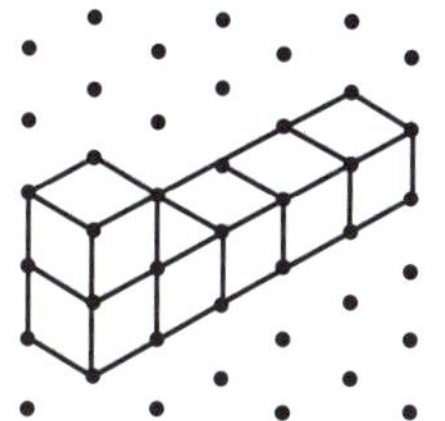

ii

iii

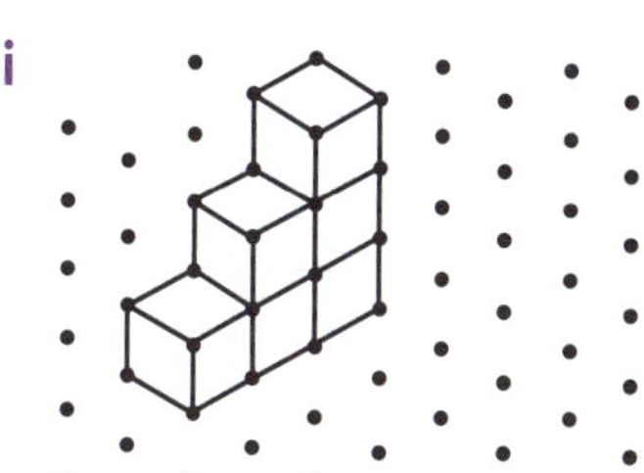

b i

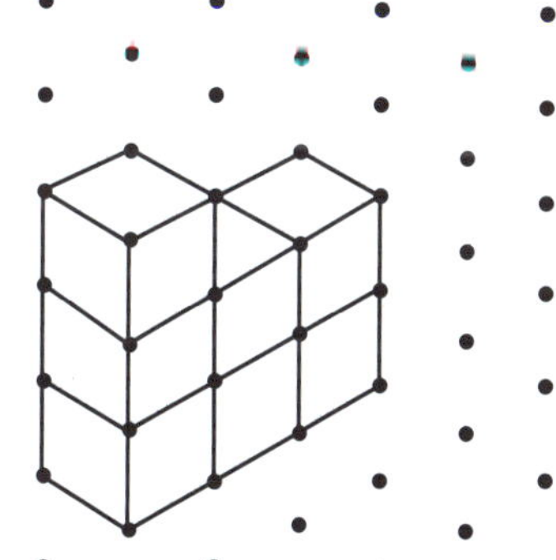

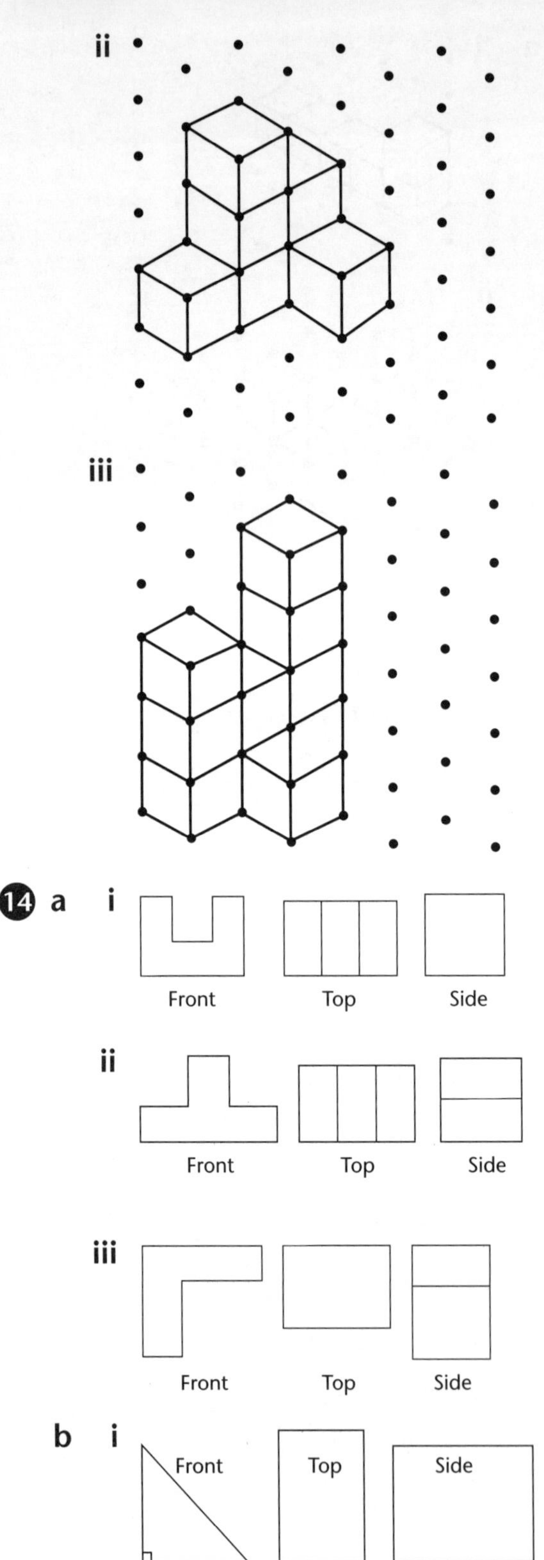

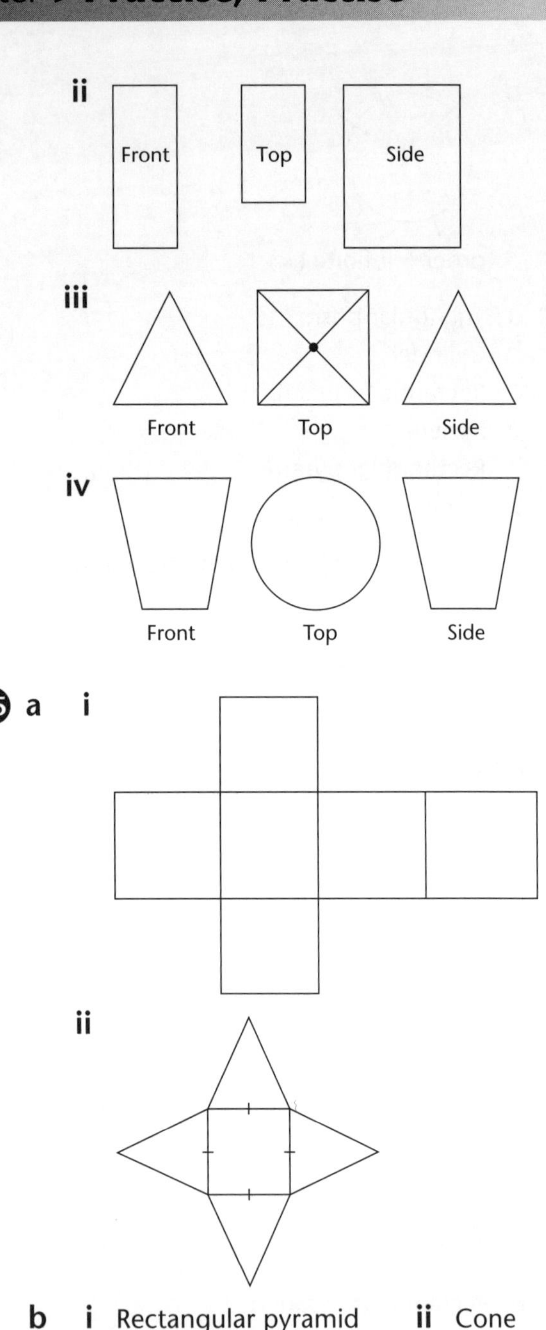

b **i** Rectangular pyramid **ii** Cone
iii Cube **iv** Cylinder
v Triangular prism

16 **a** △ABE is isosceles.
b BCDE is a trapezium.

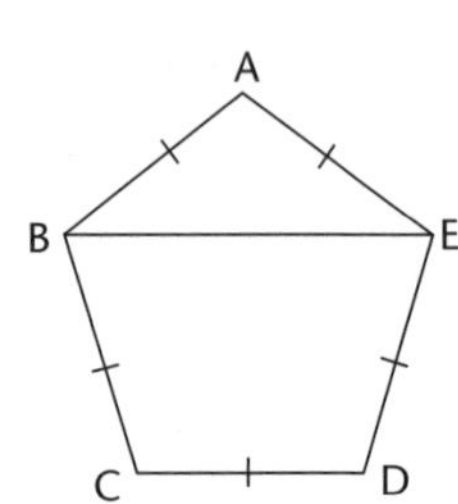

WORKED SOLUTIONS to Chapter 9 **Practise, Practise**

17 **a** This figure is called an ellipse.

b

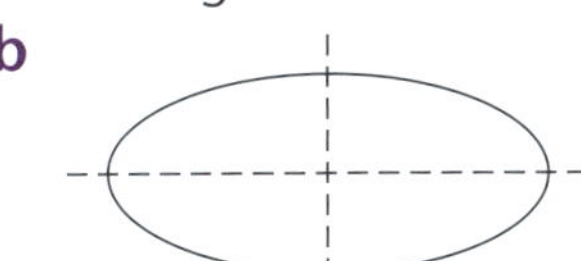

It has 2 axes of symmetry.

c No

18 **a** The vertex is point H.

b $\angle FHG$ or $\angle GHF$

c Obtuse

19 **a** $\angle AXD$ and $\angle BXC$

b $\angle DXB$, $\angle CXA$ and $\angle AXD$, $\angle BXC$

c **i** $\angle AXD + \angle DXB = 180°$

$130° + \angle DXB = 180°$

$\therefore \angle DXB = 50°$

ii $\angle CXA = \angle DXB$

(vertically opposite angles)

$\therefore \angle CXA = 50°$

d $\angle AXB$ or $\angle CXD$ are straight angles.

20

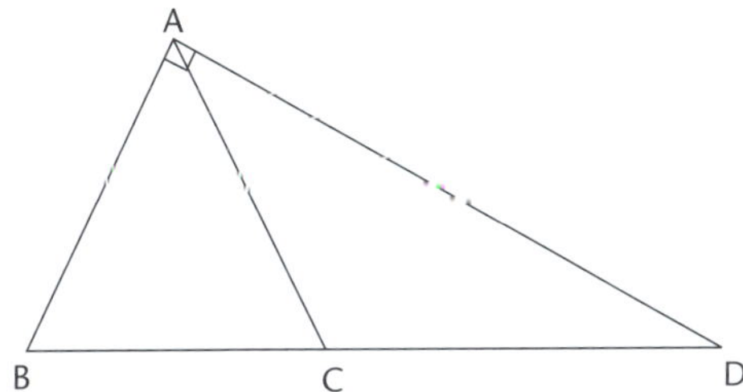

a $\angle DAB$ is a right angle.

b The following are acute angles:
$\angle ABC$, $\angle BCA$, $\angle CAB$, $\angle DAC$, $\angle CDA$.

c The angle adjacent to $\angle BCA$ is $\angle ACD$.

d The supplement of $\angle BCA$ is $\angle ACD$.

e An angle complementary to $\angle DAC$ is $\angle CAB$.

f $\angle ACD$ is obtuse.

21 **a** $90 - 42 = 48°$

b $90 - 69 = 21°$

[Complementary angles add to 90°.]

22 **a** $180 - 42 = 138°$

b $180 - 110 = 70°$

c $180 - 58 = 122°$

[Supplementary angles add to 180°.]

23 **a**

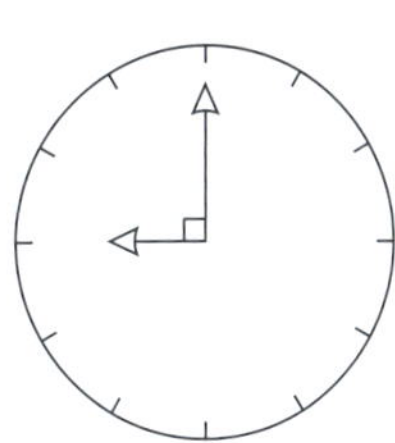

Right angle

b

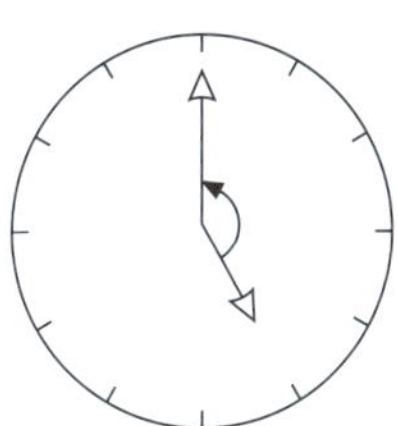

Obtuse angle

c

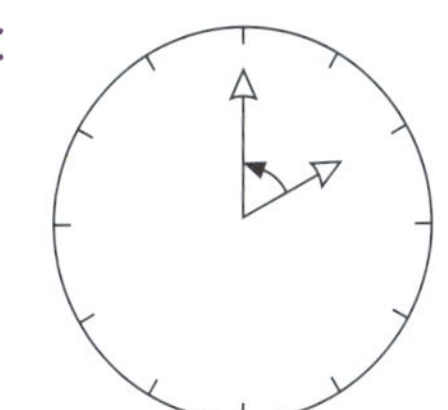

Acute angle

24 **a** $x + 68 = 90$ (complementary angles)

$x = 22$

b $x + 65 = 180$ (angles in a straight line)

$x = 115$

c $x + 90 + 54 = 180$ (angles in a straight line)

$x = 36$

d $x = 55$ (vertically opposite)

$y = 125$ (supplementary to $x = 55$)

$z = 125$ (vertically opposite to $y = 125$)

e $x + 90 + 80 + 110 = 360$ (angles at a point)

$x + 280 = 360$

$x = 80$

f $x + 63 + 48 = 180$ (angle sum of a triangle)

$x + 111 = 180$

$x = 69$

g $x + 28 + 31 = 180$ (angle sum of a triangle)

$x + 59 = 180$

$x = 121$

h

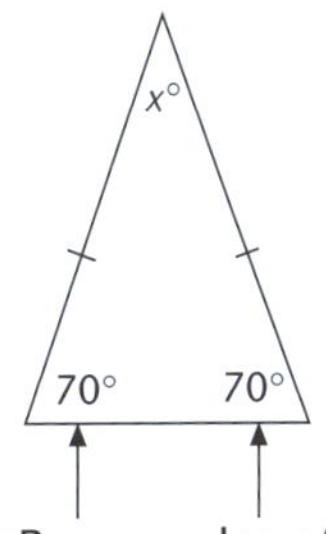

Base angles of isosceles triangle

$x + 70 + 70 = 180$ (angle sum of triangle)

$x + 140 = 180$

$x = 40$

i

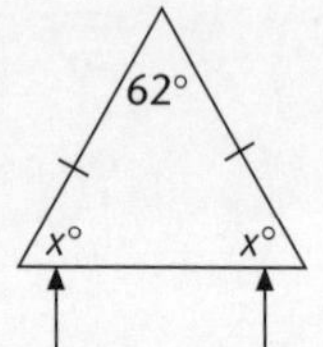

Base angles of isosceles triangle

$x + x + 62 = 180$
$x + x = 118$
$x = 59$

j $x = 112$ (corresponding angles in parallel lines)

k $x + 52 = 180$ (co-interior angles in parallel lines)
$x = 128$

l $x = 71$ (alternate angles in parallel lines)

m $x + 90 + 90 + 65 = 360$ (angle sum of a quadrilateral)
$x + 245 = 360$
$x = 115$

n $x = 60$ (angle in equilateral triangle)

o $x = 51 + 73$ (exterior angle of triangle)
$x = 124$

p $x + 90 + 120 + 85 = 360$ (angle sum of quadrilateral)
$x + 295 = 360$
$x = 65$

q

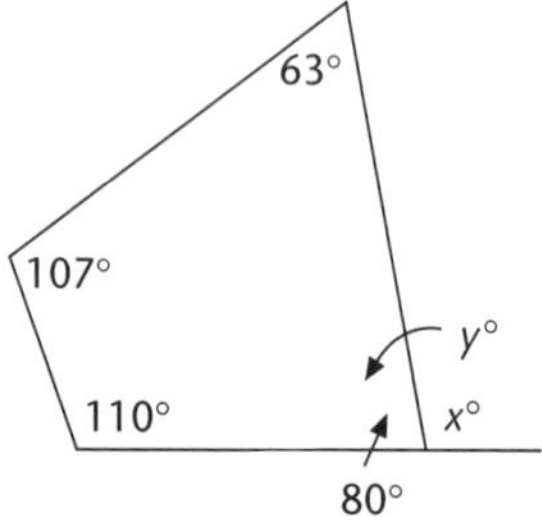

Call unknown angle in quadrilateral $y°$.

$y + 110 + 107 + 63 = 360$ (angle sum of quadrilateral)
$y + 280 = 360$
$y = 80$
$x + 80 = 180$ (angles on a straight line)
$x = 100$

r

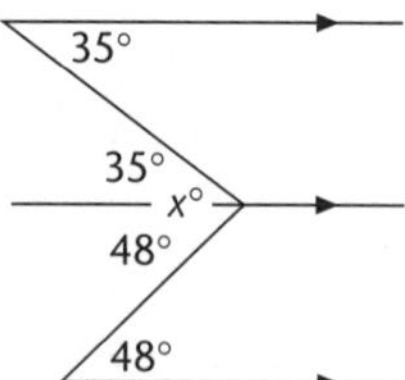

$x = 35 + 48$ (alternate ∠s in parallel lines)
$x = 83$

s

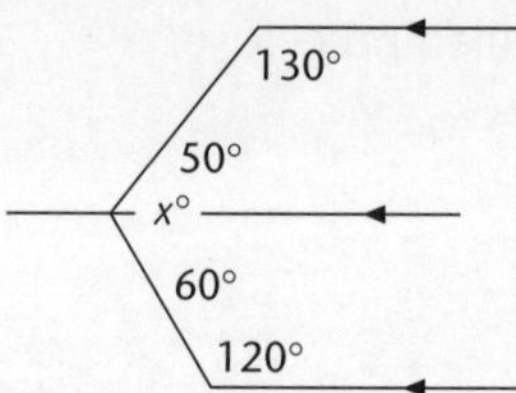

$x = 50 + 60$ (co-interior angles in parallel lines)
$x = 110$

25

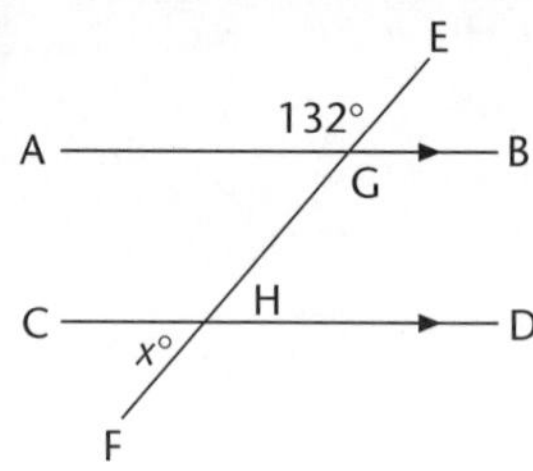

a ∠HGA = 48 (supplementary to ∠AGE)

b $x = 48$ (corresponding to ∠HGA and AB // CD)

26

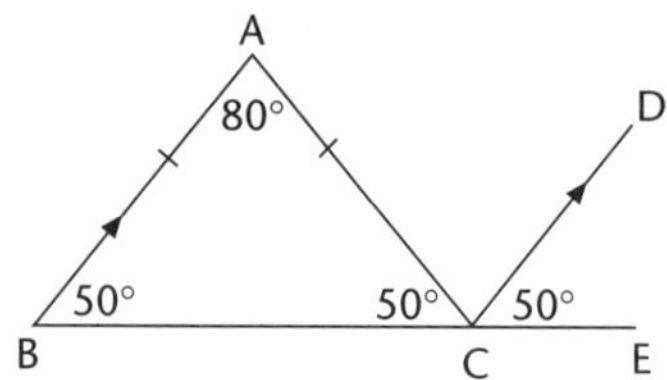

a ∠ABC = 50° (corresponding to ∠DCE and AB // DC)

b ∠BCA = 50° (base angles of isosceles △ABC)

c ∠CAB = 80° (angle sum of △ABC)

27

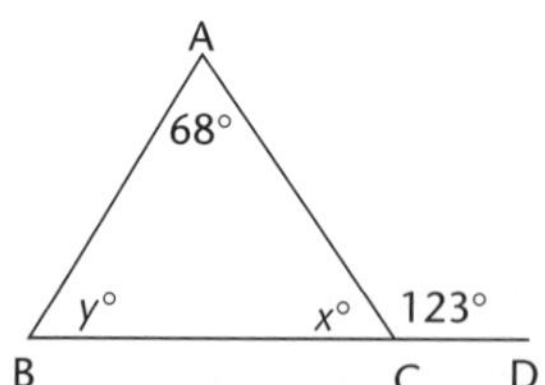

$x + 123 = 180$ (supplementary to ∠ACD)
$x = 57$

$y + 68 + 57 = 180$ (angle sum of △ABC)
$y + 125 = 180$
$y = 55$

WORKED SOLUTIONS to Chapters 9 and 10 **Practise, Practise**

28

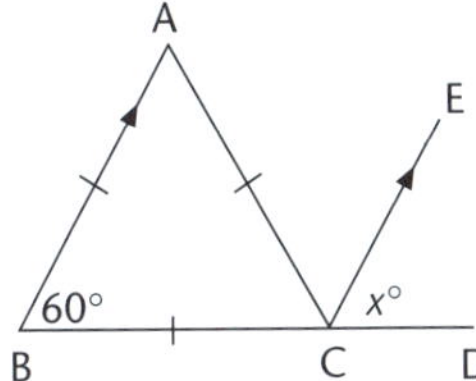

$\angle ABC = 60°$ (angle in an equilateral triangle)
$x = 60$ (corresponding to $\angle ABC$ and AB // EC)

29

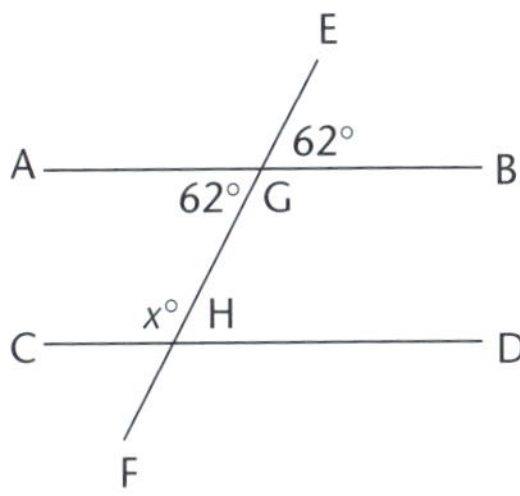

$\angle HGA = 62°$ (vertically opposite to $\angle EGB$)
$x + 62 = 180$
$\therefore x = 118$ (co-interior angles and AB // CD)

30

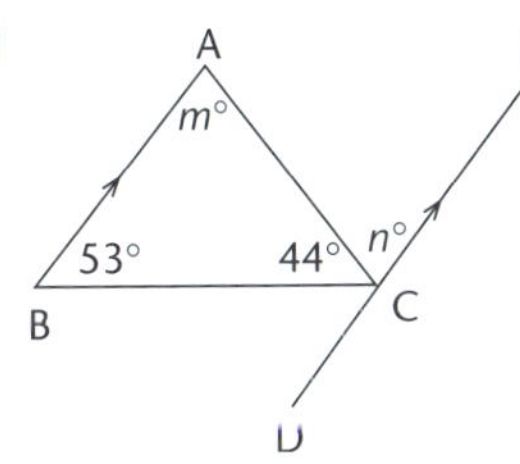

$m + 53 + 44 = 180$ (angle sum of $\triangle ABC$)
$m + 97 = 180$
$\therefore m = 83$
$m = n$ (alternate $\angle$s, AB // DE)
$\therefore n = 83$

31 $x + 60 + 70 = 180$ (angle sum of $\triangle PQR$)
$x + 130 = 180$
$x = 50$
$y = x$ (alternate $\angle$s, RS // PQ)
$\therefore y = 50$

32 $\angle ABC = 80°$ (supplementary with $\angle EBA$)
$x = 60 + 80$ (exterior $\angle$ of $\triangle ABC$)
$x = 140$

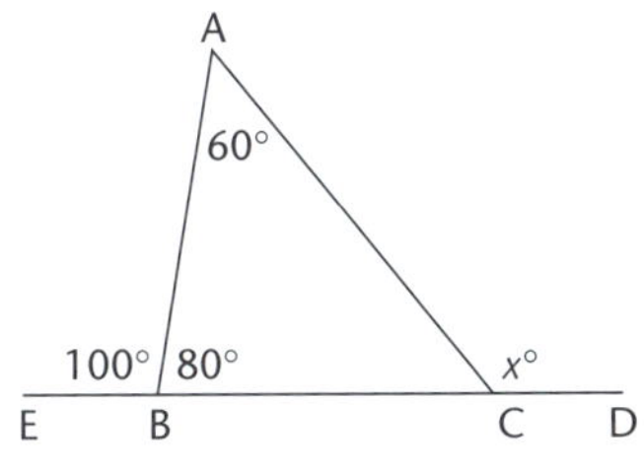

33 **a** $\angle CDB$ **b** $\angle ABD$

c $\angle ABD + \angle DBC = 90°$ (complementary $\angle$s)
$72° + \angle DBC = 90°$
$\angle DBC = 18°$

d $\angle CDB + \angle ADB = 180°$ (supplementary $\angle$s)
$126° + \angle ADB = 180°$
$\angle ADB = 54°$

34 **a** **i** A trapezium is a quadrilateral with one pair of opposite sides parallel.
ii Not necessarily.

b **i** Opposite sides are equal.
ii Opposite angles are equal.
iii Not unless it's a rhombus.

c **i** All sides are equal.
ii 2 **iii** Yes
iv Yes, a special type of parallelogram.

d **i** 4 **ii** Equal
iii 90° **iv** They are equal.

e **i** 2
ii Opposite sides are equal.
iii 90° **iv** Diagonals are equal.

f **i** 1
ii Two pairs of adjacent sides equal.

g A, C, D, E, F **h** B, C, E, F

i **i** F **ii** T **iii** T
iv T **v** F **vi** T
vii T **viii** T

j **i** Parallelogram

Chapter 10—Equations

pp. 209–210

1 **a** $x + 4 = 16$
$x = 16 - 4$
$x = 12$

b $x - 2 = 5$
$x = 5 + 2$
$x = 7$

c $x + \frac{1}{4} = \frac{1}{2}$
$x = \frac{1}{2} - \frac{1}{4}$
$x = \frac{1}{4}$

d $5x = 25$
$x = \frac{25}{5}$
$x = 5$

e $x - 3 = -18$
$x = -18 + 3$
$x = -15$

f $x + 3 = 2$
$x = 2 - 3$
$x = -1$

WORKED SOLUTIONS to Chapter 10 **Practise, Practise**

g $\frac{x}{5} = 1$
$x = 1 \times 5$
$x = 5$

h $\frac{x}{4} = 4$
$x = 4 \times 4$
$x = 16$

i $\frac{x}{3} = 2$
$x = 2 \times 3$
$x = 6$

j $2x = 5$
$x = \frac{5}{2}$
$x = 2\frac{1}{2}$

k $5x = 14$
$x = \frac{14}{5}$
$x = 2\frac{4}{5}$

l $3x = 7$
$x = \frac{7}{3}$
$x = 2\frac{1}{3}$

m $\frac{x}{2} = 3\frac{1}{2}$
$x = 3\frac{1}{2} \times 2$
$x = 7$

2 **a** $2x - 1 = 7$
$2x = 7 + 1$
$2x = 8$
$x = \frac{8}{2}$
$x = 4$

b $5x + 2 = 17$
$5x = 17 - 2$
$5x = 15$
$x = \frac{15}{5}$
$x = 3$

c $3x - 1 = 7$
$3x = 7 + 1$
$3x = 8$
$x = \frac{8}{3}$
$x = 2\frac{2}{3}$

d $2x - 1 = 9$
$2x = 9 + 1$
$2x = 10$
$\frac{2x}{2} = \frac{10}{2}$
$x = 5$

e $5x - 18 = 7$
$5x = 7 + 18$
$5x = 25$
$\frac{5x}{5} = \frac{25}{5}$
$x = 5$

f $4x + 3 = 35$
$4x = 35 - 3$
$4x = 32$
$x = \frac{32}{4}$
$x = 8$

g $3 + 2x = 7$
$2x = 7 - 3$
$2x = 4$
$x = \frac{4}{2}$
$x = 2$

h $12 + x = 16$
$x = 16 - 12$
$x = 4$

i $-5 + 3x = 16$
$3x = 16 + 5$
$3x = 21$
$x = \frac{21}{3}$
$x = 7$

j $-8 + 2x = -4$
$2x = 4 + 8$
$2x = 4$
$x = \frac{4}{2}$
$x = 2$

3 **a** $5x - 2 = 3x + 8$
$5x - 3x = 8 + 2$
$2x = 10$
$x = \frac{10}{2}$
$x = 5$

b $4x = 2x + 12$
$4x - 2x = 12$
$2x = 12$
$x = \frac{12}{2}$
$x = 6$

c $6a - 10 = 4a + 6$
$6a - 4a = 6 + 10$
$2a = 16$
$a = \frac{16}{2}$
$a = 8$

d $5b - 5 = 2b + 10$
$5b - 2b = 10 + 5$
$3b = 15$
$b = \frac{15}{3}$
$b = 5$

e $5x - 10 = 3x + 12$
$5x - 3x = 12 + 10$
$2x = 22$
$x = \frac{22}{2}$
$x = 11$

f $7x + 5 = 3x + 11$
$7x - 3x = 11 - 5$
$4x = 6$
$x = \frac{6}{4}$
$x = 1.5$

g $12x = 10x + 8$
$12x - 10x = 8$
$2x = 8$
$x = \frac{8}{2}$
$x = 4$

h $2x - 1 = 5 - x$
$2x + x = 5 + 1$
$3x = 6$
$x = \frac{6}{3}$
$x = 2$

i $4 - 3x = 5 - 4x$
$-3x + 4x = 5 - 4$
$x = 1$

j $2y = 4 - 2y$
$2y + 2y = 4$
$4y = 4$
$y = \frac{4}{4}$
$y = 1$

4 **a** $\frac{x - 2}{3} = 5$
$x - 2 = 5 \times 3$
$x - 2 = 15$
$x = 15 + 2$
$x = 17$

b $\frac{x + 2}{4} = 2$
$x + 2 = 2 \times 4$
$x + 2 = 8$
$x = 8 - 2$
$x = 6$

WORKED SOLUTIONS to Chapter 10 **Practise, Practise**

c $\dfrac{2x-4}{3} = 6$

$2x - 4 = 6 \times 3$

$2x - 4 = 18$

$2x = 18 + 4$

$2x = 22$

$x = \frac{22}{2}$

$x = 11$

d $5x = 3x + 7$

$5x - 3x = 7$

$2x = 7$

$x = \frac{7}{2}$

$x = 3\frac{1}{2}$

e $\dfrac{x}{4} = 2\frac{1}{2}$

$x = 2\frac{1}{2} \times 4$

$x = 10$

f $3x - 14 = 1$

$3x = 1 + 14$

$3x = 15$

$x = \frac{15}{3}$

$x = 5$

g $4 + a = 17$

$a = 17 - 4$

$a = 13$

h $14d - 7 = 4d + 3$

$14d - 4d = 3 + 7$

$10d = 10$

$d = \frac{10}{10}$

$d = 1$

i $4x - 12 = 2x$

$4x - 2x = 12$

$2x = 12$

$x = \frac{12}{2}$

$x = 6$

j $5x - 8 = 3x$

$5x - 3x = 8$

$\dfrac{2x}{2} = \dfrac{8}{2}$

$x = 4$

5 **a** $2x + x = 180$ [Angles on a straight line.]

$3x = 180$ [Collect like terms.]

$x = \frac{180}{3}$

$x = 60$

b $x + 63 = 180$ [Co-interior angles are supplementary when lines are parallel.]

$x = 180 - 63$

$x = 117$

c $2x + 50 + 30 = 180$ [Angle sum of a triangle.]

$2x + 80 = 180$

$2x = 180 - 80$

$2x = 100$

$x = \frac{100}{2}$

$x = 50$

d $2x - 20 = 70$ [Corresponding angles are equal when lines are parallel.]

$2x = 70 + 20$

$2x = 90$

$x = \frac{90}{2}$

$x = 45$

e $a + a + a = 180$ [Angle sum of a triangle.]

$3a = 180$ [Collect like terms.]

$a = \frac{180}{3}$

$a = 60$

f $7x = 5x + 20$ [Vertically opposite angles are equal.]

$7x - 5x = 20$

$2x = 20$

$x = \frac{20}{2}$

$x = 10$

g $2x = 60 + 70$ [Exterior angle of a triangle is equal to the sum of the opposite interior angles.]

$2x = 130$

$x = \frac{130}{2}$

$x = 65$

h $x + 90 + 90 + 70 = 360$ [Angle sum of quadrilateral.]

$x + 250 = 360$

$x = 360 - 250$

$x = 110$

i $5x - 10 = 75$ [Alternate angles are equal when lines are parallel.]

$5x = 75 + 10$

$5x = 85$

$x = \frac{85}{5}$

$x = 17$

j $3x + 2x + x + 90 = 360$ [Angles at a point; collect like terms.]

$6x + 90 = 360$

$6x = 360 - 90$

$6x = 270$

$x = \frac{270}{6}$

$x = 45$

6 **a** Let the number be x:

$x + 5 = 25$

$x = 25 - 5$

$x = 20$

Therefore, the number is 20.

b Let the number be x:

$x - 10 = 12$

$x = 12 + 10$

$x = 22$

Therefore, the number is 22.

c Let the number be x:

$3x = 19$

$x = \frac{19}{3}$

$x = 6\frac{1}{3}$

Therefore, the number is $6\frac{1}{3}$.

d Let the number be x:

$$x + 8 = 2$$
$$x = 2 - 8$$
$$x = -6$$

Therefore, the number is –6.

e Let the number be x:

$$3x + 8 = 44$$
$$3x = 44 - 8$$
$$3x = 36$$
$$x = \frac{36}{3}$$
$$x = 12$$

Therefore, the number is 12.

f Let the numbers be x and $x + 1$:

$$x + x + 1 = 43$$
$$2x + 1 = 43$$
$$2x = 43 - 1$$
$$2x = 42$$
$$x = \frac{42}{2}$$
$$x = 21$$

Therefore, the numbers are 21 and 22.

g Let the numbers be x, $x + 1$ and $x + 2$:

$$x + x + 1 + x + 2 = 39$$
$$3x + 3 = 39$$
$$3x = 39 - 3$$
$$3x = 36$$
$$x = \frac{36}{3}$$
$$x = 12$$

Therefore, the numbers are 12, 13 and 14.

h Let the number be x:

$$2x - 15 = 33$$
$$2x = 33 + 15$$
$$2x = 48$$
$$x = \frac{48}{2}$$
$$x = 24$$

Therefore, the number is 24.

i Let the number be x:

$$\frac{x}{5} = 8$$
$$x = 8 \times 5$$
$$x = 40$$

Therefore, the number is 40.

j Let the number be x:

$$2x - 3 = x + 4$$
$$2x - x = 4 + 3$$
$$x = 7$$

Therefore, the number is 7.

Chapter 11—Statistics and Probability

pp. 226–232

1

Number of ts	Tally	Frequency
2	||||	4
3	||||	4
4	~~||||~~	5
5	~~||||~~	5
6	|	1
7	|	1
	Σ	**20**

a **i** 5 **ii** 18 **iii** 2

b **i** $\frac{5}{20} = \frac{1}{4}$ **ii** $\frac{18}{20} = \frac{9}{10}$ **iii** $\frac{2}{20} = \frac{1}{10}$

c Range = 7 – 2 = 5

Mode = 4 and 5 (both have frequency of 5)

d **Number of ts**

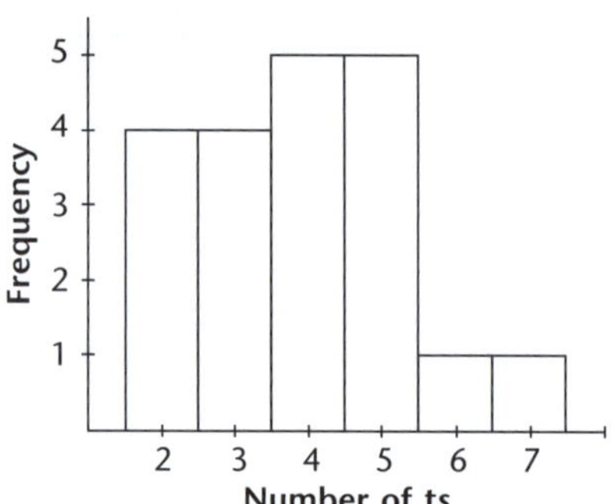

2

Class	Class centre	Tally	Frequency
141–145	143	|	1
146–150	148	~~||||~~ |||	8
151–155	153	|||	3
156–160	158	~~||||~~ |	6
161–165	163	~~||||~~ ||	7
166–170	168	|||	3
171–175	173	||	2

WORKED SOLUTIONS to Chapter 11 **Practise, Practise**

3

Stem	Leaf
5	899
6	33578999
7	111234566689
8	0111238

4

Number of lumps (x)	Tally	Frequency (f)
0	|	1
1	|	1
2	|||	3
3	|||	3
4	~~||||~~	5
5	||||	4
6	~~||||~~	5
7	|||	3
	Σ	**25**

a **i** 4 **ii** 17

iii 13 (4 or less) **iv** 8

b **i** $\frac{4}{25}$ **ii** $\frac{17}{25}$

iii $\frac{13}{25}$ **iv** $\frac{8}{25}$

c **Lumps in porridge**

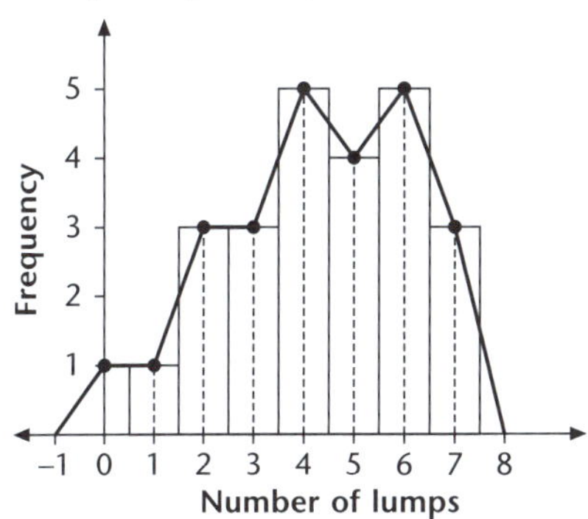

5 **Goals scored**

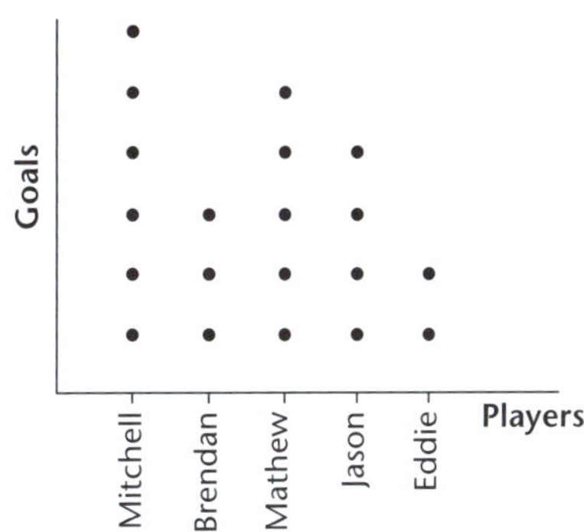

6 **Quiz results**

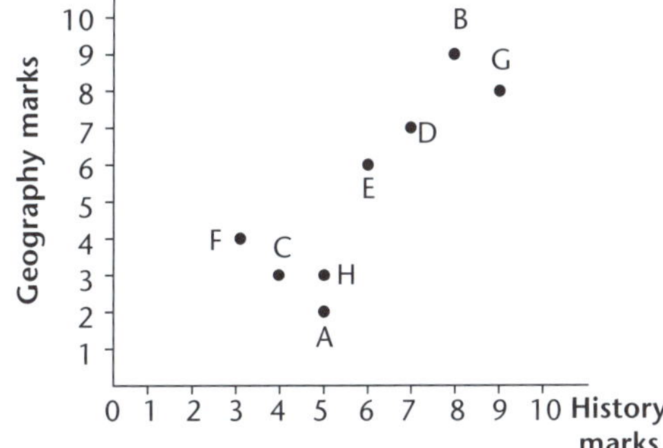

7 Range = 8 − (−1) = 8 + 1

∴ The range is 9.

8 **a** Mean = $\frac{39}{7}$ = 5.6 (1 decimal place)

Mode = 6

3 scores 3 scores

For median: 3, 4, 6, (6), 6, 7, 7

Median = 6

Range = 7 − 3 = 4

b Mean = $\frac{61}{8}$ = 7.6 (to 1 decimal place)

Mode = 5

3 scores 3 scores

For median: 5, 5, 6, (7, 8), 9, 10, 11

Median is between 7 and 8

i.e. median = $\frac{7+8}{2}$ = 7.5

Range = 6

c Mean = $\frac{36}{9}$ = 4

Mode = 1 and 4 and 7

For median: 1, 1, 1, 4, (4), 4, 7, 7, 7

Median = 4

Range = 6

d Mean = $\frac{47}{12}$ = 3.9 (to 1 decimal place)

Mode = 7

5 scores 5 scores

For median: 0, 0, 1, 2, 2, (3, 4), 5, 7, 7, 7, 9

Median = $\frac{3+4}{2}$

= 3.5 (median between 3 and 4)

Range = 9

e Mean = $\frac{105}{15}$ = 7

Mode = 7

For median: 3, 4, 4, 4, 5, 6, 7, (7), 7, 7, 9, 9, 11, 11, 11

Median = 7

Range = 8

9

Shoe size (x)	Number of children (f)	$x \times f$
3	8	24
$3\frac{1}{2}$	8	28
4	6	24
$4\frac{1}{2}$	12	54
5	10	50
$5\frac{1}{2}$	8	44
6	6	36
$6\frac{1}{2}$	4	26
7	2	14
Σ	**64**	**300**

a 64 children were surveyed.

b **i** 44 **ii** 20

c **i** $\frac{10}{64} = \frac{5}{32}$ **ii** $\frac{44}{64} = \frac{11}{16}$ **iii** $\frac{20}{64} = \frac{5}{16}$

d Range = 7 – 3 = 4 Mode = $4\frac{1}{2}$

e Mean = $\frac{300}{64}$ = 4.7 (to 1 decimal place)

f Median = $4\frac{1}{2}$ (look for 32nd and 33rd scores)

g **Children buying joggers**

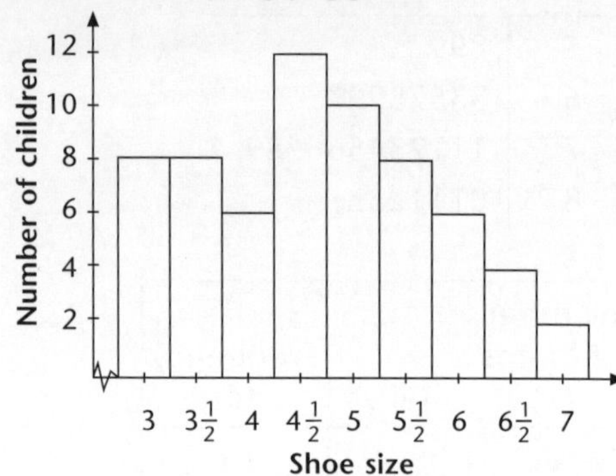

h $4\frac{1}{2}$—as this is the most common size worn. Mode.

10 **a**

Score (x)	Frequency (f)	$x \times f$
1	3	3
2	6	12
3	8	24
4	10	40
5	12	60
6	0	0
7	4	28
8	2	16
9	6	54
10	1	10
	52	**247**

b Range = 10 – 1 = 9 Mode = 5

c **i** 12 **ii** 39

iii 27 **iv** 52 – 39 = 13

d 52

e **i** $\frac{12}{52}$ or $\frac{3}{13}$ **ii** $\frac{39}{52}$ or $\frac{3}{4}$

iii $\frac{27}{52}$ **iv** $\frac{13}{52}$ or $\frac{1}{4}$

f Mean = $\frac{247}{52}$ = 4.75

g Median = 4 (26th and 27th scores)

11 **a** **i** Mean = $\frac{8+6+5+9+5+3}{6}$

= $\frac{36}{6}$

= 6

Mean number of goals = 6.

ii Range = 9 − 3 = 6 Mode = 5

iii For median: 3, 5, (5, 6), 8, 9

Median = $\frac{5+6}{2}$ = 5.5

b Mean $= \frac{8+6+5+9+5+3+13}{6}$

$= \frac{49}{7}$

= 7

For median: 3, 5, 5, (6), 8, 9, 13

Median is 6.

The mean increases by 1, while the median is 0.5 larger.

c Gai has scored (8 × 6.5) goals = 52 goals. After 7 seasons Gai had scored 49 goals. ∴ Gai scored 3 goals in her 8th season.

12 Zoran's total marks = 4 × 65 (after 4 results) = 260

To average 70 Zoran must score (6 × 70) marks, i.e. 420 marks.

Zoran must average $\frac{160}{2}$ in his other two papers, i.e. average 80.

Zoran must average 80 in each of his final papers.

13 **a** i

Number of children (score) x	Number of families (frequency) f	$x \times f$
1	2	2
2	5	10
3	4	12
4	4	16
5	3	15
6	1	6
7	1	7
Σ	**20**	**68**

ii 20 iii 68

b Range = 7 − 1 = 6 Mode = 2

c i 4 ii 11 iii 5

d i $\frac{4}{20}$ or $\frac{1}{5}$ ii $\frac{11}{20}$ iii $\frac{5}{20}$ or $\frac{1}{4}$

e Mean = $\frac{68}{20}$ = 3.4

f Median = 3 (10th and 11th scores)

14 **a** 66.76 (to 2 decimal places)

b 63.55 (to 2 decimal places)

c Mode: 63 Median: 69 Range: 27

d Median: 58

e The range of boys (27) is more than the range of the girls (22).

f The median will drop to 56.

15 Mean = $\frac{8+3+7+4+3}{5} = \frac{25}{5} = 5$

a New numbers 14, 9, 13, 10, 9

Mean $= \frac{14+9+13+10+9}{5}$

$= \frac{55}{5}$

= 11

Mean has increased by 6.

b New numbers 48, 18, 42, 24, 18

Mean $= \frac{48+18+42+24+18}{5}$

$= \frac{150}{5}$

= 30

Mean has also been multiplied by 6.

16

	Number (x)	Frequency (f)	fx	Fraction	Decimal
a	1	19	19	$\frac{19}{100}$	019
b	2	21	42	$\frac{21}{100}$	0.21
	3	20	60	$\frac{20}{100}$	0.20
	4	19	76	$\frac{19}{100}$	0.19
	5	21	105	$\frac{21}{100}$	0.21
		100	**302**		

The decimals are almost all the same value.

It shows there is an equal chance of the wheel stopping on all numbers.

c Mean = $\frac{302}{100}$ = 3.02

d Mode = 2 and 5

Median is between the 50th and 51st scores, i.e. median = 3.

WORKED SOLUTIONS to Chapter 11 **Practise, Practise**

17

$$
\begin{aligned}
\text{Average (5 innings)} &= 15.8\\
\therefore \text{Total (5 innings)} &= 15.8 \times 5\\
&= 79\\
\text{Average (next 10 innings)} &= 19.1\\
\therefore \text{Total} &= 19.1\\
&= 191\\
\text{Total over 15 innings} &= 191 + 79\\
&= 270\\
\text{Average} &= \frac{270}{15}\\
&= 18\\
\text{New total (after score of 34)} &= 270 + 34\\
&= 304\\
\text{New average (over 16 innings)} &= \frac{304}{16}\\
&= 19\\
\text{Season average} &= 19
\end{aligned}
$$

18 Total number of cards = 52 (13 of each suit—hearts, diamonds, clubs, spades).

Consider hearts and we have:

2♥, 3♥, 4♥, 5♥, 6♥, 7♥, 8♥, 9♥, 10♥, J♥, Q♥, K♥, A♥

There are four of each number.

Consider the Aces: A♥, A♦, A♠, A♣

a $P(\text{Ace}) = \frac{4}{52}$ or $\frac{1}{13}$

b $P(\text{King}) = \frac{4}{52}$ or $\frac{1}{13}$

c $P(♥) = \frac{13}{52}$ or $\frac{1}{4}$

d $P(♠) = \frac{13}{52}$ or $\frac{1}{4}$

e $P(\text{Black}) = \frac{26}{52}$ or $\frac{1}{2}$

f $P(\text{J, Q, K}) = \frac{12}{52}$ or $\frac{3}{13}$

g $P(\text{Not picture card})$ (total probability must add up to 1)

$= 1 - \frac{3}{13}$

$= \frac{10}{13}$

h $P(2, 3, 4, 5, 6) = \frac{20}{52}$ or $\frac{5}{13}$

i $P(6♥) = \frac{1}{52}$

j $P(\text{Red or 6}) = \frac{28}{52}$ or $\frac{7}{13}$

[26 red cards + 2 other 6s: 6 clubs, 6 spades.
6 diamonds and 6 hearts have already been counted.]

k $P(\text{Red 6}) = \frac{2}{52}$ or $\frac{1}{26}$

19 $P(\text{Not rain}) = 1 - \frac{7}{10}$ or $\frac{3}{10}$

20 No. of jelly babies = 5 + 6 + 4 = 15

a $P(\text{Red}) = \frac{5}{15}$ or $\frac{1}{3}$

b $P(\text{Not red}) = 1 - \frac{1}{3} = \frac{2}{3}$

c $P(\text{Black}) = \frac{6}{15}$ or $\frac{2}{5}$

d $P(\text{Red or white}) = \text{Pr(Not black)} = 1 - \frac{2}{5} = \frac{3}{5}$

e $P(\text{Green}) = 0$ (impossible as there are no green jelly babies)

f $P(\text{White}) = \frac{4}{15}$

g $P(\text{Not white}) = 1 - \frac{4}{15} = \frac{11}{15}$

21 **a** $P(\text{Female}) = 1 - 0.48 = 0.52$

b Number of male babies = 0.48 × 1400
= 672

Thus 672 male babies would be expected.

22

+	1	2	3	4	5	6
1	2	3	4	5	6	7
2	3	4	5	6	7	8
3	4	5	6	7	8	9
4	5	6	7	8	9	10
5	6	7	8	9	10	11
6	7	8	9	10	11	12

(36 possible outcomes)

a $P(8) = \frac{5}{36}$ **b** $P(5) = \frac{4}{36}$ or $\frac{1}{9}$

c $P(\text{Odd}) = \frac{18}{36}$ or $\frac{1}{2}$

d $P(>9) = \text{Pr}(10, 11, 12) = \frac{6}{36}$ or $\frac{1}{6}$

e $P(9 \text{ or greater}) = P(9, 10, 11, 12)$
$= \frac{10}{36}$ or $\frac{5}{18}$

f $P(\text{Not } 9) = 1 - P(9)$
$= 1 - \frac{4}{36}$
$= \frac{32}{36}$ or $\frac{8}{9}$

g $P(\text{Odd or less than 5}) = P(2, 3, 4, 5, 7, 9, 11)$
$= \frac{22}{36}$ or $\frac{11}{18}$

h $P(\text{Odd and less than 5}) = P(3)$
$= \frac{2}{36}$ or $\frac{1}{18}$

23

+	**1**	**2**	**3**	**4**
1	2	3	4	5
2	3	4	5	6
3	4	5	6	7
4	5	6	7	8

(16 outcomes)

a $P(5) = \frac{4}{16}$ or $\frac{1}{4}$

b $P(7) = \frac{2}{16}$ or $\frac{1}{8}$

c $P(1) = 0$

d $P(>5) = \frac{6}{16}$ or $\frac{3}{8}$

e $P(\text{At least } 5) = P(5, 6, 7, 8) = \frac{10}{16}$ or $\frac{5}{8}$

24 **a** $P(9) = \frac{1}{10}$ (only one numbered nine)

b $P(\text{Even}) = P(2, 4, 6, 8, 10) = \frac{5}{10}$ or $\frac{1}{2}$

c $P(<6) = P(1, 2, 3, 4, 5) = \frac{5}{10}$ or $\frac{1}{2}$

d $P(6 \text{ or more}) = 1 - P(<6)$

$= 1 - \frac{1}{2}$

$= \frac{1}{2}$

WORKED SOLUTIONS to Chapter 1 **Tests**

Chapter 1—Number

pp. 29–33

Level 1 Test

1. $16 + 4 + 17 = 37$ ✓ (1 mark)
2. $4 \times 25 \times 17 = 1700$ ✓ (1 mark)
3. $8 + 9 = 17$ ✓ (1 mark)
4. $8 \times 9 = 72$ ✓ (1 mark)
5. $12 + 8 = 20$ ✓ (1 mark)
6. $12 - 8 = 4$ ✓ (1 mark)
7. $12 - 8 = 4$ ✓ (1 mark)
8. $30 - 12 + 8 = 18 + 8 = 26$ ✓ (1 mark)
9. $16 + 28 = 44$ ✓ (1 mark)
10. $20 \times 7 = 140$ ✓ (1 mark)
11. 7000 ✓ (1 mark)
12. 3207, 3702, 7203, 7302 ✓ (1 mark)
13. **a** $(6 + 4) \times (7 - 5) = 20$ ✓ (1 mark)
    b $6 + [4 \times (7 - 5)] = 14$ ✓ (1 mark)
    c $(6 + 4) \times 7 - 5 = 65$ ✓ (1 mark)
14. $17 + 11 + 13 = 41$ ✓ (1 mark)
15. **a**
$$\begin{array}{r} 64\ \times \\ \underline{35} \\ 320 \\ \underline{1920} \\ \underline{2240} \end{array}$$
    Answer = 2240 ✓ (1 mark)

    b
$$\begin{array}{r} 14 \\ 27\overline{)389} \\ \underline{27} \\ 119 \\ \underline{108} \\ 11 \end{array}$$
    Answer = $14\frac{11}{27}$ ✓ (1 mark)

16. Number of nails
    $= 14 \times 75 + 5$ ✓
    $= 1050 + 5$
    $= 1055$ ✓
$$\begin{array}{r} 75\ \times \\ \underline{14} \\ 300 \\ \underline{750} \\ \underline{1050} \end{array}$$
    (2 marks)
17. **a** 14, 16, 18, 20, 22 ✓ (1 mark)
    b 4, 8, 12, 16, 20, 24, 28, 32, 36 ✓ (1 mark)
    c 1, 2, 3, 4, 6, 12 ✓ (1 mark)
    d 16, 25, 36, 49 ✓ (1 mark)
18. **a** Adding 6
    $\therefore 19 + 6 = 25$ ✓ (1 mark)
    b Multiplying by 4
    $\therefore 48 \times 4 = 192$ ✓ (1 mark)
19. Factors of 8: 1, 2, (4), 8
    Factors of 12: 1, 2, 3, (4), 6, 12
    ∴ HCF is 4 ✓ (1 mark)
20. Multiples of 8: 8, 16, (24), 32, ...
    Multiples of 12: 12, (24), 36, ...
    ∴ LCM is 24 ✓ (1 mark)
21. 12: 1, (2), (3), 4, 6, 12
    ∴ Prime factors are 2 and 3 ✓ (1 mark)
22. **a** $6 \times 6 \times 6 \times 6 \times 6 \times 6 \times 6 = 6^7$ ✓ (1 mark)
    b $(3 \times 2)^2 = 3^2 \times 2^2$
    $6^2 = 9 \times 4$
    $36 = 36$
    ∴ True ✓ (1 mark)
23. $\sqrt{81} = 9$ ✓ (1 mark)
24. **a** Factor tree: 48 → 12, 4; 12 → (3), 4; 4 → (2), (2); 4 → (2), (2) ✓
    $\therefore 48 = 3 \times 2 \times 2 \times 2 \times 2$
    $= 3 \times 2^4$ ✓ (2 marks)

    b Factor tree: 54 → 9, 6; 9 → (3), (3); 6 → (3), (2) ✓
    $\therefore 54 = 3 \times 3 \times 3 \times 2$
    $= 3^3 \times 2$ ✓ (2 marks)

WORKED SOLUTIONS to Chapter 1 **Tests**

c HCF $= 3 \times 2$
$= 6$ ✓
LCM $= 3^3 \times 2^4$
$= 27 \times 16$
$= 432$ ✓ (2 marks)

25 a 72: $7 + 2 = 9$ which is divisible by 3
∴ 72 is divisible by 3 ✓ (1 mark)
b 89: $8 + 9 = 7$ which is not divisible by 3
∴ 89 not divisible by 3 ✓ (1 mark)

(Total 40 marks)

Level 2 Test

1 a $6 + 4 + 11 + 9 = 10 + 20 = 30$ ✓ (1 mark)
b $8 \times 125 \times 19 = 1000 \times 19$
$= 19\,000$ ✓ (1 mark)

2 $42 \div 7 = 6$ ✓ (1 mark)

3 $17 - 11 = 6$ ✓ (1 mark)

4 $(9 + 15 + 8 + 6 + 12) \div 5 = 50 \div 5$
$= 10$ ✓ (1 mark)

5 Sum $= 19 + 13 = 32$ ✓
Answer $= 32 - 24 = 8$ ✓ (2 marks)

6 40 000 ✓ (1 mark)

7 62 169, 49 612, 46 129, 42 169, 21 496 ✓ (1 mark)

8 a $8 \times 6 = 48$ ✓ (1 mark)
b $5 + 60 = 65$ ✓ (1 mark)
c $56 + 84 = 140$ ✓ (1 mark)

9 a
```
  76 ×
  47
 532
3040
3572
```
Answer $= 3572$ ✓ (1 mark)

b
```
     24
36)894
    72
    174
    144
     30
```
Answer $= 24\frac{30}{36}$ or $24\frac{5}{6}$ ✓ (1 mark)

c 4700 ✓ (1 mark)

d 753 ✓ (1 mark)
e $43 \times 3 \times 10 = 1290$ ✓ (1 mark)
f $(27 \times 10) - (27 \times 1) = 270 - 27$
$= 243$ ✓ (1 mark)
g $(27 \times 10) + (27 \times 1) = 270 + 27$
$= 297$ ✓ (1 mark)

10 Cost of one $= \$14.40 \div 9 = \$1.60$ ✓
Cost of five $= \$1.60 \times 5 = \$8.00$ ✓ (2 marks)

11 a 11, 13, 15, 17, 19, 21, 23 ✓
(Implies end numbers included.) (1 mark)
b 15, 20, 25, 30, 35, 40, 45, 50, 55 ✓
(Ends not included.) (1 mark)
c 1, 2, 3, 4, 6, 9, 12, 18, 36 ✓ (1 mark)
d 1, 3, 6, 10, 15, 21, 28 ✓ (1 mark)

12 HCF $= 6$ ✓ (1 mark)

13 LCM $= 36$
(Multiples of $18 = 18, 36, 54 \ldots$) ✓ (1 mark)

14 a

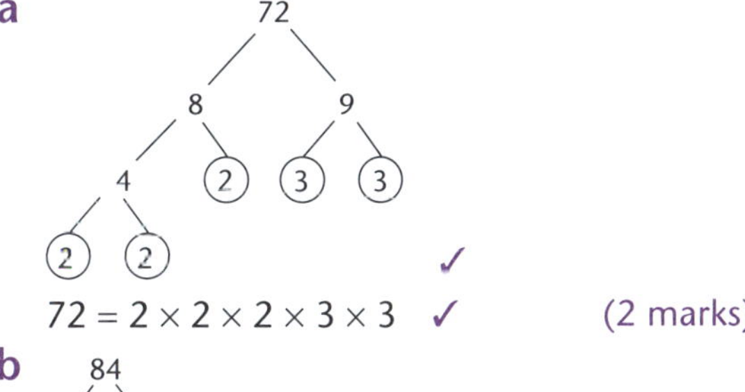

$72 = 2 \times 2 \times 2 \times 3 \times 3$ ✓ (2 marks)

b

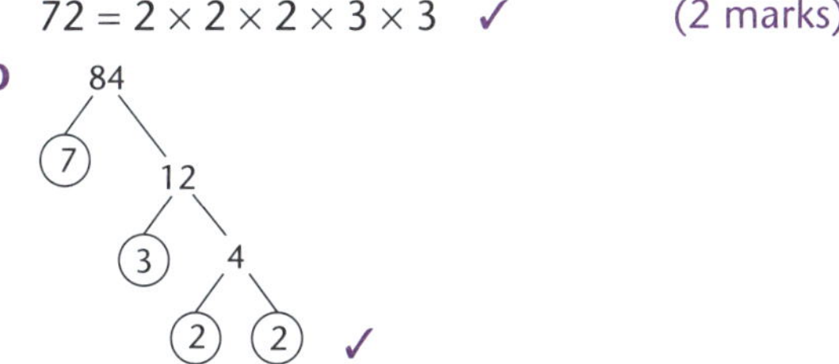

$84 = 2 \times 2 \times 3 \times 7$ ✓ (2 marks)

c HCF $= 2 \times 2 \times 3 = 12$
(All factors common to both.) ✓

LCM $= \underbrace{2 \times 2 \times 2 \times 3 \times 3}_{72} \times 7$ (84 $= 2 \times 2 \times 3 \times 7$)
$= 504$

(Includes **all** factors of both.) ✓ (2 marks)

15 $3^8 = 3 \times 3 \times 3 \times 3 \times 3 \times 3 \times 3 \times 3$
(8 factors, each 3) ✓ (1 mark)

16 $3^5 \times 4^3$ ✓ (1 mark)

WORKED SOLUTIONS to Chapter 1 **Tests**

17 $3^4 = 3 \times 3 \times 3 \times 3 = 81$ ✓ (1 mark)

18 **a** $\sqrt{49} = 7$ (as $7^2 = 49$) ✓ (1 mark)
b $\sqrt[3]{64} = 4$ (as $4^3 = 64$) ✓ (1 mark)

19 $5^2 - 4^2 = 25 - 16 = 9$ ✓
$\sqrt{5^2 - 4^2} = \sqrt{9} = 3$
$\therefore$ No, $5^2 - 4^2 \neq \sqrt{5^2 - 4^2}$ ✓ (2 marks)

(Total 40 marks)

Level 3 Test

1 $\sqrt{25 \times 4} = \sqrt{100} = 10$
$\sqrt{25} \times \sqrt{4} = 5 \times 2 = 10$
$\therefore$ True, $\sqrt{25 \times 4} = \sqrt{25} \times \sqrt{4}$ ✓ (1 mark)

2 $104 - 77 = 27$ ✓ (1 mark)

3 $17 + 177 + 117 + 717 = 1028$ ✓ (1 mark)

4 **a** $130 - 122 + 8 = 8 + 8 = 16$ ✓ (1 mark)
b $17 + 27 = 44$ ✓ (1 mark)
c $3 \times 13 = 39$ ✓ (1 mark)
d $323 + 187 = 510$ ✓ (1 mark)

5 $(4 + 8) \times (3 + 7) = 120$ ✓ (1 mark)

6 $(15 \times 8) - (15 + 8)$ ✓
$= 120 - 23 = 97$ ✓ (2 marks)

7 $\{35 - [49 - 24 - 15]\} \times 4$
$= \{35 - 10\} \times 4 = 100$ ✓ (1 mark)

8

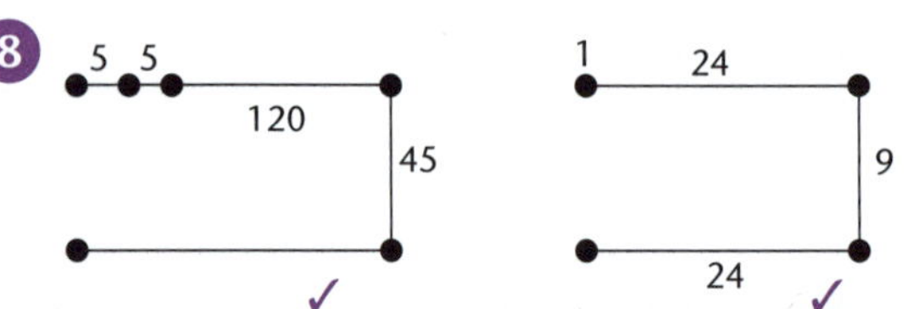

No. of posts = 1 + 24 + 9 + 24
= 58 ✓ (3 marks)

9 Total runs = $26 \times 8 = 208$ ✓
Total for average 32 = $32 \times 10 = 320$ ✓
Runs needed = $320 - 208 = 112$ ✓ (3 marks)

10 101 ✓ (1 mark)

11 HCF = 3 ✓ (1 mark)

12 LCM = 216 ✓ (1 mark)

13 **a**

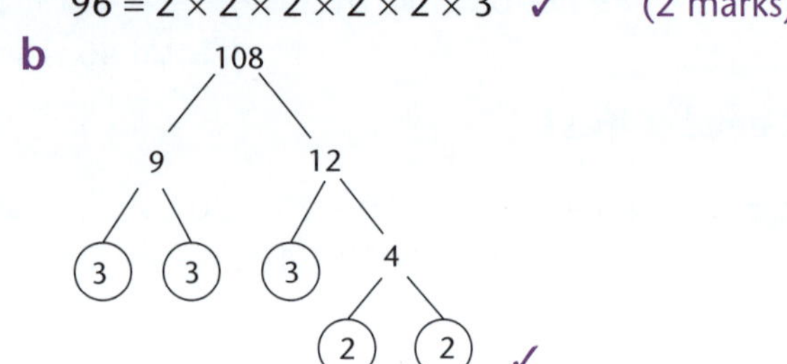

$96 = 2 \times 2 \times 2 \times 2 \times 2 \times 3$ ✓ (2 marks)

b

108
9 12
3 3 3 4
2 2 ✓

$108 = 2 \times 2 \times 3 \times 3 \times 3$ ✓ (2 marks)

c HCF = $2 \times 2 \times 3 = 12$ ✓
LCM = $\underbrace{2 \times 2 \times 2 \times 2 \times 2 \times 3}_{96} \times 3 \times 3$ (with $2 \times 2 \times 3 \times 3 \times 3$ bracketed as 108)
$= 864$ ✓ (2 marks)

14 $3 \times 3 \times 3 \times 3 \times 3 \times 5 \times 5 \times 5$ ✓ (1 mark)

15 $7^7 \times 8^5$ ✓ (1 mark)

16 **a** $\sqrt{121} = 11$ ✓ (1 mark)
b $2^8 = 2 \times 2 \times 2 \times 2 \times 2 \times 2 \times 2 \times 2$
$= 256$ ✓ (1 mark)
c $\sqrt{25 + 144} = \sqrt{169} = 13$ ✓✓ (2 marks)

17 **a** $616 \div 8 = 77$ As 8 divides last 3 digits
$\therefore$ 47 616 is divisible by 8. ✓ (1 mark)
b $4 + 7 + 6 + 1 + 6 = 24$ 3 divides 24
$\therefore$ 47 616 is divisible by 3. ✓ (1 mark)
c $4 + 6 + 6 = 16$, $7 + 1 = 8$, $16 - 8 = 8$
11 does not divide into 8
$\therefore$ 47 616 is not divisible by 11. ✓ (1 mark)
d As 47 616 is divisible by both 3 and 8
then 47 616 is divisible by (3×8) ✓
i.e. 47 616 is divisible by 24. ✓ (2 marks)

WORKED SOLUTIONS to Chapters 1 and 2 **Tests**

18

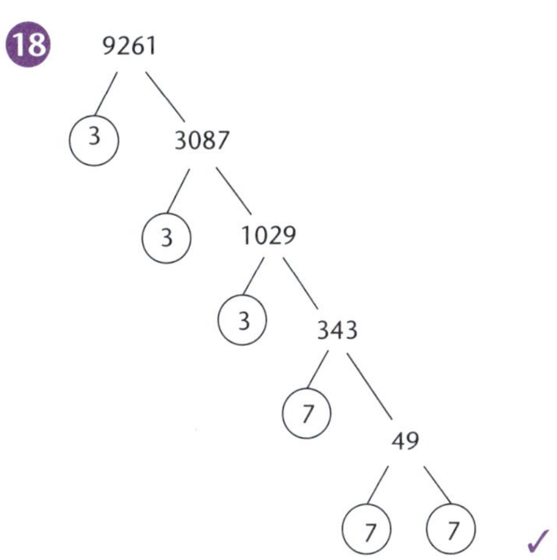

✓

$9261 = 3 \times 3 \times 3 \times 7 \times 7 \times 7$ ✓

$= (3 \times 7) \times (3 \times 7) \times (3 \times 7)$

$\sqrt[3]{9261} = 3 \times 7$

$= 21$ ✓ (3 marks)

(Total 40 marks)

Chapter 2—Fractions

pp. 50–54

Level 1 Test

1 $\frac{12}{18} = \frac{2 \times \cancel{6}^{1}}{3 \times \cancel{6}^{1}} = \frac{2}{3}$ ✓ (1 mark)

2 $\frac{5}{3} = 1\frac{2}{3}$ $(= 5 \div 3)$ ✓ (1 mark)

3 $2\frac{1}{4} = \frac{9}{4}$ $(9 = 4 \times 2 + 1)$ ✓ (1 mark)

4 **a** $\frac{1}{4} < \frac{1}{3}$ compare $\frac{3}{12} < \frac{4}{12}$ ✓ (1 mark)

b $\frac{2}{5} = \frac{4}{10}$ as $\frac{4}{10} = \frac{{}^{1}\cancel{2} \times 2}{{}^{1}\cancel{2} \times 5} = \frac{2}{5}$ ✓ (1 mark)

5 $\frac{5}{4} = 1\frac{1}{4}$ ✓ (1 mark)

6 **a** $\frac{9-7}{10} = \frac{2}{10} = \frac{1}{5}$ ✓ (1 mark)

b $\frac{15}{20} + \frac{8}{20} = \frac{23}{20} = 1\frac{3}{20}$ ✓ (1 mark)

c $4 - 2 - \frac{1}{4} = 2 - \frac{1}{4} = 1\frac{3}{4}$ ✓ (1 mark)

d $6 + 3 + \frac{1}{4} + \frac{2}{3} = 9 + \frac{3}{12} + \frac{8}{12}$ ✓

$= 9\frac{11}{12}$ ✓ (2 marks)

7 **a** $\frac{9}{20}$ ✓ (1 mark)

b $\frac{\cancel{3}^{1}}{\cancel{5}_{1}} \times \frac{\cancel{5}^{1}}{\cancel{9}_{3}} = \frac{1}{3}$ ✓ (1 mark)

c $\frac{\cancel{3}^{1}}{\cancel{2}_{1}} \times \frac{\cancel{4}^{2}}{\cancel{3}_{1}} = 2$ ✓✓ (2 marks)

d $\frac{7}{10} \times \frac{7}{5} = \frac{49}{50}$ ✓✓ (2 marks)

e $\frac{5}{\cancel{8}_{2}} \times \frac{\cancel{4}^{1}}{3} = \frac{5}{6}$ ✓✓ (2 marks)

8 $\frac{7}{10} + \left(\frac{\cancel{5}^{1}}{\cancel{10}_{5}} \times \frac{\cancel{2}^{1}}{\cancel{10}_{2}}\right) = \frac{7}{10} + \frac{1}{10}$ ✓✓ (2 marks)

$= \frac{8}{10} = \frac{4}{5}$

9 **a** $\frac{1}{\cancel{4}_{1}} \times \frac{\cancel{44}^{11}}{1} = 11$ $\therefore$ \$11 ✓ (1 mark)

b $\frac{3}{4} \times 44 = 3 \times 11$ $\therefore$ \$33 ✓ (1 mark)

10 $\frac{5}{25} = \frac{1}{5}$ ✓ (1 mark)

11 $\frac{1}{4}$, ($\frac{2}{4}$), $\frac{3}{4}$ ✓

Fraction is $\frac{2}{4} = \frac{1}{2}$ ✓ (2 marks)

12 $\frac{1}{3} \times 24 = 8$, $7 \times 8 = 56$ hours ✓✓✓ (3 marks)

13 $\frac{36}{40} = \frac{9}{10} = \frac{90}{100}$ ✓✓

90 out of 100 ✓ (3 marks)

14 $\frac{1}{10}$ of wage = \$8

$\therefore \frac{1}{10} = \$8$ ✓

$\frac{10}{10} = \$8 \times 10$ ✓

$= \$80$ ✓ (3 marks)

(Total 35 marks)

Level 2 Test

1 $\frac{27}{45} = \frac{3 \times \cancel{9}^{1}}{5 \times \cancel{9}_{1}} = \frac{3}{5}$ ✓ (1 mark)

2 $\frac{17}{4} = 4\frac{1}{4}$ $(17 \div 4)$ ✓ (1 mark)

3 $5\frac{4}{5} = \frac{29}{5}$ $(5 \times 5 + 4 = 29)$ ✓ (1 mark)

WORKED SOLUTIONS to Chapter 2 **Tests**

4 Change to 12ths:
$\frac{3}{4} = \frac{9}{12}, \frac{2}{3} = \frac{8}{12}, \frac{1}{2} = \frac{6}{12}$
Ascending order:
$\frac{6}{12}, \frac{8}{12}, \frac{9}{12} = \frac{1}{2}, \frac{2}{3}, \frac{3}{4}$ ✓ (1 mark)

5 $1\frac{2}{3} = \frac{5}{3}$ ✓
Reciprocal of $1\frac{2}{3} = \frac{3}{5}$ ✓ (2 marks)

6 **a** $\frac{4}{6} + \frac{3}{6} = \frac{7}{6} = 1\frac{1}{6}$ ✓ (1 mark)

b $5 - 3 - \frac{1}{3} = 2 - \frac{1}{3} = 1\frac{2}{3}$ ✓ (1 mark)

c $2 + 1 + \frac{1}{3} + \frac{4}{5} = 3 + \frac{5}{15} + \frac{12}{15}$
$= 3 + \frac{17}{15}$
$= 3 + 1\frac{2}{15}$
$= 4\frac{2}{15}$ ✓✓ (2 marks)

d $\frac{11}{2} - \frac{7}{10} = \frac{55}{10} - \frac{7}{10}$
$= \frac{48}{10}$
$= 4\frac{8}{10}$
$= 4\frac{4}{5}$ ✓✓ (2 marks)

7 **a** $\frac{\cancel{2}^1}{7} \times \frac{3}{\cancel{4}_2} = \frac{3}{14}$ ✓✓ (2 marks)

b $\frac{\cancel{4}^1}{\cancel{5}_1} \times \frac{\cancel{15}^3}{\cancel{8}_2}$ ✓
$= \frac{3}{2} = 1\frac{1}{2}$ ✓ (2 marks)

c $\frac{7}{\cancel{9}_3} \times \frac{\cancel{6}^2}{5}$ ✓
$= \frac{14}{15}$ ✓ (2 marks)

d $30 \div \frac{6}{5} = \frac{\cancel{30}^5}{1} \times \frac{5}{\cancel{6}_1}$ ✓
$= 25$ ✓ (2 marks)

8 $\frac{3}{4} + \frac{\cancel{7}^1}{\cancel{10}_2} \times \frac{\cancel{5}^1}{\cancel{7}_1} = \frac{3}{4} + \frac{1}{2} = 1\frac{1}{4}$ ✓✓ (2 marks)

9 $\frac{2}{\cancel{3}_1} \times \frac{\cancel{24}^8}{1} = 16$ ∴ $16 ✓✓ (2 marks)

10 Average $= \dfrac{\frac{2}{5} + \frac{7}{10}}{2}$
$= \dfrac{\frac{4}{10} + \frac{7}{10}}{2}$ ✓
$= \frac{11}{10} \times \frac{1}{2}$
$= \frac{11}{20}$ ✓ (2 marks)

11 Number of girls $= \frac{5}{8}$ of 24 ✓
$= \frac{5}{\cancel{8}_1} \times \frac{\cancel{24}^3}{1}$ ✓
$= 15$ ✓ (3 marks)

12 $\frac{18}{30} = \frac{3 \times \cancel{6}^1}{5 \times \cancel{6}_1} = \frac{3}{5}; \frac{3 \times 20}{5 \times 20} = \frac{60}{100}$ ✓
Mark is 60 out of 100. ✓ (2 marks)

13 Fraction $= \frac{40 \text{ s}}{2 \text{ min } 30 \text{ s}} = \frac{40}{150}$ ✓✓
$= \frac{4}{15}$ ✓ [1 min = 60 s; 2 min = 120 s] (3 marks)

14 Wage = $8 $\times 7\frac{1}{2}$
= ($8 × 7) + ($8 × $\frac{1}{2}$)
= $56 + $4 = $60 ✓
∴ She earns $60. (1 mark)

(Total 35 marks)

Level 3 Test

1 $\frac{38}{8} = \frac{19}{4} = 4\frac{3}{4}$ ✓ (1 mark)

2 $8\frac{4}{7} = \frac{60}{7}$ $(7 \times 8 + 4)$ ✓ (1 mark)

3 Change to 60ths
$1\frac{7}{12} = 1\frac{35}{60}, 1\frac{3}{4} = 1\frac{45}{60}, \frac{8}{5} = 1\frac{36}{60}$
Descending order:
$1\frac{45}{60} > 1\frac{36}{60} > 1\frac{35}{60}$
$1\frac{3}{4}, \frac{8}{5}, 1\frac{7}{12}$ ✓ (1 mark)

4 Sum $= \frac{5}{8} + \frac{5}{6}$ ✓
$= \frac{15}{24} + \frac{20}{24}$
$= \frac{35}{24} = 1\frac{11}{24}$ ✓ (2 marks)

5 Product $= \frac{\cancel{3}^1}{\cancel{8}_1} \times \frac{\cancel{16}^2}{\cancel{9}_3} = \frac{2}{3}$ ✓✓ (2 marks)

6 $1\frac{2}{5} - \frac{7}{10} = \frac{7}{5} - \frac{7}{10} = \frac{14}{10} - \frac{7}{10} = \frac{7}{10}$ ✓

Reciprocal of $\frac{7}{10} = \frac{10}{7} = 1\frac{3}{7}$ ✓ (2 marks)

7 **a** $\frac{25}{4} \div \frac{15}{8}$ ✓ (3 marks)

$= \frac{\cancel{25}^{5}}{\cancel{4}_{1}} \times \frac{\cancel{8}^{2}}{\cancel{15}_{3}}$ ✓

$= \frac{10}{3} = 3\frac{1}{3}$ ✓

b $5\frac{1}{3} - 3 + \frac{3}{10}$ ✓

$= 2\frac{1}{3} + \frac{3}{10}$ ✓

$= 2 + \frac{10}{30} + \frac{9}{30}$

$= 2\frac{19}{30}$ ✓ (3 marks)

8 $\frac{23}{\cancel{8}_{1}} \times \frac{\cancel{8}^{1}}{5} - \frac{\cancel{15}^{3}}{\cancel{8}_{1}} \times \frac{\cancel{8}^{1}}{\cancel{5}_{1}}$ ✓

$= \frac{23}{5} - 3$ ✓

$= 4\frac{3}{5} - 3 = 1\frac{3}{5}$ ✓ (3 marks)

9 $\frac{5}{\cancel{6}_{1}} \times \cancel{72}^{12} = 60$ ∴ \$60 ✓✓ (2 marks)

10 $(1\frac{7}{8} + 2\frac{1}{2} + 2\frac{3}{4} + \frac{7}{8}) \div 4$ ✓

$= (\frac{15}{8} + \frac{5}{2} + \frac{11}{4} + \frac{7}{8}) \div 4$ ✓

$= (\frac{15}{8} + \frac{20}{8} + \frac{22}{8} + \frac{7}{8}) \div 4$ ✓

$= \frac{64}{8} \div 4 = 8 \div 4 = 2$ ✓ (4 marks)

11 Amount per hour $= 33\frac{3}{4} \div 4\frac{1}{2}$ ✓

$= \frac{135}{4} \div \frac{9}{2}$

$= \frac{\cancel{135}^{15}}{\cancel{4}_{2}} \times \frac{\cancel{2}^{1}}{\cancel{9}_{1}}$

$= \frac{15}{2}$

$= 7.50$ ∴ \$7.50 ✓ (2 marks)

12 No. of lengths $= 200 \div 6\frac{2}{5}$ ✓

$= 200 \div \frac{32}{5}$

$= \frac{\cancel{200}^{25}}{1} \times \frac{5}{\cancel{32}_{4}}$ ✓

$= \frac{125}{4} = 31\frac{1}{4}$ ✓

No. of lengths = 31, with $\frac{1}{4}$ left over

$\frac{1}{4} \times 6\frac{2}{5} = 1\frac{3}{5}$ cm left over ✓ (4 marks)

13 **a** $\triangle = 1\frac{7}{10} - \frac{4}{5}$

$= \frac{17}{10} - \frac{4}{5}$

$= \frac{17}{10} - \frac{8}{10}$

$= \frac{9}{10}$ ✓

$\triangle = \frac{9}{10}$ (i.e. $\frac{9}{10} + \frac{4}{5} = 1\frac{7}{10}$) ✓ (2 marks)

b $\triangle + \frac{4}{5} < 1\frac{7}{10}$, then $\triangle < \frac{9}{10}$ but $\triangle > 0$

∴ $0 < \triangle < \frac{9}{10}$ ✓ (1 mark)

14 $\frac{2}{3} + \frac{1}{4} = \frac{8}{12} + \frac{3}{12} = \frac{11}{12}$ ✓

$\frac{11}{12}$ of students supported Knights or Dogs ✓

∴ $\frac{1}{12}$ had no opinion ✓

Number with no opinion

$= \frac{1}{\cancel{12}_{1}} \times \cancel{120}^{10} = 10.$ ✓ (4 marks)

15 $\frac{4}{5}$ of the class = 24 ✓

∴ $\frac{1}{5} = \frac{24}{4} = 6$ ✓

∴ $\frac{5}{5} = 6 \times 5 = 30$

Number of students = 30. ✓ (3 marks)

(Total 40 marks)

Chapter 3—Decimals

pp. 67–70

Level 1 Test

1 As 4.76 = 4.760
and 4.760 > 4.577
∴ 4.76 > 4.577 ✓ (1 mark)

2 Second decimal place
∴ 9 hundredths ✓ (1 mark)

3 0.031 ✓ (1 mark)

4 $0.04 = \frac{4}{100} = \frac{1}{25}$ ✓ (1 mark)

5 47.6 ✓ (1 mark)

6

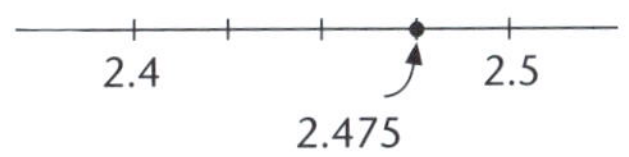

∴ X = 2.475 ✓ (1 mark)

WORKED SOLUTIONS to Chapter 3 **Tests**

7 $0.21 + 0.073 = 0.283$ ✓ (1 mark)

8 $\frac{7}{20}$ shaded

i.e. $\frac{7}{20} = \frac{35}{100} = 0.35$ ✓ (1 mark)

9 **a** $3.620 + 0.413 = 4.033$ ✓ (1 mark)
b $7.0 - 1.4 = 5.6$ ✓ (1 mark)
c $1 - 0.6 - 0.2 = 0.4 - 0.2 = 0.2$ ✓ (1 mark)

10 $3.4 + — = 7.6$
i.e. $7.6 - 3.4 = —$
∴ Missing value is 4.2. ✓ (1 mark)

11 **a** $3.41 \times 5 = 17.05$ ✓ (1 mark)
b $6.047 \times 10 = 60.47$ ✓ (1 mark)
c As $407 \times 3 = 1221$
∴ $4.07 \times 0.3 = 1.221$ ✓ (1 mark)

12 **a** $5.2 \div 2 = 2.6$ ✓ (1 mark)
b $18.43 \div 100 = 0.1843$ ✓ (1 mark)
c $1.44 \div 1.2 = 14.4 \div 12 = 1.2$ ✓ (1 mark)

13 Cost $= 0.24 \times 7 = 1.68$
∴ Pencils cost $1.68. ✓ (1 mark)

14 Remaining $= 24.60 - 3.55 = 21.05$
∴ 21.05 m remains. ✓ (1 mark)

15 $1 - 0.2 \times 0.3 = 1 - 0.06$ ✓
$= 0.94$ ✓ (2 marks)

16 Average $= \frac{1.4+0.8+3.6+0.6}{4}$ ✓
$= \frac{6.4}{4}$
$= 1.6$ ✓
∴ Average is 1.6. (2 marks)

17 Depth $= 3.40 - 0.12$ ✓
$= 3.28$ ✓
∴ New depth is 3.28 m. (2 marks)

18 Payment $= \$0.07 \times 126$ ✓
$= \$8.82$ ✓
∴ Pete is paid $8.82. (2 marks)

19 Number $= 20 \div 0.5$
$= 200 \div 5$
$= 40$ ✓✓
∴ 40 lengths of ribbon. (2 marks)

(Total 30 marks)

Level 2 Test

1 4.710, 4.701, 4.170, 4.700
∴ 4.17, 4.7, 4.701, 4.71 ✓ (1 mark)

2 $8\overline{)7.000}$ = 0.875

∴ $\frac{7}{8} = 0.875$ ✓ (1 mark)

3 $0.48 = \frac{48}{100} = \frac{12}{25}$ ✓ (1 mark)

4 'Nearest hundredth' = 2 dec. places
∴ Answer is 8.04. ✓ (1 mark)

5 Adding 0.39
∴ $2.79 + 0.39 = 3.18$ ✓ (1 mark)

6 **a**
$$\begin{array}{r} 6.015\ + \\ 2.1 \\ 3.06 \\ \hline 11.175 \\ \hline \end{array}$$
∴ Answer is 11.175. ✓ (1 mark)

b
$$\begin{array}{r} 3.05\ - \\ 0.14 \\ \hline 2.91 \\ \hline \end{array}$$
∴ Answer is 2.91. ✓ (1 mark)

7
$$\begin{array}{r} 4.05\ - \\ 1.09 \\ \hline 2.96 \\ \hline \end{array}$$
∴ Answer is 2.96. ✓ (1 mark)

8 **a** $602 \times 3 = 1806$
∴ $6.02 \times 0.3 = 1.806$ ✓ (1 mark)
b $31.486 \times 100 = 3148.6$ ✓ (1 mark)

9 **a** $3.6 \div 0.4 = 36 \div 4 = 9$ ✓ (1 mark)
b $64.76 \div 1000 = 0.06476$ ✓ (1 mark)
c $25.51 \div 0.5 = 255.1 \div 5$
$= 51.02$ ✓ (1 mark)

10 $8.5 \div ? = 1.7$
∴ $8.5 \div 1.7 = 85 \div 17 = 5$ ✓
Answer = 5. (1 mark)

11 $\square + 3.2 = 6.47$
i.e. $6.47 - 3.20$ ✓
$= 3.27$ ✓ (2 marks)

WORKED SOLUTIONS to Chapter 3 Tests

12 $104.80 \div 4$ ✓
$= 26.20$ ✓
$\therefore$ Bob pays \$26.20. (2 marks)

13 'Middle' = Average
$= \frac{1.6 + 2.8}{2}$ ✓
$= \frac{4.4}{2}$
$= 2.2$ ✓
(2 marks)

14 Amount $= 1.25 \div 5 = 0.25$ ✓
$\therefore$ 0.25 L in each glass. ✓ (2 marks)

15 As $32 \times 5 = 160$
$\therefore 3.2 \times 0.5 = 1.6$ ✓

✓ (2 marks)

16 **a** $(0.5)^2 + 2.5 = 0.25 + 2.5$ ✓
$= 2.75$ ✓ (2 marks)

b $3.2 \times 0.2 + \frac{1.5}{5}$
$= 0.64 + 0.3$ ✓
$= 0.94$ ✓ (2 marks)

17 Savings $= 68 \div 8 = 8.5$ ✓✓

$$8\overline{)68.00} = 8.50$$

$\therefore$ Laura must save \$8.50/week. (2 marks)

(Total 30 marks)

Level 3 Test

1 3.470, 3.407, 3.471, 3.074
$\therefore$ 3.074, 3.407, 3.47, 3.471 ✓ (1 mark)

2

$$11\overline{)7.0000\ \ldots} = 0.6363\ \ldots$$

$\therefore \frac{7}{11} = 0.\dot{6}\dot{3}$ ✓ (1 mark)

3 $8.055 = 8\frac{55}{1000}$
$= 8\frac{11}{200}$ ✓ (1 mark)

4 $\frac{317}{500} = \frac{634}{1000}$
$= 0.634$
$= 0.63$ (to two dec. places) ✓ (1 mark)

5 **a**
$$\begin{array}{r} 13.04 + \\ 2.9 \\ 5 \\ \hline 20.94 \\ \hline \end{array}$$
$\therefore 20.94$ ✓ (1 mark)

b
$$\begin{array}{r} 6.000 - \\ 2.031 \\ \hline 3.969 \\ \hline \end{array}$$
$\therefore 3.969$ ✓ (1 mark)

6 ____ $- 1.76 = 2.3$
$\therefore 2.3 + 1.76 = 4.06$ ✓ (1 mark)

7 **a** $1.2 \times 1.2 = 1.44$ ✓ (1 mark)
b $7.476 \times 1000 = 7476$ ✓ (1 mark)
c $0.6 \times 0.03 \times 0.1 = 0.0018$ ✓ (1 mark)

8 **a** $2.05 \div 0.05 = 205 \div 5$
$= 41$ ✓ (1 mark)

b $\frac{10.44}{0.4} = \frac{104.4}{4}$

$$\therefore 4\overline{)104.4} = 26.1$$

$\therefore 26.1$ ✓ (1 mark)

c $0.047 \div 100 = 0.000\,47$ ✓ (1 mark)

9 A: 1.25, B: 3.5
$\therefore AB = 3.5 - 1.25 = 2.25$
i.e. AB is 2.25 units ✓ (1 mark)

10 As $0.4 \times 0.4 = 0.16$
$\therefore \sqrt{0.16} = 0.4$ ✓ (1 mark)

11 As $3.4 \times 0.9 = 3.06$
$\therefore 3.06 \div 0.9 = 3.4$
i.e. $30.6 \div 0.9 = 34$ ✓ (1 mark)

12 $0.4 + 4 \times 0.5 = 0.4 + 2$
$= 2.4$ ✓✓ (2 marks)

13 $\frac{3.6}{0.4} + \frac{0.32}{0.8} = \frac{36}{4} + \frac{3.2}{8}$
$= 9 + 0.4$ ✓
$= 9.4$ ✓ (2 marks)

14 Cost $= 99 \div 6 = 16.5$
$\therefore$ One court costs \$16.50. ✓✓ (2 marks)

WORKED SOLUTIONS to Chapters 3 and 4 **Tests**

15
$$\begin{array}{r} 347\times \\ 28 \\ \hline 2776 \\ 6940 \\ \hline 9716 \end{array}$$ ✓

$\therefore 3.47 \times 2.8 = 9.716$ ✓ (2 marks)

16 Number $= 1.60 \div 0.08$
$= 160 \div 8$ ✓
$= 20$

$\therefore$ I can make 20 rissoles. ✓ (2 marks)

17
$$\begin{array}{r} 1\,699\times \\ 32 \\ \hline 3\,398 \\ 50\,970 \\ \hline 54\,368 \end{array}$$ ✓

$\therefore \$1.699 \times 32 = \$54.368$
$= \$54.37$ (to 2 dec. places)

$\therefore$ 32 L of petrol costs \$54.37. ✓ (2 marks)

18 Average of 4 days = 13.7
$\therefore$ Sum of 4 days $= 13.7 \times 4$
$= 54.8$ ✓
Now, sum of 3 days $= 14.8 + 12.7 + 14.7$
$= 42.2$
$\therefore$ Difference $= 54.8 - 42.2$
$= 12.6$ ✓
$\therefore$ Temp. on 4th day was 12.6°. (2 marks)

(Total 30 marks)

Chapter 4—Percentages pp. 82–84

Level 1 Test

1 a $\frac{3}{5} \times 100\% = 60\%$ ✓ (1 mark)
b $1\frac{1}{2} \times 100\% = 150\%$ ✓ (1 mark)
c $0.45 \times 100\% = 45\%$ ✓ (1 mark)
d $0.8 \times 100\% = 80\%$ ✓ (1 mark)
e $0.06 \times 100\% = 6\%$ ✓ (1 mark)
f $0.355 \times 100\% = 35.5\%$ ✓ (1 mark)

2 a $\frac{27}{100}$ ✓ (1 mark)
b $\frac{8}{100} = \frac{2}{25}$ ✓ (1 mark)
c $\frac{40}{100} = \frac{2}{5}$ ✓ (1 mark)

3 a 0.04 ✓ (1 mark)
b 0.075 ✓ (1 mark)
c 2.15 ✓ (1 mark)

4 a $\frac{12}{48} \times 100\% = 25\%$ ✓ (1 mark)
b $\frac{320}{400} \times 100\% = 80\%$ ✓ (1 mark)
c $\frac{12}{25} \times 100\% = 48\%$ ✓ (1 mark)

5 a $0.15 \times 20 = \$3$ ✓ (1 mark)
b $0.08 \times 600 = \$48$ ✓ (1 mark)
c $0.12 \times \$20 = \$2.40$ ✓ (1 mark)

6 90% of \$40
$= 0.9 \times \$40 = \$36$ ✓ (1 mark)

7 $\frac{15}{60} \times \frac{100}{1}\% = 25\%$ ✓ (1 mark)

8 25% of an amount = \$30
100% of amount $= \$30 \times 4 = \$120$
$\therefore$ Amount is \$120 ✓ (1 mark)

9 Amount to sister = 25% of 200
$= 0.25 \times 200$
$= 50$ ✓
$\therefore$ Amount left $= 200 - 50 = 150$
$\therefore$ Athena has \$150 left ✓ (2 marks)

10 New amount $= 1.2 \times \$450$ ✓
$= \$540$ ✓ (2 marks)

(Total 25 marks)

Level 2 Test

1 a 41%, $\frac{2}{5} = 40\%$, $0.42 = 42\%$
$\therefore \frac{2}{5}$, 41%, 0.42 ✓ (1 mark)
b 37%, $\frac{3}{8} = 37.5\%$, $0.4 = 40\%$
$\therefore$ 37%, $\frac{3}{8}$, 0.4 ✓ (1 mark)

2 a $0.04 \times \$290 = \$11.60$ ✓ (1 mark)
b $0.16 \times \$360 = \$57.60$ ✓ (1 mark)
c $0.085 \times \$7000 = \$595$ ✓ (1 mark)
d $0.0525 \times 4000 = 210$
$\therefore$ 210 mm or 21 cm ✓ (1 mark)

WORKED SOLUTIONS to Chapter 4 **Tests**

3 a $\frac{30}{300} \times 100\% = 10\%$ ✓ (1 mark)

b $\frac{5}{2000} \times 100\% = 0.25\%$ ✓ (1 mark)

c $\frac{6}{120} \times 100\% = 5\%$ ✓ (1 mark)

d $\frac{35}{2000} \times 100\% = 1.75\%$ ✓ (1 mark)

4 a 110% of \$25 = 1.1 × \$25 ✓
= \$27.50 ✓ (2 marks)

b 108% of \$650 = 1.08 × \$650 ✓
= \$702 ✓ (2 marks)

5 a 92% of \$650 = 0.92 × \$650 ✓
= \$598 ✓ (2 marks)

b $90\frac{1}{2}\%$ of \$4000 = 0.905 × \$4000 ✓
= \$3620 ✓ (2 marks)

6 a Discount = 0.15 × \$45 ✓
= \$6.75
∴ A discount of \$6.75 ✓ (2 marks)

b Discount = 0.2 × \$2480 ✓
= \$496
∴ A discount of \$496 ✓ (2 marks)

7 a 0.125 × 160 = \$20
∴ \$20 ✓ (1 mark)

b New price = \$160 − \$20
= \$140 ✓ (1 mark)

8 100% − 40% = 60% ✓
Fraction for Savvas = $\frac{60}{100} = \frac{3}{5}$ ✓ (2 marks)

9 5% of amount = 24
100% of amount = 24 × 20 = 480
∴ The amount is 480 ✓✓ (2 marks)

10 Bank amount = $\frac{1}{2} \times 2400 = 1200$
∴ Amount remaining is \$1200 ✓
Bike cost 60% of \$1200 = 0.6 × 1200
= 720
∴ Bike cost \$720 ✓ (2 marks)

(Total 30 marks)

Level 3 Test

1 a $\frac{2}{3} \times 100\% = 66.\dot{6}\%$ ✓ (1 mark)

b $\frac{7}{8} \times 100\% = 87.5\%$ ✓ (1 mark)

2 a 0.085 ✓ (1 mark)

b 0.034 ✓ (1 mark)

c 0.1275 ✓ (1 mark)

3 a 0.125 × \$70 = \$8.75 ✓ (1 mark)

b 0.008 × \$50 = \$0.40 ✓ (1 mark)

4 120 × 1.10 × 0.9 = 118.8 ✓✓ (2 marks)

5 Spent = 100% − 60%
= 40% ✓
= $\frac{2}{5}$ ✓ (2 marks)

6 a $\frac{50}{200} \times 100\% = 25\%$ ✓ (1 mark)

b $\frac{125}{1000} \times 100\% = 12.5\%$ ✓ (1 mark)

7 Percentage saved = 100 − [21 + 13 + 6]
= 100 − 40
= 60
∴ 60% ✓
Amount saved = 60% of \$800
= 0.6 × \$800
= \$480
∴ Olivia saves \$480 ✓ (2 marks)

8 Cost = 0.82 × 1200 ✓
= 984
∴ Cost \$984 ✓ (2 marks)

9 a Discount = \$25.74 − \$23.40
= \$2.34 ✓ (1 mark)

b Discount % = $\frac{234}{2340} \times 100\%$
= 10% ✓ (1 mark)

10 85% of original = \$1275 ✓
1% of original = \$1275 ÷ 85 = \$15
100% of original = \$15 × 100 = \$1500
∴ Washing machine originally
priced at \$1500 ✓ (2 marks)

11 $100 - 22\frac{1}{2} = 77\frac{1}{2}$ ✓
∴ New price = 0.775 × 205.50
= 159.2625
= 159.26 (2 dec. places)
∴ She paid \$159.26 ✓ (2 marks)

WORKED SOLUTIONS to Chapters 4 and 5 **Tests**

12 $100 - 33\frac{1}{3} = 66\frac{2}{3}$ ✓

$\therefore$ New amount $= 0.666\,666 \times 369.60$
$= 246.40$

$\therefore$ \$246.40 ✓ (2 marks)

(Total 25 marks)

Chapter 5—Integers

pp. 100–103

Level 1 Test

1 **a** $-5 < 0$ ✓ (1 mark)

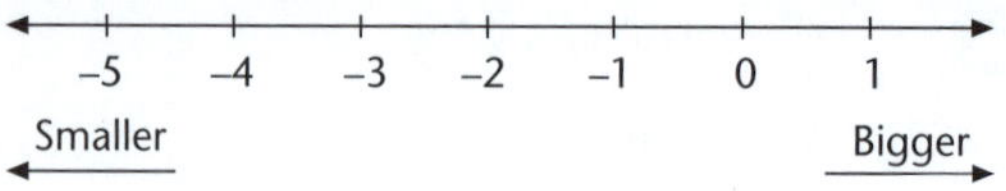

b $-3 > -4$ ✓ (1 mark)

2 **a** $-3 + 7 = 4$ ✓ (1 mark)

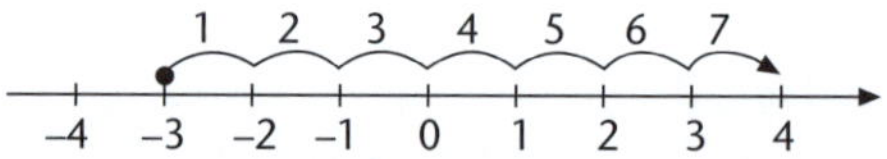

b $(-9) + (-3) = -12$ ✓ (1 mark)
c $-6 + 8 + (-4) = 2 + (-4) = -2$ ✓ (1 mark)
d $-7 - 11 = -7 + (-11) = -18$ ✓ (1 mark)
e $-3 - (-8) = -3 + 8 = 5$ ✓ (1 mark)

3 $11 + (-3) + (-5) = 8 + (-5) = 3$ ✓ (1 mark)

4 **a** $-3 + \boxed{-2} = -5$, $\square = -2$ ✓ (1 mark)
b $\boxed{-5} - 3 = -8$, $\square = -5$ ✓ (1 mark)

5 **a** $-8 + 2 = -6$ ✓ (1 mark)
b $2 - 6 = -4$ ✓ (1 mark)
c $(-5) \times (-3) = 15$ ✓ (1 mark)
d $12 \div (-4) = -3$ ✓ (1 mark)

6 **a** $3t - 5t = -2t$ ✓ (1 mark)
b $-8a + 4a = -4a$ ✓ (1 mark)
c $-2m - 3m = -5m$ ✓ (1 mark)
d $-3a \times 4 = -12a$ ✓ (1 mark)
e $15a \div (-3a) = -5$ ✓ (1 mark)

7 **a** $-7 + 5 = -2$ ✓ (1 mark)
b $-5 - (7) = 5 + 7 = 12$ ✓ (1 mark)
c $2 - (-7) = 2 + 7 = 9$ ✓ (1 mark)
d $-7 - 5 = -12$ ✓ (1 mark)
e $3 + (-7) - 5 = 3 - 7 - 5 = -9$ ✓ (1 mark)

8 $-18 + 4 + 3 = -11$ ✓
Temperature is -11°C ✓ (2 marks)

9 **a** A(–3, 1) ✓ (1 mark)
b B(1, –3) ✓ (1 mark)

10 **a** $-\frac{1}{2} + \frac{3}{4} = \frac{-2}{4} + \frac{3}{4}$
$= \frac{1}{4}$ ✓ (1 mark)

b $1.2 - 3.5 = (1.2 - 1.2) - 2.3$
$= -2.3$ ✓ (1 mark)

(Total 30 marks)

Level 2 Test

1 $-8, -1, 0$ ✓ (1 mark)

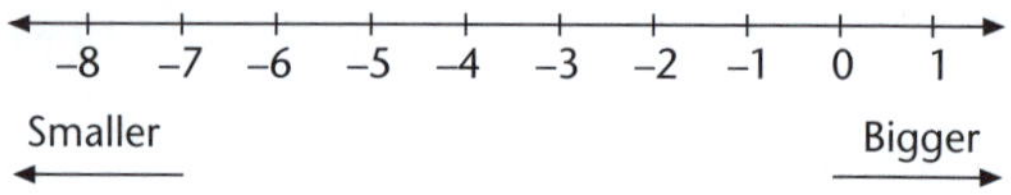

2 **a** $11 + (-8) = 3$ ✓ (1 mark)
b $-13 + 5 = -8$ ✓ (1 mark)
c $-3 - 11 = -3 + (-11) = -14$ ✓ (1 mark)
d $11 - 17 = 11 + (-17) = -6$ ✓ (1 mark)
e $5 - (-3) - 7 = 5 + 3 + (-7) = 1$ ✓ (1 mark)
f $(-9) + 5 + (-8) = -4 + (-8)$
$= -12$ ✓ (1 mark)

3 **a** $9 - \boxed{22} = -13$, $\square = 22$ ✓ (1 mark)
b $(-8) - \boxed{-12} = 4$, $\square = -12$ ✓ (1 mark)

4 $-18 + (-3) + 11 + (-5)$
$= -21 + 11 + (-5)$
$= -10 + (-5) = -15$ ✓ (1 mark)

5 **a** $(-4) \times (-6) = 24$ ✓ (1 mark)
b $40 \div (-8) = -5$ ✓ (1 mark)
c $(-100) \div 10 = -10$ ✓ (1 mark)
d $(-12) \div (-6) = 2$ ✓ (1 mark)
e $(-3) \times 2 \times (-2) = 12$ ✓ (1 mark)
f $(-6)^2 = 36$ ✓ (1 mark)

6 **a** $t - 7t = -6t$ ✓ (1 mark)
b $-8t + 2t = -6t$ ✓ (1 mark)
c $7b - b = 6b$ ✓ (1 mark)
d $-3a + 3a = 0$ ✓ (1 mark)
e $-2t + 3t - t = 0$ ✓ (1 mark)
f $6a - 8a + 12 + 4 = -2a + 16$ ✓ (1 mark)

WORKED SOLUTIONS to Chapter 5 Tests

7 a $-1\frac{1}{2} + \frac{1}{4} = -1\frac{2}{4} + \frac{1}{4}$
$= -1\frac{1}{4}$ ✓ (1 mark)

b $(5 - 9) - (-3 + 7) = -4 - 4$
$= -8$ ✓ (1 mark)

c $-12 + 3 - 2 = -11$ ✓ (1 mark)

d $1 - 7 + 2 = -4$ ✓ (1 mark)

8 a $-4 - 7 = -11$ ✓ (1 mark)

b $3 - (-7) = 3 + 7 = 10$ ✓ (1 mark)

c $-4 - (-7) + 2 = -4 + 7 + 2 = 5$ ✓ (1 mark)

d $(-7) \times (-4) = 28$ ✓ (1 mark)

e $(-7)^2 = (-7) \times (-7) = 49$ ✓ (1 mark)

f $2 \times (-4) - (-7) = -1$ ✓ (1 mark)

9 a $\frac{10}{15} - \frac{12}{15} = \frac{-2}{15}$ ✓ (1 mark)

b $0.7 + (-1.5) = 0.7 - 1.5$
$= -0.8$ ✓ (1 mark)

c $-6 + 2.3 = 2.3 - 6 = -3.7$ ✓ (1 mark)

10 Average $= [-11 + (-8) + (-5)] \div 3$ ✓
$= [-24] \div 3$
$= -8$

Average temperature is −8°C. ✓ (2 marks)

11 a $3 - 4 \times 2 = 3 - 8 = -5$ ✓ (1 mark)

b $4 \times 3 - 14 = 12 - 14 = -2$ ✓ (1 mark)

c $(-12) \div 6 - 4 = -2 - 4 = -6$ ✓ (1 mark)

(Total 40 marks)

Level 3 Test

1 5, −1, −7, −10 ✓ (1 mark)

2 a P(−3, −2) ✓
Q(4, −3) ✓ (2 marks)

b Average of −3 and 4 $= -\frac{3+4}{2} = \frac{1}{2}$

Average of −2 and −3
$= \frac{-2 + -3}{2} = \frac{-5}{2} = -2\frac{1}{2}$ ✓

Mid-point of PQ $= (\frac{1}{2}, 2\frac{1}{2})$ ✓ (2 marks)

3 a $-19 - 21 = -40$ ✓ (1 mark)

b $5 - 7 - 13 = 5 + (-7) + (-13)$
$= -15$ ✓ (1 mark)

c $-16 + 13 + (-2) = -3 + (-2)$
$= -5$ ✓ (1 mark)

d $-5 - 3 - (-8) = -5 + (-3) + 8$
$= 0$ ✓ (1 mark)

4 $-9 + (-7) = -9 - 7$
$= -16$ ✓ (1 mark)

5 $3 - (-10) = 3 + 10$
$= 13$ ✓ (1 mark)

6 a $-9 - 4 = -13$ ✓ (1 mark)

b $-9 - (-5) + 3 = -9 + 5 + 3$
$= -1$ ✓ (1 mark)

c $-7 - 3 + 10 = 0$ ✓ (1 mark)

7 a $\triangle + (-8) = -40$
$\triangle - 8 = -40$
$\therefore \triangle = -32$ ✓ (1 mark)

b $\square - (-9) = 6$
$\square + 9 = 6$
$\therefore \square = -3$ ✓ (1 mark)

8 a $10 - 16 \div (-4) = 10 + 4$
$= 14$ ✓ (1 mark)

b $(-16 \times 2) \div (-8) = -32 \div -8$
$= 4$ ✓ (1 mark)

9 $5 - \square - (-3) = 15$
$5 - \square + 3 = 15$
$8 - \square = 15$
$\therefore \square = -7$ ✓ (2 marks)

10 a $-5m - 3m = -8m$ ✓ (1 mark)

b $-8y + 2y = -6y$ ✓ (1 mark)

c $-20a - a = -21a$ ✓ (1 mark)

d $(-4y)^2 = 16y^2$ ✓ (1 mark)

e $10a - 4a \times 3 = -2a$ ✓ (1 mark)

f $(-15y) \div (-3y) \times (-2) = -10$ ✓ (1 mark)

11 a $4 - (-9) - 8 = 4 + 9 - 8 = 5$ ✓ (1 mark)

b $-9 - 8 = -17$ ✓ (1 mark)

c $8 - (-9) = 8 + 9 = 17$ ✓ (1 mark)

d $-9 + 8 = -1$ ✓ (1 mark)

e $7 - (-9) = 7 + 9 = 16$ ✓ (1 mark)

12 a $-\frac{2}{3} + 2 = 2 - \frac{2}{3} = 1\frac{1}{3}$ ✓ (1 mark)

b $-\frac{3}{4} - \frac{5}{8} = -\frac{6}{8} - \frac{5}{8}$
$= -\frac{11}{8}$
$= -1\frac{3}{8}$ ✓ (1 mark)

c $1-(-\frac{4}{5})+\frac{1}{10}=1+\frac{4}{5}+\frac{1}{10}$
$=1+\frac{8}{10}+\frac{1}{10}$
$=1\frac{9}{10}$ ✓ (1 mark)

13 $-5(-2)+3=10+3=13$ ✓ (1 mark)

14 **a** $4x-5x+2=-x+2$ ✓ (1 mark)
b $7a-3+a=8a-3$ ✓ (1 mark)
c $12-2a+3a-1=11+a$ ✓ (1 mark)

15 **a** $2x^2-3y^2=2(-4)^2-3(-3)^2$
$=2(16)-3(9)$
$=32-27$
$=5$ ✓ (1 mark)

b $x+2y=\frac{-4+2\times(-3)}{5}$
$=\frac{-4-6}{5}$
$=-\frac{10}{5}$
$=-2$ ✓ (1 mark)

(Total 40 marks)

Chapter 6—Ratios

pp. 116–118

Level 1 Test

1 **a** 1:4 ✓ (1 mark)
b 9:7 ✓ (1 mark)
c 10:1 ✓ (1 mark)

2 **a** AB: 1 unit, EF: 1 unit
∴ 1:1 ✓ (1 mark)
b CE: 2 units, AI: 8
∴ 2:8 = 1:4 ✓ (1 mark)
c BG: 5 units, AF: 5 units
∴ 5:5 = 1:1 ✓ (1 mark)

3 **a** 16:14 = 8:7 ✓ (1 mark)
b 14:30 = 7:15 ✓ (1 mark)

4 **a** 80:40 = 2:1 ✓ (1 mark)
b 120:80 = 3:2 ✓ (1 mark)

5 3 parts = 270 ✓
1 part = 90 ✓
∴ There is 90 mL of cordial in the mix. ✓ (3 marks)

6 $5+3=8$ ✓
∴ 8 parts ✓
∴ $\frac{5}{8}\times 464=290$ ✓
∴ There are 290 adults in the town. (3 marks)

7 $2+3=5$
∴ 5 parts ✓
∴ $\frac{3}{5}\times 30=18$
∴ The longer length is 18 m. ✓ (2 marks)

8 **a** 3 shaded, 6 unshaded
∴ 3:6 = 1:2 ✓ (1 mark)
b 6:9 = 2:3 ✓ (1 mark)

9 $2+3=5$
∴ 5 parts ✓
Boys = $\frac{2}{5}\times 30=12$
∴ 12 boys ✓ (2 marks)

10 5 parts = 550
1 part = 110
∴ 9 parts = 990
∴ There are 990 students. ✓✓✓ (3 marks)

(Total 25 marks)

Level 2 Test

1 **a** 320:400 = 4:5 ✓ (1 mark)
b 15:12:6 = 5:4:2 ✓ (1 mark)
c 1 metre = 10 000 metres
i.e. 1:10 000 ✓ (1 mark)

2 **a** 12:16 = 3:4
∴ $x=4$ ✓ (1 mark)
b 5:15 = 10:30
∴ $x=10$ ✓ (1 mark)

3 Unsold = 72 − 24 ✓
= 48 ✓
∴ 72:48 = 3:2 ✓ (3 marks)

4 35 parts = 280 ✓
1 part = 8
∴ There are 8 teachers. ✓ (2 marks)

WORKED SOLUTIONS to Chapter 6 **Tests**

5 $7 + 2 = 9$
$\therefore$ 9 parts ✓
$\therefore \frac{7}{9} \times 540 = 420$
$\frac{2}{9} \times 540 = 120$
$\therefore$ 420 g, 120 g ✓✓ (3 marks)

6 4 parts = 1.2 ✓
1 part = 0.3
$\therefore$ 2 parts = 0.6
$\therefore$ The shortest is 0.6 m. ✓ (2 marks)

7 $3 + 5 + 4 = 12$
$\therefore$ 12 parts ✓
i.e. $\frac{5}{12} \times 480 = 200$
$\therefore$ The area sowed with oats is 200 ha. ✓
(2 marks)

8 $4 + 5 = 9$
$\therefore$ 9 parts ✓
Black = $\frac{5}{9} \times 45 = 25$
$\therefore$ There are 25 black jellybeans. ✓ (2 marks)

9 **a** 2:6 = 1:3 ✓ (1 mark)
b 8:24 = 1:3 ✓ (1 mark)
c 4:36 = 1:9 ✓ (1 mark)

10 3 parts = 21 ✓
1 part = 7
5 parts = 35
$\therefore$ 35 kg of A ✓✓ (3 marks)
(Total 25 marks)

Level 3 Test

1 **a** $\frac{3}{4}:\frac{4}{4} = 3:4$ ✓ (1 mark)
b 5000 m^2:20 000 m^2 = 1:4 ✓ (1 mark)
c $\frac{2}{3}:\frac{3}{4} = \frac{8}{12}:\frac{9}{12} = 8:9$ ✓ (1 mark)

2 **a** 48:64 = 3:4
$\therefore x = 3$ ✓ (1 mark)
b 25:35 = 5:7
= 10:14
$\therefore x = 14$ ✓ (1 mark)

3 Angles are 60°, 40°, 80°
$\therefore$ 80°:40° = 2:1 ✓ (1 mark)

4 Discount:Original price = 15:100
= 3:20 ✓ (1 mark)

5 $2 + 3 + 5 = 10$ $\therefore$ 10 parts
$\therefore \frac{2}{10} \times 360 = 72$
$\frac{3}{10} \times 360 = 108$
$\frac{5}{10} \times 360 = 180$
$\therefore$ \$72, \$108, \$180 ✓✓ (2 marks)

6 Perimeter = 80 cm
$\therefore$ Sum of length + breadth = 40 cm
$\therefore$ 5 parts = 40 ✓
$\therefore$ $\frac{2}{5} \times 40 = 16$
$\frac{3}{5} \times 40 = 24$
$\therefore$ The sides are 16 cm × 24 cm. ✓ (2 marks)

7 4 parts = 5.6 ✓
1 part = 1.4
3 parts = 4.2
$\therefore$ 4.2 kg copper would be needed. ✓ (2 marks)

8 2 + 1 = 3 parts ✓
$\therefore \frac{1}{3}$ of \$8.70 = \$2.90
$\therefore$ Smaller share is \$2.90. ✓ (2 marks)

9 **a** 10:15 = 2:3 ✓ (1 mark)
b 6:13.5 = 12:27 = 4:9 ✓ (1 mark)

10 3 parts = 12
1 part = 4
10 parts = 40 ✓
$\therefore$ daughter 12, mother 40
$\therefore$ Total age is 52. ✓ (2 marks)

11 \$2.50:\$7.50 = 1:3
$\therefore$ 1 + 3 = 4 parts ✓
Christopher's share = $\frac{1}{4} \times 25\,000 = 6250$
$\therefore$ Chris receives \$6250. ✓ (2 marks)

12 **a** $t:b = 3:2$, $b:c = 5:2$
$\therefore t:b:c$
3:2 (multiply by 5)
5:2 (multiply by 2)
$\therefore$ 15:10:4 ✓✓ (2 marks)

b 4 parts = 48 ✓
1 part = 12
∴ 15 parts = 180
∴ 180 students went by train. ✓ (2 marks)

(Total 25 marks)

Chapter 7—Algebra

pp. 132–135

Level 1 Test

1 **a**

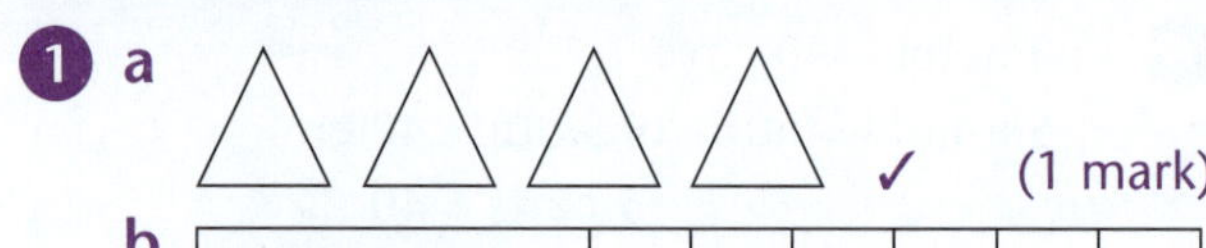

✓ (1 mark)

b

Number of △	1	2	3	4	5	6
Number of matchsticks	3	6	9	12	15	18

✓ (1 mark)

c Number of matches used
= 3 × number of triangles formed. ✓ (1 mark)

d $3 \times 15 = 45$ ✓ (1 mark)

e 3 × triangles = 30,
∴ 10 triangles ✓ (1 mark)

2 **a** 6 (adding 1) ✓ (1 mark)
b 11 (adding 2) ✓ (1 mark)
c 40 (multiply by 2) ✓ (1 mark)
d 21 (subtract 3) ✓ (1 mark)

3 **a** $y = x + 2$

x	1	2	3	4
y	3	4	5	6

✓✓ (2 marks)

b $T = 3 \times t$

t	1	2	3	4
T	3	6	9	12

✓✓ (2 marks)

4 **a** $y = x + 2$ ✓ (1 mark)
b $y = x - 1$ ✓ (1 mark)

5 **a** $5a$ ✓ (1 mark)
b $2b$ ✓ (1 mark)
c $12x$ ✓ (1 mark)
d c^2 ✓ (1 mark)

6 **a** $4 \times k$ ✓ (1 mark)
b $a \times b$ ✓ (1 mark)
c $5 \times x \times y$ ✓ (1 mark)
d $y \times y$ ✓ (1 mark)

7 **a** $3y$ ✓ (1 mark)
b $2b$ ✓ (1 mark)

8 **a** $5 + 3 = 8$ ✓ (1 mark)
b $5 - 2 = 3$ ✓ (1 mark)
c $2 \times 5 = 10$ ✓ (1 mark)
d $3 \times 5 - 10 = 5$ ✓ (1 mark)

9 **a** $4 \times 2 = 8$ ✓ (1 mark)
b $\frac{4}{2} = 2$ ✓ (1 mark)
c $3 \times 4 \times 2 = 24$ ✓ (1 mark)
d $4 \times 4 = 16$ ✓ (1 mark)

10 $3 + 5 - 2 = 6$ ✓ (1 mark)

11 $5a^2$ ✓ (1 mark)

12 **a** $5b + 3b + 2b = 10b$ ✓ (1 mark)
b $3y + 2y - y = 4y$ ✓ (1 mark)
c $2a + 3b + 5a = 7a + 3b$ ✓ (1 mark)
d $2t + 6n + 3t - 2n = 5t + 4n$ ✓ (1 mark)

13 **a** 2^3 ✓ (1 mark)
b a^4 ✓ (1 mark)
c $3a^3$ ✓ (1 mark)

14 $(3 + 1)^2 = 4^2 = 16$ ✓ (1 mark)

15 $5a \div a = 5$ ✓✓ (2 marks)

(Total 45 marks)

Level 2 Test

1 **a** ✓

(1 mark)

b

Number of △	1	2	3	4	5	6	7	8
Number of matchsticks	3	5	7	9	11	13	15	17

✓ (1 mark)

c The number of matchsticks
= 2 × number of triangles + 1 ✓ (1 mark)

WORKED SOLUTIONS to Chapter 7 **Tests**

d $M = 2T + 1$ ✓ (1 mark)
e i $2 \times 20 + 1 = 41$ ✓
ii $2T + 1 = 257$, $T = 128$ ✓ (2 marks)

2 a 6, 3 (subtract 3) ✓ (1 mark)
b 25, 36 (next square) ✓ (1 mark)
c 13, 21 (Fibonacci sequence) ✓ (1 mark)
d 18, 9 (halve previous) ✓ (1 mark)

3 a

x	1	2	3	4
y	1	3	5	7

✓ (1 mark)

b

x	2	3	4	5
y	11	9	7	5

✓ (1 mark)

4 a $y = 3x + 2$ ✓ (1 mark)
b $y = 2x - 3$ ✓ (1 mark)

5 a $10a$ ✓ (1 mark)
b $6abc$ ✓ (1 mark)
c $4a^2$ ✓ (1 mark)
d 12 ✓ (1 mark)
e 5 ✓ (1 mark)
f $10a^2b$ ✓ (1 mark)
g $9b^2$ ✓ (1 mark)
h $4a$ ✓ (1 mark)
i $12a + 4b$ ✓ (1 mark)
j $7a^2 + a$ ✓ (1 mark)

6 a $6a^3$ ✓ (1 mark)
b a^7 ✓ (1 mark)
c a^6 ✓ (1 mark)
d a^{10} ✓ (1 mark)

7 a $5 \times 4 - 3 = 17$ ✓ (1 mark)
b $2 \times 3 \times 3 = 18$ ✓ (1 mark)
c $10 - 2 \times 4 = 2$ ✓ (1 mark)
d $\frac{5 \times 3 - 3}{4} = \frac{12}{4}$, $A = 3$ ✓ (1 mark)

8 $x = 4$, $x - 2 = 2$ ✓ (1 mark)

9 $2 \times 1 \times 1 - 1 + 2 = 3$ ✓ (1 mark)

10 a $3x$ ✓ (1 mark)
b $6xy + 6x$ ✓ (1 mark)
c $21x^5y^5$ ✓ (1 mark)
d $8a^{12}$ ✓ (1 mark)
e $6ab$ ✓ (1 mark)
f $6b^2a$ ✓ (1 mark)

(Total 40 marks)

Level 3 Test

1 a 28 ✓ (1 mark)

b

Number of squares	1	3	5	7	9	11
Number of matchsticks	4	10	16	22	28	34

✓ (1 mark)

c The number of matchsticks = 3 × number of squares + 1. ✓ (1 mark)
d $m = 3T + 1$ ✓ (1 mark)
e $3 \times 57 + 1 = 172$ ✓ (1 mark)
f $3T + 1 = 601$, $T = 200$ ✓ (1 mark)

2 a 10, 12, 14, add 2 to get the next term. ✓ (1 mark)
b 15, 11, 7, subtract 4 to get the next term. ✓ (1 mark)
c 13, 21, 34, add previous two terms. ✓ (1 mark)

3 a

x	1	2	3	4
y	1	7	17	31

✓✓ (2 marks)

b

x	1	5	21	50
y	1	13	61	148

✓✓ (2 marks)

4 a $y = x^2 - 3$ ✓✓ (2 marks)
b $y = 5 - x^2$ ✓✓ (2 marks)

5 a $2xy + 4x$ ✓ (1 mark)
b $15t^3m^5$ ✓ (1 mark)
c $2x$ ✓ (1 mark)
d $5a^3$ ✓ (1 mark)
e $\frac{2x}{3}$ ✓ (1 mark)
f $8a^3$ ✓ (1 mark)
g $9a^8b^4$ ✓ (1 mark)
h $4ab$ ✓ (1 mark)

6 **a** $P = (220 - 30) \times 4 \div 5$
$= 760 \div 5$
$= 152$ ✓ (1 mark)

b $2 \times 3 \times 3 - 3 + 4 = 19$ ✓ (1 mark)

c $2x = 8, x = 4$
$20 - 3x = 20 - 3 \times 4 = 8$ ✓ (1 mark)

7 **a** $\frac{5t^2}{2}$ ✓✓ (2 marks)

b $\frac{1}{2}$ ✓✓ (2 marks)

8 **a** $c = \frac{3 \times 30}{3 + 12}$ ✓
$= \frac{90}{15}$
$= 6$ ✓ (2 marks)

b $12 = \frac{6 \times x}{6 + 12}$
$= \frac{x}{3}$ ✓
$x = 3 \times 12$
$= 36$ ✓ (2 marks)

9 **a** $3a^2 = 3 \times a^2$
$= 3 \times (2)^2$
$= 3 \times 4$
$= 12$ ✓ (1 mark)

b $x(x^2 - 2) = 5 \times (5^2 - 2)$
$= 5 \times (25 - 2)$
$= 5 \times 23$
$= 115$ ✓ (1 mark)

10 **a** $(5a)^3 - (5a)^0 = 5a \times 5a \times 5a - 1$
$= 125a^3 - 1$ ✓ (1 mark)

b $8a^2 \div 4ab = \frac{8a^2}{4ab}$
$= \frac{\cancel{8}^2 \times \cancel{a}^1 \times a}{\cancel{4}_1 \times \cancel{a}_1 \times b}$
$= \frac{2a}{b}$ ✓ (1 mark)

(Total 40 marks)

Chapter 8—Measurement pp. 158–165

Level 1 Test

1 Perimeter by measurement = 2.4 + 3.2 + 4.0
= 9.6
i.e. perimeter is 9.6 cm ✓ (1 mark)

2 As 23 000 ÷ 1000 = 23
∴ 23 000 m = 23 km ✓ (1 mark)

3 **a** Per. = 14 × 4 = 56
i.e. perimeter is 56 m ✓ (1 mark)

b Per. = 2(9 + 2) = 2(11) = 22
i.e. perimeter is 22 m ✓ (1 mark)

c Per = 7 + 7 + 12 + 12 + 8 = 46
i.e. perimeter is 46 m ✓ (1 mark)

4 434.5 cm to 435.5 cm
(i.e. 434 ± 0.5) ✓ (1 mark)

5 As perimeter = 14
∴ 5 + 5 + side = 14
∴ side = 4
i.e. missing side is 4 cm ✓ (1 mark)

6 Per. = 16 × 4 = 64 m ✓
∴ Cost = $2 × 64 = $128
∴ Cost is $128. ✓ (2 marks)

7

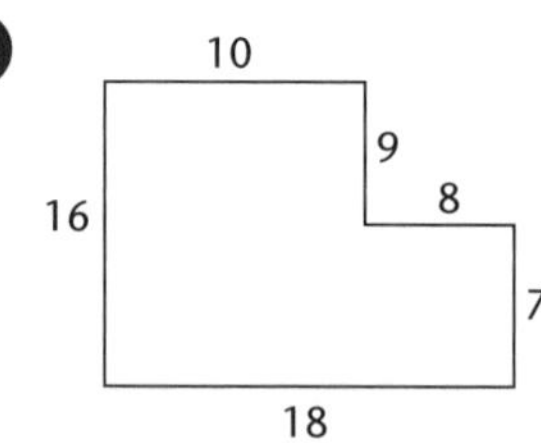

∴ Per. = 16 + 18 + 7 + 8 + 9 + 10 = 68 ✓
∴ Perimeter is 68 cm. ✓ (2 marks)

8 Length 4 × 16 + 2 × 15 + 5 ✓
= 64 + 30 + 5
= 99
∴ Requires 99 m tape. ✓ (2 marks)

9 **a** Area = 12 × 7 = 84
∴ The area is 84 cm^2. ✓ (1 mark)

b Area = 20^2 = 400
∴ The area is 400 mm^2. ✓ (1 mark)

c Area = 14 × 6 = 84
∴ The area is 84 cm^2. ✓ (1 mark)

WORKED SOLUTIONS to Chapter 8 **Tests**

10 **a** Vol. $= 120 \times 12 = 1440$

$\therefore$ Volume is 1440 cm^3. ✓ (1 mark)

b Vol. $= 8 \times 7 \times 5 = 280$

$\therefore$ Volume is 280 cm^3. ✓ (1 mark)

11 **a**

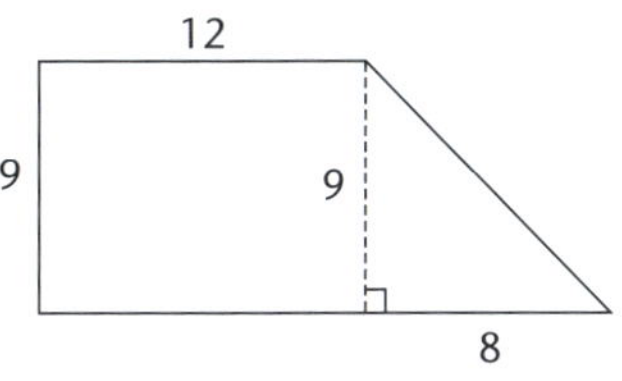

$\therefore$ Area $= 12 \times 9 + \frac{1}{2} \times 8 \times 9$ ✓

$= 108 + 36$

$= 144$

$\therefore$ Area is 144 cm^2. ✓ (2 marks)

b

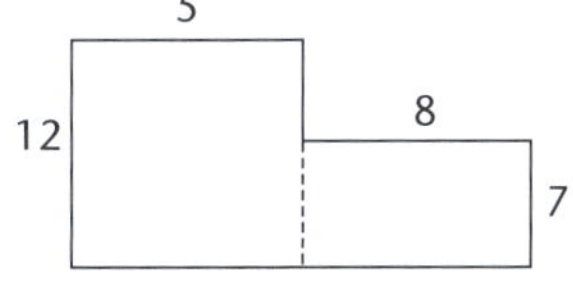

$\therefore$ Area $= 12 \times 5 + 8 \times 7$ ✓

$= 60 + 56$

$= 116$

$\therefore$ The area is 116 cm^2. ✓ (2 marks)

12 Area $= 16 \times 9 - 7 \times 4$ ✓

$= 144 - 28$

$= 116$

$\therefore$ The area is 116 cm^2. ✓ (2 marks)

13 As area = 48 cm^2

$\therefore$ length $\times$ width $= 48$

i.e. $10 \times$ width $= 48$ ✓

$\therefore$ width $= 48 \div 10 = 4.8$

$\therefore$ The width is 4.8 cm. ✓ (2 marks)

14 Vol. $= \frac{1}{2} \times 12 \times 8 \times 20$ ✓

$= 48 \times 20$

$= 960$

$\therefore$ Volume is 960 cm^3. ✓ (2 marks)

15 Vol. $= 8^3$

$= 64 \times 8$

$= 512$ ✓

i.e. volume is 512 cm^3

$\therefore$ The capacity is 512 mL. ✓ (2 marks)

(Total 30 marks)

Level 2 Test

1 5 cm by measurement ✓ (1 mark)

2 $2400 \div 100 = 24$

i.e. 24 m ✓ (1 mark)

3 **a** Per. $= 2 \times (18 + 13)$

$= 2 \times 31$

$= 62$

i.e. perimeter is 62 m ✓ (1 mark)

b Per. $= 25 \times 4$

$= 100$

$\therefore$ Perimeter is 100 mm. ✓ (1 mark)

4 68 ± 0.5

$= 67.5$ cm and 68.5 cm ✓ (1 mark)

5 Perimeter = 40 cm

$\therefore 2 \times$ (length + width) $= 40$

i.e. length + width $= 20$ ✓

$\therefore 12 +$ width $= 20$

i.e. width $= 8$

$\therefore$ Width is 8 cm. ✓ (2 marks)

6

24
13
23
14
10
10

Per. $= 23 + 10 + 10 + 14 + 13 + 24$ ✓

$= 94$

$\therefore$ Perimeter is 94 cm. ✓ (2 marks)

7

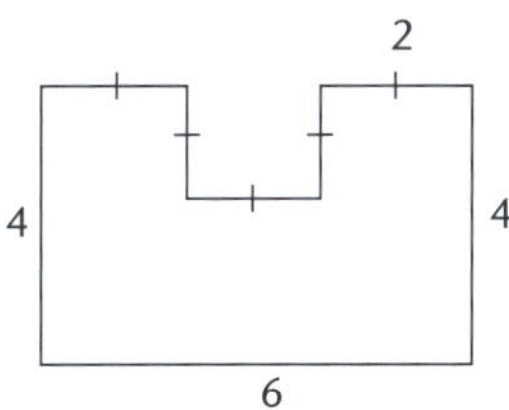

$\therefore$ Per. $= 4 + 6 + 4 + 5 \times 2$ ✓

$= 14 + 10$

$= 24$

$\therefore$ Perimeter is 24 cm. ✓ (2 marks)

8 As $310\,000 \div 10\,000 = 31$

$\therefore 310\,000$ m^2 $= 31$ ha ✓ (1 mark)

WORKED SOLUTIONS to Chapter 8 **Tests**

9 **a** Area = $48 \times 11 = 528$
$\therefore$ Area is 528 cm^2. ✓ (1 mark)

b Area = $\frac{1}{2} \times 30 \times 15 = \frac{1}{2} \times 450 = 225$
$\therefore$ Area is 225 cm^2. ✓ (1 mark)

c Area = $20 \times 12.2 = 244$
$\therefore$ Area is 244 cm^2. ✓ (1 mark)

10

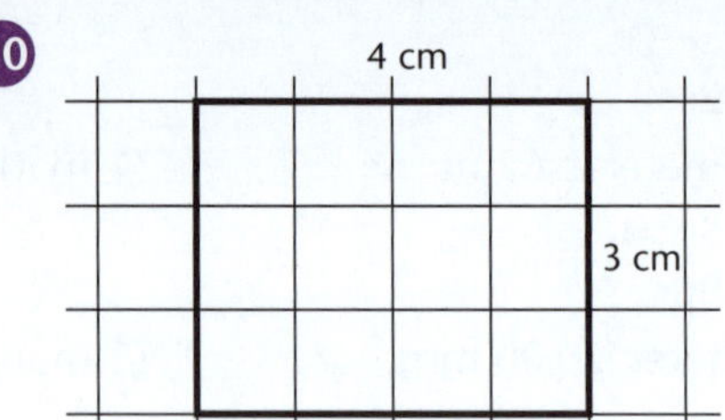

$\therefore$ A rectangle measuring 4 cm by 3 cm. ✓ (1 mark)

11 **a**

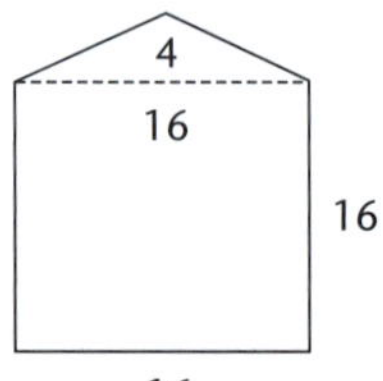

$\therefore$ Area = $16 \times 16 + \frac{1}{2} \times 16 \times 4$ ✓
= 256 + 32
= 288
$\therefore$ Area is 288 m^2. ✓

16 ×
16
96
160
256

(2 marks)

b

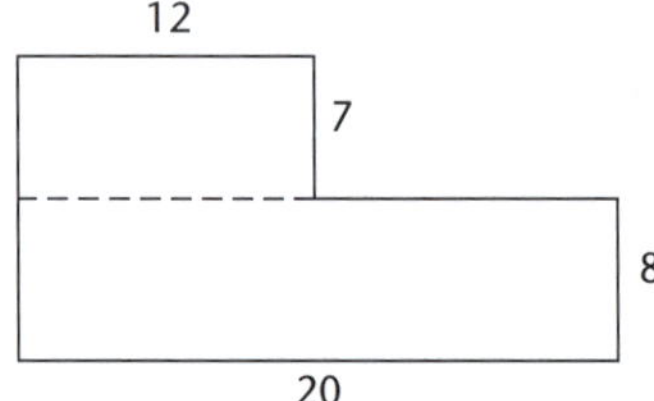

Area = $20 \times 8 + 12 \times 7$ ✓
= 160 + 84
= 244
$\therefore$ Area is 244 m^2. ✓ (2 marks)

12

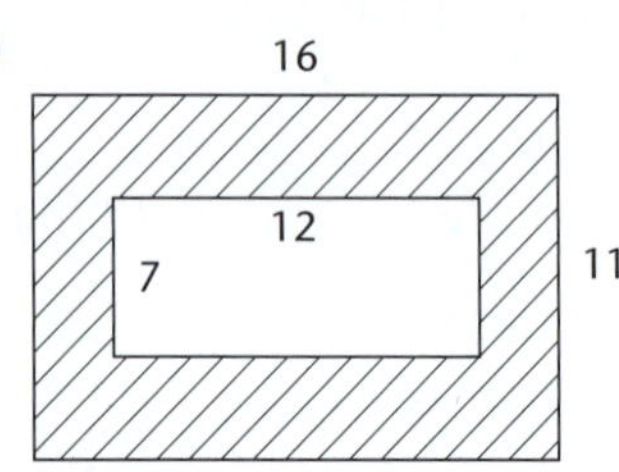

Area = $16 \times 11 - 12 \times 7$ ✓
= 176 − 84 = 92
$\therefore$ Area is 92 cm^2. ✓ (2 marks)

13 Area = $30 \times 14 = 420$ ✓
$\therefore$ Fertiliser = $420 \times 80 = 33\,600$
$\therefore$ 33 600 g fertiliser is required ✓
i.e. 33.6 kg (2 marks)

14 Area = 4.2×2.8 ✓
= 11.76
$\therefore$ Area is 11.76 m^2. ✓

42 ×
28
336
840
1176

(2 marks)

15 **a** Vol. = $\frac{1}{2} \times 12 \times 6 \times 10$ ✓
= 360
$\therefore$ Volume is 360 m^3. ✓ (2 marks)

b Vol. = $(20 \times 10 - 4 \times 2) \times 6$ ✓
= $(200 - 8) \times 6$
= $192 \times 6 = 1152$
$\therefore$ Volume is 1152 cm^3. ✓ (2 marks)

(Total 30 marks)

Level 3 Test

1 $4\,000\,000 \div 1000 = 4000$
i.e. 4000 m
$\therefore$ $4000 \div 1000 = 4$
i.e. 4 km
$\therefore$ 4 000 000 mm = 4 km ✓ (1 mark)

2 Per. = $125 \times 8 = 1000$
$\therefore$ perimeter is 1000 cm
i.e. 10 m ✓ (1 mark)

3 Fence = width + width + length
= width + width + 3 widths
= 5 widths ✓

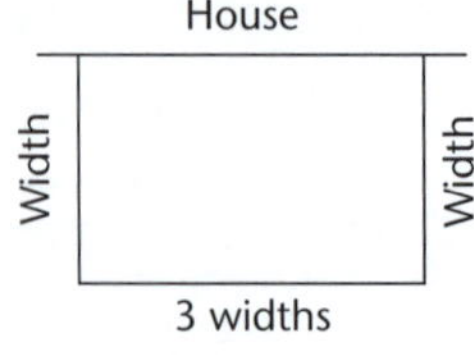

$\therefore$ 5 widths = 25
$\therefore$ width = 5
i.e. width 5 m, length 15 m ✓ (2 marks)

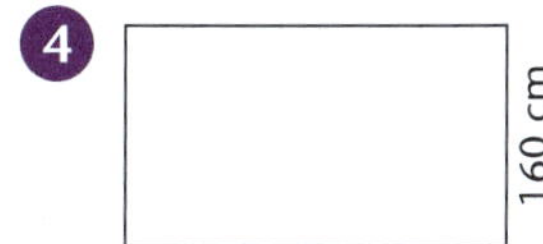

WORKED SOLUTIONS to Chapter 8 **Tests**

4

160 cm

300 cm

$\therefore$ Per. $= 2 \times (300 + 160)$ ✓
$= 2 \times 460$
$= 920$

$\therefore$ Perimeter is 920 cm. ✓ (2 marks)

5

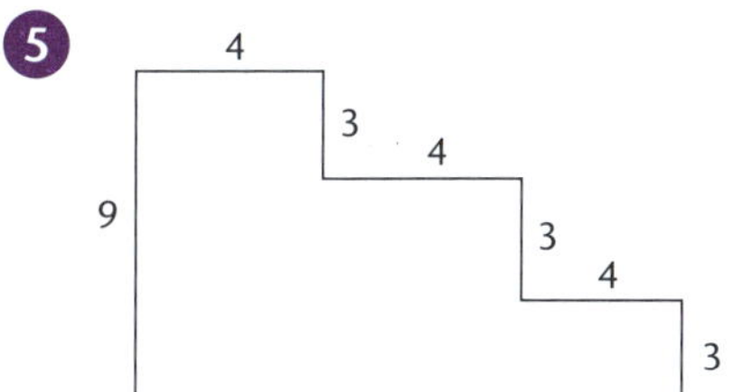

Per. $= 2 \times (9 + 12)$ ✓
$= 2(21)$
$= 42$

$\therefore$ Perimeter is 42 cm. ✓ (2 marks)

6

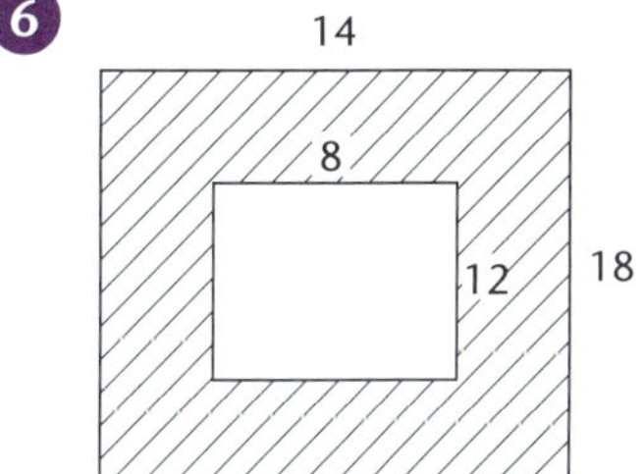

$\therefore$ Per. $= 2 \times (18 + 14)$ ✓
$= 64$

$\therefore$ Length of frame is 64 cm. ✓ (2 marks)

7

Area = base $\times$ height
$800 = \text{base} \times 32$ ✓
$\therefore$ Base $= 800 \div 32$
$= 25$

$\therefore$ Base is 25 cm. ✓ (2 marks)

8

4 cm

4 cm

✓ (1 mark)

9 $6.814 \times 1000 \times 1000$
$\therefore$ 6.814 km = 6 814 000 mm ✓ (1 mark)

10 Area $= 2 \times 1.2 = 2.4$ ✓
Area is 2.4 m^2. ✓ (2 marks)

11 Area $= 160 \times 60 = 9600$ ✓
$\therefore$ Area is 9600 m^2
Area is 0.96 ha. ✓ (2 marks)

12

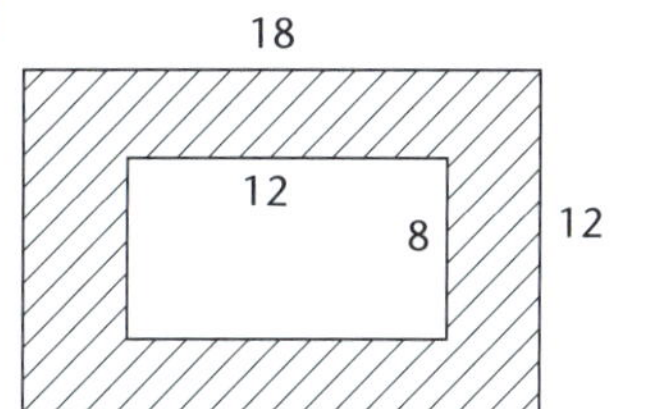

Area $= 18 \times 12 - 12 \times 8$ ✓
$= 216 - 96$
$= 120$

$\therefore$ Area is 120 cm^2. ✓ (2 marks)

13 Area $= 18 \times 12 - \frac{1}{2} \times 18 \times 6$ ✓
$= 216 - 54$
$= 162$

$\therefore$ Area is 162 cm^2. ✓ (2 marks)

14 Number of rolls $= 4 \times 5 = 20$ ✓
$\therefore$ Cost $= \$3 \times 20 = \$60$

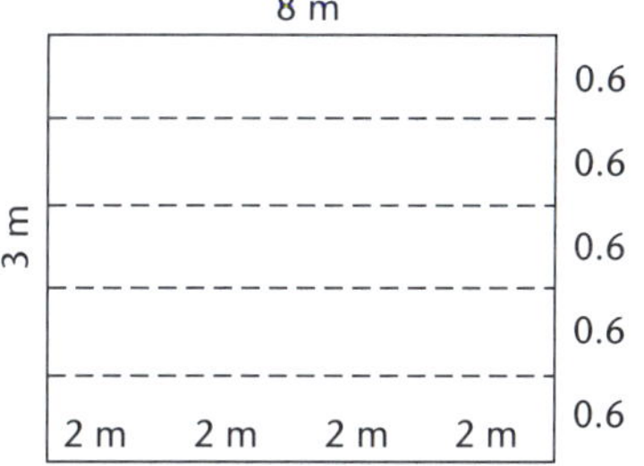

$\therefore$ The turf will cost \$60. ✓ (2 marks)

15 As volume = 27 cm^3
$\therefore$ Side = 3 cm long ✓
$\therefore$ Area $= 3 \times 3 = 9$
Area is 9 cm^2. ✓ (2 marks)

WORKED SOLUTIONS to Chapters 8 and 9 **Tests**

16 **a**

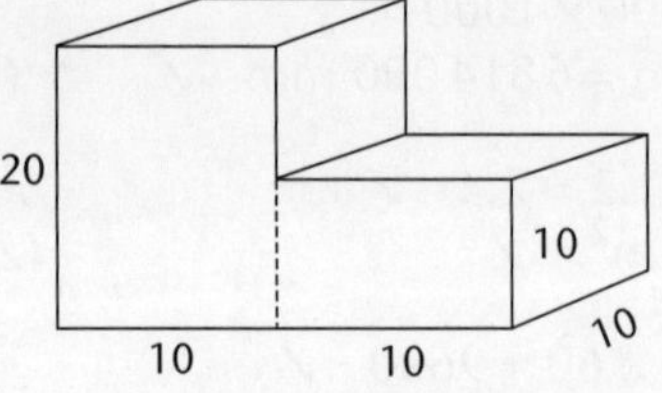

$\therefore$ Vol. $= (20 \times 10 + 10 \times 10) \times 10$ ✓
$= (200 + 100) \times 10$
$= 300 \times 10$
$= 3000$
$\therefore$ Volume is 3000 cm^3. ✓ (2 marks)

b

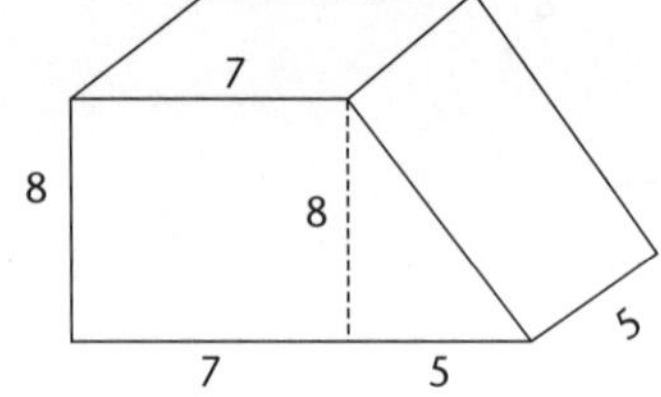

Vol. $= (8 \times 7 + \frac{1}{2} \times 5 \times 8) \times 5$ ✓
$= (56 + 20) \times 5$
$= 76 \times 5$
$= 380$
$\therefore$ Volume is 380 cm^3. ✓ (2 marks)

(Total 30 marks)

Chapter 9—Geometry

pp. 193–202

Level 1 Test

1 **a** pentagon ✓ (1 mark)
b circle ✓ (1 mark)
c parallelogram ✓ (1 mark)
d trapezium ✓ (1 mark)

2 **a** sphere ✓ (1 mark)
b cone ✓ (1 mark)
c rectangular prism ✓ (1 mark)
d triangular prism ✓ (1 mark)

3 **a** T ✓ (1 mark)
b F ✓ (1 mark)
c T ✓ (1 mark)
d T ✓ (1 mark)
e T ✓ (1 mark)
f F ✓ (1 mark)

4 **a** triangle ✓ (1 mark)
b triangular prism ✓ (1 mark)
c yes ✓ (1 mark)
d 5 ✓ (1 mark)
e 9 ✓ (1 mark)
f 6 ✓ (1 mark)

5 **a** $\angle MPN$ or $\angle NPM$ ✓ (1 mark)
b acute ✓ (1 mark)

6 octagon ✓ (1 mark)

7 4 ✓ (1 mark)

8 **a** 180° ✓ (1 mark)
b 90° ✓ (1 mark)

9 **a** D ✓ (1 mark)
b G ✓ (1 mark)
c I ✓ (1 mark)
d L ✓ (1 mark)
e K ✓ (1 mark)
f B ✓ (1 mark)
g C ✓ (1 mark)
h F ✓ (1 mark)
i A ✓ (1 mark)
j E ✓ (1 mark)

10 **a** 180° ✓ (1 mark)
b 360° ✓ (1 mark)

11 **a** $x + 115 = 180$ (supplementary angles)
$x = 65$ ✓ (1 mark)
b $x = 72$ (vertically opposite angles) ✓ (1 mark)
c $x + 60 + 70 + 90 = 360$ (angles at a point)
$x = 140$ ✓ (1 mark)

12 **a** $a + 72 + 54 = 180$ (angles in a triangle)
$a = 54$ ✓ (1 mark)
b $2x + 70 = 180$ (isosceles triangle)
$x = 55$ ✓ (1 mark)
c $m = 60$ (equilateral triangle) ✓ (1 mark)
d $x + 55 + 75 + 110 = 360$
(angles in a quadrilateral)
$x = 120$ ✓ (1 mark)

13 **a** $x = 65$ (corresponding angles) ✓ (1 mark)

b $x + 48 = 180$ (co-interior angles)
$x = 132$ ✓ (1 mark)

c $x = 117$ (alternate angles) ✓ (1 mark)

14 $x + 55 + 42 = 180$ (angles in a triangle)
$x = 83$ ✓
$y = 55 + 42$ (sum of opposite interior angles)
$y = 97$
or $x + y = 180$ (supplementary angles)
$83 + y = 180$
$y = 97$ ✓ (2 marks)

(Total 50 marks)

Level 2 Test

1 **a** Hexagon ✓ (1 mark)
b Ellipse ✓ (1 mark)
c Isosceles triangle ✓ (1 mark)
d Rhombus ✓ (1 mark)

2 **a** Cylinder ✓ (1 mark)
b Square pyramid ✓ (1 mark)
c Pentagonal prism ✓ (1 mark)
d Hemisphere ✓ (1 mark)

3 **a** F ✓ (1 mark)
b T ✓ (1 mark)
c [illegible] ✓ (1 mark)
d T ✓ (1 mark)
e T ✓ (1 mark)
f F ✓ (1 mark)
g T ✓ (1 mark)
h T ✓ (1 mark)

4

5 ✓ (1 mark)

5

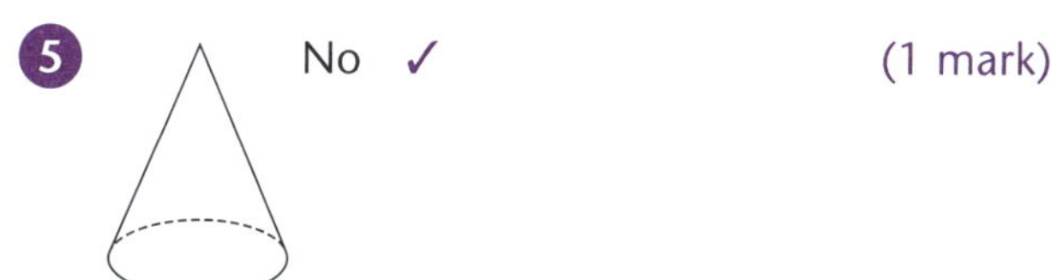

No ✓ (1 mark)

6 **a** Yes ✓ (1 mark)
b Hexagonal prism ✓ (1 mark)
c 8 ✓ (1 mark)
d Yes ✓ (1 mark)

7

✓ (1 mark)

8 **a** $x = 360 - (100 + 125 + 75)$
$= 360 - 300$
$= 60$ ✓
$y = 180 - 60$
$= 120$ ✓ (2 marks)

b $x = 38$ ✓
$y = 180 - (38 + 42)$
$= 180 - 80$
$= 100$ ✓ (2 marks)

9

✓ (1 mark)

10 **a** 15° ✓ (1 mark)
b 105° ✓ (1 mark)

11 **a** Equal ✓ (1 mark)
b **i** Supplementary ✓
ii Equal ✓
iii Equal ✓ (3 marks)

12 **a** $x + 20 = 80$ (vertically opposite angles)
$x = 60$ ✓ (1 mark)

b $x + 90 + 62 = 180$ (angles on a line)
$x = 28$ ✓ (1 mark)

c $x + 56 = 180$ (co-interior angles)
$x = 124$ ✓ (1 mark)

d $x + 2 \times 55 = 180$ (angles in isosceles triangle)
$x = 70$ ✓ (1 mark)

e Angle vertically opposite to x
$= 180 - 54$ (co-interior angles)
$\therefore x = 180 - 54$ (vertically opposite angles)
$x = 126$ ✓ (1 mark)

f $2x + 90 + 70 = 360$ (angles at a point)
$2x = 200$
$\therefore x = 100$ ✓ (1 mark)

13 $a = 60$ (angle in an equilateral triangle) ✓
$b + a = 180$ (supplementary angles)
$b = 120$ ✓ (2 marks)

14 $y + 45 + 55 = 180$ (angles in a triangle)
$y = 80$ ✓
$x + y = 180$ (supplementary angles)
$x = 100$
or $x = 55 + 45$ (opposite interior angles)
$x = 100$ ✓ (2 marks)

15 **a** ∠BCA = 52° = ∠ABC ✓
(base angles of isosceles △ABC) ✓ (2 marks)
b ∠CAB = 76° ✓
(angle sum of △ABC) ✓ (2 marks)
c ∠ACE = 76° ✓
(alternate to ∠CAB and AB || EC) ✓ (2 marks)

(Total 50 marks)

Level 3 Test

1 B ✓ (1 mark)

2 D ✓ (1 mark)

3 C ✓ (1 mark)

4 B ✓ (1 mark)

5 D ✓ (1 mark)

6 B ✓ (1 mark)

7 B ✓ (1 mark)

8 B ✓ (1 mark)

9 D ✓ (1 mark)

10 A ✓ (1 mark)

11 **a** ∠ABC or ∠CBA ✓ (1 mark)
b ∠CDB or ∠BDC ✓ (1 mark)
c ∠BDA or ∠ADB ✓ (1 mark)
d ∠DBC or ∠CBD ✓ (1 mark)

12

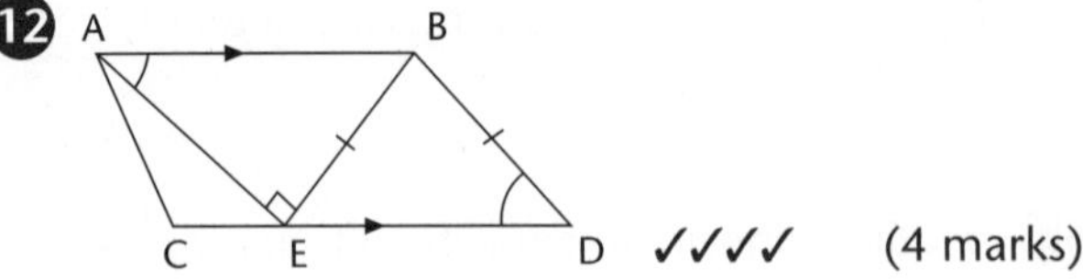

✓✓✓✓ (4 marks)

13 **a**

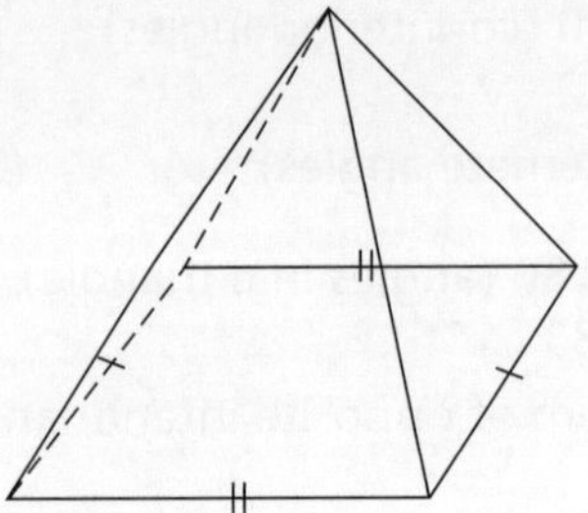

✓ (1 mark)
b No ✓ (1 mark)
c Yes ✓ (1 mark)

14 **a** $x = 55$ ✓
$y = 75$ ✓ (2 marks)
b $x = 115$ ✓
$y = 115$ ✓ (2 marks)
c $x = 119$ ✓
$y = 61$ ✓ (2 marks)

15 **a** Yes ✓ (1 mark)
b No ✓ (1 mark)

16 **a** $a = 54$ ✓
$b = 54$ ✓
$c = 54$ ✓
$x = 72$ ✓ (4 marks)
b $x = 72$ ✓
$y = 118$ ✓
$z = 118$ ✓ (3 marks)

17 **a** $x = 54$ ✓✓
b $x = 132$ ✓✓ (4 marks)

18 $x = 54$ ✓
$y = 72$ ✓
$z = 72 - 54 = 18$ ✓ (3 marks)

19 $x = 65$ (base angles of isoceles equal) ✓
$y = 115$ (angles in straight line) ✓ (2 marks)

(Total 45 marks)

Chapter 10—Equations

pp. 212–215

Level 1 Test

1 **a** $14 + 4 \neq 10$ ∴ No ✓ (1 mark)
b $8 - 5 = 3$ ∴ Yes ✓ (1 mark)
c $5 \times 9 = 45 \neq 4.5$ ∴ No ✓ (1 mark)
d $-\frac{12}{4} = -3$ ∴ Yes ✓ (1 mark)

WORKED SOLUTIONS to Chapter 10 **Tests**

e $3 \times 7 - 1 = 21 - 1 = 20 \therefore$ Yes ✓ (1 mark)

f $\frac{4+2}{3} = \frac{6}{3} \neq 9 \therefore$ No ✓ (1 mark)

2 **a** $p + 3 = 12$
$p = 12 - 3$
$p = 9$ ✓ (1 mark)

b $a - \frac{1}{2} = 3$
$a = 3 + \frac{1}{2}$
$a = 3\frac{1}{2}$ ✓ (1 mark)

c $\frac{m}{10} = 1.23$
$m = 1.23 \times 10$
$m = 12.3$ ✓ (1 mark)

d $7p = 21$
$p = \frac{21}{7}$
$p = 3$ ✓ (1 mark)

e $2y - 1 = 11$
$2y = 11 + 1$
$2y = 12$
$y = \frac{12}{2}$
$y = 6$ ✓ (1 mark)

3 **a** $a + 8 = 23$
$a = 23 - 8$
$a = 15$ ✓ (1 mark)

b $p - 7 = 15$
$p = 15 + 7$
$p = 22$ ✓ (1 mark)

c $\frac{a-1}{2} = 9$
$a - 1 = 9 \times 2$
$a - 1 = 18$
$a = 18 + 1$
$a = 19$ ✓ (1 mark)

d $\frac{x}{5} = 6$
$x = 6 \times 5$
$x = 30$ ✓ (1 mark)

e $y + 7 = 2$
$y = 2 - 7$
$y = -5$ ✓ (1 mark)

4 **a** $2x + 1 = 7$
$2x = 7 - 1$
$2x = 6$ ✓
$x = \frac{6}{2}$
$x = 3$ ✓ (2 marks)

b $2x - 3 = 7$
$2x = 7 + 3$
$2x = 10$ ✓
$x = \frac{10}{2}$
$x = 5$ ✓ (2 marks)

c $\frac{a-2}{3} = 6$
$a - 2 = 18$ ✓
$a = 18 + 2$
$a = 20$ ✓ (2 marks)

d $2x = x + 3$
$2x - x = 3$ ✓
$x = 3$ ✓ (2 marks)

e $8 + 3a = 14$
$3a = 14 - 8$
$\frac{3a}{3} = \frac{6}{3}$ ✓
$a = 2$ ✓ (2 marks)

5 **a** **i** Perimeter = 20
$4x = 20$
$x = \frac{20}{4}$
$x = 5$ ✓ (1 mark)

ii $A = 5 \times 5 = 25$ cm^2 ✓ (1 mark)

b **i** Peter = $(x - 4)$ years ✓ (1 mark)

ii $x + x - 4 = 10$
$2x - 4 = 10$
$2x = 10 + 4$
$2x = 14$
$x = 7$
$\therefore$ Michael is 7 years old. ✓ (1 mark)

(Total 30 marks)

Level 2 Test

1 $\text{LHS} = \frac{3x}{2}$
$= \frac{3 \times 5}{2}$
$= \frac{15}{2}$
$= 7\frac{1}{2}$
$\neq$ RHS
$\therefore x = 5$ is not a solution. ✓ (1 mark)

WORKED SOLUTIONS to Chapter 10 **Tests**

2 **a** $x + 2.3 = 7.4$
$x = 7.4 - 2.3$
$x = 5.1$ ✓ (1 mark)

b $x - 2\frac{1}{2} = \frac{1}{4}$
$x = \frac{1}{4} + 2\frac{1}{2}$
$x = 2\frac{3}{4}$ ✓ (1 mark)

c $x + 8 = 2$
$x = 2 - 8$
$x = -6$ ✓ (1 mark)

d $\frac{x+1}{2} = 4$
$x + 1 = 8$
$x = 7$ ✓ (1 mark)

e $5x = 6$
$x = \frac{6}{5}$
$x = 1\frac{1}{5}$ ✓ (1 mark)

f $x - 1 = -9$
$x = -9 + 1$
$x = -8$ ✓ (1 mark)

3 **a** $5x - 3 = 27$
$5x = 27 + 3$
$5x = 30$ ✓
$x = \frac{30}{5}$
$x = 6$ ✓ (2 marks)

b $3x + 5 = -7$
$3x = -7 - 5$
$3x = -12$ ✓
$x = -\frac{12}{3}$
$x = -4$ ✓ (2 marks)

c $\frac{2x - 2}{3} = 4$
$2x - 2 = 12$ ✓
$2x = 14$
$x = 7$ ✓ (2 marks)

d $5x = 3x + 12$
$5x - 3x = 12$ ✓
$2x = 12$
$x = 6$ ✓ (2 marks)

e $4x - 3 = 2x + 18$
$4x - 2x = 18 + 3$
$2x = 21$ ✓
$x = 10\frac{1}{2}$ ✓ (2 marks)

f $\frac{2x - 1}{3} = 5$
$2x - 1 = 3 \times 5$
$2x = 15 + 1$
$\frac{2x}{2} = \frac{16}{2}$ ✓
$x = 8$ ✓ (2 marks)

4 **a** $2x + x + 90 = 180$ ✓
$3x + 90 = 180$ (angle sum of a triangle)
$3x = 90$
$x = 30$ ✓ (2 marks)

b $3x = 2x + 20$ ✓
$3x - 2x = 20$ (vertically opposite angles)
$x = 20$ ✓ (2 marks)

5 **a** $v = u + at$
$= 5 + 2 \times 3$
$= 11$ ✓ (1 mark)

b $v = u + at$
$20 = 5 + a \times 3$
$20 = 5 + 3a$
$3a = 20 - 5$
$3a = 15$
$a = 5$ ✓ (1 mark)

6 $2x - 3 = 21$ ✓
$2x = 24$
$x = 12$
∴ Number is 12. ✓ (2 marks)

7 **a** Perimeter = 16
$x - 2 + x - 2 + x + x = 16$ ✓
$4x - 4 = 16$
$4x = 20$
$x = 5$ ✓ (2 marks)

b Breadth = (5 − 2) = 3 cm
Length = 5 cm
Area = $3 \times 5 = 15$ cm^2 ✓ (1 mark)

(Total 30 marks)

Level 3 Test

1 $2x - 1 = 7$
$2x = 7 + 1$
$2x = 8$
$x = 4$
Answer is A. ✓ (1 mark)

WORKED SOLUTIONS to Chapter 10 **Tests**

2 $\frac{2x}{3} - 1 = 7$

$\frac{2x}{3} = 8$

$2x = 24$

$x = 12$

Answer is C. ✓ (1 mark)

3 $x = \frac{y}{z}$

$4 = \frac{32}{z}$

$z = \frac{32}{4}$

$z = 8$

Answer is A. ✓ (1 mark)

4 $3x + 2 = 6 - x$

Lucy should move the $-x$ over and change its sign, similarly the + 2. Thus line (1) should be:

$3x + x = 6 - 2$

Lines (2) and (3) are correct for the mistaken line (1).

∴ Answer is B. ✓ (1 mark)

5 $4x = -2x + 6$

$4x + 2x = 6$

$6x = 6$

$x = 1$

Answer is A. ✓ (1 mark)

6 **a** $a - 3 = -15$

$a = -15 + 3$

$a = -12$ ✓ (1 mark)

b $x + 17 = 5$

$x = 5 - 17$

$x = -12$ ✓ (1 mark)

c $\frac{x}{0.4} = 1.2$

$x = 1.2 \times 0.4$

$x = 0.48$ ✓ (1 mark)

d $7x = 23$

$x = \frac{23}{7}$

$x = 3\frac{2}{7}$ ✓ (1 mark)

e $\frac{3x}{2} - 1 = 12$

$\frac{3x}{2} = 13$

$3x = 26$

$x = 8\frac{2}{3}$ ✓ (1 mark)

7 **a** $5x + 17 = 22$

$5x = 22 - 17$

$5x = 5$ ✓

$x = 1$ ✓ (2 marks)

b $\frac{2x - 3}{5} = 3$

$2x - 3 = 15$ ✓

$2x = 15 + 3$

$2x = 18$

$x = 9$ ✓ (2 marks)

c $12x - 7 = 4x + 49$

$12x - 4x = 49 + 7$ ✓

$8x = 56$

$x = 7$ ✓ (2 marks)

d $2x - 6 = -4x$

$2x + 4x = 6$ ✓

$6x = 6$

$x = 1$ ✓ (2 marks)

8 $K = \frac{2B - 1}{A}$

$12.5 = \frac{2B - 1}{4}$ ✓

$2B - 1 = 50$

$2B = 51$

$B = 25.5$ ✓ (2 marks)

9 Let number be x:

$17 - 3x = 2$ ✓

$-3x = -15$

$x = 5$

∴ Number is 5. ✓ (2 marks)

10 Let consecutive odd integers be n, $n + 2$: ✓

$n + n + 2 = 68$

$2n + 2 = 68$ ✓

$2n = 66$

$n = 33$

∴ Consecutive odd integers are 33 and 35. ✓ (3 marks)

11 $2x + 2x + 52 = 180$

$4x + 52 = 180$ ✓

$4x = 128$

$x = 32$

Base angles = $2x° = 64°$ ✓ (2 marks)

WORKED SOLUTIONS to Chapters 10 and 11 **Tests**

12 Perimeter = 18

$x + x + 2x + 2x = 18$ ✓

$6x = 18$

$x = 3$

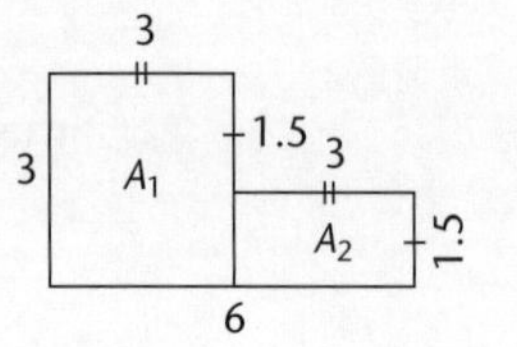

✓

$A_1 = 3 \times 3 = 9\text{ cm}^2$

$A_2 = 3 \times 1.5 = 4.5\text{ cm}^2$

$A = A_1 + A_2$

$= 9 + 4.5$

$= 13.5\text{ cm}^2$ ✓ (3 marks)

(Total 30 marks)

Chapter 11—Statistics and Probability

pp. 234–238

Level 1 Test

1 **a**

Score (x)	Tally	Frequency
11	||	2
12	||	2
13	~~||||~~	5
14	||||	4
15	||||	4
16	~~||||~~ |	6
17	~~||||~~ |	6
18	~~||||~~ |	6
19	~~||||~~	5

✓✓✓ (3 marks)

b

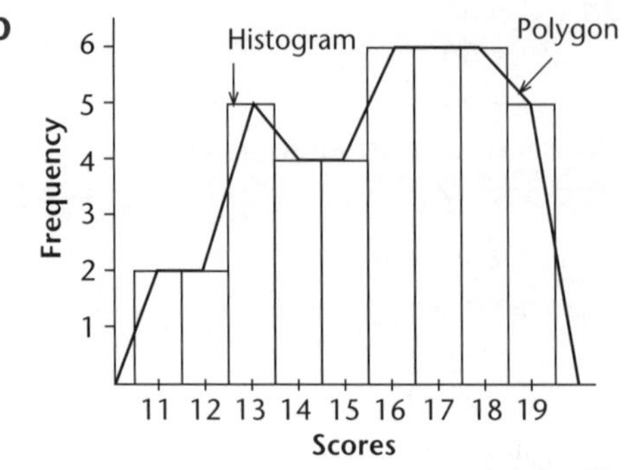

✓✓✓ (3 marks)

2 **a**

Score (x)	Frequency (f)	$x \times f$
22	5	110
23	7	161
24	10	240
25	8	200
26	5	130
Total	**35**	**841**

✓✓ (2 marks)

b Mean $= \frac{\Sigma fx}{\Sigma f}$

$= \frac{841}{35}$

$= 24.03$ (to 2 decimal places) ✓ (1 mark)

c Mode: 24 ✓
Median = 18th score
i.e. 24 ✓
Range = 4 ✓ (3 marks)

d 12 ✓ (1 mark)

e $5 + 10 + 5 = 20$ ✓ (1 mark)

3 **a** Mean = 12.5 (by calculator) ✓

or $\bar{x} = (12 + 18 + 14 + 3 + 5 + 19 + 9 + 11 + 14 + 20) \div 10$

$= \frac{125}{10}$

$= 12.5$ (1 mark)

b Median ~~3~~, ~~5~~, ~~9~~, ~~11~~, 12, 14, ~~14~~, ~~18~~, ~~19~~, ~~20~~
i.e. middle of 12 and 14
∴ The median is 13. ✓ (1 mark)

c Mode = 14 ✓ (1 mark)

d Range is $20 - 3 = 17$ ✓ (1 mark)

4 **a** The outlier is 45. ✓ (1 mark)

b $\bar{x} = 19.85$ (to 2 decimal places) ✓ (1 mark)

c

Stem	Leaf
0	~~69~~
1	~~244~~7~~89~~
2	~~3489~~
3	
4	~~5~~

∴ The median is 18. ✓ (1 mark)

d The mode is 14. ✓ (1 mark)

e The range is $45 - 6 = 39$. ✓ (1 mark)

WORKED SOLUTIONS to Chapter 11 **Tests**

5 a $P(5) = \frac{1}{6}$ ✓ (1 mark)

b $P(\text{Less than } 3) = \frac{2}{6} = \frac{1}{3}$ ✓ (1 mark)

(Total 25 marks)

Level 2 Test

1 a

Class	Class centre (x)	Tally	Frequency (f)	fx
51–55	53	||||	4	212
56–60	58	|||	3	174
61–65	63	~~||||~~ |||	8	504
66–70	68	~~||||~~ |	6	408
71–75	73	~~||||~~ |||	8	584
76–80	78	|	1	78
		Sum	30	1960

✓✓✓ (3 marks)

b $\bar{x} = \frac{\Sigma fx}{\Sigma f} = \frac{1960}{30}$

$= 65.\dot{3}$ ✓ (1 mark)

c 61–65 and 71–75 ✓ (1 mark)

d

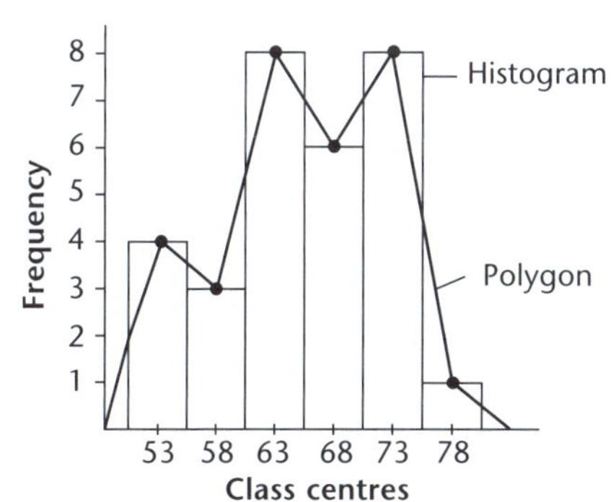

✓✓✓ (3 marks)

2 a $\bar{x} = 7.\dot{2}\dot{7}$ ✓ (1 mark)

b ~~2~~, ~~3~~, ~~4~~, ~~5~~, ~~5~~, 6, ~~7~~, ~~8~~, ~~11~~, ~~13~~, ~~16~~

∴ The median is 6. ✓ (1 mark)

c The mode is 5. ✓ (1 mark)

d The range is 16 – 2 = 14. ✓ (1 mark)

3 a

People	Houses
1	2
2	4
3	5
4	4
5	2
6	3

✓✓ (2 marks)

b Number of people

$= 1 \times 2 + 2 \times 4 + 3 \times 5 + 4 \times 4 + 5 \times 2 + 6 \times 3$

$= 2 + 8 + 15 + 16 + 10 + 18$

$= 69$ ✓✓ (2 marks)

c Number of houses $= 2 + 4 + 5 + 4 + 2 + 3$

$= 20$ ✓ (1 mark)

d Mean $= \frac{69}{20} = 3.45$ ✓ (1 mark)

4 a Sum of five scores $= 5 \times 8 = 40$ ✓ (1 mark)

b Sum of six scores $= 6 \times 10 = 60$ ✓

∴ New score $= 60 - 40 = 20$

∴ The new score is 20. ✓ (2 marks)

5 a

Stem	Leaf
2	0269
3	24578
4	1234688
5	19

✓✓ (2 marks)

b i $\bar{x} = 38.6\dot{1}$ ✓✓ (2 marks)

ii Mode = 48 ✓ (1 mark)

iii Middle of 38 and 41

∴ The median is 39.5. ✓ (1 mark)

iv 59 – 20 = 39 ✓ (1 mark)

6 a $\frac{1}{10}$ ✓ (1 mark)

b $\frac{5}{10} = \frac{1}{2}$ ✓ (1 mark)

c $\frac{7}{10}$ ✓ (1 mark)

d $P(\text{Composite}) = P(4, 6, 8, 9, 10)$

$= \frac{1}{2}$ ✓ (1 mark)

e $P(\text{Factor of } 12) = P(1, 2, 3, 4, 6)$

$= \frac{5}{10}$

$= \frac{1}{2}$ ✓ (1 mark)

f $P(\text{Multiple of } 3) = P(3, 6, 9)$

$= \frac{3}{10}$ ✓ (1 mark)

g $P(\text{Not a } 7) = 1 - P(7)$

$= 1 - \frac{1}{10}$

$= \frac{9}{10}$ ✓ (1 mark)

(Total 35 marks)

Level 3 Test

 **a**

Score (x)	Frequency (f)	fx
3	4	12
5	4	20
8	6	48
10	4	40
12	5	60
Sum	**23**	**180**

✓✓ (2 marks)

b Mean $= \frac{180}{23} = 7.83$ (to 2 decimal places) ✓

Mode = 8 ✓

Median is the 12th score, which is 8 ✓

Range is $12 - 3 = 9$ ✓ (4 marks)

c

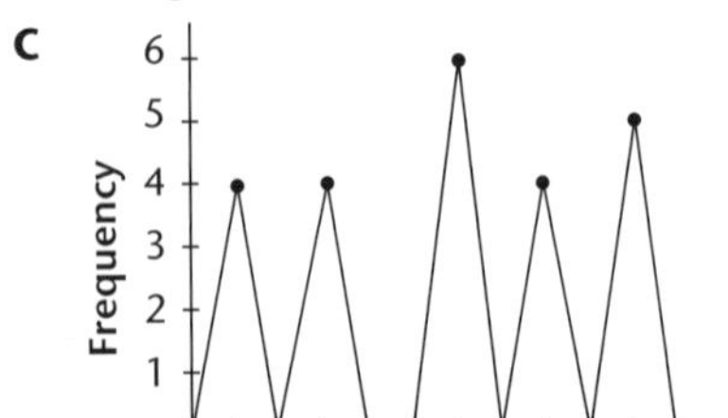

✓✓✓ (3 marks)

2 a

Stem	Leaf
3	24789
4	00246
5	258
6	36
7	☐

The median = 43 ✓ (1 mark)

b ∴ Range = 43

∴ 32 + 43 = 75

∴ ☐ = 5 ✓ (1 mark)

c

Class	Class centre	Frequency (f)	fx
31–35	33	2	66
36–40	38	5	190
41–45	43	2	86
46–50	48	1	48
51–55	53	2	106
56–60	58	1	58
61–65	63	1	63
66–70	68	1	68
71–75	73	1	73
	Sum	**16**	**758**

✓✓✓ (3 marks)

d Mean from stem-and-leaf is 47.5625

Mean from frequency table is $\frac{\Sigma fx}{\Sigma f}$

$= \frac{758}{16}$

$= 47.375$

The mean using the frequency table is less than the mean using the stem-and-leaf plot. ✓✓ (2 marks)

Score (x)	Frequency (f)	fx
2	4	8
4	6	24
Sum	**10**	**32**

∴ Mean $= \frac{32}{10} = 3.2$ ✓✓ (2 marks)

Or

Score (x)	Frequency (f)	fx
6	4	24
4	6	24
Sum	**10**	**48**

∴ The mean is 4.8.

WORKED SOLUTIONS to Chapter 11 **Tests**

4 Sum of goals in 5 games = $5 \times 22 = 110$
Sum of goals in 6 games = $6 \times 19 = 114$
$\therefore$ There were 4 goals scored in the 6th game. ✓✓ (2 marks)

5

Males	Stem	Leaf
	12	089
874	13	34478
98200	14	25
764	15	

Location: The mean of the males (145) is higher than the female mean (134). The mode for the males (140) is higher than the female mode (134), and the median for the males (142) is higher than the females median (134). ✓✓✓

Spread: The range of the females (25) is higher than the males (23). ✓ (4 marks)

6

Items (x)	Customers (f)	fx
1	1	1
2	3	6
3	6	18
4	6	24
5	5	25
6	3	18
7	2	14
8	4	32
Sum	**30**	**138**

a Median: middle of 15th, 16th customer, = 4 ✓
Mode: 3 and 4 ✓
Range: 7 ✓ (3 marks)

b Mean = $\frac{138}{30}$ = 4.6 ✓ (2 marks)

7 **a** $\frac{1}{2}$ ✓ (1 mark)
b $\frac{1}{2}$ ✓ (1 mark)
c $\frac{12}{13}$ ✓ (1 mark)
d $\frac{8}{52} = \frac{2}{13}$ ✓ (1 mark)
e $\frac{2}{52} = \frac{1}{26}$ ✓ (1 mark)
f $\frac{50}{52} = \frac{25}{26}$ ✓ (1 mark)

(Total 35 marks)

WORKED SOLUTIONS to Sample Examination 1

Sample Examination 1 pp. 240–245

Part A (50 marks)

1. 1700 ✓ (1 mark)
2. –4, –3, –2 ✓ (1 mark)
3. $5a$ ✓ (1 mark)
4. $\frac{11}{3}$ ✓ (1 mark)
5. 21 ✓ (1 mark)
6. Trapezium ✓ (1 mark)
7. $36 ✓ (1 mark)
8. T ✓ (1 mark)
9. 4.6 ✓ (1 mark)
10. 30 ✓ (1 mark)
11. 0.11 ✓ (1 mark)
12. 4.5 ✓ (1 mark)
13. $28y$ ✓ (1 mark)
14. 1 ✓ (1 mark)
15. 0.65 ✓ (1 mark)
16. 5 ✓ (1 mark)
17. $27a^2$ ✓ (1 mark)
18. 610 048 ✓ (1 mark)
19. 4.15 cm ✓ (1 mark)
20. 3 ✓ (1 mark)
21. 53° ✓ (1 mark)
22. 650 ✓ (1 mark)
23. True ✓ (1 mark)
24. $0.70 ✓ (1 mark)
25. 6 ✓ (1 mark)
26. 8 ✓ (1 mark)
27. 49 ✓ (1 mark)
28. 15 ✓ (1 mark)
29. $\frac{5}{8}$ ✓ (1 mark)
30. $\frac{1}{4}$ ✓ (1 mark)
31. 68 ✓ (1 mark)
32. 49 ✓ (1 mark)
33. 124 ✓ (1 mark)
34. 85 ✓ (1 mark)
35. 98 ✓ (1 mark)
36. 1.1 ✓ (1 mark)
37. 0.36 ✓ (1 mark)
38. 16 ✓ (1 mark)
39. 370 mm ✓ (1 mark)
40. 4 cm ✓ (1 mark)
41. –2 ✓ (1 mark)
42. 1.8 ✓ (1 mark)
43. 3:5 ✓ (1 mark)
44. 620.8 ✓ (1 mark)
45. 80 ✓ (1 mark)
46. –13 ✓ (1 mark)
47. 45 ✓ (1 mark)
48. 9 ✓ (1 mark)
49. 0.07 ✓ (1 mark)
50. $2\frac{1}{2}$ ✓ (1 mark)

WORKED SOLUTIONS to Sample Examination 1

Part B (20 marks)

1 $5 - 0.73 = 4.27$ ✓
Length remaining is 4.27 m. ✓ (2 marks)

2

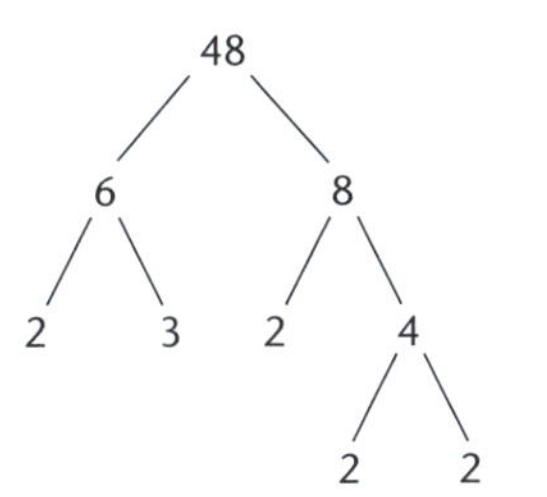

✓

$48 = 2 \times 2 \times 2 \times 2 \times 3$ or $2^4 \times 3$ ✓ (2 marks)

3 $x = 70$, $y = 70$ ✓✓ (2 marks)

4 6 pencils cost \$4.32
1 pencil costs $\$4.32 \div 6 = \$0.72$ ✓
$\therefore$ 4 pencils cost $\$0.72 \times 4 = \$2.88$ ✓
(2 marks)

5 Runs over season $= 17 \times 12 = 204$
Runs in finals $= 2 \times 21 = 42$ ✓
Total runs $= 204 + 42 = 246$ ✓ (2 marks)

6

Score (x)	Frequency (f)	$f \times x$
4	7	28
5	6	30
6	2	12
	15	70

✓

Mean $= \frac{70}{15} = 4\frac{2}{3}$ ✓ (2 marks)

7 Number of posts $= 85 \div 2\frac{1}{2} + 1$
(the + 1 is for the first post) ✓
$= 85 \div \frac{5}{2} + 1$
$= \frac{\cancel{85}^{17}}{1} \times \frac{2}{\cancel{5}_1} + 1$
$= 34 + 1 = 35$
Number of posts is 35. ✓ (2 marks)

8 $3\frac{1}{3} - 2\frac{4}{5} = 3\frac{1}{3} - 3 + \frac{1}{5}$ ✓
$= \frac{1}{3} + \frac{1}{5}$
$= \frac{5}{15} + \frac{3}{15}$
$= \frac{8}{15}$ ✓ (2 marks)

9 $9a^2 = 9 \times a \times a$
$= 9 \times 1\frac{1}{3} \times 1\frac{1}{3}$ ✓ $[a = 1\frac{1}{3}]$
$= \frac{\cancel{9}^1}{1} \times \frac{4}{\cancel{3}_1} \times \frac{4}{\cancel{3}_1}$
$= 16$ ✓ (2 marks)

10 There is one winner, so there are 17 losers. There have to be 17 matches for there to be 17 losers. ✓
The number of games is 17. ✓ (2 marks)

Part C (30 marks)

Question One

a $6x - 17 = 55$
$\therefore 6x = 55 + 17$
$6x = 72$ ✓
$\therefore x = \frac{72}{6}$
i.e. $x = 12$ ✓ (2 marks)

b Let Berend's age $= b$ years
Then Marie's age $= 4b + 3$ ✓
But Marie is 35 years old.
$\therefore 4b + 3 = 35$ ✓
$\therefore 4b = 35 - 3$
$4b = 32$
$b = \frac{32}{4}$
$\therefore b = 8$
Berend is 8 years old. ✓ (3 marks)

c **i** $3x + 5 + x + 27 + 28 = 180$
($\angle$sum of a $\triangle$) ✓
$\therefore 4x + 60 = 180$
(collecting like terms) ✓

ii From i $\therefore 4x = 180 - 60$
$4x = 120$
$\therefore x = \frac{120}{4}$
$x = 30$ ✓

Then $3x + 5 = 3 \times 30 + 5$
$= 95$

$x + 27 = 30 + 27$ ✓
$= 57$

Angles are 95°, 57° and 28°. ✓ (4 marks)

d $2\frac{1}{4} \times 1\frac{3}{4} \times 1\frac{1}{3} = \frac{9^3}{4} \times \frac{7}{4_1} \times \frac{4^1}{3_1}$ ✓
$= \frac{21}{4}$ ✓
$= 5\frac{1}{4}$ ✓ (3 marks)

e Length needed = $(3 \times 1.5) + (2 \times 1.8) + 1.9$ ✓
$= 4.5 + 3.6 + 1.9$
$= 10$ $\left[1\frac{1}{2} = 1.5\right]$ ✓
Length needed is 10 metres.
Cost = $10 \times \$3.75 = \$37.50$
The cost to Noah is \$37.50. ✓ (3 marks)

Question Two

a **i** $A = \frac{1}{2}bh$
$= \frac{1}{2} \times 15 \times 7$ ✓
$= \frac{1}{2} \times 105$
$= 52\frac{1}{2}$

Area is $52\frac{1}{2}$ cm^2. ✓

ii $A = bh$
$= 20 \times 13.2$
$= 264$ ✓
Area is 264 cm^2. ✓ (4 marks)

b

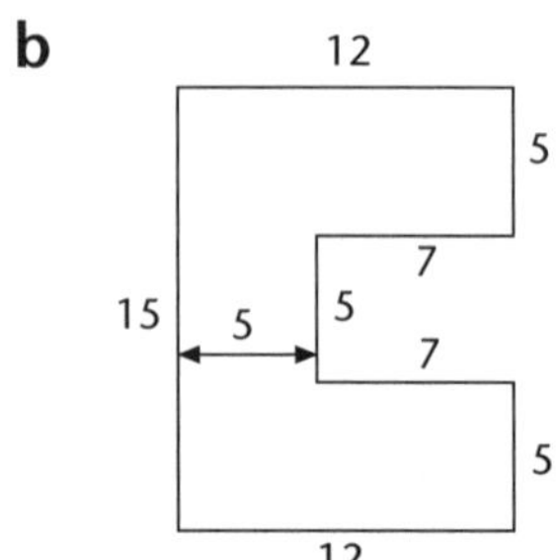

i Perimeter = Total length around the shape
$= 12 + 15 + 12 + 5 + 7 + 5 + 7 + 5$ ✓
$= 68$
Perimeter of the patio is 68 m. ✓

ii Length to be fenced = $12 + 15 + 12$
(part not next to house)
$= 39$ ✓
Length is 39 metres.
Cost = $39 \times \$27$ ✓
$= \$1053$ ✓ (5 marks)

c

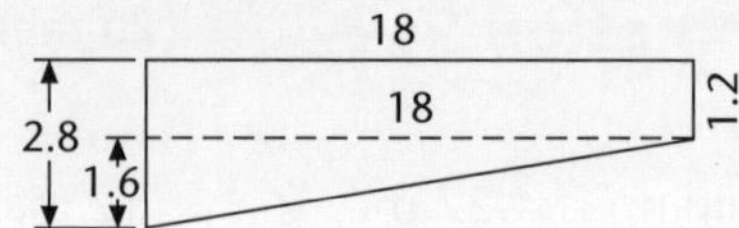

i A rectangle = lb
$= 18 \times 1.2$
$= 21.6$

A triangle = $\frac{1}{2}bh$
$= \frac{1}{2} \times 18 \times 1.6$
$= 14.4$

Total $A = 21.6 + 14.4 = 36$ ✓
Area is 36 m^2. ✓

ii

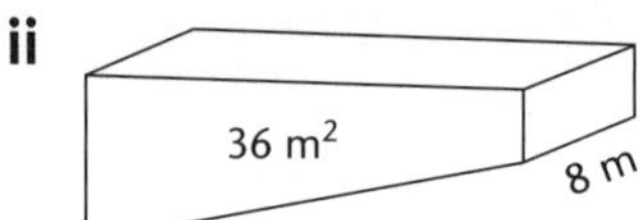

$V = Ah = 36 \times 8 = 288$
Volume of water is 288 m^3. ✓

iii

18 m

Surface of pool

8 m

Surface area of pool = $LB = 18 \times 8 = 144$
Surface area is 144 m^2. ✓
Now 7200 L = 7 200 000 mL
[1 L = 1000 mL]
= 7 200 000 cm^3
But 1 m^3 = $100 \times 100 \times 100$ cm^3
[1 cm^3 = 1 mL]
= 1 000 000 cm^3
Then 7200 L = $\frac{7200000}{1000000}$ m^3
= 7.2 m^3
Volume of water removed = 7.2 m^3. ✓

Now, let depth = d
Now $V = Ah$
$= 144 \times d$
$= 144d$ m^3

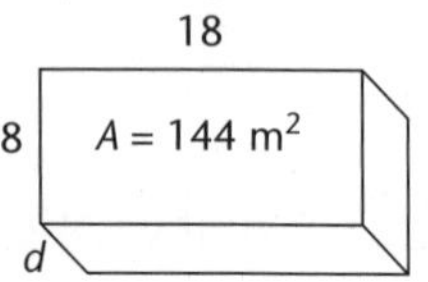

But volume = 7.2 m^3
$\therefore 144d = 7.2$ (in m^3)
$\therefore d = \frac{7.2}{144}$ m = 0.05 m
i.e. Depth is 0.05 m = 5 cm
Depth of water falls by 5 cm. ✓ (6 marks)

WORKED SOLUTIONS to Sample Examination 2

Sample Examination 2 pp. 246–252

Part A (50 marks)

1. 11 ✓ (1 mark)
2. 0 ✓ (1 mark)
3. $6a$ ✓ (1 mark)
4. $12 + 8 = 20$ ✓ (1 mark)
5. 50 ✓ (1 mark)
6. $\frac{7}{100}$ ✓ (1 mark)
7. 0.875 ✓ (1 mark)
8. 1:2 ✓ (1 mark)
9. $7x - 3y$ ✓ (1 mark)
10. $6x^2$ ✓ (1 mark)
11. 8 ✓ (1 mark)
12. $x = 9$ ✓ (1 mark)
13. 2 ✓ (1 mark)
14. $-9°$ ✓ (1 mark)
15. $8ab$ ✓ (1 mark)
16. 8 ✓ (1 mark)
17. $\frac{5}{9}$ ✓ (1 mark)
18. 1 ✓ (1 mark)
19. 23° ✓ (1 mark)
20. $a + 12$ ✓ (1 mark)
21. -4 ✓ (1 mark)
22. \$9 ✓ (1 mark)
23. 54 ✓ (1 mark)
24. $3b$ ✓ (1 mark)
25. $\frac{1}{2}$ ✓ (1 mark)
26. 0.16 ✓ (1 mark)
27. $2x^2 + 6x$ ✓ (1 mark)
28. -7 ✓ (1 mark)
29. 20 ✓ (1 mark)
30. 240 cm ✓ (1 mark)
31. \$36 ✓ (1 mark)
32. 65 ✓ (1 mark)
33. 80% ✓ (1 mark)
34. $\frac{1}{3}$ ✓ (1 mark)
35. C ✓ (1 mark)
36. D: obtuse ✓ (1 mark)
37. 8 cm ✓ (1 mark)
38. 0.2 ✓ (1 mark)
39. ∠ACB ✓ (1 mark)
40. 12 ✓ (1 mark)
41. 400 m ✓ (1 mark)
42. 30 cm^2 ✓ (1 mark)
43. 60 days ✓ (1 mark)
44. 6 axes ✓ (1 mark)
45. 11 ✓ (1 mark)
46. $a + 2b$ ✓ (1 mark)
47. $x + 1$ ✓ (1 mark)
48. 45 ✓ (1 mark)
49. $y = 2x + 3$ ✓ (1 mark)
50. 6 hours ✓ (1 mark)

WORKED SOLUTIONS to Sample Examination 2

Part B (32 marks)

1 $7.00 + 2.50 + 1.50 + 2.40 ✓

Total = $13.40 ✓ (2 marks)

2 12 litres costs $25.80

1 litre costs $\frac{\$25.80}{12}$ = $2.15 ✓

7 litres costs $2.15 × 7 = $15.05 ✓ (2 marks)

3 Cost = $1.70 + $8.35 + $0.85 + $2.35

= $13.25 ✓

Change = $20 – $13.25 = $6.75 ✓ (2 marks)

4

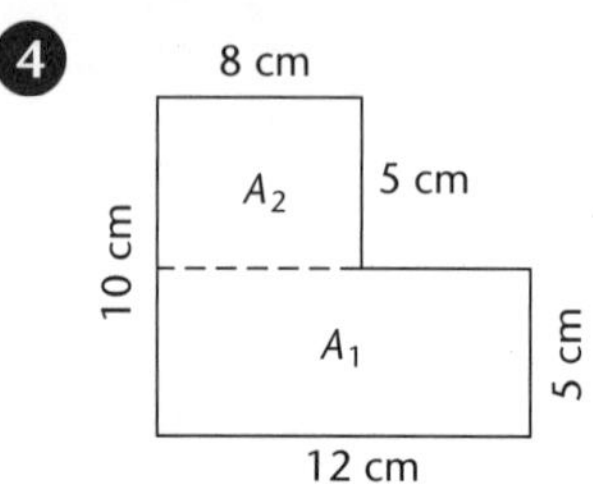

$A_1 = lb$
$= 12 \times 5$
$= 60$

$A_2 = lb$
$= 8 \times 5$
$= 40$

Area = 60 + 40 ✓
= 100

∴ Area is 100 cm^2. ✓ (2 marks)

5 Whites: 2, 4, 6, 8

Blacks: 2, 3, 4, 5 ✓

Whites = 2 × Diagram number

∴ Whites = 2 × 17 = 34

∴ 34 whites ✓ (2 marks)

6 No. of minutes = 1.75 L ÷ 5 mL

= 1750 ÷ 5

= 350 min ✓

Time taken = 350 ÷ 60 h

= 5 h 50 min ✓ (2 marks)

7 **a** Factors of 27 = 1, 3, 9, 27

Factors of 36 = 1, 2, 3, 4, 6, 9, 12, 18, 36

∴ The common factors are 1, 3, 9. ✓ (1 mark)

b HCF = 9 ✓ (1 mark)

8 $x + 2x + 3x = 180$

$6x = 180$

$x = 30$ ✓

Largest angle = $3x$

$= 3 \times 30°$

$= 90°$ ✓ (2 marks)

9 $2x - 1 = -7$

$2x = -7 + 1$

$2x = -6$ ✓

$x = \frac{-6}{2}$

$x = -3$ ✓ (2 marks)

10

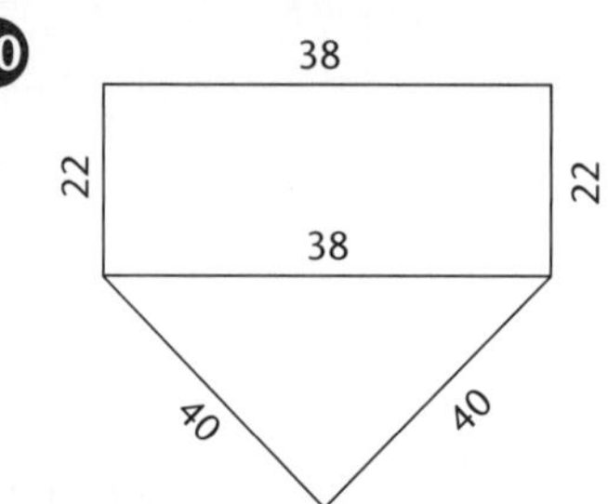

Length of fence = 2 × 22 + 2 × 38 + 2 × 40

= 200 ✓

Cost = 200 × $14.25

= $2850 ✓ (2 marks)

11 **a** $a = -6$ ✓ (1 mark)

b $\frac{x + 2}{3} = 5$

$x + 2 = 15$

$x = 13$ ✓ (1 mark)

12 ∠AGF = 60° ✓ (vertically opposite angles)

∴ $x = 180 - 60$ (co-interior angles AB // CD)

$= 120$ ✓ (2 marks)

13

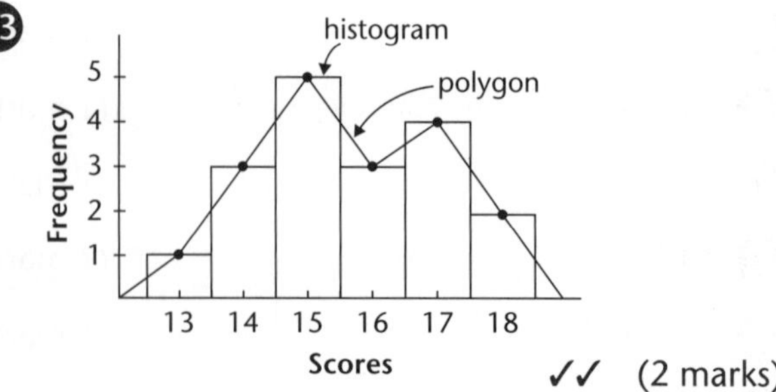

✓✓ (2 marks)

14 **a** –4, –1, $2\frac{1}{2}$, 4, 6.3, 8 ✓ (1 mark)

b $2\frac{1}{2}$ ✓ (1 mark)

WORKED SOLUTIONS to Sample Examination 2

15 Time taken = $\frac{50}{20}$ h = $2\frac{1}{2}$ h ✓

Time of arrival = 7.30 a.m. + $2\frac{1}{2}$ h
= 10.00 a.m. ✓ (2 marks)

16 **a** $\angle ABC = \frac{1}{2}(180° - 110°)$
$= \frac{1}{2} \times 70°$
$= 35°$
(base angles of isosceles △) ✓ (1 mark)

b $\angle ECD = 35°$
(corresponding to ∠ABC, AB // CE) ✓
(1 mark)

Part C (18 marks)

1 **a** Paul's share = $\frac{1}{3} \times \$42 = \$14$ ✓ (1 mark)

b Jill's share = $\frac{1}{6} \times \$42 = \$7$ ✓ (1 mark)

c Fraction remaining = $1 - (\frac{1}{3} + \frac{1}{6})$
$= \frac{6}{6} - (\frac{2}{6} + \frac{1}{6})$
$= \frac{6}{6} - \frac{3}{6}$
$= \frac{3}{6}$
$= \frac{1}{2}$

Jack has $\frac{1}{2}$ remaining. ✓ (1 mark)

2 **a** Area around picture
= Area frame – Area picture
= 60 × 60 – 40 × 26
= 2560 ✓
Area around picture is 2560 cm^2. (1 mark)

b 60 – 40 = 20
$\frac{1}{2}$ of 20 = 10

60 – 26 = 34
$\frac{1}{2}$ of 34 = 17
Width of border is 10 cm on sides and 17 cm above and below picture. ✓✓
(2 marks)

3 **a** (48 – 16) ÷ 4 = 32 ÷ 4 = 8
∴ She has 8 sports friends. ✓

b 16:48 = 1:3
Ratio is 1:3. ✓

c 0.25 × 48 = 12
There are 12 males. ✓ (3 marks)

4 **a** $a^2 - b = (-2)^2 - 3$
$= 1$ ✓

b $cb - a = 3(-5) - (-2)$
$= -15 + 2$
$= -13$ ✓

c $\frac{4a - b + 1}{c} = \frac{4(-2) - 3 + 1}{-5}$
$= \frac{-10}{-5}$
$= 2$ ✓ (3 marks)

5 $a° + a° + b° + b° = 180°$ ✓
(angle sum of straight line)
∴ $a° + b° = 90°$ ✓
But $\angle DBF = a° + b°$
i.e. $\angle DBF = 90°$ ✓ (3 marks)

6 **a** Total boys = 12 × 4 = 48 ✓ (1 mark)

b Total girls = 10 × 4 = 40 ✓
Missing girl's age = 40 – (18 + 8 + 8)
= 40 – 34
= 6 ✓
∴ Girls are 6, 8, 8, 18. ✓ (3 marks)

c Average age = $\frac{48 + 40}{8}$
$= \frac{88}{8}$
$= 11$
∴ Average age is 11 years. ✓ (1 marks)

INDEX

E

F

G

H

I

J

K

L

M

N

O

P

Q

R

S

T

U

V

W

NOTES

NOTES